Contaminants in agricultural soils

Challenges and solutions

It is widely recognised that agriculture is a significant contributor to global warming and climate change. Agriculture needs to reduce its environmental impact and adapt to current climate change whilst still feeding a growing population, i.e. become more 'climate-smart'. Burleigh Dodds Science Publishing is playing its part in achieving this by bringing together key research on making the production of the world's most important crops and livestock products more sustainable.

Based on extensive research, our publications specifically target the challenge of climate-smart agriculture. In this way we are using 'smart publishing' to help achieve climate-smart agriculture.

Burleigh Dodds Science Publishing is an independent and innovative publisher delivering high quality customer-focused agricultural science content in both print and online formats for the academic and research communities. Our aim is to build a foundation of knowledge on which researchers can build to meet the challenge of climate-smart agriculture.

For more information about Burleigh Dodds Science Publishing simply call us on +44 (0) 1223 839365, email info@bdspublishing.com or alternatively please visit our website at www.bdspublishing.com.

Related titles:

Advances in measuring soil health
Print (ISBN 978-1-78676-426-3); Online (ISBN 978-1-78676-428-7, 978-1-78676-429-4)

Developing circular agricultural production systems
Print (ISBN 978-1-80146-256-3); Online (ISBN 978-1-80146-257-0, 978-1-80146-258-7)

Improving soil health
Print (ISBN 978-1-78676-670-0); Online (ISBN 978-1-78676-672-4, 978-1-78676-673-1)

Understanding and preventing soil erosion
Print (ISBN 978-1-80146-379-9); Online (ISBN 978-1-80146-380-5, 978-1-80146-381-2)

Understanding and utilising soil microbiomes for a more sustainable agriculture
Print (ISBN 978-1-80146-474-1); Online (ISBN 978-1-80146-475-8, 978-1-80146-476-5)

Chapters are available individually from our online bookshop:
https://shop.bdspublishing.com

climate•SMART•publishing
IN AGRICULTURAL SCIENCE

BURLEIGH DODDS SERIES IN AGRICULTURAL SCIENCE
NUMBER 166

Contaminants in agricultural soils

Challenges and solutions

Edited by Distinguished Laureate Professor Ravi Naidu, University of Newcastle, Australia and CRC for Contamination Assessment and Remediation of the Environment (crcCARE), Australia

burleigh dodds
SCIENCE PUBLISHING

Published by Burleigh Dodds Science Publishing Limited
82 High Street, Sawston, Cambridge CB22 3HJ, UK
www.bdspublishing.com

Burleigh Dodds Science Publishing, 1518 Walnut Street, Suite 900, Philadelphia, PA 19102-3406, USA

First published 2025 by Burleigh Dodds Science Publishing Limited

Library of Congress Control Number: 2025939883

British Library Cataloguing in Publication Data
A catalogue record for this book is available from the British Library

ISBN 978-1-80146-878-7 (print)
ISBN 978-1-80146-879-4 (ePub)
ISBN 978-1-80146-880-0 (PDF)
ISSN 2059-6936 (print)
ISSN 2059-6944 (online)

DOI: 10.19103/AS.2025.0151

Typeset by Deanta Global Publishing Services, Dublin, Ireland

Contents

Series list

Title	Series number
Achieving sustainable cultivation of maize - Vol 1 From improved varieties to local applications Edited by: Dr Dave Watson, CGIAR Maize Research Program Manager, CIMMYT, Mexico	001
Achieving sustainable cultivation of maize - Vol 2 Cultivation techniques, pest and disease control Edited by: Dr Dave Watson, CGIAR Maize Research Program Manager, CIMMYT, Mexico	002
Achieving sustainable cultivation of rice - Vol 1 Breeding for higher yield and quality Edited by: Prof. Takuji Sasaki, Tokyo University of Agriculture, Japan	003
Achieving sustainable cultivation of rice - Vol 2 Cultivation, pest and disease management Edited by: Prof. Takuji Sasaki, Tokyo University of Agriculture, Japan	004
Achieving sustainable cultivation of wheat - Vol 1 Breeding, quality traits, pests and diseases Edited by: Prof. Peter Langridge, The University of Adelaide, Australia	005
Achieving sustainable cultivation of wheat - Vol 2 Cultivation techniques Edited by: Prof. Peter Langridge, The University of Adelaide, Australia	006
Achieving sustainable cultivation of tomatoes Edited by: Dr Autar Mattoo, USDA-ARS, USA and Prof. Avtar Handa, Purdue University, USA	007
Achieving sustainable production of milk - Vol 1 Milk composition, genetics and breeding Edited by: Dr Nico van Belzen, International Dairy Federation (IDF), Belgium	008
Achieving sustainable production of milk - Vol 2 Safety, quality and sustainability Edited by: Dr Nico van Belzen, International Dairy Federation (IDF), Belgium	009
Achieving sustainable production of milk - Vol 3 Dairy herd management and welfare Edited by: Prof. John Webster, University of Bristol, UK	010
Ensuring safety and quality in the production of beef - Vol 1 Safety Edited by: Prof. Gary Acuff, Texas A&M University, USA and Prof. James Dickson, Iowa State University, USA	011
Ensuring safety and quality in the production of beef - Vol 2 Quality Edited by: Prof. Michael Dikeman, Kansas State University, USA	012
Achieving sustainable production of poultry meat - Vol 1 Safety, quality and sustainability Edited by: Prof. Steven C. Ricke, University of Arkansas, USA	013
Achieving sustainable production of poultry meat - Vol 2 Breeding and nutrition Edited by: Prof. Todd Applegate, University of Georgia, USA	014

Introduction

With increasing pressure on land as a resource, there is a pressing need to make better use of existing agricultural land and to remediate contaminated sites for sustainable farming practices. Many soils used for agriculture already contain some degree of contamination, stemming from historical contamination, the application of agrochemicals during farming, or proximity to industrial activities. Additionally, atmospheric deposition and other routes contribute to the accumulation of contaminants. The European Union (EU) Soil Observatory e.g. reports that 61 % of EU soils are in an unhealthy state, with contamination being a key factor.

Contaminants in soils not only pose direct risks to human health and crop safety but also significantly impact ecosystem services–the benefits that ecosystems provide to humans. These impacts include reduced soil fertility, altered nutrient cycling, and diminished water quality, all of which are essential for maintaining agricultural productivity and ecological balance. Contaminated soils can impair biodiversity, disrupt pollination, and decrease the resilience of ecosystems to environmental stresses. Therefore, addressing soil contamination is critical not only for safeguarding human health but also for preserving the integrity and functionality of ecosystem services that underpin sustainable agriculture and environmental health.

This collection reviews recent research on key contaminants in agricultural soils, covering topics such as contamination pathways; challenges and techniques in identification and tracking; persistence and bioaccumulation in soils; risks to and effects on human health; and effects on soil health. Additionally, it highlights advances in bioremediation and phytoremediation techniques for managing soil contaminants.

Part 1 Types of contaminants in agricultural soils

The first group of chapters primarily focuses on various types of contamination in agricultural soil. Chapter 1 examines polycyclic aromatic hydrocarbons (PAHs) and pathways of contamination, methods for detection and tracking, their persistence and accumulation in soil, and the potential risks and impacts on human and soil health. It also reviews recent advances in managing PAH-contaminated soils and discuss associated challenges. The chapter highlights that nano remediation techniques can improve soil cleanup efforts by enhancing microbial activity and facilitating biodegradaion of contaminants. This approach addresses soil pollution caused by PAHs, accelerates cleanup process, and reduces environmental impact, in an eco-friendly and sustainable solution.

Chapter 2 addresses the challenges and solutions associated with trace element contaminants in agricultural soils. Trace elements are naturally occurring components of the environment. However, human activities such as industrial emissions, mining, agricultural practices, and waste disposal have significantly altered their distribution, and concentrations in soils. This anthropogenic influence often leads to contamination, resulting in elevated levels of certain trace elements that can persist in the soil for extended periods. Once accumulated, trace elements tend to remain in soil due to ther chemical stability, posing risks to plant growth, food safety, and human health. These elements can influence soil chemistry and biological activity, potentially causing toxicity in plants and entering the food chain. Soil plays a dual role as both a sink and a source of trace elements: it can immobilize elements, reduce immediate bioavailability, or release them over time, affecting surrounding environments. The mobility and bioavailability of trace elements in soils depend on various factors, including soil pH, organic matter content, redox conditions, and the chemical form of the elements. Plants can absorb these elements through root uptake and to a lesser extent, via atmospheric deposition on leaves. However, plants exhibit various regulatory mechanisms to mitigate their impact. Various remediation strategies can help reduce their bioavailability. Additionally, establishing and enforcing regulatory standards for acceptable trace element concentrations in soils and agricultural products is crucial. Such standards serve as vital tools for monitoring, managing, and preventing contamination, thereby indirectly controlling environmental and health impacts.

Chapter 3 examines the challenges posed by per and polyfluoroalkyl substances (PFAS) presence in water used for agriculture. It begins with a brief overview of key PFAS characteristics, including their persistence, mobility, and bioaccumulation potential. The chapter then explores the pathways through which PFAS enter the environment–such as contaminated water sources, soil application, and runoff–and discusses their environmental fate, including transport, transformation, and sequestration within soil profiles. Additionally, the chapter highlights the implications of PFAS contamination for human health and ecosystems, emphasizing potential risks associated with exposure through food and water. To address these challenges, it reviews potential mitigation strategies in agricultural settings, such as water treatment options, soil remediation techniques, and best management practices designed to reduce PFAS uptake and maintain farm viability. Overall, the chapter aims to provide a comprehensive understanding of PFAS issues in agriculture and propose approaches to minimize their impact on the food system.

Chapter 4 centers on soil-pesticide monitoring data, highlighting key findings and challenges like legacy contamination and the interpretation of results. It also explores soils' dual role, as both a sink and source of pesticide residues, stressing the importance of regular, harmonized and holistic monitoring programs. Furthermore, the chapter provides recommendations for improved assessment and prevention of soil contamination by pesticide residues, including an initial effort at defining relative benchmarks for monitoring pesticide residues in agricultural land. The emphasis is mainly on synthetic pesticides, which are used on most agricultural land, and most relevant for regulatory and pesticide reduction strategies.

Chapter 5 provides a detailed overview of the pathways through which antibiotics reach the soil, the processes that influence their fate within the soil, and the methods used to predict their behavior and longevity. Antibiotics, widely used to treat or prevent microbial diseases in animals and humans, can pose significant environmental risks when they enter soil ecosystems. It also presents insights from specific case studies and explores future research directions to address environmental challenges. The chapter highlights the importance of understanding how antibiotics interact with soil components under various environmental conditions. Finally, it stresses the need for interdisciplinary approaches and collaborative efforts among scientists, policymakers, and agricultural practitioners to develop strategies for managing antibiotic residues in the environment and mitigating potential negative impacts.

Chapter 6 addresses the use of wastewater in agricultural soils as a possible alternative to fresh water. Water recycling and the use of treated domestic sewage effluents (referred to as treated wastewater (TWW)) for agriculture, industry and non-potable urban and environmental purposes can be a highly effective strategy to exploit a sustainable water resource in water-scarce areas. It can also contribute significantly to nutrient conservation and environmental protection. However, TWW is considered to be of inferior quality compared with fresh water due to elevated salinity and sodicity, presence of toxic inorganic ions and anthropogenic organic pollutants susceptible to migration towards surface and ground water; all of which may result in contamination of irrigated lands. Understanding the processes by which irrigation with TWW may lead to contamination of the soil ecosystem is crucial. Developing management practices that mitigate the hazards involved in TWW irrigation is essential.

Part 2 Remediation techniques

Part 2 covers a range of assessment and remediation techniques for contaminants in agricultural soils. Chapter 7 examines recent advances in analytical techniques for detecting soil contaminants. A major category of soil contaminants is diffuse contaminants, alternatively termed non-point source pollutants. Another aspect of this contamination spectrum is atmospheric deposition, where airborne pollutants, gradually settle onto land and aquatic environments. Given the numerous and varied sources of contaminants, developing effective methods for their assessment, monitoring, and management is challenging. This requires the use of highly sensitive analytical instruments to ensure accurate detection.

As discussed in the chapter, two analytical techniques, spectroscopy and mass spectrometry, have consistently addressed this challenge, enabling a more detailed nuanced understanding of the nature of soil contamination.

Chapter 8 highlights recent advances in risk assessment of contaminated agricultural soils. It examines key issues in hazard identification and emphasizes the use of conceptual site models to better understand site contamination complexity. The chapter covers ecological exposure analysis and toxicological risk assessment related to human health. Special focus is given to the potential dangers of mixed contaminants for both ecological systems and humans, underscoring the importance of holistic risk evaluations. Additionally, the chapter explores future trends in this field.

Chapter 9 examines how organic amendments in agricultural soils can immobilize toxic trace elements and promote the biodegradation of organic pollutants by supporting microbiological degraders. The most commonly used organic amendments are compost, biochar, and chitosan. The chapter also highlights that many new remediation approaches are based on chemical modifications of traditional amendments and/or combining different amendments. Even conventional oxidation methods like persulfate applications benefit from being combined with organic amendments, which can act as catalysts. As discussed, complex contaminant mixtures demand innovative development of customized amendment combinations.

Chapter 10 examines clay mineral-based additives used in the remediation of hydrocarbons in agricultural soils. These soils may be contaminated with a range of pollutants, including heavy metal(loid)s, petroleum hydrocarbons (PHCs), pesticides, and microplastics. Among these, PHCs are commonly associated with oil handling, nearby mining activities, used tires, surface runoff, and agrochemicals. Bioremediation methods such as microbial degradation or phytoremediation can break down these PHCs. However, active microorganisms often require additional support, like additives that boost their degradation potential. As highlighted in the chapter, certain mineral-based additives, both in their raw and modified states, have garnered considerable research interest for enhancing the bioremediation of contaminants such as PHCs.

The goal of Chapter 11 is to explain the significance of immobilisation/phytostabilisation and recent developments that may enhance its application, including in agricultural soils. Soil remediation of heavy metal and metalloid contamination can be accomplished through immobilisation/phytostabilization techniques, which involve locking contaminants within the root zone. This process can be facilitated by plants, soil amendments, and rhizosphere microorganisms, thereby reducing their bioavailability. Selecting appropriate soil amendments is crucial for effective contaminant immobilisation. The chapter emphasizes the use of non-edible crops, which hold potential for biomass valorization, along with the importance of soil quality indicators to assess risk reduction and improvements in soil health.

Chapter 12 reviews recent advances in soil contaminant phytoextraction using accumulator and hyper-accumulator plants. Phytoremediation, especially the subset known as phytoextraction, offers an appealing alternative for remediating agricultural soils. Phytoextraction involves plants that can extract heavy metal contaminants from soil and sequester them within their tissues. The chapter discusses recent research in phytoextraction, focusing particularly on the removal of heavy metals like cadmium (Cd) and copper (Cu) from agricultural and horticultural soils.

As the concluding chapter of the book, Chapter 13 explores the scope of agricultural soil contamination in China, emphasizing the widespread occurrence of pollutants like heavy metals, organic contaminants, and emerging pollutants such as microplastics. Soil pollution has emerged as a significant concern in China, fueled by rapid industrial growth and urban expansion, posing serious risks to agroecosystems and human well-being. The chapter reviews government efforts, including the "Soil Pollution Prevention and Control Action Plan," as well as the implementation and success of various remediation projects to address this issue, offering a valuable comparison with other nations.

Part 1

Types of contaminants in agricultural soils

Chapter 1

Polycyclic aromatic hydrocarbon (PAH) contaminants in agricultural soils: occurrence and remediation

Teklit Gebregiorgis Ambaye, University of Brescia, Italy/Technical University of Denmark, Denmark; and Mentore Vaccari[1], University of Brescia, Italy

1 Introduction

Polycyclic aromatic hydrocarbons (PAHs) are hydrophobic pollutants that are nonpolar and classified as persistent organic pollutants (Ambaye et al., 2022a). They are a major cause of soil pollution and pose a substantial environmental threat (Barathi et al., 2023). Polycyclic aromatic hydrocarbons originate from natural and human activity, including the incomplete burning of diagenesis and hydrocarbons (Wei et al., 2024). Sources are industrial activity, fossil fuel combustion, sewage sludge application, volcanic activity, and forest fires (Wei et al., 2024).

Polycyclic aromatic hydrocarbons are present in diverse habitats such as agricultural products, soils, aerosols, marine sediments, and marine organisms (Ambaye et al., 2023a). They are extremely carcinogenic because

1 Corresponding author: Mentore Vaccari, email: mentore.vaccari@unibs.it

http://dx.doi.org/10.19103/AS.2025.0151.01

of their intricate nature and are resistant to degradation (Zhang et al., 2024). They gather in living and non-living organisms, impacting agricultural soils and crop production as time passes. Elevated PAH levels can result in decreased presence, variety, and functions of agricultural soil organisms that play a role in soil improvement and nutrient circulation (Zhang et al., 2024). Polycyclic aromatic hydrocarbons, due to their intricate nature, are difficult to break down and have a long-lasting presence in both living and non-living organisms, making it crucial to address them for environmental protection (Patel et al., 2020).

Polycyclic aromatic hydrocarbons, including naphthalene, anthracene, fluorene, fluoranthene, and phenanthrene, are used in commercial products but are now viewed as carcinogenic and toxic in the environment. These mutagenic and teratogenic compounds can cause adverse health effects, including brain, bladder, leukemia, DNA mutations, lung, and bone cancers. They can also cause cognitive impairments, genetic changes, oxidative harm, and cell death. Chronic exposure can lead to compromised immune function, respiratory issues, metabolic imbalances, and cancer in marine organisms, potentially affecting humans via the food chain (Rengarajan et al., 2015; Patel et al., 2020; Drwal et al., 2019). Moreover, they are frequently present in the soil at levels below fatal doses, leading to adverse effects on soil plants and animals. Ecotoxicological studies have assessed the impact of PAH exposure on soil-dwelling animals (Kumari et al., 2023; Sverdrup et al., 2002). Therefore, it is essential to address PAH pollution to ensure its environmental sustainability.

Various remediation methods have been used to remove PAHs from polluted soil; however, they are often costly, time-consuming, and require extensive operation (Ansari et al., 2023). Advanced oxidation methods may generate harmful by-products, while microbial bioremediation and phytoremediation are cost-effective but may take longer and produce poor PAH removal effectiveness (Thacharodi et al., 2023). However, most treatment methods are expensive, environmentally unfriendly, or sometimes ineffective. Policymakers and scientists are now focusing on risk-based analyses to mitigate the negative effects of PAHs on the environment and human health, thereby achieving sustainable development goals.

This chapter discussed new techniques for addressing soil diffuse pollution with PAH, their contamination routes, challenges in identification and tracking, persistence and bioaccumulation in soils, risks to human health, and soil health effects. It also discusses recent advancements in managing PAH contamination, highlighting future prospects for sustainable development and environmentally friendly management.

2 Characteristics, sources, transport and impact of polycyclic aromatic hydrocarbons

Polycyclic aromatic hydrocarbons are ecotoxic pollutants that can be categorized into two groups: low-molecular-weight and high-molecular-weight (HMW) chemicals (Skic et al., 2023). The toxicity, genotoxicity, mutagenicity, and carcinogenicity of these substances pose substantial hazards to humans and other species (Peng et al., 2023). Polycyclic aromatic hydrocarbons are subject to stringent regulations in industrialized nations owing to their inherent carcinogenicity, genotoxicity, toxicity, and mutagenicity (Ziyaei et al., 2024). The United States Environmental Protection Agency has identified 16 priority pollutants: naphthalene, acenaphthylene, acenaphthene, fluorene, phenanthrene, anthracene, fluoranthene, pyrene, benzo[a]anthracene, chrysene, benzo[b]fluoranthene, benzo[k]fluoranthene, benzo[a]pyrene, benzo[g,h,i]perylene, indeno[1,2,3-c,d]pyrene, and dibenz[a,h]anthracene (Masjedi et al., 2023). Polycyclic aromatic hydrocarbons, possessing high molecular weights and aromatic rings, have elevated levels of chemical stability, toxicity, and environmental duration. The movement and distribution of organisms within their surroundings are influenced by a range of physicochemical characteristics (Vijayanand et al., 2023). Polycyclic aromatic hydrocarbons undergo weathering mechanisms such as soil adsorption, atmospheric volatilization, and water dissolution, and their fate, behavior, and transport are influenced by their sources and physicochemical properties (Alegbeleye et al., 2017; Srogi, 2007; Eldos et al., 2022).

2.1 Source of polycyclic aromatic hydrocarbons

Polycyclic aromatic hydrocarbons are organic pollutants derived from human activities, including coal combustion, fossil fuels, and food products (Mallah et al., 2022). These pollutants can be discharged into the environment through urbanization and rapid industrialization, contaminating food, soil, and water. Soil ecology is highly vulnerable due to the significant uptake of PAHs by soil particles (Kim et al., 2013). Polycyclic aromatic hydrocarbons are primarily produced domestically, with high-temperature food preparation techniques contributing significantly. Over 75% of individuals in India, China, and Southeast Asia use high-emission resources frequently (Kumar et al., 2016). Smoked and grilled meals increase exposure to PAHs, while oils and fats are the primary sources of daily intake and are frequently found in surface waters due to industrial and urban runoff activities (Sun et al., 2019). The presence of PAHs in drinking water is typically linked to coal tar linings in distribution pipes. Polycyclic aromatic hydrocarbons are also present in surface soils due to atmospheric deposition or urban runoff (Venkatraman et al., 2024). Polycyclic aromatic hydrocarbon exposure is common in various industries, primarily from industrial operations, domestic activities, human habits, automobiles, and natural phenomena, with atmospheric conditions determining PAH levels in air and soil. (Venkatraman et al., 2024; Balmer et al., 2019). Further research and investigation into the exposure pathways of PAHs is necessary to mitigate their adverse effects.

2.2 Exposure pathways of polycyclic aromatic hydrocarbons

Polycyclic aromatic hydrocarbons can be ingested in various forms, including inhalation, ingestion of contaminated food, and contact with contaminated soil or water (Sun et al., 2021). Polycyclic aromatic hydrocarbons can infiltrate the respiratory system and pose health hazards through inhalation, while ingestion, especially from contaminated food and water, accumulates due to environmental contamination (Priyadarshanee et al., 2022). Direct exposure to flames or high temperatures during cooking can increase the concentration of PAHs. Contaminated water sources can also accumulate PAHs, thereby causing exposure when consumed. Handling tobacco products also exposes workers to PAHs (Singh et al., 2020). Dermal contact, particularly in professional sectors such as asphalt paving, roofing, and coal tar manufacturing, allows PAHs to be directly absorbed by the skin. The factors influencing absorption include high concentrations, extended contact time, skin condition, solubility, and contact area (Kriech et al., 2011). Occupational exposure to PAHs is a critical concern for workers in various industries, with certain professions having increased susceptibility to PAHs due to their work settings and substance management (Petit et al., 2019). Further research and investigation are needed into the environment's fate and transport to mitigate its detrimental effects.

2.3 The impact of polycyclic aromatic hydrocarbons contamination on health

Polycyclic aromatic hydrocarbons pose a significant global health challenge, causing cancer, respiratory issues, metabolic irregularities, and genetic alterations (Udom et al., 2023). Antibiotic resistance (AR) is a major global health issue, affecting healthcare systems and leading to higher rates of illness, death, and healthcare expenses (Gorovtsov et al., 2018). Antibiotic resistance, present in healthcare and environmental compartments, causes 1.27 million annual fatalities, and could increase to ten million within 25 years if not addressed (Das et al., 2021). The continuous use of antibiotics in agriculture and aquatic environments can promote antibiotic-resistant bacteria, and wastewater treatment facilities can facilitate the exchange of genetic materials containing antibiotic-resistant genes (Mutuku et al., 2022). Understanding the complex relationship between AR and PAHs is vital for devising effective approaches to reduce antibiotic resistance in the environment (Ondon et al., 2021).

2.4 The fate and impact of polycyclic aromatic hydrocarbons on the environment

Polycyclic aromatic hydrocarbons are a significant environmental concern because of their affinity for surfaces and ability to accumulate in terrestrial invertebrates (Sun et al., 2021). Exposure of aquatic microbes and birds toultraviolet (UV) radiation and contaminants can lead to dangerous or catastrophic impacts (Williamson et al., 2019). Human activities in cities have increased the concentration of PAHs in the soil, with coal tar used for wood preservation and roofing. Polycyclic aromatic hydrocarbons have been emitted into the atmosphere through burning coal, wood, and petroleum for heating, and ashes containing PAHs have been dispersed across urban areas. Polycyclic aromatic hydrocarbons are present in various environments and exhibit varying flow patterns, making it crucial to control and reduce air pollution in densely populated areas for public health and ecological monitoring (Sun et al., 2023). The use and spread of PAHs can be influenced by physicochemical reactions and transformations, including photochemical alterations, sensitivity to harmful toxins, and interactions with other PAHs (Tarigholizadeh et al., 2023).

Polycyclic aromatic hydrocarbons are hydrophobic compounds that are strongly attracted to organic matter in water and sediments. This attraction is more pronounced for PAHs with more than five rings, which have high Kow values (Kumar et al., 2021). The cleanup of PAHs in terrestrial and marine environments is influenced by many atmospheric processes, including air-water vapor exchange, dry deposition, and wet deposition (Balmer et al., 2019). The processes under consideration are influenced by several factors, such as salinity, physicochemical properties of the pollutant, temperature, wind speed, density, and precipitation rates (Balmer et al., 2019).

High-molecular-weight PAHs, characterized by their low vapor pressures, exhibit a higher propensity for adsorption onto sorbents in the solid/water phase than low-molecular-weight PAHs. The concentration distribution in the vapor and sorbent phases is influenced by a range of factors, including wind speed, particle size, composition, humidity, and density (Hussain et al., 2018). Wet deposition occurs when atmospheric vapors dissolve in raindrops, causing PAHs to precipitate and adhere to the rain-absorbed soil particles (Birgül et al., 2011; Eldos et al., 2022). The process is influenced by the intensity of rainfall, as well as geographical and seasonal fluctuations (Li and Duan, 2023).

Polycyclic aromatic hydrocarbons accumulate in soil and sediment, with hydrophobic high-molecular-weight PAHs as the primary constituents. The diffusion of PAHs between air and water surfaces occurs when PAHs disperse at the interface between air and water, either as atmospheric vapor or as dissolved particles (Baran et al., 2021; Abdel-Shafy and Mansour, 2016). The fugacity of a substance, which refers to its tendency to escape from a phase, is directly influenced by the temperature, thereby affecting the rate of evaporation and absorption (Lawal, 2017). Atmospheric concentrations of PAHs are influenced by two primary mechanisms: thermal degradation or deposition, and the presence of oxidizing agents such as hydroxyl radicals (Rubio-Clemente et al., 2014).

Generally, environment and human health are significantly affected by PAHs, thus necessitating the implementation of sustainable remediation strategies. Several factors must be considered, including the composition of contaminants, environment, biological conditions, microbial species, time, cost, regulatory requirements, and methods. It is imperative to embrace innovative, sustainable methodologies encompassing physical, chemical, biological, and comparable technologies.

3 Techniques for remediation of polycyclic aromatic hydrocarbons

The optimal approach for treating PAH-contaminated sites is a sustainable, efficient, and cost-effective method that utilizes adaptable, site-suitable, and sustainable technologies to ensure environmental sustainability and industrial growth (Gitipour et al., 2018). Physical treatments, like solvent extraction and electrokinetic remediation, are simple but require additional steps, while chemical processes like catalytic degradation and chemical oxidation are effective but may produce toxic by-products (Kuppusamy et al., 2017). Biological treatment is environmentally friendly but requires a prolonged duration. Thermal processes, such as thermal desorption and microwave heating, can break down high-molecular-weight hydrocarbons but require significant energy and additional treatment (Mohan et al., 2006). Chemical oxidation is a widely used remediation approach because of its simplicity, effectiveness, eco-friendliness, and speed. These processes include ozone, Fenton reactions, permanganate oxidation, and activated persulfate reactions. Fenton reactions decrease the content of PAHs in the soil by producing potent oxidizing agents such as •OH. Nevertheless, the maintenance of these conditions necessitates the adjustment of soil pH to a range of 2.8–3.0, a task that can prove to be arduous owing to the substantial buffer capacity of the soil (Sun et al., 2020). Therefore, the demand for versatile methods to efficiently eliminate PAHs from contaminated sites is increasing, necessitating the adoption of innovative sustainable approaches involving physical, chemical, biological, and comparable technologies.

3.1 Soil washing and solvent extraction

Soil washing/solvent extraction is used to remediate soils contaminated with high-molecular-weight PAHs (Von Lau et al., 2014). Polycyclic aromatic hydrocarbons have low bioavailability, slow desorption, and high hydrophobicity, which make their removal difficult (Lamichhane et al., 2016). Solid-state extraction utilizes water, organic solvents, surfactants, complexing agents, and non-toxic flushing agents such as humic acids or vegetable oils to extract PAHs from binding sites (Trellu et al., 2016). Polycyclic aromatic hydrocarbons that have been decomposed are either disposed of or subjected to biological or chemical processes for detoxification (Sakshi et al., 2019). Vegetable oils have potential applications in PAH remediation as eco-friendly solvents for physical extraction and as soil amendments to enhance biological removal (Kariyawasam et al., 2022).

Generally, PAHs can be extracted using vegetable oils, biosurfactants, surfactants, and organic solvents. Their extraction rate and efficiency depend on their solubility and interactions with sample matrix sites. The physicochemical properties of the agent enhance solvent-soil interactions, making HMW PAHs more effective when hydrophobic agents such as surfactants and vegetable oil are used (Von Lau et al., 2014). Further research is needed to explore the economic benefits of solvent extraction and soil washing, as well as the residual material produced during these processes, which can be disposed of in prescribed landfills or require further treatment.

3.2 Biological treatment techniques

Polycyclic aromatic hydrocarbons can undergo aerobic and anaerobic biodegradation using diverse biological treatment techniques (Haritash and Kaushik, 2009). Membrane bioreactors and sequencing bioreactors are commonly used to remove organic contaminants. Biological approaches facilitate the biodegradation of organic contaminants with low molecular weights. According to Sher et al. (2023), the combination of Flavobacterium, Pseudomonas aeruginosa, and Pseudomonas putida exhibited the highest efficacy in removing water-soluble PAHs compared with pure culture strains. Microbial consortia were also evaluated for their ability to stimulate the breakdown of phenanthrene and other PAHs (Sher et al., 2023).

In general, bioremediation research has made considerable progress, but further development is needed to develop more effective methods (Manian et al., 2023). Advances in molecular biology and genetic engineering have identified the genes responsible for PAH metabolism, thereby enhancing the biotechnological effectiveness of microorganisms (Banerjee et al., 2024). Multi-omics allows for the creation of tailored strains capable of metabolizing diverse hydrocarbons and adapting to polluted environments (Thacharodi et al., 2023).

Phytoremediation

Plant-assisted bioremediation, also known as phytoremediation, can effectively mitigate PAH contamination (Ansari et al., 2023). Plants can absorb, sequester, and convert toxins into soil, producing enzymes that improve the bioavailability of pollutants and soil nutrition. Plants with high adaptability and extensive root surface area are favored (Sarma and Prasad, 2015). Intercropping alfalfa with tall fescue can enhance the dissipation of PAHs through phytoremediation because of distinct microbial communities in the rhizosphere (Guo et al., 2018). A field experiment showed that the presence of PAH in intercropping was significantly higher than that in monoculture (20%) and unplanted soils (0%) (Matsodoum Nguemté et al., 2023). Multispecies mixes, such as celery, white clover, and ryegrass, have been found to enhance the phytoremediation of soils contaminated with PAHs compared with monocultures. Plant growth-promoting rhizobacteria (PGPR) can enhance phytoremediation by mitigating environmental stressors and enhancing plant survival under suboptimal conditions (Chigbo, 2013). Plant growth-promoting rhizobacteria promote plant root growth and development through processes such as nitrogen fixation, siderophore synthesis, phytohormone generation, and mineral solubilization, enhancing soil fertility and overall plant growth. The use of PGPR has been empirically validated to enhance plant growth, particularly in soils with high contamination levels (Kuppusamy et al., 2017).

Bioaugmentation

Bioaugmentation is a highly effective approach for soils with limited natural PAH degraders, using aerobic or anaerobic organisms (Haleyur et al., 2019). Aerobic bioremediation involves the use of fungi, Gram-negative bacteria, or aerobic Gram-positive (Aydin et al., 2017). Li et al. (2008) found that microbial consortia from aged oil-contaminated soil had a significant PAH degradation capacity of 45–56%. They suggested the need for chemical pretreatments to improve the bioavailability and removal effectiveness of high-molecular-weight PAHs in field soils. The extensive surface area of the fungal mycelium increases PAH breakdown. Pinelli et al. (1997) demonstrated that the slurry-phase bioreactor system, which incorporates native aerobic specialized bacteria, efficiently eliminates PAHs and organic extractable substances from polluted industrial soil. This was achieved through improved pollutant desorption, uniformity, and oxygen mass transfer.

Land farming

The conventional agricultural method of landfarming involves the utilization of tilling, bulking, irrigation, and fertilizer application techniques to enhance soil uniformity and facilitate the biodegradation of PAHs (Aparicio et al., 2022). Although it promotes the growth of native microorganisms, its effectiveness in breaking down PAHs is limited because of its low availability, scarcity of PAH degraders, inadequate absorption of contaminants, and superficial treatment (Kaur et al., 2023). Hansen et al. (2004) conducted a pilot-scale study that revealed that conventional landfarming can serve as a cost-effective treatment option for locations contaminated with PAHs if there are no time constraints. In general, landfarming has certain drawbacks, including the requirement for large workspaces, diminished microbial activities, elevated excavation expenses, and prolonged incubation periods.

Biostimulation

Biostimulation is employed to expedite the remediation of soils that contain coal tar. Nevertheless, the inclusion of nutrients alone did not improve the biodegradation of coal tar PAH (Babu et al., 2019). The breakdown of PAH components increases with the addition of biodiesel with inorganic nutrients over 55 days (Taylor and Jones, 2001). The observed enhancement in the biodegradability of PAHs can be related to tar solubilization and dispersion, which leads to an increase in the bioavailability and uptake of PAHs by native biodegrading organisms. This implies that the utilization of biodiesel treatments may be a feasible approach for addressing nutritional deficiencies in the soil.

3.3 Thermal technologies

Thermal technologies have shown effectiveness in remediating sites contaminated with PAHs owing to their efficiency and rapid elimination of total petroleum hydrocarbons (Vidonish et al., 2016). However, high temperatures require significant energy inputs and can lead to structural degradation, posing challenges for the restoration of ecological systems (Vidonish et al., 2016). A holistic approach to thermal treatment is needed, emphasizing energy preservation and the protection of the ecological system (Dai et al., 2022). Microwave irradiation is cost-effective, time-efficient, and effective for treating PAHs by dissipating electromagnetic field energy into the irradiated material (Eldos et al., 2020). This energy is converted into heat for organic pollutant removal through thermal desorption. The effectiveness of microwave irradiation depends on the dielectric characteristics of the chemical. The vaporization of PAHs is influenced by sediment permeability and desorption interactions (Eldos et al., 2020). Microwave irradiation heating reduces energy costs and remediation duration in wet areas, such as marine sediments, making it more efficient and environmentally friendly than traditional thermal remediation methods. (Falciglia et al. 2018). Thermal treatment, specifically kiln incineration, can completely degrade priority PAHs; however, its high energy cost and operating temperature of 820–1600°C raise concerns. According to reports, thermal treatment for PAH-contaminated materials and soil costs US$475–1350 per cubic yard or US$328–843 per m^3, with additives like CaO potentially lowering the temperature, saving energy, and minimizing site damage. Microwave irradiation heating is particularly suitable for the treatment of PAH-contaminated sediments in areas with abundant water resources (Dai et al., 2022).

3.4 Chemical degradation

The degradation of PAHs in contaminated environments can be achieved using chemical oxidation processes, such as ozonation, photolysis, and chlorination (Ji et al., 2024). Nevertheless, this approach has the potential to produce harmful by-products such as halogenated hydrocarbons (Ji et al., 2024). Ozonation is considered a means of eliminating PAHs from the soil. The oxidation of pollutants by ozone occurs through two distinct mechanisms: direct interaction with chemically bonded or atomic nuclei, and indirect interaction with organic matter or metal oxides (Tamadoni and Qaderi, 2019). Under basic pH conditions, the indirect process assists the pathways, whereas the direct method is favored under acidic pH conditions (Li et al., 2021). The benzene ring is broken down by O_3 and PAHs, forming intermediate molecules, such as hydroxylated benzene and carboxylate. Medium-sized molecules compete with the contaminants for O_3 and OH radicals. However, oxidation limitations, including the existence of harmful by-products and OH scavengers in contaminated water, can negatively affect PAH breakdown (Maranzana et al., 2005). To improve the efficiency of the O_3 oxidation process, ozonation in conjunction with UV light and/or H_2O_2 has been used (Kumari and Kumar, 2023).

Photolysis

Photolysis is a chemical oxidation mechanism that occurs naturally in aqueous systems and leads to PAH degradation (El-Saeid et al., 2015). Photolysis occurs when PAHs absorb light, resulting in the formation of a highly reactive form of the desired PAH (Plata et al., 2008). The degradation of PAHs occurs when their emission spectra coincide with the absorption spectrum of the target molecule, which ranges from 200 to 386 nm. The form has the potential to either return to its initial condition or undergo significant alteration. (Wang et al., 2007). Photolysis, influenced by temperature, pH, radiation, and soil composition, can generate hazardous intermediate chemicals from the direct photolysis of PAH, potentially reducing the effectiveness of chemical treatments (Jacobs et al., 2008).

3.5 Advanced oxidation processes

Advanced oxidation processes (AOPs) are chemical oxidative techniques that use combined treatment systems with catalysts, oxidants, and energy sources to oxidize PAHs and generate intermediate free radicals, hydroxyl radicals (Rayaroth et al., 2023). The effectiveness of AOPs depends on factors such as the oxidant type, light intensity, pH level, free radical generation rate, pollutants, and water composition (Cheng et al., 2016). Ozone-based AOPs using UV radiation or oxidation (H_2O_2 or UV coupled with O_3) have shown enhanced efficacy in breaking down resistant organic contaminants (Yadav et al., 2020). Polycyclic aromatic hydrocarbons can be oxidized by $O_{3,}$ $HO_{2,}$ or OH– radicals or by direct ozonation in hybrid systems, such as H_2O_2/O_3 or UV/O_3. However, these systems have lower energy efficiencies due to O_3 limited solubility in liquids (Ji et al., 2024). In general, coupled approaches of advanced oxidation processes with biological methods have demonstrated efficacy in the remediation of soils contaminated with PAH, resulting in cost reduction and enhanced operational efficiency. Further investigations should prioritize the identification of intermediate by-products, evaluation of soil quality, and determination of scaled-up parameters for practical implementation in the field.

Photocatalysis

In photocatalysis, a semiconductor absorbs light and transfers it to a higher energy level, thereby facilitating chemical reactions in substances such as organic pollutants (Chatterjee and Dasgupta, 2005). Titanium dioxide (TiO_2) is a promising semiconductor material owing to its exceptional absorption characteristics, photocatalytic degradation capabilities, and cost-effectiveness (Nasr et al., 2018). Graphitic carbon nitride is another promising semiconductor material because of its unique electronic band structure (Bharathi et al., 2023). Photocatalysts trigger redox processes when exposed to irradiation from semiconductors, and researchers are exploring various forms of photocatalyst hybridization, such as heterojunction photocatalysts, for metal loading and doping applications (Ahmad et al., 2023).

Photocatalytic treatment is a popular wastewater treatment method because of its benefits, including hybridization of photocatalysts, appropriate preparation techniques, and degradation of contaminants (Peleyeju et al., 2018). Recent studies by Da Silva Medeiros et al. (2022) have shown that benzene, toluene, ethylbenzene, xylene, and naphthalene can be successfully removed from contaminated water and soil. However, total organic carbon removal effectiveness varied across treatments. Ozonation using H_2O_2 yielded the most favorable results, effectively eliminating all organic compounds with 70% acetic acid (Xiao et al., 2015). The use of microwave radiation and UV light for TiO_2

photocatalytic reactions holds significant potential in Advanced Oxidation Processes for remediating organic contaminants in polluted soil and water (Cheng et al., 2016).

In general, TiO_2 photocatalysts have undergone thorough investigation owing to their safety, high efficacy, and cost-effectiveness in the remediation of polluted soils. Nevertheless, its primary limitation is that its breakdown only occurs on the surface of the soil, resulting in a reduction in removal efficiency as the soil layers become thicker. Moreover, the absence of visible light emission and the restricted lifespan of the bulb contributed to the high expense of using UV lamps.

3.6 Combined approaches

The efficacy of biological, chemical, and physical treatments in the remediation of PAHs has been demonstrated; however, their applicability in field soils is limited. These factors include diminished efficacy, buildup of transformation by-products, extended process duration, and higher costs. Various treatments, including physical-chemical-biological, physicochemical, chemical-biological, and physical-biological methodologies, have been proposed to address these issues (Kuppusamy et al., 2017). For instance, Dadkhah and Akgerman (2002) demonstrated the utilization of subcritical hot water for a combination of extraction/in situ oxidation. This method was implemented sequentially, resulting in minimal levels of PAHs in soil through oxidation. Nevertheless, the integrated strategy exhibited a modest restorative performance, ranging from 20% to 49%, when the soil/oil contact duration was 2–3 h and Fenton treatment lasted for 5–14 days.

Lee et al. (2001) showed that the application of 100% acetone or ethanol resulted in enhanced bioavailability of PAHs in soil contaminated with coal tar. This pretreatment led to a degradation rate of 90% within 17 days, whereas soil without pretreatment required approximately 35 days to achieve comparable levels of efficiency.

Haapea and Tuhkanen (2006) conducted a study in which biodegradation was applied as the last stage after chemical oxidation, following an initial physical treatment. Before introducing the Pseudomonas strain and artificial nutrients, the field soil was subjected to pretreatment through washing and ozonation. The results of the integrated treatments showed a reduction in oxidant consumption by 5–10 times and an elimination of PAH by 90% compared to the individual procedures. Nevertheless, a constraint arises from the augmented expenses associated with the pH correction. A study on the effects of pretreatments on intermediate products and soil degradation was not conducted, suggesting a potential avenue for future research.

4 Emerging remediation technologies

Nanotechnology, nanoremediation, electrokinetic remediation, microbial fuel cell remediation, and vermiremediation have been employed to treat soils contaminated with PAHs. Nevertheless, additional studies are required to incorporate these techniques into biological, physical, and chemical treatments to achieve maximum efficacy (Shahsavari et al., 2019).

4.1 Nanomaterial-based technologies

Nanomaterial-based technologies are utilized to remove contaminants from contaminated areas, with various nanomaterials being introduced as nanosorbents to improve the treatment of PAH-contaminated soil and water, ensuring their future reuse (Mazarji et al.,2021). Adsorbent materials for treatment were evaluated based on their structural characteristics, adsorption kinetics, specificity, and isotherms. This helps us to understand pollutant adsorption mechanics. The most common kinetic models for PAH adsorption are the pseudo-second-order and pseudo-first-order models. Adsorption processes are evaluated using equilibrium data and isotherm models, which quantify the number of pollutants adsorbed under constant-temperature conditions (Awad et al., 2020; Revellame et al., 2020; Majd et al., 2022). There are two main classifications of adsorbent materials: natural adsorbents, which include synthetic adsorbents, and organic or inorganic adsorbents. Natural sorbents are easily accessible and cost-effective substances; however, their utilization is hindered by the intricacy involved in the collection of adsorbents for subsequent disposal or reuse in oil-spill water (De Gisi et al., 2016). Synthetic sorbents encompass both laboratory-synthesized adsorbents and commercially accessible adsorbents, including activated carbon, nanotechnology-based adsorbents, zeolites, and carbon nanotubes (Eldos et al., 2022). In general, nanotechnology has been used to remediate soils contaminated with PAH; however, the issue of sequestering PAHs in aged soils has received limited attention. The propensity of PAHs to migrate to subterranean environments depends on nanotechnology and operational conditions, necessitating a more prudent approach. Large-scale preparation of nanotechnology is expensive and time-consuming, making the 'more is better' strategy unnecessary. Implementing nanotechnology for the efficient, cost-effective, and timely removal of PAHs presents challenges, including the retrieval of nanomaterials post-treatment. A comprehensive understanding of the long-term environmental safety of nanotechnology is crucial because uncontrolled soil exposure can lead to human exposure (Gulia et al., 2023).

4.2 Bioelectrokinetic remediation

The bioelectrokinetic remediation approach, alternatively referred to as electrokinetic-enhanced bioremediation, involves the use of a weak electric field to facilitate the migration and conversion of microorganisms, nutrients, and organic pollutants in soils (Beretta et al., 2019). This method promotes the breakdown of organic pollutants in the dissolved phase by utilizing electrokinetic transport processes such as electrophoresis, electroosmosis, and electromigration (Han et al., 2021). Electrokinetic processes can enhance the interactions between microorganisms and pollutants in soils, particularly clay soils, by overcoming the low bioavailability of toxins (Li et al., 2023). Bioelectrokinetic methods for treating PAH-contaminated soil have been studied on a laboratory scale using artificially contaminated soils. However, the actual soil properties and pollutants are more complex than those in experimental studies. The factors to consider when scaling up include system configuration, contaminated soil properties, site area, electrode arrangement, and electrolyte characteristics. The practical engineering parameters for efficiency and cost-effectiveness remain unknown (Lan et al., 2023).

4.3 Microbial fuel cell remediation

A microbial fuel cell (MFC) is a device that produces power by breaking down organic or inorganic substances through microbial processes. The MFC anode serves as an insoluble receptor for microorganisms, in addition to natural electron receptors such as nitrate or oxygen. This enhances electron transfer and promotes the breakdown of organic waste. Soil microbial fuel cells are a novel type of microbial fuel cell that has garnered considerable interest in recent times. Soil microbial fuel cells comprise an anode located in the soil phase and a cathode situated in an oxygen-rich water phase. Research has demonstrated that soil microbial fuel cells can produce energy and eliminate persistent organic compounds from soil and sediment, thus potentially addressing the issue of PAH-contaminated soils. The efficacy of a microbial fuel cell is contingent upon the specific characteristics and composition of the microbial population present at the anode. The utilization of molecular ecology and high-throughput sequencing, a methodology that does not depend on microbial culture, is essential for elucidating the makeup and structure of microbial communities in contaminated environments as well as enhancing the process of pollutant biodegradation (Ambaye et al., 2022a, 2023a, 2023b, 2022c).

4.4 Nanoremediation

The development of functionalized nanoparticles with proteins, DNA, enzymes, humic acids, and biosurfactants has led to an increased focus on nanoremediation techniques for contaminated environments, enabling the catalytic degradation and adsorption of PAHs (Rajput et al., 2023). A combination of bioremediation and nanotechnology can effectively remove PAH from the environment. Nanotechnology can be applied directly or through adsorption, enhancing microorganism growth and immobilizing contaminants through microbial enzymes. This process increases biosurfactant production in microorganisms, enhances hydrocarbon solubility, and facilitates the degradation of pollutants in the environment (Sam et al., 2023). Nevertheless, achieving successful integration necessitates a thorough examination, evaluation of potential risks, and establishment of regulatory frameworks. The safe and responsible deployment of innovative technologies necessitates imperative collaboration among scientists, policymakers, and stakeholders (Yu et al., 2017).

4.5 Vermiremediation

Vermiremediation is a remediation technique in which soil invertebrates such as earthworms consume detrimental substances originating from oil-contaminated soil, thereby mitigating soil pollution (Ameen and Al-Homaidan, 2024). Earthworms can alter the chemical, physical, and biological constituents of soil, thereby mitigating the dispersion of toxins. Organisms possess a substantial quantity of bacteria that are stored within their digestive tracts, initiating the process of breakdown. Earthworms, including *Eisenia fetida*, can survive in soils with elevated levels of oil, including oil polluted with hydrocarbons. The mortality rate and hydrocarbon removal capacity of earthworms are directly influenced by the duration of the experiment (Thakur et al., 2023). The survival rate of earthworms is contingent upon the presence of food in experimental containers. A range of organic amendments can be used in the treatment of oil-contaminated soils to promote the elimination of contaminants and improve the growth parameters of earthworms (Rorat et al., 2017). Vermiremediation, which is a self-sustaining process with minimal energy consumption and simple maintenance, has the potential to be a cost-effective and ecologically friendly method for addressing PAH contamination at sites within a year. However, warms may not be suitable for field applications and are limited to low-to-medium levels of contamination in controlled environments (Datta et al., 2023).

5 Monitoring of polycyclic aromatic hydrocarbons

Remediation technologies' effectiveness, predictability, and dependability are crucial. Consistent surveillance is necessary for wider acceptance. Monitoring includes treatment-specific methods, replication, appropriate sampling, and data processing (Ambaye, et al., 2022a). Biological and analytical methodologies can be used to monitor and analyze soil remediation activities.

5.1 Analytical methods

The characterization of PAHs in environmental samples utilizes advanced extraction techniques, such as miniaturized and microextraction methods, to overcome traditional limitations (Manousi and Zachariadis, 2020; Patel et al., 2020). Microextraction techniques include solid-phase microextraction and liquid-phase microextraction, whereas miniaturized extraction methods include magnetic solid-phase extraction, pipette tip solid-phase extraction, dispersive solid-phase extraction, fabric-phase sorptive extraction, and stir bar/rod/plate sorptive extraction (Chormey and Bakırdere, 2018). Advanced PAH extraction methods have been enhanced by incorporating novel sorbents such as graphene, molecularly imprinted polymers, metal-organic frameworks, graphene oxide, carbon nanotubes, and zeolitic imidazole frameworks (Patel et al., 2020). Liquid-phase microextraction uses ionic liquids instead of hazardous organic solvents. Advanced extraction methods offer advantages such as simplicity, reduced time consumption, decreased solvent requirements, smaller sample volumes, and straightforward handling (Sharma et al., 2023). Gas chromatography-mass spectrometry, gas chromatography, high-performance liquid chromatography, ultra-high-pressure liquid chromatography, UV detectors, fluorescence detectors, flame ionization detectors, tandem mass detectors, and diode-array detectors are the most used detection methods (Beyer et al., 2010; Patel et al., 2020).

5.2 Microbial characterization

Soil microorganisms are ideal for investigating the bioremediation of PAHs, but their cultivation is limited to less than 1%. To address this, molecular microbial ecology methodologies, such aspolymerase chain reaction (PCR) based community profiling have been implemented (Gupta et al., 2015). Techniques such as temperature gradient gel electrophoresis and denaturing gradient gel electrophoresis are used for environmental analysis, allowing for segregation of DNA fragments and faster identification. Stable isotope probing has significantly improved our understanding of soil microbial communities, providing a culture-independent approach for examining the impact of environmental changes on microbiota (Dash and Das, 2018). Next-generation sequencing technology and bioinformatics tools have facilitated efficient, environment-friendly, and cost-effective metagenomic examinations (Shahsavari et al., 2019). Genomics has enabled the identification of functional genes and metabolic pathways in diverse microbial communities, reducing the sequencing effort and time-consuming bioinformatics analysis. Various technologies, such as metabolomics, metaproteomics, and metatranscriptomics, have also been developed, making it increasingly important to use a combination of these technologies to fully understand complex microbial populations (Shahsavari et al., 2019).

6 Implementing sustainable remediation technologies: trends and challenges

Bioremediation is increasingly being used to clean long-standing PAH-contaminated soils using techniques such as solid-phase treatments, phytoremediation, bioreactors, and composting. Common treatments include biological, chemical, physical, and thermal (Hussain et al., 2021). However, soils with high PAH levels require pretreatment to reduce PAH concentrations and mitigate toxicity. Alternatives, such as chemical oxidation, solvent washing, and thermal desorption, are viable or necessary (Kumar et al., 2021).

Scientists are becoming more interested in environmentally friendly and sustainable methods to clean soils contaminated with PAHs. Despite the long-standing success of most green or bioremediation technologies, their failure at the field scale is primarily attributed to technical and nontechnical issues (Kariyawasam et al., 2022). This highlights the need for more efficient remediation strategies. The remediation process is influenced by four technical factors: microbial, contaminant, soil, weather, and cost, with site-specific limitations affecting the positioning and spatial arrangement of PAHs (Bhatti et al., 2024). The effectiveness of a remedial system is significantly influenced by factors such as the soil water content, temperature, environmental conditions, aeration levels, PAH concentration, soil nutrient deficiency, microorganism competition, secondary carbon bioavailability, presence of inhibitory pollutants, and water activity (Singh and Mishra, 2021). The remedial efficiency is determined by the limiting factor of the contaminant bioavailability. High-molecular-weight PAHs exhibit limited solubility in water, significant solid-water distribution ratios, and enhanced stability owing to their angular ring structure. Nevertheless, existing methods have not achieved the full degradation of HMW PAHs under various soil conditions (Eldos et al., 2022). Surfactants can effectively address bioavailability concerns by enhancing the bioavailability of PAHs for microbial breakdown. Studies have shown that organisms that produce biosurfactants, such as bacteria and fungi, along with their active chemicals on the surface, can accelerate the process of cleaning up soils contaminated with a combination of substances. Nevertheless, limited research has been conducted on the application of biosurfactants for the remediation of PAH-contaminated soils in the field, potentially owing to the significant expenses associated with their manufacturing (Ambaye et al., 2021).

Another constraint is the challenge of quantifying and regulating metabolic pathways within intricate soil ecosystems. To attain favorable outcomes on a significant scale, it is vital to assess biochemical transformations that can be regulated solely by optimal parameters (Banerjee et al., 2024). The selection between ex situ and in situ remediation, as well as the selection of pre- and/or post-treatment alternatives, necessitates meticulous deliberation. The efficacy of bioremediation is contingent on the selection of appropriate microorganisms (Wang et al., 2024). The bioremediation of soils contaminated with PAHs has historically encountered challenges, including the presence of indigenous PAH degraders, effectiveness of both Indigenous and introduced strains, tolerance of co-contaminants, self-assimilation of crucial nutrients, and capacity to tolerate varying weather conditions (Ambaye et al., 2022b). To overcome this drawback, it is necessary to establish consistent growth conditions for PAH degraders in small-scale laboratories.

Furthermore, the remediation of polluted soils is influenced by technical and nontechnical factors, public and regulatory attitudes, time constraints, and residual contaminants. Regulations govern cleaning processes and impose restrictions on methods such as genetically modified organisms (Kuppusamy et al., 2017). Regulation and environmental laws control remediation processes and market value. While more rigorous investigation is needed, funding for fundamental research decreases (Kuppusamy et al., 2017). An effective remediation program necessitates a comprehensive approach, which is frequently lacking. Universities and commercial institutes provide additional training opportunities to facilitate the acquisition of remediation skills (Kuppusamy et al., 2017). The absence of value-added products resulting from remediation hinders venture capital investment and hampers commercial research and development activities. The process of site remediation is characterized by its nontechnical nature, necessitating meticulous preparation and precise execution (Kuppusamy et al., 2017). Site characterization and treatability studies necessitate the examination of various factors, including natural degradation, biodegradability, environmental conditions, post-remediation risks, pre- and post-treatment considerations, appropriate technologies, adequate funding, and the potential placement of pollutants that may not undergo complete degradation (Kuppusamy et al., 2017).

7 Future trends

The increase in PAHs due to industrialization, urbanization, and motor vehicle exhaust necessitates the development of efficient bioremediation approaches. Advancements in molecular technology have revealed mechanisms for PAH degradation; however, understanding the interplay between microorganisms is crucial for effective management strategies (Patel et al., 2020). Regulatory licensing and risk evaluation are necessary to ensure the widespread use of GMOs. Future studies should focus on monitoring effluents, their health implications, and degradation strategies. Advancements in biotechnology such as nanoremediation and transgenic methods are expected to address other pollutants; however, their capacity to resolve PAH-contaminated soils remains unexplored (Punetha et al., 2022).

The design of an efficient, dependable, cost-effective, and risk-aware technique for removing PAHs requires the collaboration of groups of bacteria, fungi, and algae. The implementation of engineered microbial consortia has the potential to address the challenges associated with microbes in contaminated locations characterized by high levels of mixed contaminants. Algal-bacterial consortia offer greater advantages than bacterial consortia because of their ability to release lightweight molecules and extrapolymeric compounds. These chemicals enhance the capacity of bacteria to degrade aliphatic and aromatic pollutants. Microalgae also play a crucial role in providing oxygen to facilitate the aerobic breakdown of pollutants. Subsequent investigations should prioritize the development of algal-bacterial consortia strains capable of breaking down HMW PAH, exhibiting tolerance towards co-contaminants, and generating biosurfactants (Patel et al., 2020).

Nanoremediation is an emerging field that offers a low-cost, reliable, rapid, and risk-based solution for removing PAHs from polluted sites (Rajput et al., 2024). Their compact dimensions and innovative surface coatings make them suitable for in situ use in deeper soils. Combining nanoremediation with other technologies, such as bioremediation, can help remove contaminants. The use of nanofertilizers, nanomaterials, and green-produced nanooxidizers can enhance the effectiveness of PAHs nanoremediation (Mazarji et al., 2021). Integrating nanoremediation with established bioremediation methods can facilitate rapid breakdowns in real-world settings (Rajput et al., 2024).

8 Conclusion

Polycyclic aromatic hydrocarbons are carcinogenic micropollutants that pose significant risks to human health. Successful interventions require a study of transport and exposure routes. Despite advancements in cleaning techniques, most are still inadequate or in their early stages. Several technologies are currently in the prototype stage or being implemented to address PAH contamination, including heating, extraction, oxidation, and bioremediation systems. However, every site has unique characteristics, necessitating a careful evaluation of remediation policies, financial backing, and public reception.

Further research is needed to understand the collaboration between microbial communities and to develop function-directed risk-based green remediation strategies. To mitigate pollution levels, comprehensive strategies should prioritize advanced methodologies such as co-culture degradation, enzyme-mediated breakdown, and nanotechnologies. Examining the specific effects of PAH compounds on human health and the transmission of these chemicals through the food chain is crucial. Moreover, public awareness campaigns and regular evaluations of bioremediation technologies are essential.

9 References

Abdel-Shafy, H.I. and Mansour, M.S., 2016. A review on polycyclic aromatic hydrocarbons: source, environmental impact, effect on human health and remediation. *Egyptian Journal of Petroleum*, 25(1), pp.107–123.

Ahmad, I., et al., 2023. Semiconductor photocatalysts: a critical review highlighting the various strategies to boost the photocatalytic performances for diverse applications. *Advances in Colloid and Interface Science*, 311, p.102830.

Alegbeleye, O.O., et al., 2017. Polycyclic aromatic hydrocarbons: a critical review of environmental occurrence and bioremediation. *Environmental Management*, 60, pp.758–783.

Ambaye, T.G., et al., 2021. Preparation, characterization and application of biosurfactant in various industries: a critical review on progress, challenges and perspectives. *Environmental Technology & Innovation*, 24, p.102090.

Ambaye, T.G., et al., 2022a. Remediation of soil polluted with petroleum hydrocarbons and its reuse for agriculture: Recent progress, challenges, and perspectives. *Chemosphere*, 293, p.133572.

Ambaye, T.G., et al., 2022b. Ex-situ bioremediation of petroleum hydrocarbon contaminated soil using mixed stimulants: Response and dynamics of bacterial community and phytotoxicity. *Journal of Environmental Chemical Engineering*, 10(6), p.108814.

Ambaye, T.G., et al., 2022c. Insights into rhamnolipid amendment towards enhancing microbial electrochemical treatment of petroleum hydrocarbon contaminated soil. *Chemosphere*, 307, p.136126.

Ambaye, T.G., et al., 2023a. Effect of biochar on petroleum hydrocarbon degradation and energy production in microbial electrochemical treatment. *Journal of Environmental Chemical Engineering*, 11(5), p.110817.

Ambaye, T.G., et al., 2023b. Microbial electrochemical bioremediation of petroleum hydrocarbons (PHCs) pollution: recent advances and outlook. *Chemical Engineering Journal*, 452, p.139372.

Ameen, F. and Al-Homaidan, A.A., 2024. Combined impacts of bioaugmentation and vermiremediation on crude oil-contaminated soil: mitigating strategies for prospective environmental management. *Emerging Contaminants*, 10(2), p.100302.

Ansari, F., et al., 2023. Review on bioremediation technologies of polycyclic aromatic hydrocarbons (PAHs) from soil: mechanisms and future perspective. *International Biodeterioration & Biodegradation*, 179, p.105582.

Aparicio, J.D., et al., 2022. The current approach to soil remediation: a review of physicochemical and biological technologies, and the potential of their strategic combination. *Journal of Environmental Chemical Engineering*, 10(2), p.107141.

Awad, A.M., et al., 2020. Adsorption of organic pollutants by nanomaterial-based adsorbents: an overview. *Journal of Molecular Liquids*, 301, p.112335.

Aydin, S., et al., 2017. Aerobic and anaerobic fungal metabolism and Omics insights for increasing polycyclic aromatic hydrocarbons biodegradation. *Fungal Biology Reviews*, 31(2), pp.61–72.

Babu, A.G., et al., 2019. Bioremediation of polycyclic aromatic hydrocarbons (PAHs): current practices and outlook. In Arora, P.K. (ed.). *Microbial metabolism of xenobiotic compounds* (pp. 189–216). Singapore: Springer.

Balmer, J.E., et al., 2019. Sources and environmental fate of pyrogenic polycyclic aromatic hydrocarbons (PAHs) in the Arctic. *Emerging Contaminants*, 5, pp.128–142.

Banerjee, S., et al., 2024. Microbes and microbial strategies in carcinogenic polycyclic aromatic hydrocarbons remediation: a systematic review. *Environmental Science and Pollution Research*, 31(2), pp.1811–1840.

Baran, A., et al., 2021. Distribution of polycyclic aromatic hydrocarbons (PAHs) in the bottom sediments of a dam reservoir, their interaction with organic matter and risk to benthic fauna. *Journal of Soils and Sediments*, 21(6), pp.2418–2431.

Barathi, S., et al., 2023. Recent trends in polycyclic aromatic hydrocarbons pollution distribution and counteracting bio-remediation strategies. *Chemosphere*, 320, p.139396.

Beretta, G., et al., 2019. The effects of electric, magnetic and electromagnetic fields on microorganisms in the perspective of bioremediation. *Reviews in Environmental Science and Biotechnology*, 18, pp.29–75.

Beyer, J., et al., 2010. Analytical methods for determining metabolites of polycyclic aromatic hydrocarbon (PAH) pollutants in fish bile: a review. *Environmental Toxicology and Pharmacology*, 30(3), pp.224–244.

Bharathi, D., et al., 2023. Enhanced photocatalytic degradation of polycyclic aromatic hydrocarbon by graphitic carbonitride-nickel (g–C3N4–Ni) nanocomposite. *Chemosphere*, 345, p.140464.

Bhatti, S.S., et al., 2024. PAHs in terrestrial environment and their phytoremediation. In Malik, A. and Garg, V.K. (eds.) *Bioremediation for sustainable environmental cleanup* (pp. 67–85). Boca Raton: CRC Press.

Birgül, A., et al., 2011. Atmospheric wet and dry deposition of polycyclic aromatic hydrocarbons (PAHs) determined using a modified sampler. *Atmospheric Research*, 101(1–2), pp.341–353.

Chatterjee, D. and Dasgupta, S., 2005. Visible light induced photocatalytic degradation of organic pollutants. *Journal of Photochemistry and Photobiology C: Photochemistry Reviews*, 6(2–3), pp.186–205.

Cheng, M., et al., 2016. Hydroxyl radicals based advanced oxidation processes (AOPs) for remediation of soils contaminated with organic compounds: a review. *Chemical Engineering Journal*, 284, pp.582–598.

Chigbo, C.O., 2013. *Phytoremediation potential for co-contaminated soils* (Doctoral dissertation, University of Birmingham).

Chormey, D.S. and Bakırdere, S., 2018. Principles and recent advancements in microextraction techniques. In Leeburn, W.C (ed.) *Comprehensive analytical chemistry* (Vol. 81, pp. 257–294). Amsterdam: Elsevier.

Dadkhah, A.A. and Akgerman, A., 2002. Hot water extraction with in situ wet oxidation: PAHs removal from soil. *Journal of Hazardous Materials*, 93(3), pp.307–320.

Dai, C., et al., 2022. Review on the contamination and remediation of polycyclic aromatic hydrocarbons (PAHs) in coastal soil and sediments. *Environmental Research*, 205, p.112423.

Das, N., et al., 2021. Paradigm shift in antibiotic-resistome of petroleum hydrocarbon contaminated soil. *Science of The Total Environment*, 757, p.143777.

Dash, H.R. and Das, S., 2018. Molecular methods for studying microorganisms from atypical environments. In Cook, A.M. and McGenity, A.J. (eds.) *Methods in microbiology* (Vol. 45, pp. 89–122). London: Academic Press.

da Silva Medeiros, D.C.C., et al., 2022. Review on carbon-based adsorbents from organic feedstocks for removal of organic contaminants from oil and gas industry process water: production, adsorption performance and research gaps. *Journal of Environmental Management*, 320, p.115739.

Datta, S., et al., 2023. Vermiremediation of agrochemicals, PAHs, and crude oil polluted land. In Bharagava, S. (ed.) *Bio-inspired land remediation* (pp. 287–315). Cham: Springer International Publishing.

De Gisi, S., et al., 2016. Characteristics and adsorption capacities of low-cost sorbents for wastewater treatment: a review. *Sustainable Materials and Technologies*, 9, pp.10–40.

Drwal, E., et al., 2019. Polycyclic aromatic hydrocarbons (PAHs)–Action on placental function and health risks in future life of newborns. *Toxicology*, 411, pp.133–142.

Eldos, H.I., et al., 2020. Rapid assessment of the impact of microwave heating coupled with UV-C radiation on the degradation of PAHs from contaminated soil using FTIR and multivariate analysis. *Arabian Journal of Chemistry*, 13(11), pp.7609–7625.

Eldos, H.I., et al., 2022. Recent advances in the treatment of PAHs in the environment: application of nanomaterial-based technologies. *Arabian Journal of Chemistry*, 15(7), p.103918.

El-Saeid, M.H., et al., 2015. Photolysis degradation of polyaromatic hydrocarbons (PAHs) on surface sandy soil. *Environmental Science and Pollution Research*, 22, pp.9603–9616.

Falciglia, P.P., et al., 2018. A review on the microwave heating as a sustainable technique for environmental remediation/detoxification applications. *Renewable and Sustainable Energy Reviews*, 95, pp.147–170.

Gitipour, S., et al., 2018. Treatment technologies for PAH-contaminated sites: a critical review. *Environmental Monitoring and Assessment*, 190, pp.1–17.

Gorovtsov, A.V., et al., 2018. The influence of heavy metals, polyaromatic hydrocarbons, and polychlorinated biphenyls pollution on the development of antibiotic resistance in soils. *Environmental Science and Pollution Research*, 25, pp.9283–9292.

Gulia, S., et al., 2023. Bioremediation of PAHs using nanotechnology. In Bhargava, H.N. and Saxena, A. (eds.) *Nanobiotechnology for bioremediation* (pp. 249–271). Amsterdam: Elsevier.

Guo, M., et al., 2018. Enhanced polycyclic aromatic hydrocarbons degradation in rhizosphere soil planted with tall fescue: bacterial community and functional gene expression mechanisms. *Chemosphere*, 212, pp.15–23.

Gupta, S., et al., 2015. Molecular approaches for biodegradation of polycyclic aromatic hydrocarbon compounds: a review. *Reviews in Environmental Science and Bio/Technology*, 14, pp.241–269.

Haapea, P. and Tuhkanen, T., 2006. Integrated treatment of PAH contaminated soil by soil washing, ozonation and biological treatment. *Journal of Hazardous Materials*, 136(2), pp.244–250.

Haleyur, N., et al., 2019. Influence of bioaugmentation and biostimulation on PAH degradation in aged, contaminated soils: response and dynamics of the bacterial community. *Journal of Environmental Management*, 238, pp.49–58.

Han, D., et al., 2021. Critical review of electro-kinetic remediation of contaminated soils and sediments: mechanisms, performances and technologies. *Water, Air, & Soil Pollution*, 232(8), p.335.

Hansen, L.D., et al., 2004. Extended bioremediation of PAH/PCP contaminated soils from the POPILE wood treatment facility. *Chemosphere*, 54(10), pp.1481–1493.

Haritash, A.K. and Kaushik, C.P., 2009. Biodegradation aspects of polycyclic aromatic hydrocarbons (PAHs): a review. *Journal of Hazardous Materials*, 169(1–3), pp.1–15.

Hussain, K., et al., 2018. Monitoring and risk analysis of PAHs in the environment. In Hussain, C.M. (ed.) Handbook of environmental materials management (pp. 1–35). Cham: Springer International Publishing.

Hussain, K., et al., 2021. Bioremediation of waste gases and polluted soils. *Microbial Rejuvenation of Polluted Environment*, 2, pp.111–137.

Jacobs, L.E., et al., 2008. Direct and indirect photolysis of polycyclic aromatic hydrocarbons in nitrate-rich surface waters. *Environmental Toxicology and Chemistry: An International Journal*, 27(8), pp.1643–1648.

Ji, Z., et al., 2024. The degradation of polycyclic aromatic hydrocarbons (PAHs) by ozone-based advanced oxidation processes: a review. *Ozone: Science & Engineering*, 46(1), pp.26–42.

Kariyawasam, T., et al., 2022. Polycyclic aromatic hydrocarbon contamination in soils and sediments: sustainable approaches for extraction and remediation. *Chemosphere*, 291, p.132981.

Kaur, R., et al., 2023. Microbiome based approaches for the degradation of polycyclic aromatic hydrocarbons (PAHs): a current perception. *Chemosphere*, 341, p.139951.

Kim, K.H., et al., 2013. A review of airborne polycyclic aromatic hydrocarbons (PAHs) and their human health effects. *Environment International*, 60, pp.71–80.

Kriech, A.J., et al., 2011. Study design and methods to investigate inhalation and dermal exposure to polycyclic aromatic compounds and urinary metabolites from asphalt paving workers: research conducted through partnership. *Polycyclic Aromatic Compounds*, 31(4), pp.243–269.

Kumar, M., et al., 2021. Remediation of soils and sediments polluted with polycyclic aromatic hydrocarbons: to immobilize, mobilize, or degrade? *Journal of Hazardous Materials*, 420, p.126534.

Kumar, V., et al., 2016. Sources, distribution, and health effect of carcinogenic polycyclic aromatic hydrocarbons (PAHs)–current knowledge and future directions. *Journal of the Chinese Advanced Materials Society*, 4(4), pp.302–321.

Kumari, P. and Kumar, A., 2023. Advanced oxidation process: a remediation technique for organic and non-biodegradable pollutant. *Results in Surfaces and Interfaces*, 11, p.100122.

Kumari, S., et al., 2023. Bioremediation of polycyclic aromatic hydrocarbons in crude oil by bacterial consortium in soil amended with Eisenia f etida and rhamnolipid. *Environmental Science and Pollution Research*, 30, pp.1–15.

Kuppusamy, S., et al., 2017. Remediation approaches for polycyclic aromatic hydrocarbons (PAHs) contaminated soils: technological constraints, emerging trends and future directions. *Chemosphere*, 168, pp.944–968.

Lamichhane, S., et al., 2016. Polycyclic aromatic hydrocarbons (PAHs) removal by sorption: a review. *Chemosphere*, 148, pp.336–353.

Lan, J., et al., 2023. An overview of bioelectrokinetic and bioelectrochemical remediation of petroleum-contaminated soils. *Environmental Science and Ecotechnology*, 15, p.100278.

Lawal, A.T., 2017. Polycyclic aromatic hydrocarbons: a review. *Cogent Environmental Science*, 3(1), p.1339841.

Lee, P.H., et al., 2001. Use of solvents to enhance PAH biodegradation of coal tar. *Water Research*, 35(16), pp.3941–3949.

Li, J., et al., 2023. Treatment of PAHs contaminated soil in abandoned industrial site using combined method of improved in situ capping and electrokinetic enhanced-bioremediation. *Journal of Hazardous Materials*, 455, p.131606.

Li, W., et al., 2021. Influence of sludge organic matter on elimination of polycyclic aromatic hydrocarbons (PAHs) from waste activated sludge by ozonation: controversy over aromatic compounds. *Science of The Total Environment*, 797, p.149232.

Li, X., et al., 2008. Biodegradation of aged polycyclic aromatic hydrocarbons (PAHs) by microbial consortia in soil and slurry phases. *Journal of Hazardous Materials*, 150(1), pp.21–26.

Li, Y. and Duan, X., 2023. Wet deposition of polycyclic aromatic hydrocarbons in a remote area of Central South China from 2014 to 2017. *Atmospheric Science Letters*, p.e1201.

Majd, M.M., et al., 2022. Adsorption isotherm models: a comprehensive and systematic review (2010– 2020). *Science of The Total Environment*, 812, p.151334.

Mallah, M.A., et al., 2022. Polycyclic aromatic hydrocarbon and its effects on human health: an overeview. *Chemosphere*, 296, p.133948.

Manian, R., 2023. Bioremediation of polycyclic aromatic hydrocarbons contaminated soils: recent progress, perspectives and challenges. *Environmental Monitoring and Assessment*, 195(12), pp.1–17.

Manousi, N. and Zachariadis, G.A., 2020. Recent advances in the extraction of polycyclic aromatic hydrocarbons from environmental samples. *Molecules*, 25(9), p.2182.

Maranzana, A., et al., 2005. Ozone interaction with polycyclic aromatic hydrocarbons and soot in atmospheric processes: theoretical density functional study by molecular and periodic methodologies. *The Journal of Physical Chemistry A*, 109(48), pp.10929–10939.

Masjedi, M.R., et al., 2023. Polycyclic aromatic hydrocarbon (PAH) leachates from post-consumption waterpipe tobacco waste (PWTW) into aquatic environment-a primary study. *Environmental Pollution*, 327, p.121500.

Matsodoum Nguemté, P., et al., 2023. Effects of varying PAHs-polluted soils on the morpho-anatomy and physiology of Bermuda grass. *Soil and Sediment Contamination: An International Journal*, 32(6), pp.652–664.

Mazarji, M., et al., 2021. Effect of nanomaterials on remediation of polycyclic aromatic hydrocarbons-contaminated soils: a review. *Journal of Environmental Management*, 284, p.112023.

Mohan, S.V., et al., 2006. Bioremediation technologies for treatment of PAH-contaminated soil and strategies to enhance process efficiency. *Reviews in Environmental Science and Bio/Technology*, 5, pp.347–374.

Mutuku, C., et al., 2022. Occurrence of antibiotics and bacterial resistance genes in wastewater: resistance mechanisms and antimicrobial resistance control approaches. *World Journal of Microbiology and Biotechnology*, 38(9), p.152.

Nasr, M., et al., 2018. Recent progress on titanium dioxide nanomaterials for photocatalytic applications. *ChemSusChem*, 11(18), pp.3023–3047.

Ondon, B.S., et al., 2021. Sources of antibiotic resistant bacteria (ARB) and antibiotic resistance genes (ARGs) in the soil: a review of the spreading mechanism and human health risks. *Reviews of Environmental Contamination and Toxicology*, 256, pp.121–153.

Patel, A.B., et al., 2020. Polycyclic aromatic hydrocarbons: sources, toxicity, and remediation approaches. *Frontiers in Microbiology*, 11, p.562813.

Peleyeju, M.G. and Arotiba, O.A., 2018. Recent trend in visible-light photoelectrocatalytic systems for degradation of organic contaminants in water/wastewater. *Environmental Science: Water Research & Technology*, 4(10), pp.1389–1411.

Peng, B., et al., 2023. A systematic review of polycyclic aromatic hydrocarbon derivatives: occurrences, levels, biotransformation, exposure biomarkers, and toxicity. *Environmental Science & Technology*, 57(41), pp.15314–15335.

Petit, P., et al., 2019. Lung cancer risk assessment for workers exposed to polycyclic aromatic hydrocarbons in various industries. *Environment international*, 124, pp.109–120.

Pinelli, D.F.M.G., et al., 1997. Bioremediation of a polycyclic aromatic hydrocarbon-contaminated soil by using different aerobic batch bioreactor systems. *Soil and Sediment Contamination*, 6(3), pp.243–256.

Plata, D.L., et al., 2008. Photochemical degradation of polycyclic aromatic hydrocarbons in oil films. *Environmental Science & Technology*, 42(7), pp.2432–2438.

Priyadarshanee, M., et al., 2022. Mechanism of toxicity and adverse health effects of environmental pollutants. In Das, S. (ed.) *Microbial biodegradation and bioremediation* (pp. 33–53). Amsterdam: Elsevier.

Punetha, A., et al., 2022. An insight on microbial degradation of benzo [a] pyrene: current status and advances in research. *World Journal of Microbiology and Biotechnology*, 38(4), p.61.

Rajput, V.D., et al., 2023. Nano-enhanced microbial remediation of PAHs contaminated soil. *Air, Soil and Water Research*, 16, p.11786221231170099.

Rajput, V.D., et al. eds., 2024. *Nano-bioremediation for water and soil treatment: an eco-friendly approach*. Boca Raton: CRC Press.

Rayaroth, M.P., et al., 2023. Advanced oxidation processes for the removal of mono and polycyclic aromatic hydrocarbons: a review. *Science of The Total Environment*, 857, p.159043.

Rengarajan, T., et al., 2015. Exposure to polycyclic aromatic hydrocarbons with special focus on cancer. *Asian Pacific Journal of Tropical Biomedicine*, 5(3), pp.182–189.

Revellame, E.D., et al., 2020. Adsorption kinetic modeling using pseudo-first order and pseudo-second order rate laws: a review. *Cleaner Engineering and Technology*, 1, p.100032.

Rorat, A., et al., 2017. Vermiremediation of polycyclic aromatic hydrocarbons and heavy metals in sewage sludge composting process. *Journal of Environmental Management*, 187, pp.347–353.

Rubio-Clemente, A., et al., 2014. Removal of polycyclic aromatic hydrocarbons in aqueous environment by chemical treatments: a review. *Science of the Total Environment*, 478, pp.201–225.

Sakshi, S.K. and Haritash, A.K., 2019. Polycyclic aromatic hydrocarbons: soil pollution and remediation. *International Journal of Environmental Science and Technology*, 16, pp.6489–6512.

Sam, K., et al., 2023. Prospects of emerging PAH sources and remediation technologies: insights from Africa. *Environmental Science and Pollution Research*, 30(14), pp.39451–39473.

Sarma, H. and Prasad, M.N.V., 2015. Plant-microbe association-assisted removal of heavy metals and degradation of polycyclic aromatic hydrocarbons. In Mazumder, M., et al. (eds.) Petroleum geosciences: Indian contexts (pp. 219–236). Cham: Springer.

Shahsavari, E., et al., 2019. Biological degradation of polycyclic aromatic compounds (PAHs) in soil: a current perspective. *Current Pollution Reports*, 5, pp.84–92.

Sharma, K., et al., 2023. Characterization of polycyclic aromatic hydrocarbons (PAHs) associated with fine aerosols in ambient atmosphere of high-altitude urban environment in Sikkim Himalaya. *Science of The Total Environment*, 870, p.161987.

Sher, S., et al., 2023. Review of techniques for the removal of polycyclic aromatic hydrocarbons from produced water. *Environments*, 10(3), p.40.

Singh, A. and Mishra, V.K., 2021. Biodegradation of organic pollutants for its effective remediation from the environment and the role of various factors affecting the biodegradation process. In Prasad, M.N.V. (Ed.) *Sustainable environmental clean-up* (pp. 1–27). Amsterdam: Elsevier.

Singh, L., et al., 2020. PAHs, diet and cancer prevention: cooking process driven-strategies. *Trends in Food Science & Technology*, 99, pp.487–506.

Skic, K., et al., 2023. Effect of sorption properties on the content, ecotoxicity, and bioaccumulation of polycyclic aromatic hydrocarbons (PAHs) in bottom sediments. *Journal of Hazardous Materials*, 442, p.130073.

Srogi, K., 2007. Monitoring of environmental exposure to polycyclic aromatic hydrocarbons: a review. *Environmental Chemistry Letters*, 5, pp.169–195.

Sun, K., et al., 2021. A review of human and animals exposure to polycyclic aromatic hydrocarbons: health risk and adverse effects, photo-induced toxicity and regulating effect of microplastics. *Science of The Total Environment*, 773, p.145403.

Sun, L., et al., 2023. Relationship between atmospheric pollution and polycyclic aromatic hydrocarbons in fresh snow during heavy pollution episodes in a cold city, northeast China. *Ecotoxicology and Environmental Safety*, 260, p.115091.

Sun, Y., et al., 2019. Trends of research on polycyclic aromatic hydrocarbons in food: a 20-year perspective from 1997 to 2017. *Trends in Food Science & Technology*, 83, pp.86–98.

Sun, Y., et al., 2020. Oxidative degradation of polycyclic aromatic hydrocarbons in contaminated industrial soil using chlorine dioxide. *Chemical Engineering Journal*, 394, p.124857.

Sverdrup, L.E., et al., 2002. Soil ecotoxicity of polycyclic aromatic hydrocarbons in relation to soil sorption, lipophilicity, and water solubility. *Environmental Science & Technology*, 36(11), pp.2429–2435.

Tamadoni, A. and Qaderi, F., 2019. Optimization of soil remediation by ozonation for PAHs contaminated soils. *Ozone: Science & Engineering*, 41(5), pp.454–472.

Tarigholizadeh, S., et al., 2023. Transfer and degradation of PAHs in the soil–plant System: a review. *Journal of Agricultural and Food Chemistry*, 72(1), pp.46–64.

Taylor, L.T. and Jones, D.M., 2001. Bioremediation of coal tar PAH in soils using biodiesel. *Chemosphere*, 44(5), pp.1131–1136.

Thacharodi, A., et al., 2023. Bioremediation of polycyclic aromatic hydrocarbons: an updated microbiological review. *Chemosphere*, 324, p.138498.

Thakur, S.S., et al., 2023. Polycyclic aromatic hydrocarbon (PAH)–contaminated soil decontamination through vermiremediation. *Water, Air, & Soil Pollution*, 234(4), p.247.

Trellu, C., et al., 2016. Removal of hydrophobic organic pollutants from soil washing/flushing solutions: a critical review. *Journal of Hazardous Materials*, 306, pp.149–174.

Udom, G.J., et al., 2023. Pervasiveness, bioaccumulation and subduing environmental health challenges posed by polycyclic aromatic hydrocarbons (PAHs): a systematic review to settle a one health strategy in Niger Delta, Nigeria. *Environmental Research*, 229, p.115620.

Venkatraman, G., et al., 2024. Environmental impact and human health effects of polycyclic aromatic hydrocarbons and remedial strategies: a detailed review. *Chemosphere*, p.141227.

Vidonish, J.E., et al., 2016. Thermal treatment of hydrocarbon-impacted soils: a review of technology innovation for sustainable remediation. *Engineering*, 2(4), pp.426–437.

Vijayanand, M., et al., 2023. Polyaromatic hydrocarbons (PAHs) in the water environment: a review on toxicity, microbial biodegradation, systematic biological advancements, and environmental fate. *Environmental Research*, 229, p.115716.

Von Lau, E., et al., 2014. Extraction agents for the removal of polycyclic aromatic hydrocarbons (PAHs) from soil in soil washing technologies. *Environmental Pollution*, 184, pp.640–649.

Wang, F., et al., 2024. Bio-slurry-based biodegradation technology for organically contaminated soils: current work and future directions. *Journal of Environmental Chemical Engineering*, 12, p.112033.

Wang, S., et al., 2007. Light-induced cytotoxicity of 16 polycyclic aromatic hydrocarbons on the US EPA priority pollutant list in human skin HaCaT keratinocytes: relationship between phototoxicity and excited state properties. *Environmental Toxicology: An International Journal*, 22(3), pp.318–327.

Wei, L., et al., 2024. The occurrence and sources of PAHs, oxygenated PAHs (OPAHs), and nitrated PAHs (NPAHs) in soil and vegetation from the Antarctic, Arctic, and Tibetan Plateau. *Science of The Total Environment*, 912, p.169394.

Williamson, C.E., et al., 2019. The interactive effects of stratospheric ozone depletion, UV radiation, and climate change on aquatic ecosystems. *Photochemical & Photobiological Sciences*, 18(3), pp.717–746.

Xiao, J., et al., 2015. Organic pollutants removal in wastewater by heterogeneous photocatalytic ozonation. *Chemosphere*, 121, pp.1–17.

Yadav, B., et al., 2020. Degradation of organic micro-pollutants by ultraviolet radiation. In Singh, R.D., et al. (eds.) *Current developments in biotechnology and bioengineering* (pp. 427–455). Amsterdam: Elsevier.

Yu, B., et al., 2017. Remediation of PAH polluted soils using a soil microbial fuel cell: influence of electrode interval and role of microbial community. *Journal of Hazardous Materials*, 336, pp.110–118.

Zhang, X., et al., 2024. Distribution, sources, partition behavior and risk assessment of polycyclic aromatic hydrocarbons (PAHs) in the waters and sediments of Lake Ulansuhai, China. *Marine Pollution Bulletin*, 200, p.116072.

Ziyaei, K., et al., 2024. Association between exposure to water sources contaminated with polycyclic aromatic hydrocarbons and cancer risk: a systematic review. *Science of The Total Environment*, 924 p.171261.

Chapter 2

Trace element contaminants in agricultural soils: challenges and solutions

Chandima Wekumbura, Kansas State University, USA; Chammi P. Attanayake, University of Illinois Urbana–Champaign/Kansas State University, USA; and Ganga M. Hettiarachchi, Kansas State University, USA

1 Introduction

Most trace elements are naturally occurring and are part of the environment. Human activities tend to redistribute trace elements and contaminate other environmental compartments through the production, use, and disposal of products containing potentially toxic trace elements (PTTEs) (e.g. electronics, paints, dyes, batteries, and weapons used in armed conflict). Land application of waste residue and wastewater, pesticides, and certain fertilizers can contaminate agricultural soils. Once accumulated in soils, most trace elements persist in soils. Therefore, we need to create a safe living environment that minimizes the impact of contaminants while taking precautions to prevent or reduce further addition of potentially toxic contaminants to the environment. Soil is a resourceful sink for many contaminants. Soil can naturally detoxify and retain contaminants, restricting their bioavailability. This ability of soil to detoxify and

http://dx.doi.org/10.19103/AS.2025.0151.02

retain contaminants is also referred to as natural attenuation, which means that natural soil processes can alter contaminants in ways that reduce their toxicity and mobility, limiting their ability to transfer into water bodies or accumulate in plants and animals, including humans. However, it should be noted that exceeding the threshold or capacity of natural attenuation in soil will result in contaminants becoming more readily available for transfer and transport within and from the soil. Managing such soils will therefore require interventions to mitigate their impact. Various chemical and biological reactions in soil transform contaminants into stable compounds that are less likely to dissolve in water or be taken up by plants. In addition to transforming contaminants, soil can also physically and chemically bind them, restricting their movement through the environment. Soil's capacity to detoxify or retain contaminants is contaminant-specific and depends on soil properties such as organic matter content, clay content, soil pH, and soil mineralogy. Managing soil to maintain a safe level of the transfer of contaminants into living beings, either directly or indirectly via the food chain, is of utmost importance. By immobilizing and detoxifying contaminants, soil acts as a critical barrier that reduces the risk of these substances entering food chains, such as the crops we eat, or contaminating drinking water. Our goal should be to manage soil, minimize transfer, and maintain a safe level of contaminant transfer into living beings.

Most PTTEs in soil do not freely transport into plants, and even if plants take them up, they do not easily transport from roots to the aboveground plant parts. This phenomenon, known as the 'soil–plant barrier,' limits food chain transmission of heavy metals (Chaney & Giordano, 1977) and other toxic trace elements. The 'barrier' limits plant uptake of trace elements either by chemical process or by limiting plant growth before uptake reaches levels that would cause injury to consumers. For example, Pb uptake is limited by the former mechanism, whereas the latter limits Zn uptake. For elements such as Cd, Se, and Mo, however, the 'soil–plant barrier' is not effective, and their content in plants can reach levels that are potentially harmful to humans or livestock. The potential harm of these elements underscores the urgency of understanding and addressing trace element contamination. In this chapter, we discuss sources, challenges, and harmful effects of common trace element contaminants in agricultural soils and solutions for managing agricultural soils contaminated with trace elements using examples from the literature and case studies.

2 What are the common trace element contaminants and their sources?

2.1 Common trace element contaminants

Trace elements naturally present in the Earth's crust (soil, plants, water) are generally found in soil in relatively low concentrations (<100 mg kg^{-1}).

A variety of elements can be considered as trace elements: metals such as copper (Cu), zinc (Zn), lead (Pb), cadmium (Cd), cobalt (Co), chromium (Cr), manganese (Mn), nickel (Ni), molybdenum (Mo), mercury (Hg), silver (Ag), and tin (Sn), metalloids such as arsenic (As) and antimony (Sb), radionuclides (radioactive elements), such as uranium (U), and nonmetals such as selenium (Se). Some of the metallic trace elements are often called "heavy metals" due to their density being higher than 5 g cm^{-3} or atomic weight being higher than that of iron (Fe) (Antoniadis et al., 2019; Pierzynski et al., 2005; USGS, 2019). Some of the trace elements (B, Zn, Cu, Mn, Mo, and Ni) are essential for plant and microbial growth, while some do not have any role in metabolic activities (As, Cd, Pb, and Hg) (Mehri, 2020; Mishra et al., 2023). Some trace elements can even be carcinogenic (e.g. As) and/or bioaccumulate in tissues (e.g. Hg), leading to harmful effects over time. Another term, PTTEs, refers to trace elements that can have harmful effects on plants, animals, or humans when their concentration exceeds a certain level. However, Zhang et al. (2022) critically evaluated the term potentially toxic elements in various contexts and usage in literature as a 'Something that Means Everything Means Nothing.'

With the rapid socioeconomic development of the last century following the Industrial Revolution, human activities have introduced more trace elements into the atmosphere, water, and soil, resulting in significant environmental pollution (Naidu et al., 2021; Turdi & Yang, 2016). Soil serves as an important repository for all the trace elements released into the environment by human activities. Contamination of agricultural soils is a concern because the quality of agricultural produce directly influences the health and safety of both humans and animals. If soil is contaminated with pollutants, such as trace elements, these harmful substances can be taken up by plants and enter the food chain. In addition, contaminated soil can directly expose individuals through ingestion, inhalation, or dermal application of contaminated soil.

2.2 Sources of trace element contamination of agricultural soils

Agricultural soils can be primarily contaminated by trace elements from agricultural inputs, residential sources, and trace elements released from industries, either directly or indirectly, based on their location (Fig. 1) (Angon et al., 2024; Antoniadis et al., 2019; FAO & UNEP, 2021; Rashid et al., 2023; Taghavi et al., 2023). Atmospheric deposition, flooding, and sediment deposition are other methods that can carry unknown sources of trace elements into the agricultural fields as nonpoint sources (Alloway, 2013).

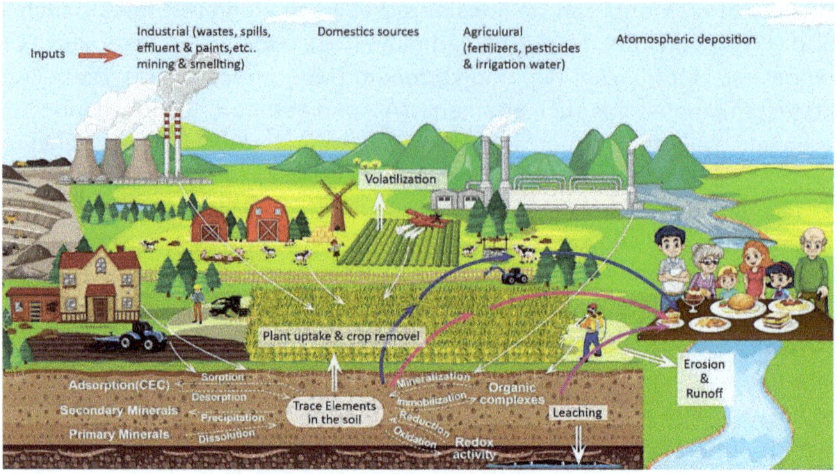

Figure 1 Trace element inputs and transformation in soils.

2.2.1 Agricultural sources

Agricultural inputs have many potential sources of trace elements, including fertilizers, other agrochemicals, manures, reclaimed wastewater, and irrigation water (Alengebawy et al., 2021; Alloway, 2013; Benke et al., 2008). Phosphate fertilizers, liming materials, manures, industrial compost, and biosolids (mostly unregulated sewage sludge) are major sources of trace elements, such as As, Cd, Cu, Hg, Mo, Ni, Pb, Se, and Zn (Bloem et al., 2017; Chen et al., 2020). Agrochemicals such as pesticides, fungicides, and herbicides could contain trace elements such as As, Hg, Cr, Cu, and Pb, which can accumulate in the soil with continuous application (Campos, 2003; FAO & UNEP, 2021). Agricultural wastewater and impure or contaminated irrigation water also contributed to the accumulation of toxic trace elements in agricultural soil (Amjad et al., 2024; Angon et al., 2024).

Application of phosphate fertilizer and Fe, Mn, and Zn micronutrient amendments has been found to be a significant contributor of trace elements such as As, Cd, and Pb in agricultural soils (Chen et al., 2008; Jiao et al., 2012). Many counties have established regulatory limits of Cd in P fertilizers (Table 1). Strict regulations on limits of trace element concentrations implemented by various governments have led to a reduction in Cd levels in phosphate fertilizers. Fertilizer manufacturing companies have invested significant resources in developing fertilizer products low in Cd. The accumulation of Cd in New Zealand and Australian top soils has historically been attributed to long-term application of Cd-rich phosphatic fertilizers (3–17.8 g Cd ha^{-1} $year^{-1}$) (Chaney, 2011; Mann et al., 2002; McDowell et al., 2013), produced from phosphate rock sourced from Nauru and Christmas Island, which contained

Table 1 The maximum acceptable limits for Cd in P fertilizers, the composition of P fertilizers expressed as ratio of Cd:P, Cd:P_2O_5, and Cd concentration in fertilizers across various countries

Country/region	Cd limit	Cd:P mg Cd kg^{-1} P	Cd:P_2O_5 mg Cd kg^{-1} P_2O_5	Cd in product containing 45% P_2O_5 mg Cd kg^{-1}
USA-Washington*	0.0889 kg Cd ha^{-1} $year^{-1}$	2040	889	400
USA-Oregon	338 mg Cd kg^{-1} (P_2O_5)	774	338	152
USA-California	180 mg Cd kg^{-1} (P_2O_5)	412	180	81
AAPFCO	450 mg Cd kg^{-1} (P_2O_5)	1030	450	202
Australia	131 mg Cd kg^{-1} (P_2O_5)	300	131	59
New Zealand	122 mg Cd kg^{-1} (P_2O_5)	280	122	–
Canada*	0.0889 kg Cd ha^{-1} $year^{-1}$	2040	889	400
China	23 mg Cd kg^{-1} (P_2O_5)	10	23	–
Japan	-	340	148	67
Denmark	48 mg Cd kg^{-1} (P_2O_5)	110	48	21.6
the Netherlands	31 mg Cd kg^{-1} (P_2O_5)	71	31	–
Finland	21.5 mg Cd kg^{-1} (P_2O_5)	49	21.5	9.7
Sweden	44 mg Cd kg^{-1} (P_2O_5)	100	43.7	19.7
EU countries from 2020 onwards	60 mg Cd kg^{-1} (P_2O_5)	137	60	27

§Annual application rate; *AAPFCO: American Association of Plant Food Control Officials.*
Source: Aadapted from: Chaney (2011); Fertilizer Association (2024); Li et al. (2020); Marini et al. (2020).

high Cd concentrations (>600 mg Cd kg^{-1} P^{-1}) (Mamun et al., 2017). Studies have indicated that soil Cd enrichment has slowed or stopped due to use of low-Cd fertilizers in these two countries (McDowell & Gray, 2022; De Vries & McLaughlin, 2013). Similarly, the historical (pre-1955) use of lead arsenate ($PbHAsO_4$) pesticides in former orchards across different parts of the USA has resulted in legacy Pb and As soil contamination (Fritz et al., 2022; McBride, 2013).

Different states in the US began adopting limits for trace elements in fertilizers, and the American Association of Plant Food Control Officials (AAPFCO) established recommended limits (Chaney, 2011). Internationally, several countries introduced regulations, and the EU considered progressively lowering Cd limits in phosphate fertilizers from 60 to 40 to 20 mg Cd per kg P_2O_5. However, in 2019, the EU adopted a revised regulation (EU) 2019/1009 with a limit value of 60 mg kg^{-1} P_2O_5 without approving a continued reduction plan (Marini, 2020).

Supplements used in animal feed and chemical additives used in organic amendments can contribute to contaminating soils with trace elements (Capar et al., 1978). Animal feeds are routinely supplemented with additives to minimize mineral deficiencies, enhance feed efficiency, or prevent diseases. Manure is increasingly being treated with chemical additives to control odor, adjust pH, precipitate suspended solids, or enhance biological treatments (Day & Funk, 1998). Examples are liquid dairy manure has 9.59 mg Se L^{-1}. Swine and poultry manure have 1.20–2.40 mg Cr L^{-1}, while liquid dairy has 10 times less (Mayland et al., 1989; NRC, 1988).

The reuse of various residual materials, such as biosolids, manures, and wastewater, in agriculture replaces the disposal costs with environmentally beneficial agronomic uses. Schlosser et al. (2014) predicted that 52% of the world's population will live in regions under water stress by 2050, creating a pressing need for wastewater recycling and reuse for agricultural production. These materials could contain trace elements that potentially contribute to contaminating agricultural soils.

Land application of Class A and B biosolids is a common practice in many states in the USA. Biosolids that meet both pathogen reduction requirements and metal limits (pollutant concentrations) are considered Class A, while biosolids that meet metal limits are considered Class B. 'Exceptional Quality' (EQ) biosolids also meet high levels of vector control requirements. Exceptional quality biosolids can be used in agriculture with a few site restrictions except, following normal agronomic practices. Land application of biosolids require to follow best management practices, consider site location (e.g. proximity to water bodies, buffer zones around water bodies), and not exceed ceiling levels established for each metal(loid)s.

There are growing concerns related to trace elements in irrigation water (Gillispie et al., 2015; Heikens et al., 2007; Mora et al., 2022; Stroud et al., 2011). Therefore, irrigation water quality guidelines usually have set limits on trace element concentrations (Table 2) (FAO, 1999), and irrigation water management plants require testing for trace elements (CLSI, 2014). Arsenic and Cr-contaminated irrigation water has recently been the focus of numerous studies due to both prevalence and the associated risk of food crop contamination (Alam et al., 2021; Mahfooz et al., 2020; Marks et al., 2021; Shahid et al., 2021).

2.2.2 Residential sources

Residential sources of trace elements are directly linked to several primary contributors: paints and other decorative materials, the corrosion of metal structures like roofing and fencing, coal burning, and disposal of sewage (Alloway, 2013; Clarke et al., 2015; Malone & Shakya, 2024; Mitchell et al., 2014).

Table 2 Regulations for irrigation water quality – FAO standards and reclaimed water for irrigation in short- and long-term applications.

Trace elements	FAO standards for irrigation[a] water (mg L^{-1})	Reclaimed water for irrigation	
		Long-term use (mg L^{-1})	Short-term use (mg L^{-1})
Arsenic	0.1	0.1	2
Cadmium	0.01	0.01	0.05
Chromium	0.1	0.1	1
Cobalt	0.05	0.05	0.5
Copper	0.2	0.2	5
Lead	2	5	10
Mercury	0.02	–	–
Molybdenum	0.01	0.01	0.05
Nickel	0.2	0.2	2
Selenium	0.02	0.02	0.2
Vanadium	–	0.1	1
Zinc	2	2	10

[a]Guidelines for irrigation water quality (G.N. No. 617 of 1999).
Source: Adapted from: Rowe and Abdel-Magid, 1995 in USEPA (2004).

2.2.3 Industrial sources

Industrial activities such as mining and smelting, burning fossil fuels, metal processing in refineries, microelectronics, wood, textile, and leather processing, improper handling and incineration of industrial waste (both solid and effluent water), and the production of paints, pigments, dust, and fumes can directly or indirectly mobilize PTTEs into soils. These industrial activities are the primary source of Zn, Pb, Ni, As, Hg, Cu, Cd, and Cr accumulation in agricultural soils (Angon et al., 2024; Dudka et al., 1995; Ephraim et al., 2013; Wu et al., 2018). Dyeing, glass coloring, and textile industrial effluent used for irrigation can contribute to the contamination of agricultural lands with Pb, Cd, Ni, Cu, and Cr (Riza et al., 2021; Tusher et al., 2017). Wastewater from the leather processing industry is rich in Cr and Na. In addition, As used as a preservative in leather tanning, and Pb and Cd are essential parts of pigments used for tanning. In some countries, it is common for small-scale leather industries to discharge this wastewater into open fields or ditches, rendering the land unsuitable for cultivation due to contamination with these trace elements (Afzal, 2014; Khalid et al., 2021; Mizan et al., 2023). Copper, Cr, As, Pb, Hg, and Zn are common trace elements found in preservatives used in industries such as wood preservation (USEPA, 1992), the production of alloys, steel, pigments, and batteries (Tchounwou et al., 2012).

3 Transformation of trace elements in soils

The introduction of trace elements into agricultural soil, currently either natural or anthropogenic sources, is almost impossible to avoid. Soil possesses various mechanisms for retaining, eliminating, or reducing toxicity of trace elements. These mechanisms have significant implications for agriculture, as they can affect the availability of essential nutrients for plant growth and the potential for trace element accumulation in crops. The main processes that can eliminate trace elements are leaching, absorption by plants, subsequent removal through harvest, soil erosion, and volatilization (Fig. 1). Primary mechanisms for retention of trace elements in soils are sorption (adsorption, absorption, and surface precipitation) onto organic and mineral colloids and complexes (through specific, such as chemisorption, or nonspecific mechanisms, such as ion exchange), and co-precipitation (Fig. 1) (Alloway, 1995; Pierzynski et al., 2005). Chemical form (i.e. speciation) in soil solid and soil solution determines the fate, behavior, and toxicity of trace elements. Although relative toxicity can also depend on the specific organisms and environmental conditions, some species of trace elements can be more toxic than others. For example, among As(III) (arsenite), As(V) (arsenate), Se(IV) (selenite), Se(VI) (selenate), Cr(III) chromite and Cr(VI) chromate, arsenite (As^{III}), selenite (Se^{IV}), and chromate (Cr^{VI}) being more toxic (Pierzynski et al., 2005).

Soil solution is the primary reaction medium that interacts with soluble species and the solid phase. Trace elements also exist in soil solution primarily as free cations (e.g. Ag^{+}, Cd^{2+}, Co^{2+}, Cr^{3+}, Cu^{2+}, Pb^{2+}, Zn^{2+}, Hg^{2+}, Ni^{2+}) or oxyanions (e.g. $AsO4^{3-}$, CrO_4^{2-}, MoO_4^{2-}, SeO_4^{2-}). The trace elemental cations and oxyanions often display similar properties and behavior to other major elements, such as Ca^{2+}, Mg^{2+}, PO_4^{3-}, etc. Soil solid surfaces are highly reactive due to their large surface area and functional groups and reactivity depends on the type of mineral or organic functional group present. The ionic form of trace elements first forms soluble complexes with other elements or with dissolved organic carbon (DOC) in solution. Cationic or anionic trace elements can undergo hydrolysis or complexation in soil solution. Complexation can occur through weak electrostatic interactions known as outer-sphere complexation, where a charged surface functional group of a solid interacts with an ionic form of a trace element across one intervening water molecule. Outer surface-bound ions are easily replaced by other ions in solution, playing a key role in retaining both cations and anions or exchanging them with other ions. Strong covalent bonding forms inner-sphere complexation when there is no water molecule between the adsorbed ionic trace element and the solid (Fig. 2). The term 'monodentate' indicates that one surface functional group bonds to the adsorbed species, while 'bidentate' indicates two bonding positions are involved in inner-sphere complexation. The term 'mono/binuclear' shows the number of metal atoms

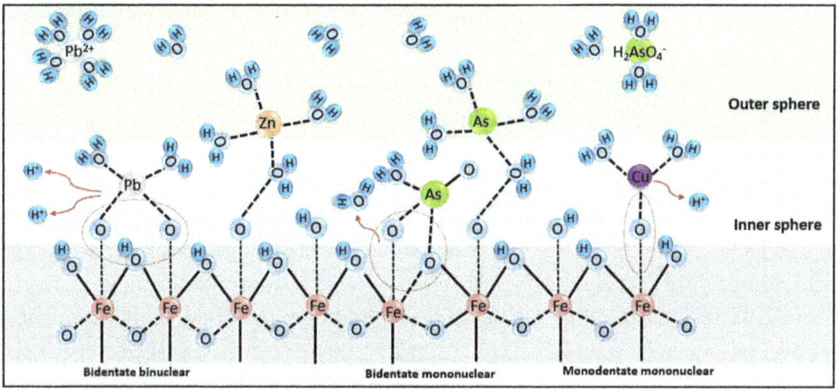

Figure 2 The complexation and ligand exchange of cations and oxyanion in soil solution with soil solid surface by releasing protons and water molecules.

involved in the bonding. Later, trace elements can undergo sorption processes or form precipitates (Pierzynski et al., 2005; Essington, 2015).

An increase in DOC in soil solution can impact trace element bioavailability and mobility. Reactivity of organic functional groups in DOC is controlled by pH-dependent charges and hydrophilic and hydrophobic character. Organic functional groups can ionize and transform into their anionic or cationic forms through proton dissociation or association reactions (Essington, 2015). The adsorption of trace elements onto soil surfaces can be reduced by DOC in the soil solution due to competition for sorption sites by DOC. On the other hand, DOC contains a variety of functional groups, capable of forming soluble complexes or ion pairs with cations or anions (Essington, 2015). The formation of these soluble organo–metallic complexes can enhance the risk of increased bioavailability and mobility of trace elements. Enhanced plant uptake, availability, and leachability of trace elements such as Cd, Pb, Ni, and Zn have been observed in many studies with increasing DOC in the soil (Antoniadis & Alloway, 2002; Hernandez-Soriano & Jimenez-Lopez, 2012).

Soil processes, such as sorption/desorption and precipitation/dissolution, take place under specific soil conditions. Soil pH, organic matter content, clay content, redox status, and mineralogy primarily influence and control these soil processes (Neina, 2019; Palansooriya et al., 2020; Rieuwerts et al., 1998). The presence of more organic matter and clay minerals can enhance the adsorption of trace elements, and carbonates or phosphates in soils can precipitate trace elements, resulting in highly stable secondary minerals (Bradl, 2004; Sipos et al., 2008; Yu et al., 2023). Under low soil pH, cationic trace elements adsorbed onto cation exchange sites of soil colloids (i.e. clay and organic matter) are released into the soil solution, while the anionic trace elements in the soil solution adsorb onto soil colloids. Conversely, at high soil pH, cationic trace elements tend to adsorb onto the colloids and anionic trace elements are released into the soil solution.

The trace elements that exist in multiple oxidation states at natural pH and redox range show more complex transformations in soils, forming complexes that vary in mobility and toxicity. The dominant inorganic species of As and Se in soil solutions are As(III) (arsenite), As(V) (arsenate), Se(IV) (selenite), and Se(VI) (selenate). The toxicity of these elements depends on their oxidation state, with the lower redox states (arsenite and selenite) being more bioavailable and mobile and arsenite being more toxic. Under the most common natural pH range (6–8), arsenite and selenite are predominantly present in solution as $H_3AsO_3^0$ and $HSeO_3^-$, respectively. In contrast, arsenate primarily occurs as $H_2AsO_4^-$ and $HAsO_4^{2-}$, while selenate is present as SeO_4^{2-}. Redox transformations of As and Se occur relatively slowly, both oxidation states are often found coexisting in soil solutions (Goldberg, 2011). Selenite and arsenate sorbed to soil colloids inner-sphere mechanisms and thereby reducing Se^{IV} and As^{V} mobility (Goldberg et al., 2008; Goh & Lim, 2004). Among the two most stable and common oxidation states of Cr, Cr(VI) is readily available in soil solution (pH > 6) as chromate (CrO_4^{2-}) or dichromate ($Cr_2O_7^{2-}$) anions and is considered the more toxic form prevalent in oxidizing conditions and in alkaline soils. Less mobile Cr(III) under acidic or neutral soils strongly adsorbed onto soil particles (Chaney et al., 1996a; Smith et al., 1989). Molybdenum in soil solution under most environmental conditions is prevalent as molybdate (MoO_4^{2-}) anions, which increase with rising soil pH, particularly in alkaline soils (Goldberg et al., 1996, 2002; Kaiser et al., 2005).

4 Transfer of trace elements to plants and animals and regulatory mechanisms

4.1 Total and bioavailable concentrations of trace elements in soils

Numerous studies have been conducted to explore the toxicity of trace elements to macro and microorganisms, plants, and humans. A wide variability in the total concentrations of trace elements in soils, causing toxicity in these organisms, has been observed because of the variable speciation of trace elements in soils depending on soil properties. Total elemental concentration, commonly estimated by 'fully' dissolving these materials in strong extractants like concentrated nitric and perchloric acids, along with the application of high heat and pressure, typically represents trace elements associated with all soil components. In addition, several direct methods or analytical techniques (portable X-ray fluorescence spectroscopy devices (pXRF), near-infrared (NIR), and mid-infrared (MIR) methods) have been developed to measure total trace element concentrations in these materials (Rieuwerts et al., 1998).

Measuring the total concentration of trace elements alone is insufficient for a comprehensive understanding of fate and transport of trace elements in the environment and their health impact on organisms. The bioavailable fraction of trace elements is crucial for plant and human health. Bioavailability refers to the portion of the total concentration of trace elements in the soil that is accessible to organisms (Hettiarachchi & Pierzynski, 2004). Generally, trace elements that exist in soil solution or potential to readily transfer in soil solution as a free ion or complexes become bioavailable.

4.2 Transfer of trace elements to plants

Plants can be contaminated with trace elements by absorption of trace elements in soils from roots or by deposition of trace elements bound to particulates on plant surfaces. While plants require certain trace elements for growth, even essential elements become toxic at higher concentrations. Crops growing in areas with industrial activities that release trace elements into the atmosphere can have a significant proportion of trace elements deposited on the above-ground plant parts. Recent studies have revealed that some trace elements deposited on leaf surfaces from atmospheric sources can be absorbed primarily through stomata, cuticles, and epidermal pores. Dong et al. (2024) and Wang et al. (2021) reported a strong correlation between excess Cd received from atmospheric deposition and increased Cd accumulation in rice plants. Ma et al. (2022) found that wheat leaves could directly uptake atmospheric fine particulate matter Cd ($PM_{2.5}$) through stomata using scanning electron microscopy and energy dispersive X-ray spectroscopy (SEM-EDS) techniques. Uzu et al. (2010) suggested two mechanisms of leaf Pb transfer, based on their observation, that Pb-containing nanoparticles are found in stomata and cuticles on lettuce leaves. Lead particles may penetrate the apoplast as solid compounds and/or dissolved Pb from source particles can diffuse through aqueous pores in the cuticle and stomata via a hydrophilic pathway, potentially causing necrotic lesions enriched with Pb. Plants growing in contaminated agricultural soil can absorb trace elements through their roots. Plant roots are able to absorb trace elements from root rhizosphere, which may then accumulate in plant tissues. The uptake and movement of trace elements within plants are influenced by various factors. Soil characteristics, including pH, organic matter, clay content, cation-exchange capacity, root surface area, root exudates that solubilize trace elements, and the presence of mycorrhizae, represent external factors. Internal factors include plant genetics, plant root structure, physiological processes such as transpiration and photosynthesis, and the plant's life stage (Chen et al., 2021; Ma et al., 2008; Ur Rahman et al., 2024; Zhao et al., 2010).

4.2.1 Plant regulatory mechanisms of potentially toxic trace elements

Plants have two main strategies to manage potentially toxic trace elements: external exclusion and internal tolerance. In the exclusion strategy, trace elemental ions are prevented from entering plant cells by being immobilized in the rhizosphere through interactions with soil microorganisms, mucilage, and root exudates containing organic acids, pectins, and other metal-binding compounds (Krzesłowska et al., 2016; Pasricha et al., 2021). Traina and Laperche (1999) reported the precipitation of pyromorphite ($Pb_5(PO_4)_3X$, where X = F, Cl, OH^-, Br) particles on the surface of roots treated with apatite in plants grown in Pb-contaminated soil. Localized acidity within the rhizosphere enhances P availability, while another mechanism, root-exudate phosphatase, causes increased local phosphate concentrations in the rhizosphere. The internal tolerance mechanism involves neutralizing metal toxicity within cells and removing harmful ions from the protoplasts (Krzesłowska, 2011). This is achieved through the formation of organic ligands like cysteine, glutathione, phytochelatins, and metallothioneins, which reduce toxicity. Plants that employ an accumulation strategy storing metals in extracytoplasmic compartments (compartmentalization), such as cell walls and vacuoles, thereby protecting sensitive sites in the protoplast. Kopittke et al. (2011) reported that Cu accumulates within the cell wall region (rhizodermis), primarily bound to polygalacturonic acid, the main component of pectin in the cell wall. Accumulation of Zn as Zn phytate within vacuoles of root cells (Kopittke et al., 2011) and mature root parts accumulated Ni in the endodermis and pericycle also reported (Seregin et al., 2007). Metal ions are bound in the cytosol by low-molecular-weight compounds, proteins or metabolites and transported across the tonoplast into the vacuole (Krzesłowska et al., 2016; Lange et al., 2017). The Casparian strip, a specialized structure in the cell walls of endodermal cells in plant roots acts as a physical barrier, ensuring that harmful substances, such as trace elements, are filtered out or regulated while allowing selective entry of essential substances into the plant's vascular system (Chen et al., 2011; Collin et al., 2022; Seregin et al., 2004; Zhao et al., 2022). Indicators are another type of plant response, in which proportional relationships are observed between the concentration of trace elements in soil and their uptake and accumulation in plant parts via transport regulation. The concentration of the element in the plant tissue can be directly related to the concentration in the soil due to plant absorption and accumulation (Baker, 1981; Pidlisnyuk et al., 2021).

Uptake and accumulation of PTTEs vary widely between plant species and even among cultivars within a species. Understanding how plants transport and tolerate trace elements is crucial for breeding nutrient-rich crops that can withstand PTTE exposure. However, much remains to be learned about how plants sense and regulate metal levels in response to deficiencies or toxicities. The ability of plants to accumulate trace elements from the soil can be estimated by the enrichment coefficient (EC), or bioconcentration factor (BCF), or bioaccumulation factor (BAF), which is based on the ratio of metal in the plant to the metal in the soil. Plants are categorized based on EC of trace elements as high-accumulator plants (EC 1–10), medium-accumulator plants (EC 0.1–1), low-accumulator plants (EC 0.01–0.1), and non-accumulator plants (EC < 0.01.) (Pidlisnyuk et al., 2021; Zgorelec et al., 2020). Hyperaccumulators are unique plants capable of accumulating specific metals or metalloids in their tissues at concentrations that are 100–1000 times higher than what is typical for most plants (Reeves et al., 2018).

A meta-analysis of 19 studies on metal accumulation in *Amaranthus* spp. revealed that, based on BCFs (ratio of metal in the plant to the metal in the soil), the accumulation potential for Cd in the leaf was high (BCF > 1, critical value), while it was moderate in the root and stem. All 19 studies included both field (37%) and greenhouse pot experiments (63%), using soils from various contaminated sources such as industrial sites, mines, and sewage-impacted areas (47%), as well as artificially contaminated (laboratory-spiked) soils (53%) (Tőzsér et al., 2023). Plant uptake, as determined by field experiments, is considered a more accurate representation of the uptake of trace elements in soil. There can be differences in uptake observed in soil pot experiments, especially with small amount of soil (<1 kg). A pot experiment conducted by Ahmed et al. (2022) reported that bioconcentration values <0.5 for Cr, Zn, Cd, Ni, and Cu for five different species of leafy vegetables grown in tannery effluent-contaminated soils. Experimental soil was collected from canal side at downstream of a tannery area with a history of discharging approximately 20 000 m^3 of tannery effluents daily for over 50 years (Ahmed et al., 2022). A field study conducted by Zhou et al. (2016) reported that <0.5 bioconcentration values for Pb, Cd, Cu, Zn, and As in five types of vegetables representing root, stem, fruit, and seeds as edible portions. This study reported ~1.0 bioconcentration value in Cd in leafy vegetables. This farmland soil was contaminated with multiple elements as a result of mine tailings deposited by floodwaters. Similarly, Samsøe-Petersen et al. (2002), in their field study, found that Cr, Cu, Ni, Pb, Zn, and As had <0.5 bioconcentration value in moderately to highly contaminated soil grown with root and legume crops while >1 bioconcentration value for Cd in lettuce and carrot with peel grown in uncontaminated soil. Soil piles, approximately 12 m × 12 m and 0.5 m deep, were prepared from uncontaminated agricultural soil, moderately contaminated soil from diffuse sources, and highly contaminated soil from a former landfill, as used in Samsøe-Petersen et al. (2002) for their study. These studies utilized soils with varying sources of contamination and experimental setups. This highlights the significance of the origin or source of contaminants, as well as factors such as the age of contaminants in the soil, in influencing the availability of trace elements in the soil. The differences were evident in experiments conducted with freshly spiked versus field-contaminated/aged soils. Studies based on freshly spiked soils may overestimate the risk of plant uptake and human exposure compared to real-world conditions with aged contaminants. Freshly spiked contaminants are more bioavailable as they have not undergone any attenuation processes in the soil. Their limited interactions with soil constituents result in higher mobility and availability, making them more readily absorbed by plants and other organisms due to their immediate presence in the soil solution (Oorts et al., 2006; Smolders et al., 2015).

Growing conditions can also impact plant uptake of trace elements. For example, excessive accumulation of Cd and As in rice grains can occur without affecting yield in contaminated fields, particularly under specific soil and water management conditions. Flooded, anaerobic environments favor As mobilization through the reduction of arsenate to arsenite and the reductive dissolution of iron oxyhydroxides, releasing adsorbed or coprecipitated As into the solution. Managing soil contamination and adopting practices such as intermittent flooding can help reduce the risk of As accumulation in rice grains (Williams et al., 2014; Xu et al., 2008). Intermittent flooding reduces the risk of arsenite formation and As uptake by limiting anaerobic conditions that favor arsenite production. However, this practice can increase Cd availability in the soil, leading to higher Cd accumulation in rice grains. Soil Cd exhibits rapid mobilization during drying and immobilization during flooding, with near-complete transitions occurring within 2 days of aerobic or anaerobic conditions (Carrijo et al., 2022; Zhao & Wang, 2020). In rice soils, field drainage during the grain-filling stage, intended to optimize yields and facilitate harvesting, lowers soil pH to pre-flooding levels, increasing Cd phytoavailability. This also enhances the solubility of secondary Al precipitates to release Cd and promotes the desorption of Cd from soil organic matter. Conversely, the pH levels alter toward near neutrality as the system transition from oxidizing to reducing conditions, and oxidation leads to the formation of Fe and Mn oxides, which co-precipitate Cd, reducing its availability (Chaney et al. 1996b; Simmons et al., 2008). Tang et al. (2023) reported that root uptake and further transport of Cd within plants follow divalent cationic nutrient transporters such as Fe(II), Mn(II), and Zn(II). Chen et al. (2021) reported that in moderately Cd-contaminated acidic soils, strategies such as liming the soil to a pH of 6.5 or continuous flooding are effective in reducing rice grain Cd concentrations. However, these approaches are ineffective in highly Cd-contaminated acidic soils. For heavily contaminated alkaline soils with very low extractable Cd concentrations, continuous flooding alone is sufficient, but it is ineffective when Cd concentrations are highly extractable. Genetic engineering of rice plants, such as knocking out *OsNramp5* (transporter responsible for Cd uptake into root cells) or overexpressing *OsHMA3* (tonoplast Cd transporter sequestering Cd into the vacuoles), has proven to be highly effective in producing low-Cd rice grains in any conditions (Chen et al., 2021). Recent breeding programs have focused on increasing Zn and Fe content and reducing As, Cd, and Cu levels in rice grains (Chen et al., 2017; Jing et al., 2023; Sun et al., 2023). Depending on the type and level of contaminants in the field, selecting specific plant species or cultivars, such as low accumulators, can help reduce the transfer of potentially toxic trace elements to humans.

The availability of other nutrients can impact the uptake of potentially toxic trace elements. Arsenate uptake in plants occurs through phosphate transporters due to its chemical similarity to phosphate. In contrast, arsenite is taken up via a highly efficient Si pathway into root cells, followed by efflux into the xylem in rice plants. Additionally, several aquaporin channels (facilitated by nodulin26-like intrinsic proteins) are capable of facilitating arsenite transport (Ma et al., 2008; Zhao et al., 2009; 2010, 2022). Therefore, deficiencies in Fe, Si, and P in the soil can enhance the adsorption of As by plants, as As competes with these essential nutrients for uptake pathways.

4.2.2 Transfer of trace elements to humans and animals

Trace elements can enter the human and animal body by direct contact with contaminated soil through soil ingestion, inhalation, or dermal absorption. Individuals can be exposed to trace elements in soil and dust by consuming fresh fruits, vegetables, grains, or other produce that has come into contact with contaminated soil (i.e. surface contamination). When these crops are grown in contaminated soil, dust containing trace elements and other harmful substances may adhere to plant surfaces and by consuming those plants trace elements can enter the body. This combination of inhalation, dermal, and ingestion represents a significant route of exposure to soil contaminants for both humans and animals (Tchounwou et al., 2012; Wang et al., 2022).

Indirect exposure to trace elements occurs when contaminants move through environmental pathways before reaching humans or animals. When humans consume contaminated food crops, livestock, and their products, the trace elements are transferred up into the food chain (Fig. 3). Through food chain contamination, concentration of contaminants in higher trophic levels can increase (biomagnification) (Barwick & Maher, 2003; Mann et al., 2011).

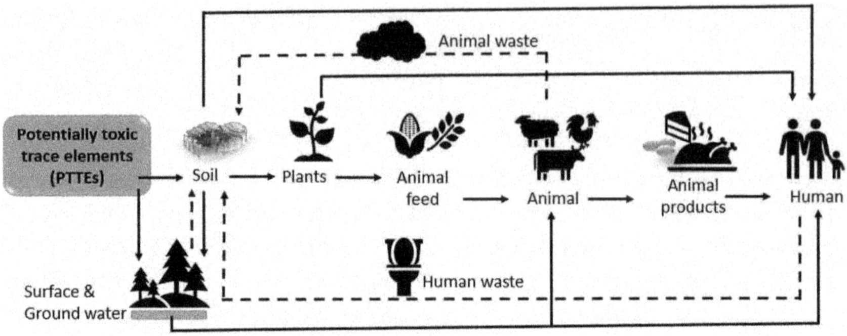

Figure 3 Direct and indirect transfer of potentially toxic trace elements into human. Source: Adapted from: Pierzynski et al. (2005).

5 Adverse effects of potentially toxic trace elements in soil

5.1 Effects on animals and human

Trace elements, which fall under the category of micronutrients, are crucial for the physiological functions, growth, and development of all living organisms, as well as for enhancing immunity (Mertz, 1998). The deficiency in these elements can lead to poor growth and development, and increased vulnerability to diseases. In contrast, exposure to excessive concentrations of essential trace elements can lead to toxicity in living organisms. When present in amounts beyond what is necessary, these elements can disrupt normal biological processes and cause significant health issues. Moreover, trace elements that have no biological function in living organisms can pose severe health risks when they accumulate. In animals, the toxic effects of potentially toxic trace elements begin with the disruption of cellular homeostasis through the generation of free radicals, which leads to oxidative stress and impaired cellular functions (Afzal & Mahreen, 2024). This can cause a decline in overall body condition scores and a significant reduction in the production potential of animals, especially in farm animals (Raikwar et al., 2008). In both animals and humans, prolonged exposure to elevated levels of these elements can result in various health issues, including organ damage, neurological disorders, and an increased risk of cancer (Antoniadis et al., 2019; Gupta & Gupta, 1998; He et al., 2005; Mitra et al., 2022; Nielsen, 2003; NRC, 1989). Well-known toxic elements, such as Cd, disrupt processes involving other essential elements, Ca, leading to health issues such as osteoporosis, renal tubular dysfunction, heart disease, and cancer in humans. Elevated blood Pb levels in young children have been linked to impaired neurological development (Seyfferth et al., 2024). Acute exposure to high doses of As has been associated with clinical signs such as lethargy, ataxia, anorexia, and diarrhea in calves (Faires, 2004).

5.2 Effects on plants

Accumulation of excessive concentrations in the plant, exceeding the optimal levels of essential trace elements, causes direct and indirect damage to plants. Direct effects include inhibition of cytoplasmic enzymes and oxidative stress-induced damage to cell structures. Indirect effects involve the replacement of essential nutrients at cation exchange sites in plants (Schützendübel & Polle, 2002). Lead, Cd, Hg, and As do not play any beneficial role in plant growth and exhibit toxic effects even at low concentrations in the growth medium. A significant reduction in plant height, reduced tiller and panicle formation, decreased shoot and root growth, inhibited seed germination, and reduced biomass were observed in many studies (Tang et al., 2023). Reduced photosynthetic activities, disrupted plant mineral nutrition uptake and translocation, and decreased enzyme activities are the key metabolic effects on plants caused by trace elements (Chibuike & Obiora, 2014; Zhao et al., 2022). Additionally, an excess of certain essential micronutrients, such as Zn, Cu, and Ni, can cause plant toxicity, resulting in stunted growth, reduced yield, and even plant death (McCauley et al., 2011).

5.3 Soil microbial communities and other soil biota

Microbial communities are far more sensitive to trace element stress than soil macrofauna and flora. Trace element contamination in soil inhibits microbial activity, causing both acute toxicity and long-term effects (Giller et al., 1998). Studies of soil contaminated with Cu, Zn, Pb, and other toxic trace elements have shown a reduction in microbial biomass, indicating that the contamination source has a direct impact on microbial health and activity. Trace element contamination imposes stress on sensitive microorganisms, altering their diversity, population, and representation within groups. Higher concentrations of Cd, Cu, Zn, and Pb in mine-affected soil have been found to decrease microbial density and inhibit enzymatic activities. (Giller et al., 1998; Okereafor et al., 2020; Šmejkalov et al., 2003; Xie et al., 2016). Aponte et al. (2020) reported that there was a two times greater reduction in endoenzymes than exoenzymes with potentially toxic trace elements contamination. The overall reduction order of different soil enzymes was reported as arylsulfatase > dehydrogenase > β-glucosidase > urease > acid phosphatase > alkaline phosphatase > catalase, based on a meta-analysis (Aponte et al., 2020). The mechanisms behind the reduction of these enzyme activities by potentially toxic trace elements are primarily attributed to their toxicity to microorganisms. This occurs mainly due to a few factors: (1) trace elements have greater binding affinity than essential elements, and (2) they displace essential elements from their native binding sites or interfere with ligand interactions, (3) trace elements masking ability over catalytically active groups, (4) the denaturing of protein conformation, and (5) trace elements competing with metal ions that are needed to form enzyme–substrate complexes. As a result, the cycling of C, N, P, and S in the soil may be slowed by contamination of potentially toxic trace elements, resulting in potential C and S limitation for microorganisms living in the soil (Aponte et al., 2020).

Earthworms are key indicators of healthy agricultural soil and play a crucial role in maintaining soil health and promoting plant growth by enhancing aeration, nutrient cycling, and soil structure. However, exposure to trace elements such as Zn, Pb, Cd, Hg, and Cu can lead to their bioaccumulation in earthworms. The toxic effects of these trace elements include reduced growth rates, impaired reproduction, DNA damage, and disruptions to their physiological functions (Liang et al., 2011; Uwizeyimana et al., 2017; Wang et al., 2018; Yadav et al., 2023). Since earthworms and soil microorganisms are essential for organic matter decomposition, a decline in their diversity or abundance can disrupt nutrient cycling in the soil. Many arthropod species can detect metal-contaminated areas, but others cannot, leading to reproduction challenges, including damage to the life cycle and reduced fitness. Some insects attempt to escape polluted sites by increasing movement, smaller populations, lower species diversity, and dominance of PTTEs-tolerant species in contaminated environments (Mogren & Trumble, 2010; Tóth et al., 2023).

6 Managing contaminated soils

6.1 Screening for contaminants

Because the sources of contaminants are widespread, screening soils for contaminants is essential. The need for such screening can be narrowed down by assessing potential sources of contaminants. The history of land use, current and past surrounding land uses, and industrial and residential activities in the surrounding areas provide clues for potential sources of contaminants. For example, if the land is located in an urban area, we assume that residential or industrial activities could contaminate the land, and the type of contaminants can be identified based on the specific activities. In large-scale corn fields, agricultural inputs (e.g. fertilizers) can be a source of trace elements.

Once the need for soil screening for trace elements is identified, initial screening for contaminants can be conducted by using in situ field equipment such as handheld XRF. To verify the level of contamination, we could collect surface soils (i.e. 0–15 cm and 15–30 cm or soils from the topmost horizon) based on a sampling scheme and analyze for contaminant concentrations in the laboratories using more robust analytical methods. The results of these analyses would provide critical data to assess the extent of soil contamination.

This information could further guide making decisions on requirements for further testing for different attributes (source identification, water analysis, bioavailability testing), remediation strategies, land-use decisions, and risk assessments.

6.2 Regulatory limits of potentially toxic trace elements

Maximum permissible levels of potentially toxic trace elements in agricultural soil can refer to the highest concentrations of specific trace elements that are legally allowed in soil without posing significant risks to human health, ecosystems, or agriculture (Table 3). These levels vary across countries and regions, depending on regulatory frameworks and environmental conditions (Antoniadis et al., 2020). Gardening and agricultural land use Cd concentration limit set by the US EPA was 0.48 mg kg^{-1} and WHO was 0.003 mg kg^{-1}. However, these values are different from other soil uses. Table 3 shows the target or permissible values of potentially toxic trace elements in agricultural soils set by different countries or organizations. There is a significant discrepancy in soil limits between regulatory bodies, primarily due to variations in how toxicity is evaluated, whether total or bioavailability of trace elements is considered, or the targeted endpoint, such as human health and soil microbial processes.

Table 3 Target or permissible values of trace elements in agricultural soils from different countries

Elements	Common range (average concentration) (mg kg^{-1}) (Lindsay, 1979)	USA[a] (mg kg^{-1})	European Union (Tóth et al., 2023)	China (Monchanin et al., 2021)	India (Srivastava et al., 2017)
As	1–50 (5)	21	5	20–40	–
Cd	0.01–0.7 (0.06)	20	1–3	0.3–0.6	3-6
Cr	1–1000 (100)	1500	50–150	–	–
Cu	2–100 (30)	750	100	–	135-270
Co	–	13–230	20–100	–	–
Hg	0.01–0.3 (0.03)	8	0.5–1.5	0.3–1	–
Ni	5–500 (40)	210	30–75	–	75-150
Pb	2–200 (10)	150	50–300	200	250–500
Sb	–	–	2–10		–
V	–	–	100–150		–
Zn	10–300 (50)	1400	100–200		300–600

[a]Total metal concentration allowed in agricultural soils treated with biosolids; Calculated from maximum cumulative pollutant loading limits, assuming incorporation of biosolids to 15 cm depth and average soil bulk density of 1.33 g cm^{-3} but not including background concentrations of these elements in soils (McGrath et al., 1995).

The maximum permissible concentrations of trace elements in biosolids-applied soils have been evaluated by the USEPA, EU, and other regulatory bodies. The land application of biosolids in the US is regulated by 40-CFR Part 503. It limits the total mass of nine different elements that can be applied to a unit land area and concentration limits in risk analyses (USEPA, 1994). However, there are no specific guidelines for levels of contamination or remediation action. We may use maximum permissible concentrations specified for biosolids-amended soils, as a guide to categorize the levels of contamination and identify the need for further actions, with some caution, considering that soils could have variable bioavailabilities of trace elements depending on soil properties (Strawn & Hettiarachchi, 2021).

Extension publications produced by universities, government, and nongovernment organizations often provide guidelines for gardeners and farmers on safe gardening practices in contaminated soils. These resources focus on soil testing facilities and methods to identify contaminated soils and management of those soils (Brown & Hettiarachchi, 2025). In addition, regulations for fertilizers and other agricultural inputs set maximum permissible levels for toxic elements like Cd, Pb, and As to minimize contaminants input into soils. The EU Fertilizer Regulation establishes strict limits on contaminants like Cd, requiring that fertilizers contain no more than 60 mg kg^{-1} of P_2O_5 to reduce environmental contamination and health risks (Ballabio et al., 2024; Huygens et al., 2019). Regulations for trace elements in food and feed are essential to protect public health and ensure food safety. These regulations establish permissible levels for various trace elements in foods to ensure that trace element intake through food is within safe limits for consumers. The Codex Alimentarius Commission, established by the FAO and the WHO in 1995, is updated yearly or as needed. It provides international food standards, guidelines, and codes of practice to limit toxic trace elements in food. For example, Codex sets limits for Pb concentration in various foods, such as 0.1 mg kg^{-1} fresh weight in most fruits and vegetables, 0.05 mg kg^{-1} fresh weight in milk, and 0.1 mg kg^{-1} dry weight in cereals (CODEX, 2023).

The United States Food and Drug Administration (FDA) regulates some trace elements in foods to ensure that concentrations of these elements are within safe limits. In 2020, the FDA launched the Closer to Zero action plan to reduce exposure to toxic elements such as As, Pb, Cd, and Hg in foods commonly consumed by infants and young children (Woodcock, 2021). The EU countries have the most comprehensive food safety regulations, with specified limits depending on the food type. For example, EU regulations limit Cd to 0.1 mg kg^{-1} in cereals (excluding rice), 0.04 mg kg^{-1} wet weight in vegetables (excluding leafy vegetables), and 0.05 mg kg^{-1} in wet weight meat; the maximum levels for inorganic As are set at 0.2 mg kg^{-1} wet weight for white rice and 0.1 mg kg^{-1} wet weight for products intended for infants (European Commission, 2023).

6.3 Soil amendments for managing agricultural soils contaminated with trace elements

6.3.1 Organic amendments

Incorporation of organic amendments into soils has been tested extensively as a means to reduce the bioavailability of trace elements in soils. Use of organic matter amendment in agricultural lands is becoming popular, especially with increasing interest in urban gardening in potentially contaminated sites that could have low-to-moderate levels of contamination. The immediate beneficial effect of incorporating organic amendments into contaminated soil is the physical dilution of soil matrix and hence, the dilution of the total concentration of the contaminants, which may not directly influence the bioavailability. It is possible to dilute the concentrations of trace elements by 10–53% depending on the amounts of amendments, depth of incorporation, moisture content, and particle size of the amendments (Attanayake et al., 2014, 2015, 2021; Defoe et al., 2014; Farfel et al., 2005). The US EPA recommends incorporating organic amendments to the ratio of 50:50 (soil:amendment in volume-based) in urban gardening to have a significant dilution of contaminants (US EPA, 2011). One argument against relying on organic amendments for dilution of trace element concentration is the length of effectiveness as the organic material decompose with time (Beckett et al., 1979). The regular incorporation of organic amendments every 3–5 years could be a practical solution, offering a new avenue for growers to enhance soil health and crop productivity (Larney & Angers, 2012; Pitumpe Arachchige et al., 2024).

In addition to the dilution of total concentrations of trace elements, organic amendments have the potential to chemically reduce the bioavailability of trace elements in soils. The most common organic amendments used in agricultural soils are compost, biosolids, and animal manure. The properties and composition of these amendments vary widely depending on the processing methods and resources used to produce compost and biosolids, and the type of animal manure. Some key factors of the amendments that affect the speciation and, consequently, bioavailability of trace elements in soils are pH, content, and composition of organic C, concentrations and type of Fe and Mn oxy(hydr)oxides, and concentration of soluble inorganic P (Fig. 4). Soil pH could change according to the amendment's pH after incorporation. The higher the pH, higher the sorption of cationic trace elements (e.g. Cu^{2+}, Zn^{2+}, Pb^{2+}, Cd^{2+}) and the lower the sorption of anionic trace elements (e.g. arsenate (AsO_4^{3-}), chromate (CrO_4^{2-}), and molybdate (MoO_4^{2-})). DOC can increase

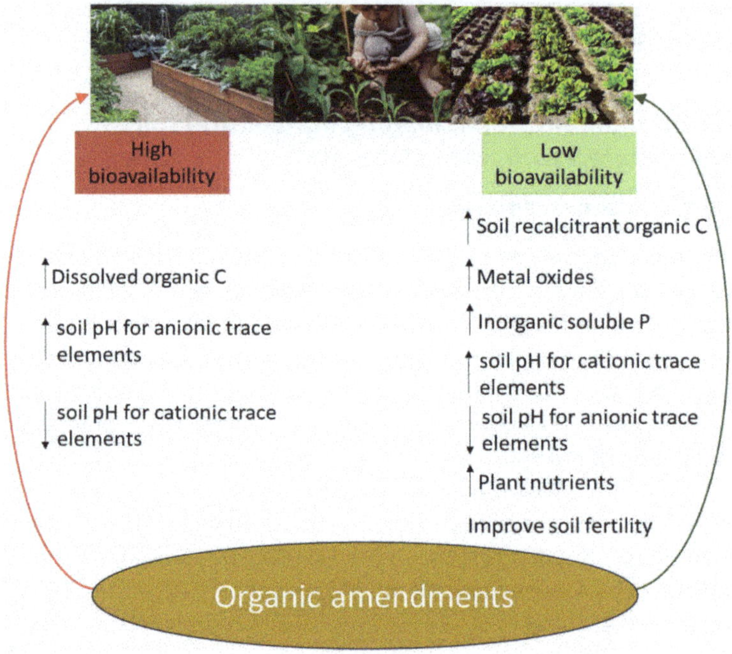

Figure 4 Alterations of soil properties and consequent direction of change of bioavailability of trace elements as a result of incorporating organic amendments into contaminated soils.

the mobility of trace elements in soil upon compost addition (Antoniadis & Alloway, 2002; Ashworth & Alloway, 2008; Attanayake et al., 2015). Metal oxy(hydr)oxides, such as Fe, Al, and Mn oxides and hydroxides, play a key role in adsorbing trace elements in soils. These compounds are highly reactive and possess a large surface area with a variety of surface sites capable of binding trace elements (Li et al., 2024; Suda & Makino, 2016). Biosolids contain more Fe oxy(hydr)oxides compared to other organic amendments (Hettiarachchi et al., 2003; 2006; Li et al., 2001; Mahler et al., 1987; Singh, 1981). Soluble inorganic P forms sparingly soluble phosphate minerals with trace elements such as lead pyromorphite and cadmium phosphate, reducing the bioavailability of trace elements (Palansooriya et al., 2020; Zwonitzer et al., 2003). The bioavailability resulting from the incorporation of an amendment is a combination of the effects of all these factors. Therefore, the effect of incorporation of organic amendments has shown mixed effects on reducing bioavailability of trace elements, as explained in the below examples. Table 4 presents the results of several field studies that examined the application of organic amendments to reduce the bioavailability of trace element contamination.

Table 4 Effect of application of organic amendments to reduce the bioavailability of trace element contamination in field studies

Type of amendment	Trace element concentration	Effect of amendments	Reasoning of mechanism	Reference
Leaf-based compost	Pb: 60–300 mg kg^{-1}	Positive Compost reduced Pb in Swiss chard and carrots and bioaccessibility of Pb	Dilution of soil total Pb and increase in soil available P on compost addition. Average available P concentrations were 68 mg kg^{-1} in the no-compost soils and 438 mg kg^{-1} in the compost-added soils. The dissolution of Pb-Ferrihydrite during the extraction was low to negligible as soil–compost mixture matured in the field. Compost amendment showed potential to contribute to reduced bioaccessibility of Pb as compost aged in the soil by increasing Pb-Organic fraction in the field	Attanayake et al., 2014 Attanayake et al., 2017
Composted biosolids	Pb: 475 mg kg^{-1} As: 95 mg kg^{-1}	Composted biosolids reduced the bioaccessibility of soil Pb by ~17% temporarily increased the bioaccessibility of As by ~ 69% compared with the control when soluble inorganic P concentration in soil was elevated by P fertilizer application No change in concentrations of Pb and As in collard greens, tomato, and carrot	Composted biosolids seemed to be effective at reducing bioaccessibility of soil Pb; this could have been due to the form of Fe in composted biosolids. Increased soluble inorganic P concentration via fertilizer application seemed to induce sparingly soluble Pb-phosphate formation. In contrast, the P fertilizer application increased As bioaccessibility in composted biosolids-amended soils likely via desorbing the adsorbed As from active Fe oxides.	Attanayake et al., 2015
Non-composted biosolids	Pb: 475 mg kg^{-1} As: 95 mg kg^{-1}	No change in Pb and As bioaccessibility No change in concentrations of Pb and As in collard greens, tomato, and carrot	High Pb sorption capacity of soil with high Fe oxides (active Fe 28 g kg^{-1}) and organic C (~30 g kg^{-1})	

(*Continued*)

Table 4 (*Contiuned*)

Type of amendment	Trace element concentration	Effect of amendments	Reasoning of mechanism	Reference
Leaf compost	Pb: 475 mg kg^{-1} As: 95 mg kg^{-1}	No change in Pb and As bioaccessibility No change in concentrations of Pb and As in collard greens, tomato, and carrot	High Pb sorption capacity of soil with high Fe oxides (active Fe 28 g kg^{-1}) and organic C (~30 g kg^{-1})	
Mushroom compost	Pb: 475 mg kg^{-1} As: 95 mg kg^{-1}	Mushroom compost initially increased the percentage of bioaccessible As. No change in Pb bioaccessibility No change in concentrations of Pb and As in collard greens, tomato, and carrot	Mushroom compost initially increased the percentage of bioaccessible As, which could have been a result of increased DOC in mushroom compost – amended soils. Increasing DOC concentration enhances desorption of As from Fe oxides	
Horse manure-based compost	Pb: 200 mg kg^{-1}	No change in bioaccessibility or concentrations of Pb in lettuce, carrot, radish, and beetroots	The increase in the levels of P and organic C in this study was not as large as in some previous studies, where organic amendments were effective in reducing the Pb concentration in plants in urban gardens The high C:N ratio may have resulted in the low reactivity of C in compost due to slow mineralization and humification, 66 which may have been partly responsible for not increasing the Pb sorption capacity of compost-amended soils.	Attanayake et al., 2021
50% biosolids, 25% sand, and 25% sawdust with dolomite	Pb: 51–312 mg kg^{-1} As: 39–146 mg kg^{-1}	reduced As concentrations in lettuce and tomato (46–80%) and carrot (<10%). Pb in carrot, lettuce, and tomato by about 25–80% Reduced bioaccessible Pb by ~30–50% Did not reduce bioaccessible As	Dilution of total concentration of Pb and As. Increase of P	Defoe et al., 2014

Recycled yard and food waste with dolomite	Pb: 506–2022 mg kg^{-1} As: <20 mg kg^{-1}	Pb in carrot, lettuce and tomato were reduced by 20–60% Reduced bioaccessible Pb by 13–15%	Dilution of total concentration of Pb and As. Increase of P	
High Fe (~100 g kg^{-1}) composted biosolids	Pb: 1700 mg kg^{-1} As: 240 mg kg^{-1}	No change in bioaccessibility of Pb as a percent of total Pb. 23% reduction in As bioaccessibility %	It would seem that the $FeCl_2$ addition resulted in some changes in Pb minerals but was insufficient to result in reduced bioaccessibility. Results from the μ-XANES[a] and bioaccessibility extracts suggest that the Fe may have altered the form of As in the Fe wastewater treatment residual compost soil and bioaccessibility of As in the control	Brown et al., 2012
Manure compost	Pb: 816 mg kg^{-1} As: 88 mg kg^{-1}	Pb and As concentrations in root (onion), leaf (kale, cabbage), fruit (eggplant, tomato) crops reduced but not significant. Bioaccessibility of Pb was significantly reduced by ~30%. No change in As bioaccessibility.	The dilution of the total concentrations of Pb and As could be the primary reason, although the paper does not discuss this in detail.	Paltseva et al., 2020

[a]μ-XANES: micro-X-ray absorption near edge structure.

Adding organic amendments to trace element-contaminated agricultural soils offers numerous benefits, including enhancing soil fertility and structure, increasing microbial activity, stabilizing soil pH, and reducing erosion (Lwin et al., 2018; Palansooriya et al., 2020) besides immobilizing and reducing uptake and toxicity of trace elements. Organic amendments provide essential nutrients, including N, P, K, and micronutrients, which contribute to overall soil fertility. They promote aggregation of soil particles, leading to improved soil structure. This leads to increased porosity, reduced compaction, and enhanced aeration and water infiltration. As a result, the soil's capacity to retain moisture improves, reducing drought stress, and enabling more efficient water use. Organic matter serves as an energy source for soil microorganisms, fostering microbial diversity and activity. Many organic amendments act as buffers, helping to stabilize soil pH such as composted materials can neutralize acidic soils, making the environment more conducive to plant growth and reducing the solubility and mobility of toxic trace elements. Additionally, organic amendments enhance the soil's cation exchange capacity, suppress soil-borne pathogens, carbon sequestration and provide a cost-effective means of managing contaminated soils while promoting sustainable agriculture. Finally, crops grown in amended soils often display better growth characteristics benefiting from diluting the total concentrations of trace elements in plant biomass (Figure 5) and are more resilient to environmental stressors.

6.3.2 P fertilizers

The nature of phosphorus compounds and specific trace elements can influence whether their application results in the mobilization or immobilization of metals. Introduction of a multipurpose anion, such as phosphate, which can enhance metal sorption by increasing cation exchange capacity and promoting metal precipitation (Bolan & Duraisamy, 2003). The addition of P fertilizers for P-contaminated soils has been proven to reduce bioaccessibility of Pb in soils (Mayer et al., 2022). Phosphorus forms sparingly soluble Pb–P precipitates (e.g. lead chloropyromorphite and lead hydroxy pyromorphite) in field leading to lowering bioaccessibility (Park et al., 2011; Scheckel et al., 2013). Also, the sparingly soluble Pb–P form in human digestive tract leads to less absorption of Pb into the bloodstream upon direct ingestion of Pb-contaminated soils (Tang et al., 2004; Karna et al., 2018).

Arsenic mobility in soils is primarily controlled by positively charged Fe and Al oxides. Inorganic P fertilizers specifically increase As mobility either through the exchange of orthophosphate for arsenic in soil or by competing for adsorption sites (Campos, 2002; Gupta et al., 2014). Similarly, the application of inorganic P fertilizer has been shown to increase Cd concentrations in crops such as carrots, lettuce, oats, ryegrass, durum wheat, and flax. However,

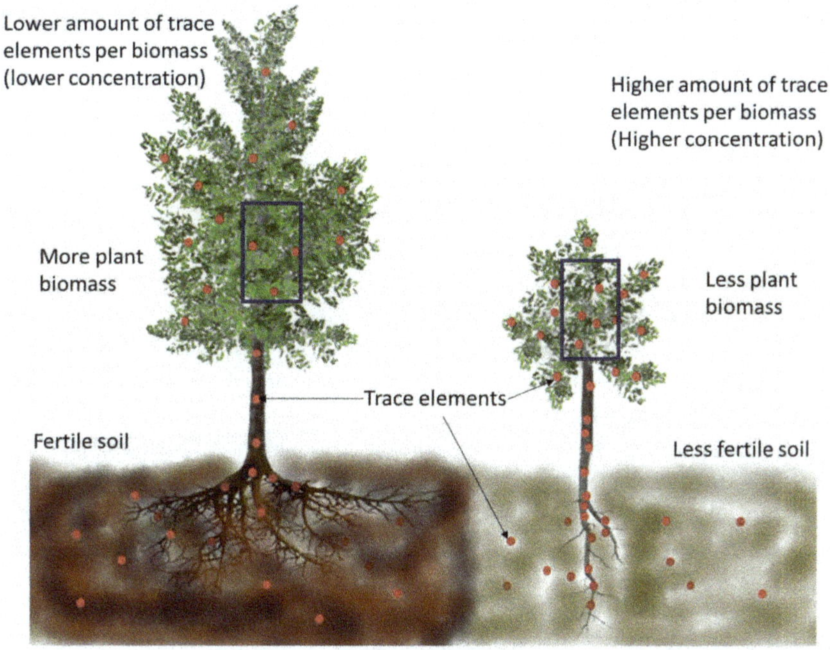

Figure 5 Healthy plant growth dilutes the total concentrations of trace elements in plant biomass. The red dots denote trace elements.

some studies have reported a decrease in Zn concentrations in plant tissues, specifically in durum wheat and flax when applied P fertilizers having Cd in them (Grant et al., 2010; He & Singh, 1995; Jiao et al., 2004).

In contrast, Chen and Lai (2024) found that the application of P-containing organic amendments, such as vermicompost, led to decreased Cd and Ni concentrations in lettuce crops compared to inorganic P fertilizers. Zhang et al. (2021) found that P-enriched modified biochar reduced the available Cd content by 81% via the formation of P–Cd complexes or organometallic chelates. Another field study utilized an amendment composed of limestone, iron powder, silicon fertilizer ($NaSiO_3$), and Ca–Mg–P fertilizer. This combination was found to significantly reduce the absorption and accumulation of Cd and As in rice, bringing levels of both Cd and inorganic As in brown rice within safe edible standards in a moderately co-contaminated paddy field (Jiang et al., 2022). Similarly, Wang et al. (2020) and Gu et al. (2018) reported that the application of hydroxyapatite combined with other amendments such as biochar, manure, and zeolite promoted the transformation of soil Cd and As from soluble to insoluble species, thereby reducing the bioaccumulation capacity of maize and rice. Phosphorus-containing amendment mixtures can raise soil pH and provide more adsorption sites and facilitate co-precipitation, forming surface complexes. Organic matter dissociation creates additional ion-exchange centers and increases surface area, while both hydroxyapatite and biochar, which contain Ca, facilitate the formation of Ca–P–As/Cd precipitates (Gu et al., 2018).

However, the large-scale use of phosphorus compounds raises concerns about potential surface and groundwater contamination. Therefore, future research should focus on understanding the role of phosphorus compounds in in situ remediation and natural attenuation at metal-contaminated sites while minimizing their impact on water quality.

Most of the research on evaluating the effectiveness of P addition in reducing Pb and As bioaccessibility has been conducted for mine and smelter-impacted soils. The species of Pb in mine-impacted soils is mostly sulfur based (PbS) and the Pb concentration is extremely high (100–20 000 mg kg^{-1}). Therefore, the applicability of the findings to moderately contaminated soils that have adsorbed Pb (Fleming et al., 2013; Haque et al., 2021) as dominant Pb species is questionable. Mine-impacted soils tend to have less clay and low organic matter content due to their origin with ore-based parent material and low plant habitat. To form Pb–P precipitates the concentrations of soluble Pb and P should be high. The concentration of soluble Pb and P required for precipitation depends on several factors, including soil pH, presence of competing ions, and organic ligands (Martínez et al., 2004). Therefore, it is possible that agricultural soils do not meet the high Pb and P requirements for the formation of Pb–P precipitates.

Soils that have low-to-moderate levels of contamination can be managed through rather simple approaches without implementing high-cost and high-tech remediation approaches. For example, in urban gardens (1) use of raised beds, (2) mixing top soils within the same land diluting the concentrations of contaminants in relatively more contaminated areas, (3) gardening only in areas where the concentrations of the contaminants are safe, (4) incorporate organic amendments into soil to dilute the contaminants in soils, and (5) maintain good plant growth via providing enough plant nutrients.

6.3.3 Lime and other acid-neutralizing materials

Soil pH is a crucial determinant of the chemical environment, as most soil reactions and biological processes are pH-sensitive (Bolan et al., 2003). In agricultural lands, soil acidification occurs due to the release of H^+ ions from activities such as the application of fertilizers containing C, N, P, and S, irrigation, and the excessive removal of base cations caused by improper cultivation practices, including monocropping and intensive tillage (Bolan et al., 2023;

Goulding & Blake, 1998). This leads to reduced pH, diminished nutrient availability, increased solubility of trace elements, and negative effects on plant growth and microbial activity. Bolan et al. (2003) discussed the effect of low soil pH on the transformation of trace elements as a result of surface charge modification, changing speciation and chemical forms, and influencing the redox reactions that control the mobility and availability of metals. The lime application increases pH, causing an increase in surface negative charge and resulting in an increase in cationic trace elements adsorption and the formation of hydroxy species of cations such as $CdOH^+$, which are adsorbed preferentially over the metal cation Cd^{2+} (Naidu et al., 1994). In addition, lime addition can alter soil pH, trigger metal hydrolysis reactions, and promote co-precipitation with carbonates. These processes enable lime to function as a precipitating agent, reducing the mobility and availability of metals in the soil. Lime also increases the Ca concentration in soil solution. When adding liming materials to Ni-contaminated soil, Ca competes with Ni for plant uptake, thereby reducing the uptake of Ni. (Bolan et al., 2003; Lwin et al., 2018).

Amendments such as lime, dolomite, slag, alkaline biosolids, biochar, and compost are commonly used to alter soil acidity (Bolan et al., 2023). These materials neutralize acidity, improve soil structure, enhance nutrient availability, immobilize trace elements, and support beneficial microbial processes. A pot experiment tested the impact of lime and limestone enhanced pH to 7.3 from 5.7 on Cd, Pb, and Zn availability in contaminated soil. This study found that the mobile fractions of Zn, Cd, and Pb decreased by 80%, 50%, and 20%, respectively, while their availability in wheat straw and grains was lower in limed soil compared to the control (Tlustoš et al., 2006). Similarly, Gong et al. (2022) found lower Cd concentrations in potatoes grown in a pot experiment with lime and calcite. Lim et al. (2013) found that lime-based waste materials (Oyster shells and eggshells) reduced the maize and TCLP extractable Cd and Pb in soil. Illera et al. (2004) revealed that dolomite containing industrial by-products applied into acid soil increased the capacity of the soil to sorb Cd, Cu, and Pb using sorption isotherm and SEM images.

Biochar is another important acid-neutralizing organic amendment. However, its acid-neutralizing potential and other properties largely depend on feedstock materials and pyrolysis conditions (Bolan et al., 2023). The alkaline nature of biochar is largely attributed to $CaCO_3$, $MgCO_3$, $Mg(OH)_2$, and MgO in biochar and its original feedstock. Additionally, there is a strong correlation between the concentration of exchangeable base cations in biochar and its pH levels. Numerous studies have demonstrated that biochar acts as an effective pH-neutralizing agent while simultaneously reducing the mobility and toxicity of trace elements in plants and contaminated soil (Bilias et al., 2021; Bolan et al., 2023; Hasnain et al., 2023).

6.3.4 Acidulation materials

High pH increases the solubility and mobility of some trace elements (Mo, As, Se) in soil. In general, adsorption of As(V) and Cr(VI) decreases with increasing pH. Cationic trace elements are more likely to form hydroxide precipitates and adsorb strongly to the soil solid phase, making them difficult to release into solution in high pH soil (Strawn, 2021). Soil acidification is needed in highly alkaline agricultural soils to increase the availability of plant nutrients while reducing the abovementioned trace elements. Organic matter that releases organic acids such as biosolids, acidified biochar, sulfur-containing materials such as Ca (gypsum), Fe, and Al sulfate, some industry by-products, N- and P-containing fertilizers, and irrigation water with low pH are the potential materials for reducing or neutralizing soil pH in alkaline or calcareous soil.

Alkaline and calcareous soil can greatly impact by pH reduction by elevating the mobility and bioavailability of trace elements in the soil solution. However, reducing pH with soil amendments facilitated precipitation and formation of strongly bound metal complexes that are not readily soluble, further reducing pH. With the reduction of pH, the exchangeable fractions of metal, such as soil carbonate-bound and Fe/Mn oxide-bound fractions of trace elements, are increased. Many studies have shown that plant availability and increase in the efficiency of phytoextraction of trace elements after soil acidification in calcareous and alkaline soils (Kayser et al., 2000; Wang et al., 2020; Zeng et al., 2011). However, Yang et al. (2001) found that soil treatment with H_3PO_4 (phosphoric acid) resulted in the formation of pyromorphite and reduced Pb bioaccessibility in smelter-contaminated urban soil with slight alkalinity. This is due to the initial low pH induced by H_3PO_4 favoring the dissolution of soil Pb and the gradual pH increase favoring pyromorphite formation. These results are further discussed in other sections.

7 Case study: community garden in Kansas City (Missouri)

7.1 Background and preliminary assessments

The community garden is located in a vacant lot in the Washington Wheatley neighborhood in Kansas City, MO, at N 39° 04′ 48″ and W 94° 32′ 59″ (Fig. 6). The extent of the garden is 42 m × 37 m. The site was formerly occupied by four houses that were demolished in the late 1970s. The possible sources of soil Pb are leaded paint on the houses, burning coal for heat, and leaded gasoline from urban activities in the surroundings. A preliminary assessment was conducted to determine whether the soil and vegetables in the garden were contaminated with trace elements up to a hazard level. The people in the neighborhood had begun gardening in the site a few months prior to the preliminary assessment.

Figure 6 The community garden at Washington Wheatley neighborhood in Kansas City, MO.

The site soils were screened in situ for trace elements approximately every 3 m using an XL3T Niton handheld X-ray fluorescence analyzer (Thermo Scientific) (Fig. 7). This screening revealed that the site soils had moderately elevated Pb concentrations and none of the other trace element concentrations exceeded the permissible values for agricultural soils, except in a few samples (Table 5).

To quantify Pb contamination in vegetables, ten plant samples from the edible portions of vegetables grown by the community gardeners were collected. The concentrations of Pb in all the vegetables (Fig. 8) were below the maximum permissible levels recommended by FAO/WHO. The maximum permissible levels are 0.3 mg kg^{-1} fresh weight (5.0 mg kg^{-1} dry weight, moisture content 94%) for leafy vegetables, 0.1 mg kg^{-1} fresh weight (1.6 mg kg^{-1} dry weight, moisture content 94%) for fruiting vegetables, and 0.1 mg kg^{-1} fresh weight (1.5 mg kg^{-1} dry weight, moisture content 93%) for root vegetables (CODEX, 2023). The concentration of Pb in the carrot sample was lower but close to the maximum permissible level.

The preliminary assessment concluded that the concentrations of trace elements, including Pb, in soil and vegetables do not pose any hazard to the community, so gardening can be continued. However, to further reduce potential transfer of soil Pb to vegetable consumers, assessment of suitable crop types that accumulate lower concentrations of Pb and methods to stabilize soil Pb would be beneficial.

Figure 7 In situ screening of soils for potentially toxic trace elements using XL3T Niton handheld X-ray fluorescence analyzer.

7.2 Field experiments

The field experiment was conducted to assess the effectiveness of a selected soil amendment in reducing soil–plant transfer of Pb and to select suitable vegetable types that accumulate lower concentrations of Pb. The amendments that could stabilize soil Pb and are suitable for a gardening site include (1) liming materials, (2) P fertilizers, and (3) organic amendments. Due to the neutral soil pH and adequate availability of soil phosphorus for vegetable growth, liming materials and phosphorus fertilizers were not chosen. Instead, locally available leaf-based compost was selected as the most suitable amendment to be tested in the field experiment. Leaf-based composts tend to have low concentrations of trace elements compared to biosolids and animal waste-based compost because leafy raw materials have lower concentrations of contaminants. Two field trials were conducted at two separate locations that vary in total Pb concentrations in completely randomized block design. Trial 1 was conducted in a low Pb location in 2009 with three replicates and Trial 2 in high Pb location

Table 5 Total concentrations of potentially toxic trace elements from in situ soil analysis using handheld X-ray fluorescence analyzer

Trace element	Range	Average	LOD[a]
	---------------------mg kg^{-1}---------------------		
As	<LOD-28	12	11
Cd	<LOD-29	17	12
Co	<LOD-550	268	260
Cr	16–71	37	85
Cu	<LOD-63	37	35
Hg	<LOD-24	11	10
Mn	230–794	481	85
Pb	58–305	154	13
Sb	<LOD-56	37	30
Sn	<LOD-41	30	30
Zn	145–636	296	25

[a]LOD, limit of detection. In all samples, concentrations of Ni <LOD = 65 mg kg^{-1}, Se <LOD = 65 mg kg^{-1}, and V <LOD = 65 mg kg^{-1}.

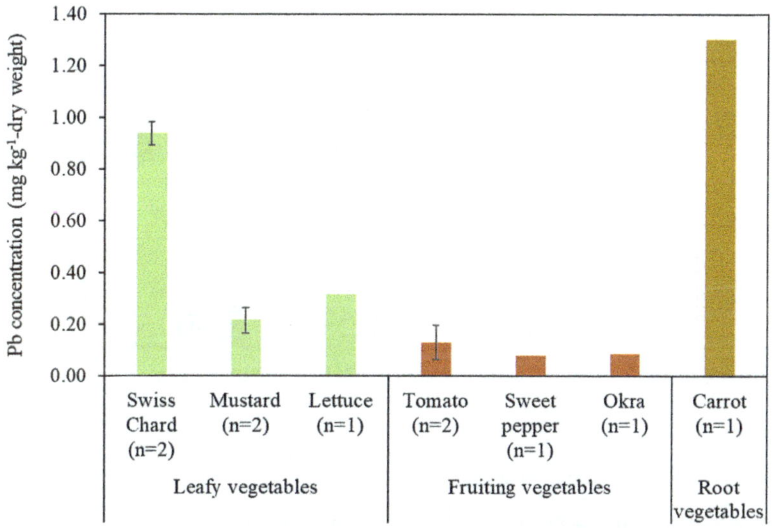

Figure 8 Concentrations of Pb in plant samples collected from the community garden. The error bars represent the standard error of independent samples.

in 2010 with four replicates. Compost was incorporated into soil up to 15 cm depth at 28 kg m^{-2} rate and control treatments were maintained without compost incorporation. A leafy vegetable (Swiss chard), a fruiting vegetable (tomato), and two root/tuber vegetables (carrot or sweet potato) were grown

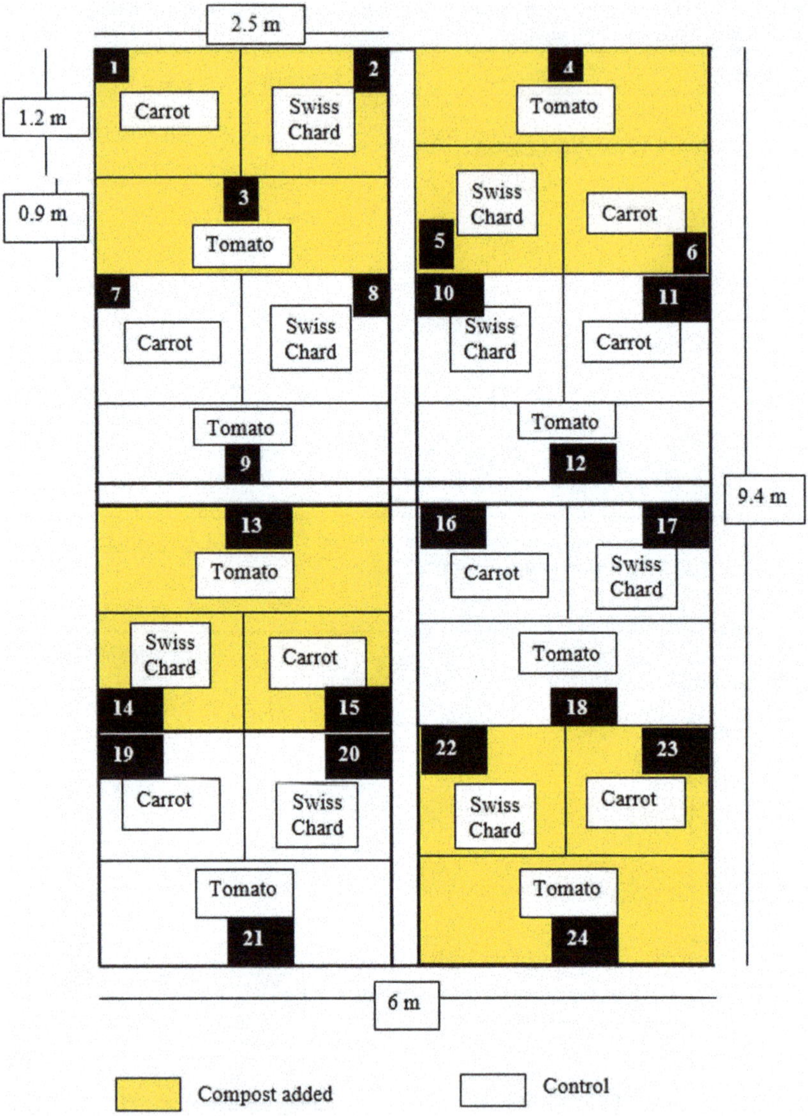

Figure 9 Plot diagram depicting the field arrangement of twenty-four plots in randomized split-plot design in four blocks. Main plot factor: compost vs. control. Sub-plot factor: vegetable type.

in sub plots in each main plot with compost/control treatments (Fig. 9). Total and extractable Pb concentrations in soils after adding amendments and Pb concentrations in edible portions of the vegetables were determined. Compost reduced the total Pb concentration in soils by approximately 30% in the trial with high Pb soils because of the physical dilution of the soil matrix with compost incorporation (Fig. 10). However, this dilution effect was not observed in the trial with low Pb soils, likely because the soil Pb concentrations were not high enough to demonstrate the dilution effect. Compost increased the extractable Pb concentrations in soils in both low and high Pb soils. The increase of extractable Pb is temporary and an effect from increased dissolved organic matter from compost (Attanayake et al., 2014). Compost reduced the concentration of Pb in Swiss chard grown in high Pb soils and did not change the Pb concentrations in any other vegetables (Fig. 11). The concentration of

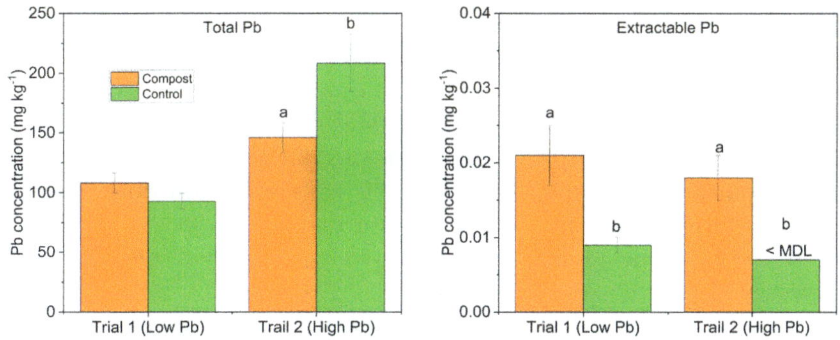

Figure 10 Total and $Sr(NO_3)_2$ extractable Pb concentrations in soils in the two field trials at high and low Pb-contaminated locations in the community garden. <MDL: Extractable Pb in control soils at Trial 2 were less than the method detection limit. The different letters on the bars represent the significant differences at $P < .05$. The error bars represent standard errors of 3 (in Trial 1) or 4 (in Trial 2) replicates.

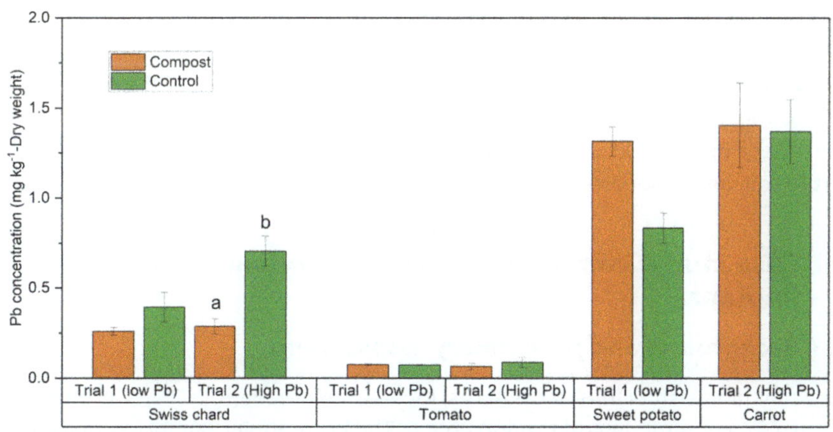

Figure 11 Concentrations of Pb in vegetables in the two field trials at high and low Pb-contaminated locations in the community garden. The different letters on the bars represent the significant differences at $P < .05$. The error bars represent standard errors of 3 (in Trial 1) or 4 (in Trial 2) replicates.

Pb was the lowest in tomato and the highest in carrot and sweet potato. All the vegetables had lower Pb concentrations than the maximum permissible levels recommended by FAO/WHO. The concentration of Pb in sweet potato and carrot were lower but close to the maximum permissible levels. Compost incorporated plots produced higher biomass than the control soils as observed visually.

7.3 Speciation experiment

Lead speciation in soil was assessed to detect whether the incorporation of compost changed the speciation of Pb in soils, which could explain the change in the bioavailability of Pb in soil. Soil samples from compost-amended and control plots 16 days and 69 days and 121 days after compost incorporation were subjected to synchrotron-based bulk X-ray absorption fine structure spectroscopy (XAFS). The XAFS data demonstrated that in both compost-amended and control soils, 70–95% of Pb were adsorbed onto Fe oxide minerals and the remainder of Pb was adsorbed onto organic C molecules. Compost did not change speciation of Pb in soils, proving that compost did not chemically stabilize Pb in soils.

7.4 Conclusions and recommendations

The soil in the urban garden was moderately contaminated with Pb. The vegetables grown in the garden did not have harmful concentrations of Pb. Root crops (carrot and sweet potato) accumulated more Pb than leafy vegetables (Swiss chard) and fruiting vegetables (tomato). Therefore, it was recommended to continue gardening. As an extra precaution, growing root crops can be avoided. Although the Pb speciation experiment demonstrated that the use of leaf-based compost did not stabilize Pb during the timespan of the experiment (121 days), it was recommended to use compost when gardening considering the other benefits that compost offer such as providing plant nutrients, maintain fertile soil, and resource recycling.

8 Case study: community garden in Indianapolis (Indiana)

8.1 Background and preliminary assessments

A vacant land parcel in Monon Acres, located in Indianapolis, IN, was intended for use as a community gardening space. The community wanted to make sure the soil did not have harmful levels of contaminants.

We conducted a preliminary assessment on soil contamination. Screening soils with XRF revealed that the soils were contaminated with Pb (475 mg kg^{-1})

and As (95 mg kg^{-1}). Because of coal burning in adjutant railroad in the past, a few soil samples were tested for polycyclic aromatic hydrocarbons (PAHs), a group of organic contaminants that are emitted at partial burning. The soils had 23–50 mg kg^{-1} of 16 PAHs listed as priority pollutants regulated by the USEPA (2014). The PAHs are naphthalene, acenaphthylene, acenaphthene, fluorene, phenanthrene, anthracene, fluoranthene, pyrene, benz[a]anthracene, chrysene, benzo[b]fluoranthene, benzo[k]fluoranthene, benzo[a]pyrene, indeno[1,2,3-cd]pyrene, dibenz[a,h]anthracene, and benzo[g,h,i]perylene.

8.1.1 Field experiment

To inform the community about ensuring safe gardening in the site, we conducted a field experiment (Fig. 12) similar to the above-described experiment in the Kansas City site.

In this experiment, we assess the effectiveness of four locally available organic amendments on reducing the accumulation of Pb, As, and PHAs in vegetables and bioaccessibilities of soil Pb and As (an estimation of bioavailability of Pb and As due to ingestion of contaminated soil). The organic amendment treatments (i.e. composted biosolids, non-composted biosolids, mushroom compost, leaf compost, and unamended-control) were arranged in four blocks as the main-plot factor and the vegetables (collard greens, carrots,

Figure 12 Field experiment in Indianapolis, IN.

and tomatoes) were grown within each main-plot as sub-plot factor in split-plot arrangement (Fig. 13). The organic amendments were incorporated to top 15 cm depth of soil at 44 kg m^{-2} rate. The vegetables were grown in two summers in two consecutive years.

Organic amendments diluted the Pb and As in soils due to physical dilution of soil matrix in amended soils due to incorporation of the amendments (Fig. 14). This is an immediate advantage of incorporation of organic amendments to contaminated soils. Although all the amount of amendments incorporated were the same for all the amendments, the degree of dilution of soil Pb and As

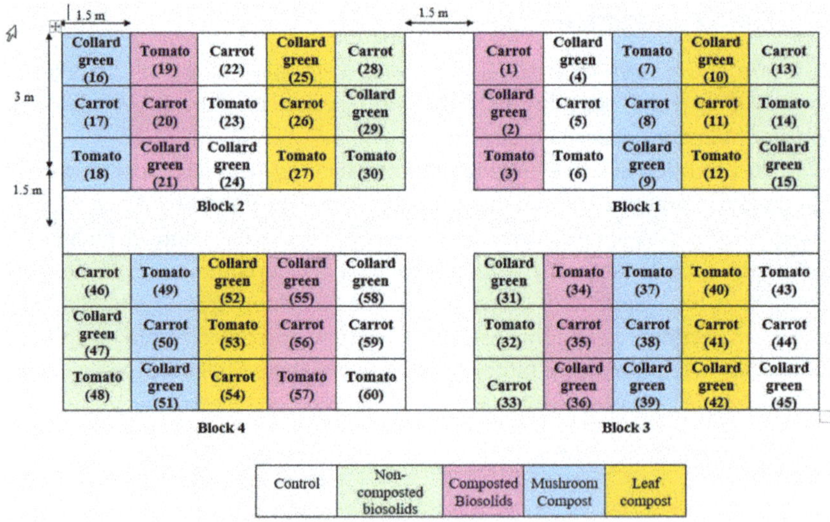

Figure 13 Plot diagram depicting the field arrangement of 60 plots in randomized split-plot design in four blocks. Organic amendment treatment is the main plot factor. Vegetables are the sub-plot factor.

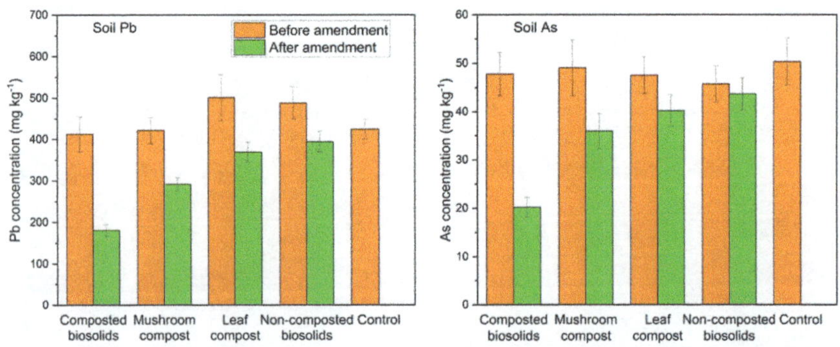

Figure 14 Concentration of Pb and As in soil before and after incorporating organic amendments. The error bars represent the standard errors of field four replicates.

varies among the amendments due to the physical nature of the amendments (e.g. particle size and moisture content). The amount of compost added was 44 kg m^{-2} (23% in mass basis (w/w), assuming 1.3 g cm^{-3} bulk density and 15-cm soil depth). This compost incorporation caused variable amounts of total Pb and As dilution depending on the properties of the compost (e.g. moisture content and particle size). Total Pb dilution ranged from 18% to 30%, and total As dilution ranged from 7% to 43% (Attanayake et al., 2015).

Immediately after incorporation into soil, composted biosolids reduced the concentration of naphthalene, noncomposted biosolids increased the concentration of benzo[g,h,i]perylene (Fig. 15) and the concentrations of other PAHs did not change. The reduction of concentration of naphthalene – the PAHs that has the lowest molecular weight among the 16 priority PAHs – could be a result of both increased volatilization and physical dilution of soil matrix. The increase of concentration of benzo[g,h,i]perylene could be attributed to desorption of the PAH molecules due to exchange with dissolved organic C with incorporation of noncomposted biosolids, that increased the dissolved organic C in soil more so than leaf compost and composted biosolids. At least one type of organic amendment seemed to increase the degradation of naphthalene, acenaphthylene or benzo [g,h,i]perylene over the time, likely through increased microbial activities due to improvement

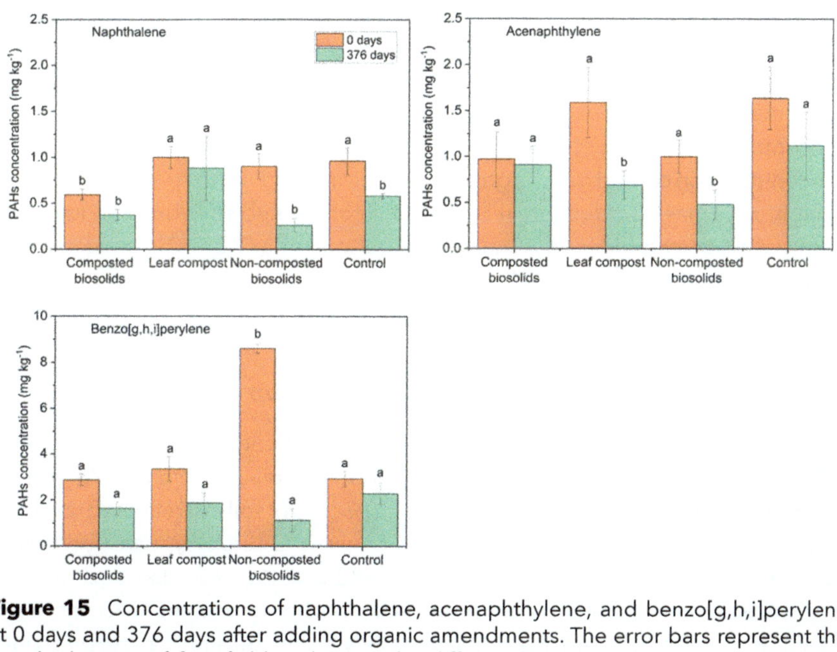

Figure 15 Concentrations of naphthalene, acenaphthylene, and benzo[g,h,i]perylene at 0 days and 376 days after adding organic amendments. The error bars represent the standard errors of four field replicates. The different letters on the bars represent the statistical differences at $P < .05$).

of soil aeration and nutrient availability for microorganisms. Further, the degradation of PAHs can be enhanced as a result of desorption of PAHs from soil colloids due to exchange with dissolved organic molecules from organic amendments. The concentration of naphthalene reduced in control soil over time probably because of improved aeration with soil preparation that facilitated volatilization.

8.2 Vegetable Pb, As, and polycyclic aromatic hydrocarbon concentrations

None of the contaminants in vegetables were found to be at a level that could exceed the maximum permissible level according to the FAO/WHO standards, except Pb in carrots. The results on concentration of Pb in vegetables were similar to what we observed in the abovementioned case study in Kansas City. Briefly, collard greens and tomato had the lowest Pb concentrations which were less than the maximum permissible levels of Pb in leafy and fruiting vegetables declared by FAO/WHO. The Pb concentration in carrots was close to the maximum permissible level for root crops, or at most, up to three times higher than the maximum permissible level indicating that carrots (or possibly the other root crops) may not be a suitable crop to grow in the garden. Organic amendments did not reduce Pb concentrations in any of the vegetables.

Arsenic was detected in collard greens but not in carrot and tomato. Composted biosolids, which was the most effective in reducing As in collard greens, lowered As concentration in collard greens by at least 68% compared to the control. The similar As BCF (the ratio of As concentration in leaves and As concentration in soil) in collard greens across all plots suggests that reduced As concentration in collard greens grown with composted biosolids were likely due to higher biomass production in biosolids-incorporated soils but not due to As stabilization by the biosolids. The concentrations of all 16 PAHs in vegetables were nondetectable (<0.01–0.1 mg kg^{-1}) suggesting that accumulating PAHs in the vegetables growing in the garden would be minimal.

8.3 Pb and As bioaccessibilities and speciation

The fraction of bioaccessible Pb and As from total concentrations in soil was 1–5% for Pb and 7–12% for As. This low bioaccessibilies of Pb and As can be attributed to inherently higher sorption capacity in soil as reflected by high amorphous Fe minerals (28 g kg^{-1}) and organic C (3%). Synchrotron-based bulk EXAFS analysis demonstrated that 70–80% of Pb was adsorbed onto Fe (oxyhydr)oxides and 20–30% of Pb was adsorbed onto organic C. Composted biosolids further reduced Pb bioaccessibility by increasing Pb sorbed onto organic C in field during the 2-year period and reducing the dissolution of Pb adsorbed onto Fe oxy(hydr)oxides during the bioaccessibility extraction that mimic acidic digestive system (pH 2.5). Further, the acidity of the extraction solution facilitated formation of pyromorphite, which is a stable Pb mineral that could contribute to low Pb bioaccessibility.

Synchrotron-based bulk X-ray absorption near-edge structure spectroscopy (XANES) demonstrated that 52-65 % of As was in As(III) form and sorbed to pyrite and 35-48 % of As was in As(V) form adsorbed on to Fe oxy(hydr)oxides. The presence of majority of As in the form of As(III) sorbed on to pyrite is a result of coal burning in the past. The above As species are known to have low bioaccessibility (Hohmann et al., 2011) and explain the observed low As bioaccessibility in soil. Mushroom compost increased As bioaccessibility in soil immediately after incorporation, by liberating As(V) from Fe oxy(hydr)oxide by exchanging As with dissolved organic C and promoting reduction of As(V) into As(III) by oxidation of dissolved organic C. The increase of As bioaccessibility by mushroom compost was temporally and later (i.e. at least within a year) the As bioaccessibility and speciation returned to that in control soil. Noncomposted biosolids and leaf compost also changed As speciation temporarily as explained above, but the speciation change did not lead to change in As bioaccessibility. The temporary liberation of As by mushroom compost, noncomposted biosolids, and leaf compost facilitated formation of lead arsenate- a stable form of As during the bioaccessibility extraction leading to counteract the potential increase of As bioaccessibility due to liberation of As adsorbed on to Fe oxy(hydr)oxide in the field. Composted biosolids did not change As speciation and bioaccessibility.

8.4 Conclusions and recommendations

The soil contained moderate concentrations of Pb, As, and PAHs. Except for Pb in carrot, the concentration of contaminants in vegetables were within the safe levels. The bioaccessibility of Pb and As were low (<10%). Speciation of Pb and As supported the low plant availability and bioaccessibility. Composted biosolids were the most effective in reducing or maintaining low plant availability and bioacessibility of Pb and As. However, gardeners could benefit from incorporation of any of the organic amendments because the amendments increase plant nutrients and soil fertility. Growing vegetables other than root crops with the incorporation of organic amendments helps to reduce contaminant transfer from soil to vegetables. Due to human health concerns of emerging contaminants, such as per and polyfluoroalkyl substances (PFAS), although contaminants in biosolids and other organic sources may not be as available as they are in other mediums, such as water, organic amendments should also be tested for potential contamination.

9 Where to look for further information

Bolan, N. S., & Duraisamy, V. (2003). Role of inorganic and organic soil amendments on immobilisation and phytoavailability of heavy metals: A review involving specific case studies. *Soil Research*, 41(4), 533–555. https://doi.org/10.1071/SR02122

Brown, S. L., Chaney, R. L., & Hettiarachchi, G. M. (2016). Lead in urban soils: A real or perceived concern for urban agriculture? *Journal of Environmental Quality*, 45(1), 26–36. https://doi.org/10.2134/jeq2015.07.0376

Carrijo, D. R., LaHue, G. T., Parikh, S. J., Chaney, R. L., & Linquist, B. A. (2022). Mitigating the accumulation of arsenic and cadmium in rice grain: A quantitative review of the role of water management. *Science of the Total Environment*, 839, 156245. https://doi.org/10.1016/j.scitotenv.2022.156245

Defoe, P. P., Hettiarachchi, G., Benedict, C., & Martin, S. (2014). Safety of gardening on lead- and arsenic-contaminated urban brownfields. *Journal of Environmental Quality*, 43(6), 2064–2078.

Seyfferth, A. L., Limmer, M. A., Runkle, B. R. K., & Chaney, R. L. (2024). Mitigating toxic metal exposure through leafy greens: A comprehensive review contrasting cadmium and lead in spinach. *GeoHealth*, 8, e2024GH001081. https://doi.org/10.1029/2024GH001081

Zhao, F. J., McGrath, S. P., & Meharg, A. A. (2010). Arsenic as a food chain contaminant: mechanisms of plant uptake and metabolism and mitigation strategies. *Annu Rev Plant Biol*. 61:535–59. doi: 10.1146/annurev-arplant-042809-112152. PMID: 20192735

10 References

Afzal, A. & Mahreen, N. (2024). Emerging insights into the impacts of heavy metals exposure on health, reproductive and productive performance of livestock. *Frontiers in Pharmacology, 15*, 1375137. https://doi.org/10.3389/fphar.2024.1375137

Afzal, M., et al. (2014). Assessment of heavy metal contamination in soil and groundwater at leather industrial area of Kasur, Pakistan. *CLEAN – Soil, Air, Water, 42*(8), 1133–1139. https://doi.org/10.1002/clen.201100715

Ahmed, S., et al. (2022). Health risk assessment for heavy metal accumulation in leafy vegetables grown on tannery effluent contaminated soil. *Toxicology Reports, 9*, 346–355. https://doi.org/10.1016/j.toxrep.2022.03.000

Alam, M. F., et al. (2021). Human arsenic exposure risk via crop consumption and global trade from groundwater-irrigated areas. *Environmental Research Letters, 16*(12), 124013. https://doi.org/10.1088/1748-9326/ac34bb

Alengebawy, A., et al. (2021). Heavy metals and pesticides toxicity in agricultural soil and plants: ecological risks and human health implications. *Toxics, 9*(3), 42. https://doi.org/10.3390/toxics9030042

Alloway, B. J. (1995). Soil processes and the behavior of metals. In B. J. Alloway (Ed.), *Heavy metals in soils* (pp. 11–38). Chapman & Hall.

Alloway, B. J. (2013). Sources of heavy metals and metalloids in soils. In B. J. Alloway (Ed.), *Heavy metals in soils: Trace metals and metalloids in soils and their bioavailability* (pp. 11–50). Springer. https://doi.org/10.1007/978-94-007-4470-7_2

Amjad, M., et al. (2024). Accumulation and translocation of lead in vegetables through intensive use of organic manure and mineral fertilizers with wastewater. *Scientific Reports, 14*, Article 63076. https://doi.org/10.1038/s41598-024-63076-x

Angon, P. B., et al. (2024). Sources, effects and present perspectives of heavy metals contamination: soil, plants and human food chain. *Heliyon, 10*(7), e28357. https://doi.org/10.1016/j.heliyon.2024.e28357

Antoniadis, V. & Alloway, B. J. (2002). The role of dissolved organic carbon in the mobility of Cd, Ni and Zn in sewage sludge-amended soils. *Environmental Pollution, 117*(3), 515–521. https://doi.org/10.1016/S0269-7491(01)00172-5

Antoniadis, V., et al. (2019). A critical prospective analysis of the potential toxicity of trace element regulation limits in soils worldwide: are they protective concerning health risk assessment? *Environment International, 127*, 819–847. https://doi.org/10.1016/j.envint.2019.03.039

Antoniadis, V., et al. (2020). International trace element regulation limits in soils. In Y. S. Ok, et al. (Eds.), *Soil and groundwater remediation technologies* (pp. 29–38). CRC Press. https://doi.org/10.1201/9780429322563

Aponte, H., et al. (2020). Meta-analysis of heavy metal effects on soil enzyme activities. *Science of the Total Environment, 737*, 139744. https://doi.org/10.1016/j.scitotenv.2020.139744

Ashworth, D. J. & Alloway, B. J. (2008). Influence of dissolved organic matter on the solubility of heavy metals in sewage-sludge-amended soils. *Communications in Soil Science and Plant Analysis, 39*(3–4), 538–550. https://doi.org/10.1080/00103620701826787

Attanayake, C. P., et al. (2014). Field evaluations on soil-plant transfer of lead from an urban garden soil. *Journal of Environmental Quality, 43*(2), 475–487. https://doi.org/10.2134/jeq2013.07.0282

Attanayake, C. P., et al. (2015). Potential bioavailability of lead, arsenic, and polycyclic aromatic hydrocarbons in compost-amended urban soils. *Journal of Environmental Quality, 44*(3), 930–944. https://doi.org/10.2134/jeq2014.09.0400

Attanayake, C. P., et al. (2021). Phytoavailability of lead for vegetables in urban garden soils. *ACS Agricultural Science & Technology, 1*(3), 173–181. https://doi.org/10.1021/acsagscitech.0c00068

Attanayake, C. P., et al. (2017). Lead speciation and in vitro bioaccessibility of compost-amended urban garden soils. *Journal of Environmental Quality, 46*(6), 1215–1224. https://doi.org/10.2134/jeq2017.02.0065

Baker, A. J. M. (1981). Accumulators and excluders: -strategies in the response of plants to heavy metals. *Journal of Plant Nutrition, 3*(1–4), 643–654. https://doi.org/10.1080/01904168109362867

Ballabio, C., et al. (2024). Cadmium in topsoils of the European Union: an analysis based on LUCAS topsoil database. *Science of the Total Environment, 912*, 168710. https://doi.org/10.1016/j.scitotenv.2023.168710

Barwick, M. & Maher, W. (2003). Biotransference and biomagnification of selenium, copper, cadmium, zinc, arsenic and lead in a temperate seagrass ecosystem from Lake Macquarie Estuary, NSW, Australia. *Marine Environmental Research, 56*(4), 471–502. https://doi.org/10.1016/S0141-1136(03)00028-X

Beckett, P. H. T., et al. (1979). The disposal of sewage sludge onto farmland:the scope of the problems of toxic elements. *Water Pollution Control (Maidstone, England), 78*, 419–445.

Benke, M. B., et al. (2008). Trace element changes in soil after long-term cattle manure applications. *Journal of Environmental Quality, 37*(3), 798–807. https://doi.org/10.2134/jeq2007.0214

Bilias, F., et al. (2021). Towards a soil remediation strategy using biochar: effects on soil chemical properties and bioavailability of potentially toxic elements. *Toxics, 9*(8), 184. https://doi.org/10.3390/toxics9080184

Bloem, E., et al. (2017). Contamination of organic nutrient sources with potentially toxic elements, antibiotics and pathogen microorganisms in relation to P fertilizer potential and treatment options for the production of sustainable fertilizers: a review. *Science of the Total Environment, 607–608*, 225–242. https://doi.org/10.1016/j.scitotenv.2017.06.274

Bolan, N. S. & Duraisamy, V. (2003). Role of inorganic and organic soil amendments on immobilisation and phytoavailability of heavy metals: a review involving specific case studies. *Soil Research, 41*(4), 533–555. https://doi.org/10.1071/SR02122

Bolan, N. S., et al. (2003). Soil acidification and liming interactions with nutrient and heavy metal transformation and bioavailability. *Advances in Agronomy, 78*, 215–272. https://doi.org/10.1016/S0065-2113(02)78006-1

Bolan, N. S., et al. (2023). Soil acidification and the liming potential of biochar. *Environmental Pollution, 317*, 120632. https://doi.org/10.1016/j.envpol.2022.120632

Bradl, H. B. (2004). Adsorption of heavy metal ions on soils and soil constituents. *Journal of Colloid and Interface Science, 277*(1), 1–18. https://doi.org/10.1016/j.jcis.2004.04.005

Brown, S. L. & Hettiarachchi, G. M. (2025). Risks and benefits associated with urban green space through the lens of environmental justice. *Soil & Environmental Health, 3*(1), 100127. https://doi.org/10.1016/j.seh.2024.100127

Brown, S. L., et al. (2012). High-iron biosolids compost–induced changes in lead and arsenic speciation and bioaccessibility in co-contaminated soils. *Journal of Environmental Quality, 41*(5), 1612–1622.

Campos, V. P. (2002). Arsenic in groundwater affected by phosphate fertilizers at São Paulo, Brazil. *Environmental Geology, 42*(1), 83–87. https://doi.org/10.1007/s00254-002-0540-0

Campos, V. P. (2003). Trace elements in pesticides. *Communications in Soil Science and Plant Analysis, 34*(9–10), 1261–1268. https://doi.org/10.1081/CSS-120020442

Capar, S. G., et al. (1978). Multielement analysis of animal feed, animal wastes, and sewage sludge. *Environmental Science & Technology, 12*(7), 785–790.

Carrijo, D. R., et al. (2022). Mitigating the accumulation of arsenic and cadmium in rice grain: a quantitative review of the role of water management. *Science of the Total Environment, 839*, 156245. https://doi.org/10.1016/j.scitotenv.2022.156245

Chaney, R. L. (2011). Food safety issues for mineral and organic fertilizers. In D. L. Sparks (Ed.), *Advances in agronomy* (Vol. 117, pp. 51–116). Academic Press. https://doi.org/10.1016/B978-0-12-394278-4.00002-7

Chaney, R. L. & Giordano, P. M. (1977). Microelements as related to plant deficiencies and toxicities. In L. F. Elliott & F. J. Stevenson (Eds.), *Soils for management of organic wastes and waste waters* (pp. 234–279). Soil Science Society of America.

Chaney, R. L., et al. (1996a). Development of the US-EPA limits for chromium in land-applied biosolids and applicability of these limits to tannery by-product derived fertilizers and other Cr-rich soil amendments. In S. Canali, F. Tittarelli, & P. Sequi (Eds.), *Chromium environmental issues* (pp. 229–295). Franco Angeli. (Proceedings of the Chromium Environmental Issues Workshop, San Miniato, Italy, April 12–13, 1996.)

Chaney, R. L., et al. (1996b). Phyto-availability and bio-availability in risk assessment of Cd in agricultural environments. In *Sources of cadmium in the environment* (pp. 49–78). OECD. (Proceedings from Stockholm, Sweden, October 15–22, 1995.)

Chen, H.-P., et al. (2021). Producing Cd-safe rice grains in moderately and seriously Cd-contaminated paddy soils. *Chemosphere, 267*, 128893. https://doi.org/10.1016/j.chemosphere.2020.128893

Chen, K. S. & Lai, H. Y. (2024). Effect of increased soil available phosphorus from vermicompost application on the bioavailability, chemical form, and bioaccessibility of heavy metals. *Environmental Geochemistry and Health, 46*(9), 343. https://doi.org/10.1007/s10653-024-02127-8

Chen, T., et al. (2011). Casparian strip development and its potential function in salt tolerance. *Plant Signaling & Behavior, 6*(10), 1499. https://doi.org/10.4161/psb.6.10.17054

Chen, X., et al. (2020). Health risk assessment associated with heavy metal accumulation in wheat after long-term phosphorus fertilizer application. *Environmental Pollution, 262*, 114348. https://doi.org/10.1016/j.envpol.2020.114348

Chen, Y., et al. (2017). Arsenic transport in rice and biological solutions to reduce arsenic risk from rice. *Frontiers in Plant Science, 8*, 245590. https://doi.org/10.3389/fpls.2017.00268

Chen, W., et al. (2008). Arsenic, cadmium, and lead in California cropland soils: role of phosphate and micronutrient fertilizers. *Journal of Environmental Quality, 37*(2), 689–695. https://doi.org/10.2134/jeq2007.0444

Chibuike, G. U. & Obiora, S. C. (2014). Heavy metal polluted soils: effect on plants and bioremediation methods. *Applied and Environmental Soil Science, 2014*, 752708. https://doi.org/10.1155/2014/752708

Clarke, L. W., et al. (2015). Urban legacies and soil management affect the concentration and speciation of trace metals in Los Angeles community garden soils. *Environmental Pollution, 197*, 1–12. https://doi.org/10.1016/j.envpol.2014.11.015

Clinical and Laboratory Standards Institute. (2014). *Performance standards for antimicrobial susceptibility testing: 24th informational supplement* (CLSI document M100-S24). CLSI.

Codex Alimentarius Commission. (2023). *General standard for contaminants and toxins in food and feed (CXS 193-1995)*. FAO & WHO. https://www.fao.org/fao-who-codexalimentarius/sh-proxy/

Collin, S., et al. (2022). Bioaccumulation of lead (Pb) and its effects in plants: a review. *Journal of Hazardous Materials Letters, 3*, 100064. https://doi.org/10.1016/j.hazl.2022.100064

Day, D. L. & Funk, T. L. (1998). Processing manure: physical, chemical, and biological treatment. In J. L. Hatfield & B. A. Stewart (Eds.), *Animal waste utilization: effective use of manure as a soil resource* (pp. 243–282). Ann Arbor Press.

De Vries, W. & McLaughlin, M. (2013). Modeling the cadmium balance in Australian agricultural systems in view of potential impacts on food and water quality. *Science of The Total Environment, 461–462*, 240–257. https://doi.org/10.1016/j.scitotenv.2013.04.069

Defoe, P. P., et al. (2014). Safety of gardening on lead- and arsenic-contaminated urban brownfields. *Journal of Environmental Quality, 43*(6), 2064–2078.

Dong, Q., et al. (2024). Revealing the sources of cadmium in rice plants under pot and field conditions from its isotopic fractionation. *ACS Environmental Au, 4*(3), 162–172. https://doi.org/10.1021/acsenvironau.3c00067

Dudka, S., et al. (1995). Trace metal contamination of soils and crop plants by the mining and smelting industry in Upper Silesia, South Poland. *Journal of Geochemical Exploration, 52*(1–2), 237–250. https://doi.org/10.1016/0375-6742(94)00047-F

Essington, M. E. (2015). Surface chemistry and adsorption reactions, organic matter in soil. In M. E. Essington (Ed.), *Soil and water chemistry* (2nd ed., pp. 155–487). CRC Press.

Ephraim, P. I., et al. (2013). Investigation of soils affected by burnt hospital wastes in Nigeria using PIXE. *SpringerPlus, 2*(1), 208. https://doi.org/10.1186/2193-1801-2-208

European Commission. (2023). Commission Regulation (EU) 2023/915 of 25 April 2023 on maximum levels for certain contaminants in food and repealing Regulation (EC) No 1881/2006. *Official Journal of the European Union, L 119*, 103–158. https://eur-lex.europa.eu/legal-content/EN/TXT/PDF/?uri=CELEX:32023R0915

Faires, M. C. (2004). Inorganic arsenic toxicosis in a beef herd. *The Canadian Veterinary Journal, 45*(4), 329. https://www.ncbi.nlm.nih.gov/pmc/articles/PMC548622/

FAO. (1999). *General Notice No. 6/7 of 999: guidelines for irrigation water quality*. Department of Environment, Food and Agricultural Organisation (FAO).

FAO & UNEP. (2021). Chapter 3: sources of soil pollution. In Natalia Rodríguez Eugenio (Ed.), *Global assessment of soil pollution: Report*. FAO. https://doi.org/10.4060/cb4894en

Farfel, M. R., et al. (2005). Biosolids compost amendment for reducing soil lead hazards: a pilot study of Orgro® amendment and grass seeding in urban yards. *Science of The Total Environment, 340*(1–3), 81–95. https://doi.org/10.1016/j.scitotenv.2004.08.018

Fertiliser Association. (2024). *Fertiliser monitoring*. The Fertiliser Association of New Zealand. https://www.fertiliser.org.nz/site/about/caring-for-our-environment/managing_contaminants

Fleming, M., et al. (2013). Extractability and bioavailability of Pb and As in historically contaminated orchard soil: effects of compost amendments. *Environmental Pollution, 177*, 90–97. https://doi.org/10.1016/j.envpol.2013.02.013

Fritz, B. G., et al. (2022). Effects of lead and arsenic in soils from former orchards on growth of three plant species. *Environmental Toxicology and Chemistry, 41*(6), 1459–1469. https://doi.org/10.1002/etc.5325

Giller, K. E., et al. (1998). Toxicity of heavy metals to microorganisms and microbial processes in agricultural soils: a review. *Soil Biology and Biochemistry, 30*(10–11), 1389–1414. https://doi.org/10.1016/S0038-0717(97)00270-8

Gillispie, E. C., et al. (2015). Soil pollution due to irrigation with arsenic-contaminated groundwater: current state of science. *Current Pollution Reports, 1*(1), 1–12. https://doi.org/10.1007/s40726-015-0001-y

Goh, K. & Lim, T. (2004). Geochemistry of inorganic arsenic and selenium in a tropical soil: effect of reaction time, pH, and competitive anions on arsenic and selenium adsorption. *Chemosphere, 55*(6), 849–859. https://doi.org/10.1016/j.chemosphere.2003.11.041

Goldberg, S. (2011). Chemical equilibrium and reaction modeling of arsenic and selenium in soils. In H. M. Selim (Ed.), *Dynamics and bioavailability of heavy metals in the rootzone* (1st ed., pp. 66–87). CRC Press. https://doi.org/10.1201/b10508

Goldberg, S., et al. (1996). Molybdenum adsorption on oxides, clay minerals, and soils. *Soil Science Society of America Journal, 60*(2), 425–432. https://www.ars.usda.gov/arsuserfiles/20361500/pdf_pubs/P1324.pdf

Goldberg, S., et al. (2002). Predicting molybdenum adsorption by soils using soil chemical parameters in the constant capacitance model. *Soil Science Society of America Journal, 66*(6), 1836–1842. https://doi.org/10.2136/sssaj2002.1836

Goldberg, S., et al. (2008). Chemical modeling of arsenic(III, V) and selenium(IV, VI) adsorption by soils surrounding ash disposal facilities. *Vadose Zone Journal, 7*(4), 1231–1238. https://doi.org/10.2136/vzj2008.0013

Gong, S., et al. (2022). Calcareous materials effectively reduce the accumulation of Cd in potatoes in acidic cadmium-contaminated farmland soils in mining areas. *International Journal of Environmental Research and Public Health, 19*(18), 11736. https://doi.org/10.3390/ijerph191811736

Goulding, K. & Blake, L. (1998). Land use, liming and the mobilization of potentially toxic metals. *Agriculture, Ecosystems & Environment, 67*(2–3), 135–144. https://doi.org/10.1016/S0167-8809(97)00111-4

Grant, C. A., et al. (2010). Preceding crop and phosphorus fertilization affect cadmium and zinc concentration of flaxseed under conventional and reduced tillage. *Plant and Soil, 333*, 337–350. https://doi.org/10.1007/s11104-010-0349-7

Gu, J. F., et al. (2018). Effects of an additive (hydroxyapatite–biochar–zeolite) on the chemical speciation of Cd and As in paddy soils and their accumulation and translocation in rice plants. *Environmental Science and Pollution Research, 25*, 8608–8619. https://doi.org/10.1007/s11356-017-0921-2

Gupta, D., et al. (2014). Role of phosphate fertilizers in heavy metal uptake and detoxification of toxic metals. *Chemosphere, 108*, 134–144. https://doi.org/10.1016/j.chemosphere.2014.01.030

Gupta, U. C. & Gupta, S. C. (1998). Trace element toxicity relationships to crop production and livestock and human health: Implications for management. *Communications in Soil Science and Plant Analysis, 29*(11–14), 1491–1522. https://doi.org/10.1080/00103629809370045

Haque, E., et al. (2021). Lead (Pb) concentrations and speciation in residential soils from an urban community impacted by multiple legacy sources. *Journal of Hazardous Materials, 416*, 125886. https://doi.org/10.1016/j.jhazmat.2021.125886

Hasnain, M., et al. (2023). Biochar-plant interaction and detoxification strategies under abiotic stresses for achieving agricultural resilience: a critical review. *Ecotoxicology and Environmental Safety, 249*, 114408. https://doi.org/10.1016/j.ecoenv.2022.114408

He, Q. B. & Singh, B. R. (1995). Cadmium availability to plants as affected by repeated applications of phosphorus fertilizers. *Acta Agriculturae Scandinavica, Section B – Soil & Plant Science, 45*(1), 22–31. https://doi.org/10.1080/09064719509410930

He, Z. L., et al. (2005). Trace elements in agroecosystems and impacts on the environment. *Journal of Trace Elements in Medicine and Biology, 19*(2–3), 125–140. https://doi.org/10.1016/j.jtemb.2005.02.010

Heikens, A., et al. (2007). Arsenic behaviour from groundwater and soil to crops: impacts on agriculture and food safety. *Reviews of Environmental Contamination and Toxicology, 189*, 43–87. https://doi.org/10.1007/978-0-387-35368-5_3

Hernandez-Soriano, M. C. & Jimenez-Lopez, J. C. (2012). Effects of soil water content and organic matter addition on the speciation and bioavailability of heavy metals. *Science of The Total Environment, 423*, 55–61. https://doi.org/10.1016/j.scitotenv.2012.02.033

Hettiarachchi, G. M. & Pierzynski, G. M. (2004). Soil lead bioavailability and in situ remediation of lead-contaminated soils: a review. *Environmental Progress, 23*(1), 78–93. https://doi.org/10.1002/ep.10004

Hettiarachchi, G. M., et al. (2003). Sorption and desorption of cadmium by different fractions of biosolids-amended soils. *Journal of Environmental Quality, 32*(5), 1684–1693. https://doi.org/10.2134/jeq2003.1684

Hettiarachchi, G. M., et al. (2006). micro-XANES and micro-XRF investigations of metal binding mechanisms in biosolids. *Journal of Environmental Quality, 35*(1), 342–351. https://doi.org/10.2134/jeq2004.0259

Hohmann, C., et al. (2011). Molecular-level modes of as binding to Fe(III) (oxyhydr)oxides precipitated by the anaerobic nitrate-reducing Fe(II)-oxidizing Acidovorax sp. strain BoFeN1. *Geochimica et Cosmochimica Acta, 75*(17), 4699–4712. https://doi.org/10.1016/j.gca.2011.02.044

Huygens, D., et al. (2019). Technical proposals for selected new fertilising materials under the Fertilising Products Regulation (Regulation (EU) 2019/1009). European Commission, Directorate B – Growth and Innovation. https://ec.europa.eu/jrc

Illera, V., et al. (2004). Immobilization of the heavy metals Cd, Cu and Pb in an acid soil amended with gypsum- and lime-rich industrial by-products. *European Journal of Soil Science, 55*(1), 135–145. https://doi.org/10.1046/j.1365-2389.2003.00583.x

Jiang, Y., et al. (2022). Combined amendment improves soil health and Brown rice quality in paddy soils moderately and highly Co-contaminated with Cd and As. *Environmental Pollution, 295*, 118590. https://doi.org/10.1016/j.envpol.2021.118590

Jiao, Y., et al. (2004). Effects of phosphorus and zinc fertilizer on cadmium uptake and distribution in flax and durum wheat. *Journal of the Science of Food and Agriculture, 84*(7), 777–785. https://doi.org/10.1002/jsfa.1648

Jiao, W., et al. (2012). Environmental risks of trace elements associated with long-term phosphate fertilizers applications: a review. *Environmental Pollution, 168*, 44–53. https://doi.org/10.1016/j.envpol.2012.03.052

Jing, H., et al. (2023). Exploring the mechanism of Cd uptake and translocation in rice: future perspectives of rice safety. *Science of The Total Environment, 897*, 165369. https://doi.org/10.1016/j.scitotenv.2023.165369

Kaiser, B. N., et al. (2005). The role of molybdenum in agricultural plant production. *Annals of Botany, 96*(5), 745. https://doi.org/10.1093/aob/mci226

Karna, R. R., et al. (2018). Point of zero charge: role in pyromorphite formation and bioaccessibility of lead and arsenic in phosphate-amended soils. *Soil Systems, 2*(2), 22. https://doi.org/10.3390/soilsystems2020022

Kayser, A., et al. (2000). Enhancement of phytoextraction of Zn, Cd, and Cu from calcareous soil: the use of NTA and sulfur amendments. *Environmental Science & Technology, 34*(9), 1778–1783. https://doi.org/10.1021/es9907198

Khalid, N., et al. (2021). Rising metals concentration in the environment: a response to effluents of leather industries in Sialkot. *Bulletin of Environmental Contamination and Toxicology, 106*(3), 493–500. https://doi.org/10.1007/s00128-021-03111-z
Kopittke, P. M., et al. (2011). In situ distribution and speciation of toxic copper, nickel, and zinc in hydrated roots of cowpea. *Plant Physiology, 156*(2), 663. https://doi.org/10.1104/pp.111.173716
Krzesłowska, M. (2011). The cell wall in plant cell response to trace metals: polysaccharide remodeling and its role in defense strategy. *Acta Physiologiae Plantarum, 33*(1), 35–51. https://doi.org/10.1007/s11738-010-0581-z
Krzesłowska, M., et al. (2016). Pectinous cell wall thickenings formation – A common defense strategy of plants to cope with Pb. *Environmental Pollution, 214*, 354–361. https://doi.org/10.1016/j.envpol.2016.04.019
Lange, B., et al. (2017). Copper and cobalt accumulation in plants: a critical assessment of the current state of knowledge. *New Phytologist, 213*(2), 537–551. https://doi.org/10.1111/nph.14175
Larney, F. J. & Angers, D. A. (2012). The role of organic amendments in soil reclamation: a review. *Canadian Journal of Soil Science, 92*(1), 19–38. https://doi.org/10.4141/cjss2010-064
Li, F., et al. (2024). Understanding the role of manganese oxides in retaining harmful metals: insights into oxidation and adsorption mechanisms at microstructure level. *Eco-Environment & Health, 3*(1), 89–106. https://doi.org/10.1016/j.eehl.2024.01.002
Li, H., et al. (2020). Input of Cd from agriculture phosphate fertilizer application in China during 2006–2016. *Science of The Total Environment, 698*, 134149. https://doi.org/10.1016/j.scitotenv.2019.134149
Li, Z., et al. (2001). Adsorption of cadmium on biosolids-amended soils. *Journal of Environmental Quality, 30*(3), 903–911. https://doi.org/10.2134/jeq2001.303903x
Liang, S., et al. (2011). Cadmium-induced earthworm metallothionein-2 is associated with metal accumulation and counteracts oxidative stress. *Pedobiologia, 54*(5–6), 333–340. https://doi.org/10.1016/j.pedobi.2011.07.011
Lim, J. E., et al. (2013). Effects of lime-based waste materials on immobilization and phytoavailability of cadmium and lead in contaminated soil. *CLEAN – Soil, Air, Water, 41*(12), 1235–1241. https://doi.org/10.1002/clen.201200169
Lindsay, W. L. (1979). *Chemical equilibria in soils*. Wiley.
Lwin, C. S., et al. (2018). Application of soil amendments to contaminated soils for heavy metal immobilization and improved soil quality–a critical review. *Soil Science and Plant Nutrition, 64*(2), 156–167. https://doi.org/10.1080/00380768.2018.1440938
Ma, C., et al. (2022). The relative contributions of different wheat leaves to the grain cadmium accumulation. *Toxics, 10*(11), 637. https://doi.org/10.3390/toxics10110637
Ma, J. F., et al. (2008). Transporters of arsenite in rice and their role in arsenic accumulation in rice grain. *Proceedings of the National Academy of Sciences, 105*(29), 9931–9935. https://doi.org/10.1073/pnas.0802361105
Mahfooz, Y., et al. (2020). Critical risk analysis of metals toxicity in wastewater irrigated soil and crops: a study of a semi-arid developing region. *Scientific Reports, 10*(1), 1–10. https://doi.org/10.1038/s41598-020-69815-0
Mahler, R. J., et al. (1987). Cadmium sulfate application to sludge-amended soils. I: effect on yield and cadmium availability to plants. *Science of The Total Environment, 67*, 117–131.

Malone, M. & Shakya, K. M. (2024). Trace metal contamination in community garden soils across the United States. *Sustainability, 16*(5), 1831. https://doi.org/10.3390/su16051831

Mamun, S. A., et al. (2017). Effects of lime and organic amendments derived from varied source materials on cadmium uptake by potato. *Journal of Environmental Quality, 46*(4), 836–844. https://doi.org/10.2134/jeq2017.02.0059

Mann, R. M., et al. (2011). Metals and metalloids in terrestrial systems: Bioaccumulation, biomagnification, and subsequent adverse effects. In F. Sánchez-Bayo, et al. (Eds.), *Ecological impacts of toxic chemicals* (pp. 43–62). Bentham Science Publishers Ltd. https://doi.org/10.2174/978160805121611101010043

Mann, S. S., et al. (2002). Cadmium accumulation in agricultural soils in Western Australia. *Water, Air, & Soil Pollution, 141*, 281–297. https://doi.org/10.1023/A:1021300228019

Marks, S. J., et al. (2021). Uptake of arsenic by irrigated vegetables and cooked food products in Burkina Faso. *Frontiers in Water, 3*, 667308. https://doi.org/10.3389/frwa.2021.667308

Marini, M., et al. (2020). The new fertilizer regulation: a starting point for cadmium control in European arable soils? *Science of The Total Environment, 745*, 140876. https://doi.org/10.1016/j.scitotenv.2020.140876

Martínez, C. E., et al. (2004). Lead phosphate minerals: solubility and dissolution by model and natural ligands. *Environmental Science & Technology, 38*(21), 5584–5590. https://doi.org/10.1021/es049617x

Mayer, M. M., et al. (2022). Using phosphate amendments to reduce bioaccessible Pb in contaminated soils: a meta-analysis. *Frontiers in Soil Science, 2*, 1028328. https://doi.org/10.3389/fsoil.2022.1028328

Mayland, H. F., et al. (1989). Selenium in seleniferous environments. In L. W. Jacobs (Ed.), *Selenium in agriculture and the environment* (SSSA Special Publication No. 23, p. 30). Soil Science Society of America.

McBride, M. B. (2013). Arsenic and lead uptake by vegetable crops grown on historically contaminated orchard soils. *Applied and Environmental Soil Science, 2013*, Article 283472. https://doi.org/10.1155/2013/283472

McCauley, A., et al. (2011). *Plant nutrient functions and deficiency and toxicity symptoms* (Nutrient Management Module No. 9). Montana State University Extension. https://apps.msuextension.org/publications/pub.html?sku=4449-9

McDowell, R. & Gray, C. (2022). Do soil cadmium concentrations decline after phosphate fertiliser application is stopped: a comparison of long-term pasture trials in New Zealand? *Science of The Total Environment, 804*, 150047. https://doi.org/10.1016/j.scitotenv.2021.150047

McDowell, R., et al. (2013). Natural background and anthropogenic contributions of cadmium to New Zealand soils. *Agriculture, Ecosystems & Environment, 165*, 80–87. https://doi.org/10.1016/j.agee.2012.12.011

McGrath, S. P., et al. (1995). Long-term effects of metals in sewage sludge on soils, microorganisms and plants. *Journal of Industrial Microbiology, 14*(2), 94–104. https://doi.org/10.1007/BF01569890

Mehri, A. (2020). Trace elements in human nutrition (II) – An update. *International Journal of Preventive Medicine, 11*, Article 2. https://doi.org/10.4103/ijpvm.IJPVM_48_19

Mertz, W. (1998). Review of the scientific basis for establishing the essentiality of trace elements. *Biological Trace Element Research, 66*(1), 185–191. https://doi.org/10.1007/BF02783137

Mishra, P., et al. (2023). Biofortification revisited: addressing the role of beneficial soil microbes for enhancing trace elements concentration in staple crops. *Microbiological Research, 275*, 127442. https://doi.org/10.1016/j.micres.2023.127442

Mitchell, R. G., et al. (2014). Lead (Pb) and other metals in New York City community garden soils: factors influencing contaminant distributions. *Environmental Pollution, 187*, 162–169. https://doi.org/10.1016/j.envpol.2014.01.007

Mitra, S., et al. (2022). Impact of heavy metals on the environment and human health: Novel therapeutic insights to counter the toxicity. *Journal of King Saud University - Science, 34*(3), 101865. https://doi.org/10.1016/j.jksus.2022.101865

Mizan, A., et al. (2023). Metal contamination in soil and vegetables around Savar tannery area, Dhaka, Bangladesh: a preliminary study for risk assessment. *Heliyon, 9*(3), e13856. https://doi.org/10.1016/j.heliyon.2023.e13856

Mogren, C. L. & Trumble, J. T. (2010). The impacts of metals and metalloids on insect behavior. *Entomologia Experimentalis et Applicata, 135*(1), 1–17. https://doi.org/10.1111/j.1570-7458.2010.00967.x

Monchanin, C., et al. (2021). Current permissible levels of metal pollutants harm terrestrial invertebrates. *Science of The Total Environment, 779*, 146398. https://doi.org/10.1016/j.scitotenv.2021.146398

Mora, A., et al. (2022). Effects of wastewater irrigation on groundwater quality: an overview. *Current Opinion in Environmental Science & Health, 25*, 100322. https://doi.org/10.1016/j.coesh.2021.100322

Naidu, R., et al. (1994). Ionic-strength and pH effects on the sorption of cadmium and the surface charge of soils. *European Journal of Soil Science, 45*(4), 419–429. https://doi.org/10.1111/j.1365-2389.1994.tb00527.x

Naidu, R., et al. (2021). Chemical pollution: a growing peril and potential catastrophic risk to humanity. *Environment International, 156*, 106616. https://doi.org/10.1016/j.envint.2021.106616

National Research Council. (1988). *Nutrient requirements of dairy cattle*. Academy Press.

National Research Council. (1989). *Diet and health: implications for reducing chronic disease risk* (Committee on Diet and Health). National Academies Press. https://www.ncbi.nlm.nih.gov/books/NBK218751/

Neina, D. (2019). The role of soil pH in plant nutrition and soil remediation. *Applied and Environmental Soil Science, 2019*, Article 5794869. https://doi.org/10.1155/2019/5794869

Nielsen, F. (2003). Trace elements. In B. Caballero, et al. (Eds.), *Encyclopedia of food sciences and nutrition* (2nd ed., pp. 5820–5828). https://doi.org/10.1016/B0-12-227055-X/01204-9

Okereafor, U., et al. (2020). Toxic metal implications on agricultural soils, plants, animals, aquatic life and human health. *International Journal of Environmental Research and Public Health, 17*(7), 2204. https://doi.org/10.3390/ijerph17072204

Oorts, K., et al. (2006). Discrepancy of the microbial response to elevated copper between freshly spiked and long-term contaminated soils. *Environmental Toxicology and Chemistry, 25*(3), 845–853. https://doi.org/10.1897/04-673R.1

Palansooriya, K. N., et al. (2020). Soil amendments for immobilization of potentially toxic elements in contaminated soils: a critical review. *Environment International, 134*, 105046. https://doi.org/10.1016/j.envint.2019.105046

Paltseva, A. A., et al. (2020). Remediation of an urban garden with elevated levels of soil contamination. *Science of The Total Environment, 722*, 137965. https://doi.org/10.1016/j.scitotenv.2020.137965

Park, J. H., et al. (2011). Comparative value of phosphate sources on the immobilization of lead, and leaching of lead and phosphorus in lead contaminated soils. *Science of The Total Environment, 409*(4), 853–860. https://doi.org/10.1016/j.scitotenv.2010.11.003

Pasricha, S., et al. (2021). Molecular mechanisms underlying heavy metal uptake, translocation and tolerance in hyperaccumulators-an analysis: heavy metal tolerance in hyperaccumulators. *Environmental Challenges, 4*, 100197. https://doi.org/10.1016/j.envc.2021.100197

Pidlisnyuk, V., et al. (2021). Phytotechnologies for site remediation. In L. E. Erickson & V. Pidlisnyuk (Eds.), *Phytotechnology with biomass production: sustainable management of contaminated sites* (1st ed., pp. 6–25). CRC Press. https://doi.org/10.1201/9781003082613

Pierzynski, G. M., et al. (2005). Trace elements. In G. M. Pierzynski, et al. (Eds.), *Soils and environmental quality* (3rd ed., pp. 331–362). CRC Press.

Pitumpe Arachchige, P. S., et al. (2024). Direct evidence on the impact of organic amendments on carbon stabilization in soil microaggregates. *Soil Science Society of America Journal, 88*(5), 1529–1544. https://doi.org/10.1002/saj2.20701

Raikwar, M. K., et al. (2008). Toxic effect of heavy metals on livestock health. *Veterinary World, 1*(1), 28–30.

Rashid, A., et al. (2023). Heavy metal contamination in agricultural soil: environmental pollutants affecting crop health. *Agronomy, 13*(6), 1521. https://doi.org/10.3390/agronomy13061521

Reeves, R. D., et al. (2018). A global database for plants that hyperaccumulate metal and metalloid trace elements. *New Phytologist, 218*(2), 407–411. https://doi.org/10.1111/nph.14907

Rieuwerts, J. S., et al. (1998). Factors influencing metal bioavailability in soils: preliminary investigations for the development of a critical 98 loads approach for metals. *Chemical Speciation & Bioavailability, 10*(2), 61–75. https://doi.org/10.3184/095422998782775835

Riza, M., et al. (2021). Portrayal of textile-based pollutants and its impact on soil, plants, and fisheries. *Nature Environment and Pollution Technology, 20*(3), 1269–1275. https://doi.org/10.46488/NEPT.2021.v20i03.038

Samsøe-Petersen, L., et al. (2002). Uptake of trace elements and PAHs by fruit and vegetables from contaminated soils. *Environmental Science & Technology, 36*(14), 3057–3063. https://doi.org/10.1021/es015691t

Scheckel, K. G., et al. (2013). Amending soils with phosphate as a means to mitigate soil lead hazard: a critical review of the state of the science. *Journal of Toxicology and Environmental Health, Part B: Critical Reviews, 16*(6), 337–380. https://doi.org/10.1080/10937404.2013.825216

Schlosser, C. A., et al. (2014). *The future of global water stress: an integrated assessment* (Joint Program Report Series No. 254, p. 30). MIT Joint Program on the Science and Policy of Global Change. https://globalchange.mit.edu/publication/15492

Schützendübel, A. & Polle, A. (2002). Plant responses to abiotic stresses: heavy metal-induced oxidative stress and protection by mycorrhization. *Journal of Experimental Botany, 53*(372), 1351–1365. https://doi.org/10.1093/jexbot/53.372.1351

Seregin, I. V., et al. (2007). Role of root and shoot tissues of excluders and hyperaccumulators in nickel transport and accumulation. *Doklady Biological Sciences, 415*(1), 295–297. https://doi.org/10.1134/S0012496607040138
Seregin, I. V., et al. (2004). Distribution and toxic effects of cadmium and lead on maize roots. *Russian Journal of Plant Physiology, 51*(5), 525–533. https://doi.org/10.1023/B:RUPP.0000035747.42399.84
Seyfferth, A. L., et al. (2024). Mitigating toxic metal exposure through leafy greens: a comprehensive review contrasting cadmium and lead in spinach. *GeoHealth, 8*, e2024GH001081. https://doi.org/10.1029/2024GH001081
Shahid, M., et al. (2021). Health risks of arsenic buildup in soil and food crops after wastewater irrigation. *Science of The Total Environment, 772*, 145266. https://doi.org/10.1016/j.scitotenv.2021.145266
Simmons, R. W., et al. (2008). Analysis of field-moist Cd-contaminated paddy soils during rice grain fill allows reliable prediction of grain Cd levels. *Plant and Soil, 302*(1–2), 125–137. https://doi.org/10.1007/s11104-007-9460-9
Singh, S. S. (1981). Uptake of cadmium by lettuce (*Lactuca sativa*) as influenced by its addition to soil as inorganic or in sewage sludge. *Canadian Journal of Soil Science, 61*, 19–28.
Sipos, P., et al. (2008). Sorption of copper, zinc and lead on soil mineral phases. *Chemosphere, 73*(4), 461–469. https://doi.org/10.1016/j.chemosphere.2008.06.046
Šmejkalová, M., et al. (2003). Effects of heavy metal concentrations on biological activity of soil micro-organisms. *Plant, Soil and Environment, 49*(7), 321–326.
Smith, S., et al. (1989). Chromium accumulation, transport, and toxicity in plants. *Toxicological & Environmental Chemistry, 24*(4), 241–251. https://doi.org/10.1080/02772248909357496
Smolders, E., et al. (2015). Toxicity in lead salt spiked soils to plants, invertebrates and microbial processes: unraveling effects of acidification, salt stress and ageing reactions. *Science of The Total Environment, 536*, 223–231. https://doi.org/10.1016/j.scitotenv.2015.07.067
Srivastava, V., et al. (2017). Agroecological responses of heavy metal pollution with special emphasis on soil health and plant performances. *Frontiers in Environmental Science, 5*, 64. https://doi.org/10.3389/fenvs.2017.00064
Strawn, D. G. (2021). Sorption mechanisms of chemicals in soils. *Soil Systems, 5*(1), 13. https://doi.org/10.3390/soilsystems5010013
Strawn, D. G. & Hettiarachchi, G. M. (2021). Fifty years of articles in JEQ on trace elements in the environment and future outlook. *Journal of Environmental Quality, 50*(6), 1266–1281. https://doi.org/10.1002/jeq2.20296
Stroud, J. L., et al. (2011). The dynamics of arsenic in four paddy fields in the Bengal Delta. *Environmental Pollution, 159*(4), 947–953. https://doi.org/10.1016/j.envpol.2010.12.020
Suda, A. & Makino, T. (2016). Functional effects of manganese and iron oxides on the dynamics of trace elements in soils with a special focus on arsenic and cadmium: a review. *Geoderma, 270*, 68–75. https://doi.org/10.1016/j.geoderma.2015.12.017
Sun, L., et al. (2023). The road toward Cd-safe rice: from mass selection to marker-assisted selection and genetic manipulation. *The Crop Journal, 11*(4), 1059–1071. https://doi.org/10.1016/j.cj.2023.01.007

Taghavi, M., et al. (2023). Ecological risk assessment of Trace Elements (TEs) pollution and human health risk exposure in agricultural soils used for saffron cultivation. *Scientific Reports, 13*(1), 1–15. https://doi.org/10.1038/s41598-023-31681-x

Tang, X., et al. (2004). Assessment of the effectiveness of different phosphorus fertilizers to remediate Pb-contaminated soil using in vitro test. *Environment International, 30*(4), 531–537. https://doi.org/10.1016/j.envint.2003.10.008

Tang, Z., et al. (2023). Molecular mechanisms underlying the toxicity and detoxification of trace metals and metalloids in plants. *Journal of Integrative Plant Biology, 65*(3), 570–593. https://doi.org/10.1111/jipb.13438

Tchounwou, P. B., et al. (2012). Heavy metals toxicity and the environment. *EXS, 101*, 133. https://doi.org/10.1007/978-3-7643-8340-4_6

Tlustoš, P., et al. (2006). The effect of liming on cadmium, lead, and zinc uptake reduction by spring wheat grown in contaminated soil. *Plant, Soil and Environment, 52*(1), 16–24. https://doi.org/10.17221/3341-PSE

Tóth, Z., et al. (2023). Urban soil quality deteriorates even with low heavy metal levels: an arthropod-based multi-indices approach. *Ecological Applications, 33*(4), e2848. https://doi.org/10.1002/eap.2848

Tőzsér, D., et al. (2023). A meta-analysis on the heavy metal uptake in *Amaranthus* species. *Environmental Science and Pollution Research International, 30*(36), 85102. https://doi.org/10.1007/s11356-023-28374-3

Traina, S. J. & Laperche, V. (1999). Contaminant bioavailability in soils, sediments, and aquatic environments. *Proceedings of the National Academy of Sciences, 96*(7), 3365–3371.

Turdi, M. & Yang, L. (2016). Trace elements contamination and human health risk assessment in drinking water from the agricultural and pastoral areas of Bay County, Xinjiang, China. *International Journal of Environmental Research and Public Health, 13*(10), 938. https://doi.org/10.3390/ijerph13100938

Tusher, T. R., et al. (2017). Soil quality and heavy metal concentrations in agricultural lands around dyeing, glass, and textile industries in Tangail District of Bangladesh. *Journal of Environmental Science and Natural Resources, 10*(2), 109–116.

Ur Rahman, S., et al. (2024). Pb uptake, accumulation, and translocation in plants: plant physiological, biochemical, and molecular response: a review. *Heliyon, 10*(6), e27724. https://doi.org/10.1016/j.heliyon.2024.e27724

U.S. Environmental Protection Agency. (1992). *Contaminants and remedial options at wood preserving sites* (EPA/600/R-92/182). Risk Reduction Engineering Laboratory, Office of Research and Development. https://clu-in.org/download/contaminantfocus/dnapl/treatment_technologies/epa-600-r-92-182.pdf

U.S. Environmental Protection Agency. (1994). *A plain English guide to the EPA Part 503 biosolids rule* (EPA/832/R-93/003). Office of Wastewater Management. https://www.epa.gov/biosolids/plain-english-guide-epa-part-503-biosolids-rule

U.S. Environmental Protection Agency. (2004). *Guidelines for water reuse* (EPA/625/R-04/108). Office of Wastewater Management, Office of Water; Office of Research and Development, National Risk Management Research Laboratory. https://www.epa.gov/sites/default/files/2015-08/documents/epa625r04108.pdf

U.S. Environmental Protection Agency. (2011). *Reusing potentially contaminated landscapes: growing gardens in urban soils* (EPA/542/F-10/011). https://www.epa.gov/sites/default/files/2014-03/documents/urban_gardening_fina_fact_sheet.pdf

U.S. Environmental Protection Agency. (2014). *Priority pollutant list*. https://www.epa.gov/sites/default/files/2015-09/documents/priority-pollutant-list-epa.pdf

U.S. Geological Survey. (2019). *Metals and other trace elements* [Fact sheet]. U.S. Geological Survey. https://www.usgs.gov/mission-areas/water-resources/science/metals-and-other-trace-elements (Accessed August 22, 2024)

Uwizeyimana, H., et al. (2017). The eco-toxic effects of pesticide and heavy metal mixtures towards earthworms in soil. *Environmental Toxicology and Pharmacology, 55*, 20–29. https://doi.org/10.1016/j.etap.2017.08.001

Uzu, G., et al. (2010). Foliar lead uptake by lettuce exposed to atmospheric fallouts. *Environmental Science & Technology, 44*(3), 851–856. https://doi.org/10.1021/es902190u

Wang, F., et al. (2021). Typical sources of Cd to paddy fields in different contaminated areas and their impacts on Cd accumulation in topsoil and rice in Changzhutan, China. *Environmental Research, 193*, 110523. https://doi.org/10.1016/j.envres.2020.110523

Wang, J., et al. (2022). Dermal bioaccessibility and cytotoxicity of heavy metals in urban soils from a typical plateau city: implication for human health. *Science of The Total Environment, 835*, 155544. https://doi.org/10.1016/j.scitotenv.2022.155544

Wang, K., et al. (2018). Bioaccumulation of heavy metals in earthworms from field contaminated soil in a subtropical area of China. *Ecotoxicology and Environmental Safety, 148*, 876–883. https://doi.org/10.1016/j.ecoenv.2017.11.058

Wang, Z., et al. (2020). Acid buffering capacity of four contrasting metal-contaminated calcareous soil types: changes in soil metals and relevance to phytoextraction. *Chemosphere, 256*, 127045. https://doi.org/10.1016/j.chemosphere.2020.127045

Williams, P. N., et al. (2014). Localized flux maxima of arsenic, lead, and iron around root apices in flooded lowland rice. *Environmental Science & Technology, 48*(15), 8498–8506. https://doi.org/10.1021/es501127k

Woodcock, J. (2021). *FDA releases action plan for reducing exposure to toxic elements from foods for babies, young children*. United States Food and Drug Administration. https://www.fda.gov/news-events/press-announcements/fda-releases-action-plan-reducing-exposure-toxic-elements-foods-babies-young-children

Wu, J., et al. (2018). Risk assessment and source identification of toxic metals in the agricultural soil around a Pb/Zn mining and smelting area in Southwest China. *International Journal of Environmental Research and Public Health, 15*(9), 1838. https://doi.org/10.3390/ijerph15091838

Xie, Y., et al. (2016). Effect of heavy metals pollution on soil microbial diversity and bermudagrass genetic variation. *Frontiers in Plant Science, 7*, 755. https://doi.org/10.3389/fpls.2016.00755

Xu, X. Y., et al. (2008). Growing rice aerobically markedly decreases arsenic accumulation. *Environmental Science & Technology, 42*(15), 5574–5579. https://doi.org/10.1021/es800324u

Yadav, R., et al. (2023). Heavy metal toxicity in earthworms and its environmental implications: a review. *Environmental Advances, 12*, 100374. https://doi.org/10.1016/j.envadv.2023.100374

Yang, J., et al. (2001). Lead immobilization using phosphoric acid in a smelter-contaminated urban soil. *Environmental Science & Technology, 35*(17), 3553–3559. https://doi.org/10.1021/es001770d

Yu, H., et al. (2023). A review on adsorption characteristics and influencing mechanism of heavy metals in farmland soil. *RSC Advances, 13*(6), 3505–3519. https://doi.org/10.1039/d2ra07095b

Zeng, F., et al. (2011). The influence of pH and organic matter content in paddy soil on heavy metal availability and their uptake by rice plants. *Environmental Pollution, 159*(1), 84–91. https://doi.org/10.1016/j.envpol.2010.09.019

Zgorelec, Z., et al. (2020). Cadmium and mercury phytostabilization from soil using *Miscanthus × giganteus*. *Scientific Reports, 10*(1), 6685. https://doi.org/10.1038/s41598-020-63488-5

Zhang, H., et al. (2021). Effects of phosphorous precursors and speciation on reducing bioavailability of heavy metal in paddy soil by engineered biochars. *Environmental Pollution, 285*, 117459. https://doi.org/10.1016/j.envpol.2021.117459

Zhang, X., et al. (2022). "Potentially toxic element"–something that means everything means nothing. *Environmental Science & Technology, 56*(17), 11922–11925. https://doi.org/10.1021/acs.est.2c03056

Zhao, F. J., et al. (2022). Toxic metals and metalloids: uptake, transport, detoxification, phytoremediation, and crop improvement for safer food. *Molecular plant*, *15*(1), 27–44. https://doi.org/10.1016/j.molp.2021.09.016

Zhao, F. J. & Wang, P. (2020). Arsenic and cadmium accumulation in rice and mitigation strategies. *Plant and Soil, 446*, 1–21. https://doi.org/10.1007/s11104-019-04374-6

Zhao, F. J., et al. (2009). Arsenic uptake and metabolism in plants. *New Phytologist, 181*(4), 777–794. https://doi.org/10.1111/j.1469-8137.2008.02716.x

Zhao, F.-J., et al. (2010). Arsenic as a food chain contaminant: mechanisms of plant uptake and metabolism and mitigation strategies. *Annual Review of Plant Biology, 61*(1), 535–559. https://doi.org/10.1146/annurev-arplant-042809-112152

Zhou, H., et al. (2016). Accumulation of heavy metals in vegetable species planted in contaminated soils and the health risk assessment. *International Journal of Environmental Research and Public Health, 13*(3), 289. https://doi.org/10.3390/ijerph13030289

Zwonitzer, J. C., et al. (2003). Effects of phosphorus additions on lead, cadmium, and zinc bioavailabilities in a metal-contaminated soil. *Water, Air, and Soil Pollution, 143*(1–4), 193–209. https://doi.org/10.1023/A:1022810310181

Chapter 3

Per- and polyfluoroalkyl substances (PFAS) as contaminants in agricultural soils

Lynda Peter and Linda S. Lee, Purdue University, USA

1 Introduction

Per- and poly-fluoroalkyl substances (PFAS) are a diverse family of several thousand synthetic compounds, now coined 'forever chemicals' due to their high thermal and chemical stability and resistance to microbial mineralization (Schymanski et al., 2023). PFAS have been found in all environmental compartments globally due to their extensive use (Espartero et al., 2022; Panieri et al., 2022). The use and number of different compounds in this family have increased exponentially since they were first discovered (Gaines, 2022). Their stability, along with other unique properties including imparting grease-, water- and stain-repelling properties, have resulted in their use in a myriad of consumer products and manufacturing processes for several decades (Gaines, 2022; Glüge et al., 2020). Familiar uses of PFAS include grease and stain-repellant coatings on food contact materials, carpets, furniture, and other textiles, as well as on non-stick cookware and construction materials (Kotthoff et al., 2015). PFAS are present in numerous personal care products including

http://dx.doi.org/10.19103/AS.2025.0151.03

lotions, shampoos, and cosmetics, as well as other cleaning products and automotive fluids (Whitehead et al., 2021). Since the 1960s, PFAS-containing aqueous-film forming foams (AFFFs) have been in use at airports and military sites because of their ability to extinguish hydrocarbon-based fires in a few seconds (Place & Field, 2012; Prevedouros et al., 2006). PFAS are also used in manufacturing electronics including semiconductor and computer chips, wire coatings, metal plating, medical devices, artificial turf, and automotive fluids among many others (Chen et al., 2024; Jacob et al., 2021; Lin et al., 2017). PFAS use in so many aspects of society, coupled to their persistence has resulted in their presence at varying scales in nearly all environmental compartments including drinking water, surface water, groundwater, soil, air, sediment, and household dust (Evich et al., 2022; Espartero et al., 2022; Panieri et al., 2022). Their ubiquitous presence and persistence, bioaccumulation potential, and linkages to numerous adverse effects in humans and animals, even at low concentrations (Fenton et al., 2021; Halldorsson et al., 2012; Dickman & Aga, 2022a; Lee et al., 2020; NASEM, 2022) have resulted in mounting concerns with regards to our water resources and food system.

PFAS presence in water used for agriculture, atmospheric deposition, and PFAS accumulation within the soil profile, where crops are grown and animals graze, are major challenges in agricultural systems. This chapter will focus on these challenges prefaced by briefly highlighting key PFAS characteristics followed by environmental entry modes and fate of PFAS, impacts on human and ecosystem health, and potential mitigation strategies in agriculture to keep PFAS-impacted farms viable and minimize PFAS impacts to our food system.

2 PFAS characteristics

PFAS occur in two broad classes, polymers and non-polymers, with the latter including the PFAS of greatest concern in terms of exposure to humans and biota. Unique to the chemical structure of most PFAS is the presence of one or more perfluorocarbons that are typically wihtin or as a linear or branched carbon chain of varying lengths (C_nF_{2n+1}-) bound to a polar functional group (e.g., carboxylic acid, sulfonate, alcohol, phosphate, amino group, or other group), which may be attached to a nonfluorinated hydrocarbon entity. The combination of chain lengths, number of perfluoroalkyl tails, polar functional groups and associated linkages govern their behavior, sub-classification, and nomenclature (Buck et al., 2011). Two classes within the PFAS family that have received the most attention are the perfluoroalkyl sulfonic acids (PFSAs) and the perfluoroalkyl carboxylic acids (PFCAs), grouped together as perfluoroalkyl acids (PFAAs), (Kwiatkowski et al., 2020; Wang et al., 2015). PFAAs are known as the terminal products of microbial degradation of larger PFAS known as precursor compounds. While PFAAs have garnered the most attention, in recent years, we have found that the bulk of the PFAS-derived organofluorine present in our sludges, often converted to biosolids for land application, are precursors (Alukkal et al., 2024b; Lee et al., 2014a; Thompson et al., 2023). Therefore, although PFAS use in some processes and products is decreasing through phaseouts and stewardship programs (Gallen et al., 2022; Ulrich et al., 2016), alternative PFAS are being produced and the capacity to quantify a wider variety of PFAS is increasing (Dickman & Aga, 2022b). Moreover, legislative response to limit the production and use of PFAS varies globally. As early as 2009, the use of perfluorooctane sulfonate (PFOS) was partially restricted through the United Nations Stockholm Convention Persistent Organic Pollutants regulation, followed by a complete ban and addition of perfluorooctanoic acid (PFOA) in 2020 (Brennan et al., 2021). More recently in 2023, the European Chemicals Agency took legislative steps through the EU Registration, Evaluation, Authorization, and Restriction of Chemicals program to restrict all legacy and novel PFAS use except in firefighting foams starting in 2026 (Thomas et al., 2023). Other countries including Australia and Canada has taken steps to regulate the use of PFAS especially long-chain PFAAs. Legislation in the United State is less cohesive with federal and state governments considering different measures. The US EPA, which is responsible for PFAS regulation at the federal level in the USA, recently approved enforceable drinking water limits for six PFAS including PFOS, PFOA, perfluorohexanesulfonic acid (PFHxS), perfluorononanoic acid (PFNA), perfluorobutanoic acid (PFBS) and Gen-X (EPA, 2024a), although the 2025 administration have put a pause on the regulations for all but PFOS and PFOA. Meanwhile, individual states in the USA have pursued legislative actions, some of which are targeting banning PFAS in fire-fighting foam and consumer products, limiting land-application of biosolids and adopting independent drinking water standards (Langenback & Wilson, 2021; Hughes, 2023; BCLP, 2025a, 2025b). Despite some successes, enforcements of PFAS policies in the US remains a challenge at the national and state levels due to political and economic constraints, public awareness, feasibility of implementation, and inadequate scientific evidence. Ultimately, the nonuniformity in worldwide response to PFAS contamination poses difficulties to reducing PFAS emissions given the global nature of markets and economies.

PFAS are both hydrophobic (perfluoroalkyl tail) and lipophilic (polar functional group), which give them their surfactant behavior, and are proteinophilic. Within each PFAS class, they are grouped as short and long chains with the longer chains having higher sorption and bioaccumulation propensity. PFAS affinity for proteins has made some PFAS highly bioaccumulative, leading to their accumulation in blood, meat, liver, kidneys, and breast milk of humans throughout the world (Jian et al., 2018; Liu et al., 2019a; Poothong et al., 2020) and most recently in brains (Xie et al., 2024). The longer the perfluoroalkyl tail length, the higher their bioaccumulation and longer their half-lives within biota

(Khan et al., 2023; Shi et al., 2018). PFAS has been found in the blood samples of nearly all human beings (Calafat et al., 2019; Göckener et al., 2020a; Haines et al., 2017). In addition, PFAS are frequently found in wildlife, domesticated animal products, and foods (Brown et al., 2020; De Silva et al., 2021; Death et al., 2021). PFAS are not acutely toxic but have been associated with numerous adverse human health effects, including some types of cancers, thyroid and heart diseases, and immunosuppression (Fenton et al., 2021; Sinclair et al., 2020; Sunderland et al., 2019). Children's vulnerability to PFAS exposure was highlighted in a systematic review where positive links were observed between PFAS serum levels and dyslipidemia, renal function, and suppression of vaccine response (Rappazzo et al., 2017). Most toxicological studies have focused on PFOS and PFOA given their early detection in human serum and environmental prevalence; however, a recent study (Xie et al., 2023) reported a possible link between perfluorooctane sulfonamide (FOSA) and glioma (a type of brain tumor). Additionally, longitudinal studies of adverse health impact from PFAS exposure have mainly been on workers and communities impacted by PFAS manufacturing (Barry et al., 2013; Darrow et al., 2013; Lopez-Espinosa et al., 2012; Steenland et al., 2013). For example, plant workers' exposure to PFOA was connected to high cholesterol, kidney and testicular cancer, thyroid disease, and ulcerative colitis. Non-occupational exposure to PFAS in communities with contaminated drinking water sources has also been linked to altered thyroid and kidney function (Blake et al., 2018). The adverse effect of PFAS exposure in agricultural communities is currently unclear due to limited evidence. A study in the Veneto region of Italy reported high serum concentrations of PFOA and PFOS in farmers that were exposed to contaminated drinking water (Ingelido et al., 2020). However, the contamination was not from agricultural activities but from a PFAS manufacturing plant. Another study (Batzella et al., 2024) examined the impact of the exposure from the same cohort over a four-year period and found a positive relationship between PFAS concentration in serum and cholesterol levels, which corroborates with an earlier report from Uppsala, Sweden (Dunder et al., 2022). In another longitudinal study investigating lung functions of adolescents and young adults of an agricultural population, no evidence of negative effects from PFAS exposure was observed (Kornher et al., 2025). Similarly, exposure to background levels of eight PFAS mixtures among farmers and other rural residents did not increase the risk of coronary heart disease over the period of 10 years (Mattsson et al., 2015). Nonetheless, the risk of adverse health effects in agricultural workers might increase over time, especially in highly contaminated situations where PFAS exposure is comparable to occupational levels. In aquatic organisms and wildlife, PFAS at environmentally relevant concentrations adversely affected metabolic processes, neurological pathways, immune system function, and overall behavior (Dickman & Aga, 2022a; Lee et al., 2020).

3 PFAS pathways into the environment and environmental concentrations

PFAS are released to the environment via a myriad of anthropogenic activities and multiple pathways, each with its own magnitude of impact, concentration, and suite of PFAS. Two of the most significant pathways of PFAS to air, land, and water include industrial discharges and aqueous fire-fighting foams (AFFF) releases during military and fire department use or training. Although these modes of entry have been dramatically reduced, they have left a long-term legacy of these persistent PFAS in the environment. Other frequent sources of PFAS into the environment are landfills (air emissions and unintended landfill leaching), wastewater discharge to surface water bodies, wastewater reuse (irrigation water), and land application of biosolids. The latter, along with AFFF-impacted surface and well water used for irrigation and livestock drinking water, are the two biggest sources directly impacting agriculture (Figure 1).

Direct PFAS sources include air emissions and liquid discharges from PFAS manufacturing, and release from PFAS-associated products with AFFFs dominating product sources of PFAS to the environment. PFAS in AFFFs enhance interfacial tension properties required to efficiently extinguish hydrocarbon fires. AFFFs have been used worldwide at military bases, airports, oil and gas refineries, chemical plants, and storage facilities, resulting in direct release of PFAS to soil during transport, storage, fire training activities, and disposal (Prevedouros et al., 2006). Concentrations up to 460 000 ng/g have been reported for PFOS and 50 000 ng/g for PFOA in soils collected from fire training sites (Brusseau et al., 2020).

Airborne-PFAS transport via wind beyond emission locations contributes to background levels of PFAS in soils, including remote areas. A global study analyzing soil samples from 62 locations with minimal to no human impact reported at least three quantifiable concentrations of PFAAs in all the sampling locations (Rankin et al., 2016). Total PFCAs and PFSAs concentrations ranged from 0.29 ng/g to 14.3 ng/g and 0.007 ng/g to 3.7 ng/g, respectively. Similarly, a study conducted in Maine, US reported background concentrations of PFCAs between 0.03 ng/g and 17.4 ng/g and PFSAs 0.16–4.35 ng/g (MDEP, 2022). Relatively higher maximum background concentrations were measured in Vermont, USA for PFCAs and PFSAs at 23 ng/g and 12 ng/g, respectively (Zhu et al., 2022).

4 PFAS routes into agriculture

Due to the extensive use of PFAS in consumer goods, waste disposal systems including landfills and wastewater treatment plants serve as the conduits for

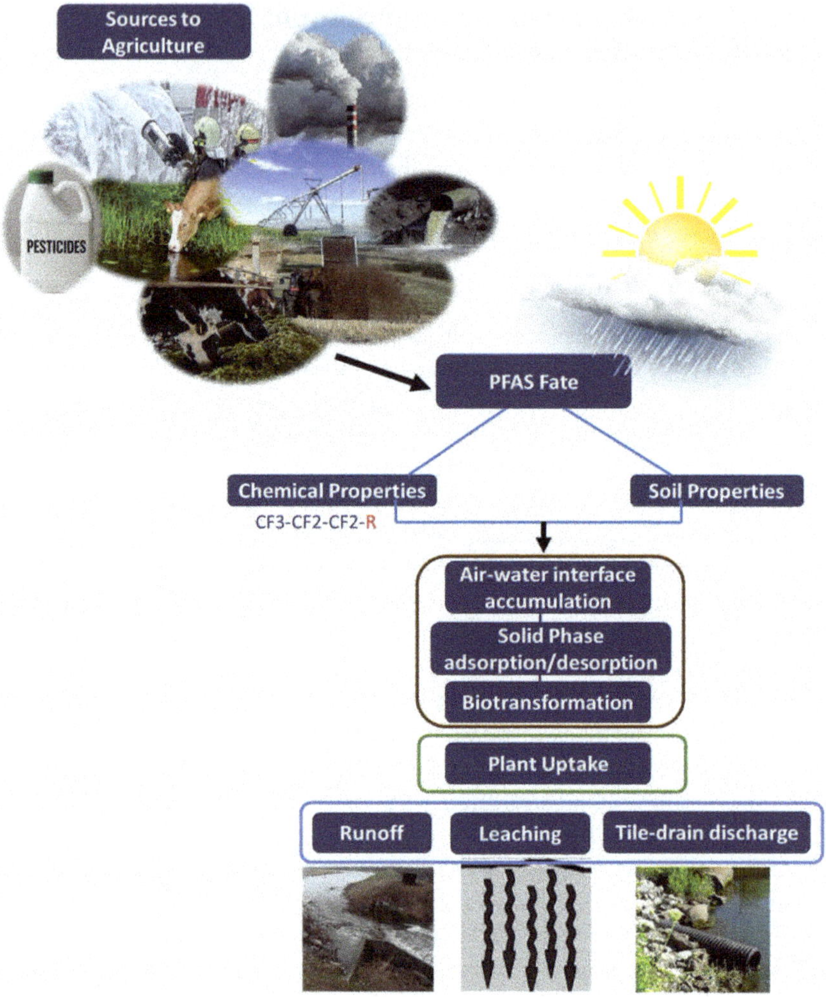

Figure 1 Schematic of primary pathways for PFAS entry into agriculture followed by subsequent fate and transport factors and processes.

PFAS release to agricultural soils in levels higher than prevailing background concentrations. The most documented PFAS sources associated with these systems are biosolids and irrigation water, which includes well water as well as surface water contaminated from a myriad of sources (e.g. wastewater effluent, AFFFs, industrial discharges) (Islam et al., 2025). Free-grazing livestock are particularly vulnerable to PFAS-contaminated surface water (Mikkonen et al., 2023). Additionally, atmospheric deposition, runoff or drainage onto reighboring properties, and agricultural inputs, e.g., pesticides are potential sources of PFAS in agricultural ecosystems.

Published by Burleigh Dodds Science Publishing Limited, 2025.

4.1 PFAS in land-applied biosolids

Biosolids are end-products of sludge generated from wastewater treatment plants that are rich in organic matter (OM), microbial biomass, and plant nutrients. Conversion of sludge to biosolids occurs through a variety of processes, including anaerobic/aerobic digestion and thermal hydrolysis targeted at reducing odor-producing compounds, vector attraction, and pathogens while improving beneficial reuse and commercial value, and in some cases, biogas production (Wang et al., 2009). The use of biosolids as a soil amendment in agriculture is a global practice. For example, an average of 44% of sewage sludge as biosolids is land-applied in the EU, 73% for Australia in 2021, and 60% in the USA in 2023 (Braine et al., 2024; USEPA, 2024b). Although treatment may alter PFAS present in biosolids through some precursor transformation, PFAS are not removed (Eriksson et al., 2017); Schaefer et al., 2022; Saliu and Sauvé, 2024; Alukkal et al., 2024a). Total PFAS loads reported for biosolids range from <100 to a few thousand µg/kg (Higgins et al., 2005; Sun et al., 2011; Moodie et al., 2021; Munoz et al., 2022; Schaefer et al., 2021), but this range is likely greatly underestimating the PFAS present. Precursor compounds (PFAS that break down to other PFAS) make up a substantial proportion of total PFAS measured in biosolids (Lee et al., 2014a; Thompson et al., 2023) and even for many of the known precursors, standards are not available. Alukkal et al. (2024b) targeted analysis of almost 60 PFAS (including many precursors), but the total calculated PFAS-associated organofluorine was less than 1%. Noteworthy is a recent study that revealed up to 75% of the organofluorine in biosolids originates from CF_n groups in pharmaceuticals, which by most definitions are not deemed PFAS (Ruyle et al., 2025), but that still leaves 25% or more unaccounted organofluorine likely associated with unquantified PFAS. Biosolids generated from industrial wastewater influent and paper fiber residuals typically have higher total PFAS loads but with fewer types of PFAS compared to biosolids derived from domestic wastewater (municipal biosolids) where PFAS concentrations are diffuse and originate from use of a variety of products. Therefore, beneficial reuse of biosolids can redistribute and elevate levels of PFAS in agricultural soils. The differences in PFAS types and concentrations in biosolids are reflected by the variable extent of contamination observed at impacted sites.

4.2 PFAS in irrigation water

In many parts of the world, irrigation with surface water, groundwater, and reclaimed wastewater is essential for crop production. Water resources impacted by landfill leachates, wastewater treatment effluent, AFFF use, and industrial discharges are potential sources of PFAS to irrigation waters, thus a pathway to agricultural soils as reported in many countries (Pepper et al., 2021; Mroczko et al., 2022; Kim et al., 2019; Shigei et al., 2020).

4.3 PFAS in pesticides

Pesticides formulated with PFAS either as an active ingredient, present in adjuvants or the "inert" ingredients used, residual impurity, or PFAS leached from containers are direct sources of contamination. Fluorination serves to improve many characteristics of the pesticide including stereospecificity and enhanced effects on target pests particularly insects, lipophilicity and residual activity, facilitating application, and reducing pesticide drift (Alexandrino et al., 2022; Ogawa et al., 2020). The single largest contribution to PFAS in pesticides overall is as an active ingredient (Donley et al., 2024). Pesticides are highly diluted before application; however, continuous use contributes to PFAS accumulation in agricultural soils. For example, N-ethyl perfluorooctane sulfonamide (EtFOSA) is as an active ingredient in sulfluramid. EtFOSA degrades into PFOS through various intermediates. Application of this pesticide in Pine and Eucalyptus plantations in Brazil to control leaf cutter ants resulted in low concentrations of PFOS (0.03–5.4 ng/g) in plantation soils (Nascimento et al., 2018). The presence of PFOS has also been reported for insecticides with the following six active ingredients: abamectin, novaluron, mineral oil, imidacloprid, spiromesifen, and malathion (Lasee et al., 2022).

4.4 PFAS in atmospheric deposition

Neutrally charged PFAS are released into the atmosphere during manufacturing and industrial use. In addition, combustion of industrial wastes and landfill waste generate PFAS-containing particles or volatile PFAS compounds to the atmosphere, respectively (Baqar et al., 2024; Lin et al., 2024). PFAS that are emitted partition between particle and gas interfaces until they are deposited onto the biosphere including agricultural soils. Atmospheric deposition of PFAS can proceed via dry and wet deposition mechanisms. Dry deposition refers to release from the atmosphere as gases or particulate matter while wet deposition involves transfer of atmospheric-PFAS in aqueous form during precipitation. Precursors such as fluorotelomer alchohols (FTOHs) and perfluoroalkyl sulfonamides (FASAs) can undergo atmospheric oxidation to PFAAs prior to land deposition (Aschmann et al., 2000; D'Eon et al., 2006Ellis et al., 2004). Soils located near PFAS emission zones are susceptible to substantial levels of contamination. Total PFAS concentrations between 79.9 ng/g and 200 ng/g have been reported in agricultural soils located 300 m from a fluorochemical manufacturing plant (Liu et al., 2019b). Nevertheless, atmospheric deposition

Published by Burleigh Dodds Science Publishing Limited, 2025.

is challenging to delineate because of long-range transport of PFAS across geopolitical boundaries.

5 PFAS fate in agricultural fields

PFAS behavior in the soil and transport outcomes depends on prevailing soil properties and PFAS class (Sharifan et al., 2021) as well as topography, climate, and management (Figure 1). Within a class, differences in perfluorocarbon chain length impact PFAS mobility and bioavailability. Short-chain PFAS are more mobile and bioavailable compared to their long-chain homologs. Both sorption of PFAS to soil surfaces as well as at the air-water interface (AWI) within the vadose zone directly affect their leaching behavior and bioavailability (Post et al., 2017; Lyu et al., 2018). For example, short-chain PFAS absorb less than longer-chain PFAS, leading them to a greater presence in porewater, thus leaching to a greater extent and being taken up more by plants than long-chain PFAS (Li et al., 2020). This leads to the inverse relationship between soil depth and the concentration of long-chain PFAS (Johnson, 2022; Pepper et al., 2023; Sepulvado et al., 2011; Washington et al., 2010; Álvarez-Ruiz et al., 2024). In addition, the presence of organic co-contaminants, nonaqueous phase liquids, or inorganic ions can influence PFAS sorption (Guelfo & Higgins, 2013; Le et al., 2022). For example, divalent cations increase sorption to soil surfaces as well as interfacial sorption of PFAS (Nickerson et al., 2023; Le et al., 2022). Once introduced into the soil, PFAS are subjected to multiple transport processes. Key processes in agricultural settings include sorption, accumulation at AWIs, transformation, plant uptake, surface runoff, and leaching.

5.1 PFAS sorption to soil surfaces

PFAS retention in the soil is controlled by a range of sorption mechanisms as well as air-water interfacial processes. The sorption (K_d, L/kg) of organic compounds to soils is determined by the properties of the compounds (chemical structure, ionic state, and surface-active charge) and the soil conditions (OM, pH, exchange capacity, and soil-solution ionic strength). In the environment, PFAS can exist as anions, cations, zwitterions, and as neutral molecules. The anionic form is more prevalent under typical environmental pH (4–9) ranges given that most PFAS, especially PFCAs and PFSAs, have relatively low acid dissociation constants (pK_a) (PFCAs, <1.6; PFSAs, <0.3) (Goss, 2008; Mei et al., 2021; Rayne & Forest, 2010; Vierke et al., 2013). The interactions between PFAS and soils include hydrophobic interactions, electrostatic interactions, and hydrogen bonding (Mei et al., 2021; Kookana et al., 2022).

Hydrophobic interaction can occur between the perfluoroalkyl chains and soil OM with longer chains having a higher affinity for OM. However, PFAS sorption is complex and sorption trends derived from laboratory and field studies are not consistent. While several studies have shown organic carbon (OC) to be a driver, others have exemplified a weak relationship between soil OC and K_d values (Li et al., 2018; Kookana et al., 2022; Uhme et al., 2021; Evich et al., 2025). For short-chain PFAAs and some of the longer chain PFCAs, electrostatic and mineralogical features appear to play a greater role in retention, whereas for long-chain PFAS, particularly PFOS, air-water interfacial sorption and sorption to OC are larger drivers. Soil protein content has also been shown to impact PFAS sorption (Li et al., 2019; Longstaffe et al., 2012). Interactions between anionic PFAS and protonated amino and amide groups in proteins are one of the reasons hypothesized for such observations (Li et al., 2019). In addition, given that PFAS commonly occur as mixtures, competitive sorption further complicates sorption processes, especially when PFAS are present at high concentrations such as may be found at military, industrial, or other heavily impacted sites (Uhme et al., 2024), For example, at a dedicated land disposal site where PFAS-laden biosolids were applied annually for 40 years, PFOA along with several short chain PFAS leached to groundwater although most of the other long-chain PFAS mass, e.g., PFOS, was retained in the 0–30 cm depth (Alverez-Ruiz et al., 2024). Substantial PFOA concentrations found in the groundwater in the absence of any other long-chain PFAS reflect the likely role of competitive processes. Lastly, literature on the desorption behavior of PFAS is also inconsistent although data are limited. Desorption hysteresis is driven by slow desorption kinetics, steric hindrances, and/or thermodynamics. Uhme et al. (2024) reported nonlinear sorption isotherms for PFAS and complete reversibility based on initial 48-h batch equilibrations of freshly spiked soils at relatively high concentrations. Conversely, for AFFF historically contaminated soils, Schaefer et al. (2021) and Klamerus et al. (2025) observed linear desorption isotherms but with significant PFAS-dependent non-labile fractions.

For cationic and zwitterionic PFAS, electrostatic interactions dominate and can occur through divalent cation bridging between the negative charges on PFAS (e.g., RF-COO- in PFCAs and RF-SO3- in PFSAs) and soil organic matter (SOM) functional groups (Mejia-Avendaña et al., 2020; Xiao et al., 2019). Electrostatic interactions are pH dependent, with pH affecting both PFAS speciation and the speciation of soil functional groups. Soil cation exchange capacity and anion exchange capacity as well as Al- and Fe-oxides play important roles in electrostatic interactions (Gao & Chorover, 2012; Oliver et al., 2019, 2020). Increasing electrolyte concentration can also increase PFAS "hydrophobic" sorption in soils through the "salting-out" effect (Wang & Shih, 2011). Hydrogen bonds can occur between the hydroxyl groups in SOM and the oxygen atom in the carboxylic (–COOH) and sulfonic (–SO_3H) groups for PFCAs and PFSAs, respectively (Mei et al., 2021). Desorption hysteresis has

Published by Burleigh Dodds Science Publishing Limited, 2025.

not been observed for cationic PFAS, whereas zwitterionic PFAS did display desorption hysteresis, which was attributed to entrapment mechanisms (Xiao et al., 2019).

5.2 Air-water interfacial sorption of PFAS

In addition to solid-phase sorption processes that impact most organic chemicals in the subsurface, PFAS with their combined hydrophilic polar head groups and hydrophobic fluorinated tails exhibit extraordinary surface-active properties. Therefore, PFAS can accumulate AWI and subsequently impact migration in the vadose zone (unsaturated) soils. Interfacial surfactant sorption is dependent on soil particle size, moisture content and PFAS affinity to the AWI, which is PFAS structure dependent. PFAS interfacial sorption increases with decreasing particle size and % moisture while increasing with increasing C-F chain length and ionic strength (Silva et al., 2019; Costanza et al., 2019; Fabregat-Palau et al., 2022; Liao et al., 2022; Li et al., 2023; Lyu & Brusseau, 2020). With the transient nature of saturation in the vadose zone, PFAS accumulation at the AWI can influence PFAS migration and reduces the surface tension of water, thus also affecting soil porewater drainage (Arshadi et al., 2024; Brusseau et al., 2021; Karagunduz et al., 2015; Silva et al., 2022). Air-water interfacial adsorption can account for over 50% of PFAS retention under unsaturated conditions (Costanza et al., 2019; Lyu et al., 2018).

PFAS sorption to the AWI relative to aqueous concentration is estimated using the air-water interfacial sorption coefficient (K_{aw}), which can be estimated indirectly through correlations with molar volume (Brusseau, 2019). The appropriate sorption model for fitting measured data to estimate K_{aw} is still debated; K_{aw} values estimated using Freundlich and Langmuir isotherm model fits vary considerably at low PFAS concentrations. Also, competitive sorption for the AWIs for PFAS mixtures has been observed further challenging predictions (Silva et al., 2021). Nevertheless, adsorption to the AWI has been integrated into models simulating steady unsaturated flow of PFAS through the vadose zone. Meanwhile, models for transient flow conditions are more challenging due to changes in the vadose zone interfacial area under natural dry and wet cycles (Guo et al., 2020). Incorporating air-water interfacial sorption into well-known transient flow models (HYDRUS-1D and LEACHM) has proven to reasonably capture observed PFOS and PFOA profiles in agricultural fields receiving wastewater effluent irrigation for four decades (Liao et al., 2024). Both PFAS sorption to soil surfaces as well as at the AWI within the vadose zone directly affect their leaching behavior and bioavailability (Post et al., 2017; Lyu et al. 2018) and sorption by both processes increases with increasing chain length. Therefore, short-chain PFAS tend to leach to a greater extent and get taken up by plants more readily than long-chain PFAS. This leads to the inverse relationship between soil depth and the concentration of long-chain PFAS (Johnson, 2022; Pepper et al., 2023; Sepulvado et al., 2011; Washington et al., 2010).

5.3 Transformation of precursors

Both in the wastewater treatment processes and after land application, PFAS that are not PFAAs can undergo biotransformation to other PFAS. The chemical structure of the precursor, which is informed largely by the PFAS manufacturing process, plays a critical role in the transformation process. For example, n:2 fluorotelomer (FT) phosphate diesters (n:2 diPAPs), n:2 FT alcohols (n:2 FTOHs), FT sulfonates, and FT alkylbetaine degrade to several PFCAs through a series of intermediates (Mejia-Avendaño & Liu, 2015; Dasu et al., 2012, 2013, 2015; Royer et al., 2015; Benskin et al., 2013). Fluorotelomers break down first to yield n:2 FTOHs which then degrade through a hypothesized "one-carbon removal" pathway to yield PFCAs (Ruan et al., 2013; Shaw et al., 2019; N. Wang et al., 2012). Electrochemical fluorination (ECF)-derived sulfonamides such as FASAs will break down through a series of intermediates to the PFSA with the same perfluoroalkyl chain length as the precursor (Mejia-Avendaño & Liu, 2015; Zhang et al., 2016b). Transformation rates are much greater under aerobic conditions compared to almost negligible in typical anaerobic conditions (Zhang et al., 2013, 2016a). Also, FT-derived PFAS degrade much quicker than the ECF-derived sulfonamides (Guelfo et al., 2021). For example, in a recent study at a dedicated biosolids land disposal site on the west coast, most of the FT-PFAS were gone within one year after biosolids application leaving intermediate FTCAs and FT unsaturated acids and terminal PFCAs while changes in the ECF precursors were small (Álvarez-Ruiz et al., 2024). Root exudates and earthworm-excreted enzymes have been reported to increase PFAS degradation (Zabaleta et al., 2018) (Munoz et al., 2020). Enzymatic activities of glutathione S-transferase and cytochrome P450 in the earthworms are proposed to be the likely mechanisms for PFAS transformation. Failure to account for precursor loads in biosolids could lead to underestimating long-term mass release of intermediates and PFAAs in biosolids-amended soils.

5.4 Plant uptake

Plant activity in the vadose zone can moderate the biogeochemical processes that impact chemical fate including PFAS. Through the process of plant-mediated energy transfer (photosynthesis, mineralization, and evapotranspiration), organic molecules can be transported to their root surfaces where they may be transformed or retarded. Molecules that are restricted in the rhizosphere or within the root tissues are prevented from further transport, thus limiting their leaching potential. In some instances, PFAS are translocated into other plant parts either through passive or active mechanisms. PFAS-specific plant uptake studies have evaluated the role of the root system in PFAS uptake and distribution. The general observation is that PFAS tends to accumulate passively in the root tissues, meaning that uptake is driven by diffusion from areas with high PFAS concentration (Blaine et al., 2014; Stahl et al., 2009; Zhao et al., 2017). The protein and lipid contents of roots can also impact PFAS accumulation in the root. Translocation from roots to shoots has also been observed especially for PFCAs in lab and greenhouse studies (Costello & Lee, 2020) primarily through transpiration-induced transport through the xylem. PFOS from biosolids-amended soils have been shown to be taken up into the leaves of forage crops (Simones et al., 2024). Plant uptake and translocation into its aerial part is inversely proportional to PFAS chain length (Wang et al., 2020; Costello & Lee, 2020). PFAS with shorter perfluorocarbon chain lengths are more mobile and available for plant uptake through transpiration. However, larger PFAS such as PFOS are also found in grasses and other plant parts. In some cases, offloading from the xylem to the phloem is required for PFAS to enter some crop parts such as the grain. Offloading may occur through non-selective substrate transporters – membrane proteins that can move chemically diverse molecules across the cell membrane (Yao et al., 2020; Gill et al., 2021). These transporters include those that transport amino acids, which have a polar group common to PFCAs (a carboxyl group). Detection of intermediate metabolites of PFAS precursors in plant tissues has suggested the possibility of PFAS biotransformation within plant cells. However, further investigation of PFAS metabolism in plants is required to understand transformation mechanisms.

5.5 Vertical transport (leaching)

PFAS transport through the soil profile to groundwater depends on multiple factors including PFAS sorption to soil and at AWIs (See section 3.3.2) as well as hydrological conditions and soil hydraulic properties. Natural cycles of weathering and drainage can release PFAS that were previously sorbed to the soil matrix. Therefore, the intensity of precipitation events such as rainfall and snowmelt as well as irrigation plays a key role in PFAS leaching. Additionally, release and vertical migration of soil colloids during infiltration can enhance PFAS leaching (Borthakur et al., 2021a, 2021b). Critical to PFAS transport are the rate and magnitude of water movement in the soil which is controlled by hydraulic properties such as texture, structure, and pore size distribution. As previously noted, short-chain PFAS ($CF_2 < 6$) tend to desorb easily from the soil, and thus susceptible to leaching compared to the longer chain homologs (Johnson, 2022; Peter et al., 2024; Sepulvado et al., 2011). This is reflected by the inverse relationship between concentration of long chain PFAS and soil depth (Álvarez-Ruiz et al., 2024; Peter & Lee, 2024a; Peter, 2025; Peter and Lee, 2025).

The presence of artificial drainage networks can alter the movement of pore water down the soil profile, and thereby PFAS leaching. In the USA where poorly drained soils dominate such as in the Midwest, tile drains are installed in agricultural fields to lower the water table and promote aeration. Agrochemicals and fertilizers percolating through the soil are redirected into tile drains and ultimately surface water, thus short-circuiting direct transport to groundwater used for drinking water (Peter & Lee, 2025). Likewise, PFOS and PFOA were detected in tiles draining from a biosolids-amended field (Gottschall et al., 2017), but not in groundwater.

5.6 Surface runoff

During precipitation events, PFAS can be released from surface soils to become part of overland flow also known as runoff. The intensity, depth, and duration of rainfall typically influence the magnitude of chemical release. Soil properties that regulate leaching, such as micro and macropore distribution, texture, and bulk density also influence surface runoff. Due to their strong association with the solid phase, PFAS are likely subjected to "first flush" removal. This means that they are carried by runoff in the form of suspended solids early in a rainfall event and delivered to surface waters (Borthakur et al., 2021c). Peter and Lee (2024b) reported substantial PFAS in runoff, particularly of PFOS and PFOA at several hundred ppt, from an agricultural field that had not received biosolids for 6 years. Results are consistent with the longevity of long-chain PFAS in the upper soil profile keeping them available for continued runoff with intense storm events.

6 Human and ecological exposure routes and risks from agriculture

6.1 Humans

Although there are several ways that humans may be exposed to PFAS in both the indoor and outdoor environment, the key exposure route is through PFAS-contaminated water given the daily intake volumes of water (De Silva et al., 2021; DeLuca et al., 2022; Sunderland et al., 2019). Given the long half-lives of PFAS within the body, even low ppt levels of intake, especially long-chain PFAS, can lead to bioaccumulation in the body. PFAS may also be consumed through contaminated food sources including commercial stores, home gardens, and recreational fishing. PFAS presence in packaging used for food products can also lead to PFAS transfer into foods (Liu et al., 2023b; Phelps et al., 2024). PFAS are known to bioaccumulate in plants, milk, meat, and particularly fish and shellfish. In addition, given the use of PFAS in so many products, it is not surprising that we also are exposed to PFAS through inhalation of airborne PFAS within our homes, cars, and workplace and dermally when using various PFAS-containing personal care products (e.g., cosmetics, shampoos, lotions, etc.).

The primary direct pathway linking PFAS exposure to agriculture includes plants and animals that have consumed PFAS-contaminated water or feed (De Silva et al., 2021; Jha et al., 2021). Animal product consumption, and particularly dairy, tends to be the largest exposure pathway of concern linked to agriculture because of PFAS bioaccumulation and biomagnification, PFAS affinity for proteins (high in meat and meal), and that children are the biggest direct consumers of milk. Cattle typically eat 25 pounds or more a day of plant matter; therefore, even low PFAS concentrations in the feed can lead to unacceptable levels in meat and milk (Fitzgerald et al., 2025). Currently, there are no formal maximum concentration levels set for PFAS in milk and meat; however, the State of Maine in the USA set action levels of <211 ppt in milk based on risk to 1–2 y old children (Maine CDC, 2017) and <3.4 ppb in meat based on risks to 1–< 6 y old children (Maine CDC, 2020). PFAS transfer from eggs via feed (Granby et al., 2024) and to laying hens free-ranging on PFAS-contaminated land including in home-produced eggs (Lasters et al., 2022; Göckener et al., 2020b). PFAS have been shown to accumulate in the yolk more than the high-protein albumen, which was attributed to PFAAs having higher binding affinities to low- and high-density lipoprotein and vitellin proteins in yolk compared to ovalbumin albumen proteins (F. Wang et al., 2019). Precursor transfer to eggs as well as in vitro precursor transformation with metabolites transferring to eggs has also been observed (Kolanczyk et al., 2020).

Indirect exposures include potential contamination of rural water supply from runoff of soil into surface waters from lands where PFAS-laden biosolids have been applied to grow crops as well as PFAS from the same source leaching to groundwater. Additionally, surface runoff into water bodies including farm ponds, streams/rivers where fishing, swimming, and other recreational activities occur can expose communities to PFAS intake. Aquatic species in closed water bodies versus streams are particularly vulnerable to bioaccumulation, although PFAS half-lives within aquatic species are relatively short compared to mammals. For example, early investigations of fish in private farm ponds located in biosolids-applied agricultural lands have shown, in some cases, PFOS at several hundred ppb in largemouth bass (Hoskins et al., 2023).

6.2 Ecological

The primary direct pathway linking PFAS exposure to wildlife with agriculture is primarily run-off or tile drain discharge from PFAS-contaminated agricultural lands into the aquatic system, although terrestrial wildlife can also be exposed through direct plant consumption grown on PFAS-impacted lands (De Silva et al., 2021). Aquatic animals can be exposed through both their respiratory system (gills) as well as dietary exposure. Dermal exposure can also be important for frogs and salamanders given their permeable skin (Abercrombie et al., 2019, 2021). PFAS have been shown to adversely impact the physical development of aquatic organisms, the endocrine and immune systems, and lead to oxidative stress, endocrine disruption, immunotoxicity, and cognitive function impairment, among others (Ankley et al., 2021; Ma et al., 2023; Pandelides et al., 2023). Some studies have indicated that long-term exposure to PFAS can affect population dynamics in organisms by reducing reproductive success or survival rates and altering their general behavior (Ankley et al., 2021; Ma et al., 2023). While long-term ecological risk from PFAS had not been fully assessed due to data limitations, their potential to persist, bioaccumulate, and biomagnify in both terrestrial and aquatic wildlife could disrupt ecosystem functions.

6.3 Effects on soil health

PFAS impacts on soil health are limited, but high concentrations of PFAS (up to 1000 ppb) in laboratory studies have been shown to alter soil properties and microbial function. Xu et al. (2023) found a decrease in water-stable aggregates with PFOS exposure and soil respiration with both PFOS and PFBS exposure. However, Xu et al. (2023) also observed an increase in litter decomposition rates, which in turn increased pH possibly due to changes in microbial community composition. Impacts on soil microbial community composition, structure, enzymes, and gene expression have been observed in some studies with some communities and enzymes but not all, and observations are inconsistent (Qiao et al., 2018; Zhang et al., 2020; Xu et al., 2023). Changes in microbial diversity or function could adversely affect soil health and productivity given the crucial role they play in the biogeochemical cycling of nutrients.

7 Techniques in PFAS identification and tracking

Although scientific studies on PFAS have burgeoned over the past decade, the exact scope of global PFAS contamination is challenging to estimate with certainty. Studies focused on tracing and fingerprinting sources of PFAS are still in their nascent stage and have focused on characterizing sources based on distinct PFAS chemical profiles. For example, chemical fingerprinting with the aid of supervised machine learning was used to differentiate water sources impacted by AFFF from other sources based on the distribution of

PFAS in contaminated sites (Kibbey et al., 2021). Likewise, contaminated sites associated with landfill leachates have been reported to be enriched with FT carboxylic acids (Benskin et al., 2012). Characterizing PFAS chemical profiles depends heavily on advancements in analytical techniques. The number of targeted PFAS compounds in commercial labs has increased over time, from focusing only on PFOS and PFOA to 18 targeted compounds based on US EPA method 537.1, and currently 40 compounds in method 1633 (US EPA, 2024c USEPA, 2020). Research and private labs have developed more comprehensive methods using high-resolution mass spectrometry to quantify over 100 targeted and non-targeted PFAS compounds (Benotti et al., 2020). Furthermore, analysis that identifies PFAS isomer patterns has provided additional diagnostic insights. Branched and linear isomers typically denote ECF and FT manufacturing processes, respectively (Buck et al., 2011). ECF process was discontinued in the early 2000s. Therefore, analytical techniques that show the ratio of branched to linear isomers can provide additional source information with respect to manufacturers and the timeline of PFAS release (Benskin et al., 2010).

Despite the progress with PFAS analytical methods, tracking contamination from comingled sources remains a challenge. For example, the major PFAS sources to agricultural soils, including biosolids and contaminated irrigation water, are aggregates of multiple PFAS inputs, which further complicates forensic analysis. Further, tracking the contribution of such agricultural non-point sources to PFAS presence in connected hydrological systems is challenging for several reasons, including differences in environmental fate of individual compounds and biotransformation (Lee et al., 2010, 2014b; Schaefer et al., 2022; Silva et al., 2022). In such scenarios, source identification and tracking integrity could be enhanced by considering site-specific history and conditions, including surface application timelines and loading rates, soil properties, and hydrogeologic features (Charbonnet et al., 2021). Additionally, multivariate statistical analysis, such as hierarchical clustering, principal component, and multiple linear regression analysis on chemical and site data, can identify patterns of PFAS occurrence and map distinct chemical profiles to PFAS sources within a location of interest (Qi et al., 2016; Salim et al., 2019; Yao et al., 2014; Zhang et al., 2016b).

8 Remediation and mitigation options for PFAS-contaminated farms

Specific to agriculture, remediation strategies are needed for both water sources used for irrigation and livestock rearing, and soil that supports food crops and forage production. Reverse osmosis (RO) and sorbent technologies are the easiest and most acceptable to deploy for removing PFAS from well water. Unfortunately, both RO and sorbent-based technologies result in PFAS-contaminated residuals (RO rejection water and spent sorbents) which will need to be managed (Tow et al., 2021; Didenko et al., 2024). Therefore, while treatment technologies that lead to PFAS destruction are becoming available, they are still far from being feasible for deployment at many farms.

The biggest challenge in agriculture is PFAS-contaminated soils given the longevity of PFAS in the soils and the scale of the problem. However, the development and testing of pragmatic approaches to minimize impacts of PFAS-contaminated farm soils on crops and livestock are underway to facilitate keeping farms viable. Approaches include the choice of crops grown, selection of crops used for feed, and sorbent addition in the till layer where crops are grown. For example, uptake of PFAS is minimal to negligible in the corn and cob (Bhattacharya et al., 2024) whereas PFAS accumulate in the corn silage (stalk and leaves) as well as grasses (Wang et al., 2020; Costello et al., 2020). Also, the long-chain and highly bioaccumulating PFAS like PFOS do not appear to be offloaded into the soybean (Lazo and Lee, 2024) similar to what has been observed for corn kernels (Bhattacharya et al., 2024). Therefore, shifting livestock feed away from silage and grasses facilitates PFAS reduction in milk and meat. Research is currently being done to optimize rotations and timing based on when the livestock commodity will go into the consumer market to take advantage of expected livestock excretion rates (PFAS half-life in the body). Also, there is some evidence that PFAS uptake into grasses changes throughout the season, thus the potential to use grass feeds when PFAS uptake is much lower. For farms in Maine, the Maine CDC (Simones et al., 2024) found PFOS uptake into grasses twice as high in late summer compared to earlier in the year. Similar trends may not be observed in other regions since PFAS uptake via transpiration will vary with function of temperature, moisture, and soil conditions. Intercropping has been proposed as a potential management strategy to alter PFAS uptake into a target crop, e.g., corn leaves and stalks used for feed, by interrow plant cover, but positive outcomes were not observed in greenhouse studies (Scearce et al., 2024) and potential reductions in yield due to nitrogen dynamics and competition for water and soil oxygen are likely (Qiu et al., 2025; Deines et al., 2022). Field studies to further examine interrow plant covers are underway (U.S.EPA, 2024d). However, such plant cover can also reduce runoff into surface water bodies used for livestock drinking or fishing as we have observed for other chemicals. Another potential PFAS management approach being investigated addresses enhancing PFAS excretion from livestock already impacted by PFAS in the feed or reducing PFAS uptake from PFAS-contaminated feed by using PFAS binders in the water or feed (U.S.EPA, 2024d).

8.1 Sorbent technologies

Sorbent mitigation at the field scale has much promise to reduce PFAS availability to porewater, thus reducing leaching and plant uptake (Mahinroosta et al., 2020; Khair Biek et al., 2024). Both waste treatment residuals (WTRs) and biochars have been used as soil amendments in agricultural settings to improve soil health, increase crop yields, regulate the loss of applied nitrogen, and control phosphorus mobilization for decades across the globe (Bayley et al., 2008; Ibrahim et al., 2023; Li et al., 2024; Ippolito et al., 2009; 2011; Turner et al., 2019; Liu et al., 2013). Over the past decade, the potential for sorbents to alter PFAS availability has been explored at the lab-scale in spiked water, soil, and soil fractions, or from PFAS-containing biosolids (Sørmo et al., 2021; 2024; Openiyi et al., 2024; Holly et al., 2024). Generally, sorbent materials with a high specific surface area, small pore size distribution, and high exchange capacity have higher sorptive capacity (Lu et al., 2016) including for PFAS (Du et al., 2014). Solution factors observed at the lab-scale to affect sorbent potential for PFAS include soil-solution pH, ionic composition, and dissolved OM (Dasu et al., 2013; Zhang et al., 2021) Limited desorption from WTRs and biochars has also been demonstrated at the lab-scale (Openiyi et al., 2024; Gravesen et al., 2023). While lab-scale studies have shown the effectiveness of various sorbent materials to sorb PFAS and reduce PFAS leaching, applying these immobilization approaches to reduce both crop uptake and leaching of PFAS in agricultural fields and associated challenges are not yet well understood. Therefore, pilot-scale and field-scale data are needed to better understand the potential and limitations. Greenhouse studies using locally sourced sorbents mixed into PFAS-contaminated soil (e.g., 400 ppb PFOS) show great promise in reducing PFAS uptake by grasses, in which even 1.5 wt% high carbon wood ash reduced PFOS uptake by 80% and 6 wt% to >95% (Openiyi and Lee, 2024; Openiyi et al., 2025; Illango et al., 2024). Mitigating PFAS uptake into plants by sorbents will be dependent on PFAS type, soil properties, and crop species, as well as sorbent amendment type and rates (Ilaango et al., 2024). In addition, weather patterns may greatly alter expected mitigation in some cases. Research on effective strategies mitigate PFAS in farming systems is ongoing.

8.2 Phytoremediation

Exploiting plant uptake (i.e., phytoextraction) has been considered for phytoremediation of soils (Kavusi et al., 2023) but has several limitations. While low-cost and perhaps visually appealing, phytoextraction will be limited in terms of magnitude and number of PFAS as well as depth of soil explored. Also, prior to harvest, there is a high potential of posing a risk to wildlife, and then the harvested plants will require subsequent disposal or destruction. In field trials with hemp at a former PFAS-contaminated Air Force Base, hemp was only able to extract 6 of 28 PFAS, mostly short-chain and PFCAs, and only at a maximum of 2% was removed (Nason et al., 2024). Harvested plants were subsequently treated using hydrothermal liquefaction, which reduced PFAS mass, but not all PFAS were removed. Huff et al. (2020) explored phytoremediation potential of eight herbaceous and seven woody plant species in a greenhouse study. They reported almost 3.8 mass % of PFOS and up to 43 mass % were taken up by *Festuca rubra* (red fescue) and *Schedonorus arundinaceus* (tall fescue), respectively. However, plant exposure to PFAS was done through weekly irrigation of PFAS-amended water, thus a vastly different scenario than trying to phytoextract PFAS adsorbed to soils that have been contaminated for a long time. In a greenhouse study , Illango et al. (2024) investigated the use of a grass-legume mix in combination with low doses of biochars (≤1%) to phytoextract PFAS from PFAS-contaminated soils amended with biosolids and additional PFAS. Phytoextraction varied with soil type and biochar type and rates, negatively correlated to PFAS chain length, and percent removal was higher for the PFCAs. Phytoextraction of PFOS, one of the PFAS of greatest concern due to its bioaccumulation and biomagnification potential, was less than 1%. It is plausible that genetic engineering techniques may aid in increasing the magnitude and number of PFAS that can be phytoextracted as well as the depth of soil that is explored by the roots (Mayakaduwage et al., 2022). Advancing endophyte-assisted phytoremediation for reducing PFAS concentrations has also been posed. Endophytes, such as bacteria and fungi, excrete enzymes that may promote organic chemical degradation (He et al., 2020); however, their value for degrading PFAS has not been explored (Mayakaduwage et al., 2022).

8.3 Bioremediation technology potential

Destruction of PFAS through biotransformation processes is highly desired due to being ecofriendly, likely cost-effective, and with high potential to be used in situ at a large scale like the agricultural landscape. PFAS degradation by microbes can occur through several reaction steps including reductive defluorination, decarboxylation, hydroxylation, and hydrogenation (Kumar et al., 2023); however, in most cases, fluorine removal has been limited. Multiple isolated bacterial species including *Pseudomonas*, *Acidimicrobium*, *Mycobacterium*, *Gordonia*, as well fungal species such as *Phanerochaete chrysosporium*, *Gloeophyllum trabeum*, and *Phanerochaete chrysosporium* have demonstrated the potential of transforming PFAS precursors, but only to terminal PFCAs (Kim et al., 2012; Merino et al., 2018; Shaw et al., 2019; Tseng et al., 2014; Van Hamme et al., 2013). Huang and Jaffé (2019) reported PFOS removal up to 33% with a pure *Acidimicrobium sp.* strain A6 and up to 50% when used in a mixed culture (Ralstonia, Bacillus, Acidimicrobium, Aciditerrimonas, and Desulfosporosinus) in 100-d incubations in which Fe (III) and $NH4^+$ were supplied as the electron acceptor and donor, respectively. Similar studies were done for PFOA but included microcosms with biosolids (Huang et al. 2022). PFCAs were generated in some cases as well as fluoride, but the high concentrations and likely impurities in the PFAS material used, as well as fluorine mole balance, introduced uncertainties into the study outcomes (Liu et al., 2023a). Other researchers have isolated microorganisms from contaminated media that used PFOS and/or PFOA as sole carbon sources when amended at high ppm levels. For example, Beskoski et al. (2018) found chemoorganoheterotrophic bacteria and a mix of yeast and molds isolated from PFAS-contaminated sediments to remove up to 69% and 36% PFOS and PFOA, respectively, in the 30-d aerobic microcosms. However, through target and nontarget screening, they only found single F-containing fatty acids. Chetverikov et al. (2017) using *Pseudomonas plecoglossicida* 2.4-D isolated from soil near a petroleum plant, observed PFOS degradation to perfluoroheptanoic acid coupled to fluoride generation. Other researchers also isolated strains from contaminated media and claimed degradation, but the weight of evidence was lacking.

While PFAS do undergo biotransformation including defluorination at some level, degradation mechanisms and the functional genes responsible for PFAS transformation are currently not well understood (Kumar et al., 2023; Shahsavari et al., 2021). Future efforts in advancing PFAS bioremediation technology potential should do what is needed to provide fluorine mole balance and exploit technologies in microbial ecology (metagenomics, transcriptomics, and metabolomics). The latter coupled to advances in biotechnology may expedite the development of PFAS bioremediation technologies that can be scaled up to address PFAS-contaminated soils and contaminated groundwater in situ.

PFAS remediation technologies typically require significant investment during and post implementation. Depending on the level of technological advancement, the cost of remediating contaminated soils can range from \$20 000 to \$90 000 per hectare, with equipment costs accounting for 40%, labor 30%, energy 20%, and other factors 10% (Valalamontes & Adamopoulos, 2025). Advanced technologies such as plasma-based and thermal desorption treatment are more expensive compared to bioaugmentation and phytoremediation. Other options such as ex-situ soil washing and landfilling fall under medium cost at the rate of \$100 to \$200/ton according to (Quinnan et al., 2022). Unlike reports of combined treatment measures for aqueous matrices (Lu et al., 2020), there is currently no field-scale demonstration for contaminated soils. Integrated strategies can reduce treatment expenses by combining low-cost approaches such as the implementation of in-situ sorbent technologies combined with phytomanagement. Still, the cost of the least expensive remediation option may be unfeasible for small-scale farms as treatment expenditures can exceed farm revenues. Nonetheless, the cost of inaction might surpass remediation costs in the long run. Contaminated farmlands can incur temporary or permanent losses from discarding impacted farm products or shutting down production. For example, farmers in several states in the USA had to dispose of milk and meat products and, in some instances, abandon the farm following incidences of high PFAS contamination (Cordner et al., 2021). Left untreated, non-productive farmlands may ultimately weaken food production and arguably pose obstacles to food security and local economies. Additionally, agricultural communities can incur an additional burden of health-related costs from exposure to PFAS, given that contaminated farmlands can become diffuse sources of PFAS to drinking water. Already, PFAS-related health costs for the United States are estimated at $37–$59 billion annually (Cordner et al., 2021), an amount that will likely increase if PFAS contamination is not contained. Other indirect costs derived from effects on the well-being of farmers and impacted communities, such as stress, depression, and low standards of living, are challenging to estimate but likely account for a sizable burden from PFAS contamination. Therefore, remediation of contaminated farmlands should be viewed as a societal issue, one that requires significant financial investment from the government and industry and cross-sector partnership between relevant stakeholders. Post remediation, long-term monitoring is required to assess the effectiveness of remediation efforts, especially for in-situ sorbent technologies. The availability of real-time sensors in the future can improve monitoring efficiency as readily available data can be coupled to predictive models to identify spatiotemporal trends of remediation outcomes.

9 Conclusion

PFAS contamination in agricultural systems is one of the biggest environmental issues of the twenty-first[t] century given the growing number of farmland contaminations and background levels in surface soil contamination and on a global scale. Sources of PFAS to the environment are diverse, but the primary sources in agricultural systems are the land application of biosolids and composted material and irrigation water contaminated by PFAS, typically from industrial, military, airport, or landfill sources. The properties of this family of several thousand compounds led to their use in multiple products and processes as well as their environmental presence, persistence, and complex retention and transport mechanisms.

Addressing PFAS contamination in agricultural systems is complex with multiple challenges. First, despite active research and ongoing monitoring efforts that have identified several contaminated farms, potential occurrence at most has not been assessed, especially in less-developed countries. There is a need for rapid quantitative methods for PFAS that have been validated and standardized and can be done at a lower cost than current analysis. Such methods will expedite needed data collection for farms and communities to minimize risks to the food system and human health.

Product and process source controls are needed, including regulating out nonessential PFAS use in industrial processes and consumer goods, including use in food packaging. Such controls can minimize direct human exposure and PFAS presence in biosolids that can be beneficially reused and effluent-treated irrigation water that is often used for irrigation, especially in water-scarce regions. In addition, pre-treatment for PFAS capture or destruction in influent streams with high concentrations of legacy PFAS, such as those typical in many landfill leachates, is required. Actions to advance source control are occurring in some regions, but still quite limited.

While source control will eventually benefit the agricultural community as well as all systems impacted by PFAS, the PFAS legacy in the soils will remain a challenge for several decades. Currently, treatment of a PFAS-contaminated water source used for irrigation or livestock drinking water can be treated with capture technologies, but the soil associated with PFAS-contaminated landscapes with depth brings forth a particularly difficult challenge. While some PFAS do get taken up into vegetables and plants consumed directly by humans, the bigger concern is uptake into feed consumed by livestock, which then enters the human food web. PFAS's high affinity for proteins leads to high bioaccumulation in meat, milk, and eggs in livestock that typically consume a high ration of feedstock, particularly in cattle. Current mitigation approaches being used include selection of crop types and species that have lower PFAS uptake, the part of the crop used for feed, and when crops are harvested (times of high transpiration will lead to higher PFAS uptake). Other approaches being explored include the use of high carbon materials to reduce PFAS availability to the crop and leaching as well as seasonal rotation of feed and feed binders. For free grazing cattle, providing clean water for drinking and minimizing access to PFAS-contaminated streams can be implemented. Given the diverse nature of PFAS chemicals and their environmental behavior, the likelihood of a one-size-fits-all solution is not feasible, rather addressing contamination will require a combination of approaches. Mitigation strategies are costly, so funds to help farms stay in business while protecting the food system and developing PFAS thresholds and action limits are needed at the local, state, and federal levels. The latter requires more refined risk assessment and management plans and data to support the assessments, which are all lagging. PFAS exposure almost always occurs in mixtures, but quantification and understanding of chemical mixture toxicity remain elusive, and scaling from lab models to humans remains a challenge, although the rapid advancements in omics and other analytical tools coupled with machine learning have promise to accelerate the gaps (Kleinstreuer & Hartung, 2024; Limbu et al., 2024; Cheng et al., 2024).

Furthermore, while progress to quantify and model the fate and transport of PFAS in agricultural systems and the environment has increased substantially in the past decade, there are still gaps in understanding. For example, the long-term impact of PFAS contamination on ecosystem health and function has not been adequately assessed. Filling these gaps will facilitate solutions for agricultural systems as well as aid in risk assessment, management, and regulatory guidance. Likewise, the development and integration of existing data networks and mapping tools are needed to aid in evaluating current and future areas of vulnerability and risk. The PFAS story is complex from occurrence to effects; therefore, transparent and targeted communication tools and educational programs for different and diverse stakeholders are needed to generate widespread awareness and public involvement. Communication and education need to surmise what is known about PFAS in layperson terms, address the growing uncertainty with regard to PFAS and food safety, reassure consumers, farm communities, and producers that PFAS in agricultural systems is being addressed, and generate assurance of alternative PFAS-safe paths.

10 Where to look for further information

Further reading: ITRC (Interstate Technology and Regulatory Council) offers free up to date information on PFAS with focused fact sheets as well as a full document capturing PFAS characteristics, occurrence, toxicity, fate, remediation, regulations, risk assessment, sampling and analysis, treatment technologies, stakeholders perspectives, and risk communication. https://pfas-1.itrcweb.org/-.

Perfluoroalkyl Substances in the Environment: Theory, Practice, and Innovation (2018). Editors: Kempisty, Racz, and Xing explores the challenges across the topical areas of regulation and management, toxicology, environmental remediation, and analytical sampling and analysis.

Per- and Polyfluorinated Alkyl Substances: Occurrence, Toxicity and Remediation of PFAS (2025). Editors: Naidu, Mallavarapu, Liu and Umeh captures PFAS dynamics in the terrestrial environment including many challenges relating to analytical science, environmental fate, toxicity, regulatory considerations, risk and remediation.

Key journals/conferences: Numerous journals include PFAS environmental fate and remediation articles including but not limited to *Environmental Science and Technology*, *Science of the Total Environment*, *Journal of Environmental Quality*, *Journal of Agriculture and Food Chemistry*, *Environment International*, *Environmental Toxicology and Chemistry*, *Chemosphere*, *Journal of Hazardous*

Published by Burleigh Dodds Science Publishing Limited, 2025.

Materials, Water Research - American Society of Agronomy, Soil Science Society of America, and Crop Sciences of America (ASA/SSSA/CSA) annual meetings in the United States and Internationally. Society of Environmental Toxicology and Chemistry has annual meetings in multiple regions/continents around the world – American Chemical Society has annual meetings in North America – International Conference on Environmental Pollution and Remediation – PFAS International Congress: Management of Environmental & Health Risks, not annual but reoccurring, one planned for 2025 and again in 2027. National PFAS Conference, held even years and typically in North America, www.nationalpfasconference.org

Major research funding initiatives: SERDP/ESTCP – Strategic Environmental Research and Development Program (SERDP) Environmental Security Technology Certification Program (ESTCP) – funds numerous multi-year projects annually focused on environmental fate, detection, and remediation of PFAS. Project abstracts, reports, and webinars can all be found at https://serdp-estcp.mil/. The United States EPA funded ten new projects in 2024 focused on varied aspects of PFAS in agriculture, https://www.epa.gov/research-grants/research-grants-understanding-pfas-uptake-and-bioaccumulation-plants-and-animals

11 References

Abercrombie, S. A., et al. (2019). Larval amphibians rapidly bioaccumulate poly- and perfluoroalkyl substances. *Ecotoxicology and Environmental Safety*, 178(March), 137–145. https://doi.org/10.1016/j.ecoenv.2019.04.022

Abercrombie, S. A., et al. (2021). Sublethal effects of dermal exposure to poly- and perfluoroalkyl substances on postmetamorphic amphibians. *Environmental Toxicology and Chemistry*, 40(3), 717–726. https://doi.org/10.1002/etc.4711

Alexandrino, D. A. M., et al. (2022). Revisiting pesticide pollution: the case of fluorinated pesticides. *Environmental Pollution*, 292(PA), 118315. https://doi.org/10.1016/j.envpol.2021.118315

Alukkal, C. R., et al. (2024a). Understanding the impact of pre digestion thermal hydrolysis process on PFAS in anaerobically digested biosolids. *Chemosphere*. https://doi.org/10.1016/j.chemosphere.2024.143406

Alukkal, C. R., et al. (2024b). Per- and polyfluoroalkyl substances behavior: insights from autothermal thermophilic aerobic digestion - storage nitrification denitrification reactors. *Chemosphere*. https://doi.org/10.1016/j.chemosphere.2024.143357

Álvarez-Ruiz, R., et al. (2024). PFAS fate at a 40-year dedicated municipal biosolids land disposal site, Science of the Total Environment. https://doi.org/10.1016/j.scitotenv.2024.176540

Ankley, G. T., et al. (2021). Assessing the ecological risks of per- and polyfluoroalkyl substances: current state-of-the science and a proposed path forward. *Environmental Toxicology and Chemistry*, 40(3), 564–605. https://doi.org/10.1002/etc.4869

Arshadi, M., et al. (2024). Modeling 1-D aqueous film forming foam transport through the vadose zone under realistic site and release conditions. *Science of the Total Environment*, 897, 170566. https://doi.org/10.1016/j.scitotenv.2024.170566

Aschmann, S. M., et al. (2000). Atmospheric chemistry of hydroxycarbonyls. *ACS National Meeting Book of Abstracts*, 40(2), 729–731.

Baqar, M., et al. (2024). Combustion of high-calorific industrial waste in conventional brick kilns: an emerging source of PFAS emissions to agricultural soils. *Science of the Total Environment*, 906, 167612. https://doi.org/10.1016/j.scitotenv.2023.167612

Barry, V., et al. (2013). Perfluorooctanoic acid (PFOA) exposures and incident cancers among adults living near a chemical plant. *Environmental Health Perspectives*, 121(11–12). http://dx.doi.org/10.1289/ehp.1306615

Batzella, E., et al. (2024). Decreases in serum PFAS are associated with decreases in serum lipids: a longitudinal study on a highly exposed population. *Science of the Total Environment*, 954, 176227. https://doi.org/10.1016/j.scitotenv.2024.176227

Bayley, R. M., et al. (2008). Water treatment residuals and biosolids co-applications affect phosphatases in a semi-arid rangeland soil. *Communications in Soil Science and Plant Analysis*, 39(19–20), 2812–2826. https://doi.org/10.1080/00103620802432733

BCLP (2025a). https://www.bclplaw.com/en-US/events-insights-news/pfas-in-consumer-products-state-by-state-regulations.html#:~:text=Dental%20Floss;-,Juvenile%20Products;,in%20all%20products%20by%202033 (last accessed September 27, 2025).

BCLP (2025b). https://www.bclplaw.com/en-US/events-insights-news/pfas-drinking-water-standards-state-by-state-regulations.html; https://www.epa.gov/pfas/us-state-resources-about-pfas (last accessed September 27, 2025).

Benotti, M. J., et al. (2020). A forensic approach for distinguishing PFAS materials. *Environmental Forensics*, 21(3–4), 319–333. https://doi.org/10.1080/15275922.2020.1771631

Benskin, J. P., et al. (2010). Perfluorinated acid isomer profiling in water and quantitative assessment of manufacturing sources. *Environmental Science and Technology*, 44(23), 9049–9054. https://doi.org/10.1021/es102582x

Benskin, J. P., et al. (2012). Per- and Poly fl uoroalkyl substances in land fi ll Leachate: patterns, time trends, and sources. Environmental Science & Technology, 46(21), 11532–11540.

Benskin, J. P., et al. (2013). Biodegradation of N-ethyl perfluorooctane sulfonamido ethanol (EtFOSE) and EtFOSE-based phosphate diester (SAmPAP diester) in marine sediments. *Environmental Science and Technology*, 47(3), 1381–1389. https://doi.org/10.1021/es304336r

Beškoski, V. P., et al. (2018). Defluorination of perfluoroalkyl acids is followed by production of monofluorinated fatty acids. *Science of The Total Environment*, 636, 355–359. https://doi.org/10.1016/j.scitotenv.2018.04.243

Bhattacharya, A., et al. (2025) Limited accumulation of perfluoroctane sulfonic acid (PFOS) in corn kernels from biosolids-impacted soils: site-specific evaluation can reduce uncertainty in exposure and risk assessment. *Journal of Environmental Quality*. pp. 1–17. https://doi.org/10.1002/jeq2.70067.

Blaine, A. C., et al. (2014). Perfluoroalkyl acid uptake in lettuce (Lactuca sativa) and Strawberry (Fragaria ananassa) irrigated with reclaimed water. *Environmental Science and Technology*, 48(24), 14361–14368. https://doi.org/10.1021/es504150h

Published by Burleigh Dodds Science Publishing Limited, 2025.

Blake, B. E. & Fenton, S. (2018, 2020). Early life exposure to per- and polyfluoroalkyl substances (PFAS) and latent health outcomes: a review including the placenta as a target tissue and possible driver of peri- and postnatal effects. *Toxicology*, 443, 15265. https://doi.org/10.1016/j.tox.2020.152565

Borthakur, A., et al. (2021a). Release of soil colloids during flow interruption increases the pore-water PFAS concentration in saturated soil. *Environmental Pollution*, 286(May), 117297. https://doi.org/10.1016/j.envpol.2021.117297

Borthakur, A., et al. (2021b). Dry-wet and freeze-thaw cycles enhance PFOA leaching from subsurface soils. *Journal of Hazardous Materials Letters*, 2: 100029. https://doi.org/10.1016/j.hazl.2021.100029

Borthakur, A., et al. (2021c). Perfluoroalkyl acids on suspended particles: Significant transport pathways in surface runoff, surface waters, and subsurface soils. *Journal of Hazardous Materials*, 417(February), 126159. https://doi.org/10.1016/j.jhazmat.2021.126159

Braine, M. F., et al. (2024). Quality and risk management frameworks for biosolids: an assessment of current international practice. *Science of the Total Environment*, 915, 169953. https://doi.org/10.1016/j.scitotenv.2024.169953

Brennan, N. M., et al. (2021). Trends in the regulation of per- and polyfluoroalkyl substances (PFAS): a scoping review. *International Journal of Research and Public Health*, 20, 10900. https://doi.org/10.3390/ijerph182010900

Brown, J. B., et al. (2020). Assessing human health risks from per- and polyfluoroalkyl substance (PFAS)-impacted vegetable consumption: a tiered modeling approach. *Environmental Science and Technology*, 54(23), 15202–15214. https://doi.org/10.1021/acs.est.0c03411

Brusseau, M. L. (2019). Estimating the relative magnitudes of adsorption to solid-water and air/oil-water interfaces for per- and poly-fluoroalkyl substances. *Environmental Pollution*, 254(Part B), 113102. https://doi.org/10.1016/j.envpol.2019.113102

Brusseau, M. L., et al. (2020). PFAS concentrations in soils: background levels versus contaminated sites. *Science of the Total Environment*, 740, 140017. https://doi.org/10.1016/j.scitotenv.2020.140017

Brusseau, M. L., et al. (2021). Ideal versus nonideal transport of PFAS in unsaturated porous media. *Water Research*, 202, 117405. https://doi.org/j.watres.2021.117405

Buck, R. C., et al. (2011). Perfluoroalkyl and polyfluoroalkyl substances in the environment: Terminology, classification, and origins. *Integrated Environmental Assessment and Management*, 7(4), 513–541. https://doi.org/10.1002/ieam.258

Calafat, A. M., et al. (2019). Legacy and alternative per- and polyfluoroalkyl substances in the U.S. general population: Paired serum-urine data from the 2013–2014 National Health and Nutrition Examination Survey. *Environment International*, 131(June), 105048. https://doi.org/10.1016/j.envint.2019.105048

Charbonnet, J. A., et al. (2021). Environmental source tracking of per- and polyfluoroalkyl substances within a forensic context: current and future techniques. *Environmental Science and Technology*, 55(11), 7237–7245. https://doi.org/10.1021/acs.est.0c08506

Chen, Y. J., et al. (2024). Emerging Perfluorobutane Sulfonamido derivatives as a new trend of surfactants used in the semiconductor industry. *Environmental Science and Technology*, 58(3), 1648–1658. https://doi.org/10.1021/acs.est.3c04435

Cheng, Y., et al. (2024). A brief review on the assessment of potential joint effects of complex mixtures of contaminants in the environment. *Environmental Science: Advances*, 3, 66. https://doi.org/10.1039/d4va00014e

Chetverikov, S. P., et al. (2017). Degradation of perfluorooctanyl sulfonate by strain *Pseudomonas plecoglossicida* 2.4-D. *Applied Biochemistry and Microbiology*, 53, 533–538. https://doi.org/10.1134/S0003683817050027

Cordner, A., et al. (2021). The true cost of PFAS and the benefits of acting now. *Environmental Science and Technology*, 55(14), 9630–9633. https://doi/epdf/10.1021/acs.est.1c03565

Costanza, J., et al. (2019). Accumulation of PFOA and PFOS at the air-water interface. *Environmental Science and Technology Letters*, 6(8), 487–491. https://doi.org/10.1021/acs.estlett.9b00355

Costello, M. C. S. & Lee, L. S. (2020). Sources, fate, and plant uptake in agricultural systems of per- and polyfluoroalkyl substances. *Current Pollution Reports*, 6(4), 179–190. https://doi.org/10.1007/s40726-020-00168-y

Darrow, L. A., et al. (2013). Serum perfluorooctanoic acid and perfluorooctane sulfonate concentrations in relation to birth outcomes in the Mid-Ohio Valley, 2005–2010. *Environmental Health Perspectives*, 121(10), 1207–1213. https://doi.org/10.1289/ehp.120637

Dasu, K. & Lee, L. S. (2015). Aerobic biodegradation of Toluene-2,4-di(8:2 fluorotelomer urethane) and hexamethylene-1,6-di(8:2 fluorotelomer urethane) monomers in soil. *Chemosphere*, 144, 2482–2488.

Dasu, K., et al. (2012). Aerobic soil biodegradation of 8:2 fluorotelomer stearate monoester degradation. *Environmental Science & Technology*, 46, 3831–3836.

Dasu, K., et al. (2013). Aerobic biodegradation of 8:2 fluorotelomer stearate monoester and 8:2 fluorotelomer citrate triester in forest soil. *Chemosphere*, 91, 399–405.

De Silva, A. O., et al. (2021). PFAS exposure pathways for humans and wildlife: a synthesis of current knowledge and key gaps in understanding. *Environmental Toxicology and Chemistry*, 40(3), 631–657. https://doi.org/10.1002/etc.4935

Death, C., et al. (2021). Per- and polyfluoroalkyl substances (PFAS) in livestock and game species: a review. *Science of the Total Environment*, 774, 144795. https://doi.org/10.1016/j.scitotenv.2020.144795

Deines, J. M., et al. (2022). Recent cover crop adoption is associated with small maize and soybean yield losses in the United States. *Global Change Biology*, 29(3), 794–807. https://doi.org/10.1111/gcb.16489

DeLuca, N. M., et al. (2022). Human exposure pathways to poly- and perfluoroalkyl substances (PFAS) from indoor media: a systematic review. *Environment International*, 162, 107149. https://doi.org/10.1016/j.envint.2022.107149

D'Eon, J. C., et al. (2006). Atmospheric chemistry of N-methyl perfluorobutane sulfonamidoethanol, $C_4F_9SO_2N(CH_3)CH_2CH_2OH$: kinetics and mechanism of reaction with OH. *Environmental Science & Technology*, 40, 1862–1868.

Dickman, R. A. & Aga, D. S. (2022a). A review of recent studies on toxicity, sequestration, and degradation of per- and polyfluoroalkyl substances (PFAS). *Journal of Hazardous Materials*, 436, 129120. https://doi.org/10.1016/j.jhazmat.2022.129120

Dickman, R. A. & Aga, D. S. (2022b). Efficient workflow for suspect screening analysis to characterize novel and legacy per- and polyfluoroalkyl substances (PFAS) in biosolids. *Analytical and Bioanalytical Chemistry*, 414, 4497–4507. https://doi.org/10.1007/s00216-022-04088-2

Didenko, T., et al. (2024). Regeneration of PFAS-laden granular activated carbon in modified supercritical CO_2 extraction. ChemRxiv. https://doi.org/10.26434/chemrxiv-2024-b1rvv-v2 (preprint, peer-review pending)

Donley, N., et al. (2024). Forever pesticides: a growing source of PFAS contamination in the environment. *Environmental Health Perspectives*, 132(7), 1–15. https://doi.org/10.1289/EHP13954

Du, Z., et al. (2014). Adsorption behavior and mechanism of perfluorinated compounds on various adsorbents: a review. *Journal of Hazardous Materials*, 274, 443–454. https://doi.org/10.1016/j.jhazmat.2014.04.038

Dunder, L., et al. (2022). Changes in plasma levels of per- and polyfluoroalkyl substances (PFAS) are associated with changes in plasma lipids: a longitudinal study over 10 years. *Environmental Research*, 211, 112903. https://doi.org/10.1016/j.envres.2022.112903

Ellis, D. A., et al. (2004). Degradation of fluorotelomer alcohols: a likely atmospheric source of perfluorinated carboxylic acids. *Environmental Science and Technology*, 38(12), 3316–3321. https://doi.org/10.1021/es049860w

Eriksson, U., et al. (2017). Contribution of precursor compounds to the release of per- and polyfluoroalkyl substances (PFASs) from wastewater treatment plants (WWTPs). *Journal of Environmental Sciences (China)*, 61, 80–90. https://doi.org/10.1016/j.jes.2017.05.004

Espartero, L. J. L., et al. (2022). Health-related toxicity of emerging per- and polyfluoroalkyl substances: comparison to legacy PFOS and PFOA. *Environmental Research*, 212(Part C), 113431. https://doi.org/10.1016/j.envres.2022.113431

Evich, M. G., et al. (2022). Per- and polyfluoroalkyl substances in the environment. *Science*, 375(6580), 380–383. https://doi.org/10.1126/science.abg9065

Evich, M. G., et al. (2025). Mineralogical controls on PFAS and anthropogenic anions in subsurface soils and aquifers. *Nature Communications*, 16, 3118. https://doi.org/10.1038/s41467-025-58040-w

Fenton, S. E., et al. (2021). Per- and polyfluoroalkyl substance toxicity and human health review: current state of knowledge and strategies for informing future research. *Environmental Toxicology and Chemistry*, 40(3), 606–630. https://doi.org/10.1002/etc.4890

Fitzgerald, C., et al. (2025). *Guide to Investigating PFAS Risk on Your Farm*. University of Maine Cooperative Extension. https://extension.umaine.edu/agriculture/guide-to-investigating-pfas-risk-on-your-farm/

Gaines, L. G. T. (2022). Historical and current usage of per - and polyfluoroalkyl substances (PFAS): a literature review. *American Journal of Industrial Medicine*, 66(5), 353–378. https://doi.org/10.1002/ajim.23362

Gallen, C., et al. (2022). Temporal trends of perfluoroalkyl substances in an Australian wastewater treatment plant: a ten-year retrospective investigation, *Science of the Total Environment*, 804, 150211. https://doi.org/10.1016/j.scitotenv.2021.150211

Gao, X. & Chorover, J. (2012). Adsorption of perfluorooctanoic acid and perfluorooctanesulfonic acid to iron oxide surfaces as studied by flow-through ATR-FTIR spectroscopy. *Environmental Chemistry*, 9(2), 148–157. https://doi.org/10.1071/EN11119

Gill, R. A., et al. (2021). The role of membrane transporters in plant growth and development, and abiotic stress tolerance. *International Journal of Molecular Sciences*, 26(23), 12792. https://doi.org/10.3390/ijms222312792.

Glüge, J., et al. (2020). An overview of the uses of per- And polyfluoroalkyl substances (PFAS). *Environmental Science: Processes and Impacts*, 22(12), 2345–2373. https://doi.org/10.1039/d0em00291g

Göckener, B., et al. (2020a). Transfer of per- and polyfluoroalkyl substances (PFAS) from feed into the eggs of laying hens. Part 1: analytical results including a modified total oxidizable precursor assay. *Journal of Agricultural and Food Chemistry*, 68(45), 12527–12538. https://doi.org/10.1021/acs.jafc.0c04456

Göckener, B., et al. (2020b). Human biomonitoring of per- and polyfluoroalkyl substances in German blood plasma samples from 1982 to 2019. *Environment International*, 145(June), 106123. https://doi.org/10.1016/j.envint.2020.106123

Goss, K. U. (2008). The pK_a values of PFOA and other highly fluorinated carboxylic acids. *Environmental Science and Technology*, 42(2), 456–458. https://doi.org/10.1021/es702192c

Gottschall, N., et al. (2017). Brominated flame retardants and perfluoroalkyl acids in groundwater, tile drainage, soil, and crop grain following a high application of municipal biosolids to a field. *Science of the Total Environment*, 574, 1345–1359. https://doi.org/10.1016/j.scitotenv.2016.08.044

Granby, K., et al. (2024). Per- and poly-fluoroalkyl substances in commercial organic eggs via fishmeal in feed. Chemosphere, 346, 140553. https://doi.org/10.1016/j.chemosphere.2023.140553

Gravesen, C., et al. (2023). Per and polyfluoroalkyl substances in water treatment residuals: occurrence and desorption. Journal of Environmental Quality, 52(5), 1291–1302. https://doi.org/10.1002/jeq2.20520

Guelfo, J. L. & Higgins, C. P. (2013). Subsurface transport potential of perfluoroalkyl acids at aqueous film-forming foam (AFFF)-impacted sites. *Environmental Science and Technology*, 47(9), 4164–4171. https://doi.org/10.1021/es3048043

Guelfo, J. L., et al. (2021). Environmental sources, chemistry, fate and transport of per- and polyfluoroalkyl substances: state of the science, key knowledge gaps, and recommendations presented at the August 2019 SETAC focus topic meeting. Environmental Toxicology and Chemistry, 40(12), 3234–3260. https://doi.org/10.1002/etc.5182

Guo, B., et al. (2020). A mathematical model for the release, transport, and retention of per- and polyfluoroalkyl substances (PFAS) in the Vadose Zone. *Water Resources Research*, 56(2), 1–21. https://doi.org/10.1029/2019WR026667

Haines, D. A., et al. (2017). An overview of human biomonitoring of environmental chemicals in the Canadian Health Measures Survey: 2007–2019. *International Journal of Hygiene and Environmental Health*, 220(2), 13–28. https://doi.org/10.1016/j.ijheh.2016.08.002

Halldorsson, T. I., et al. (2012). Prenatal exposure to perfluorooctanoate and risk of overweight at 20 years of age: a prospective cohort study. *Environ Health Perspect*, 120(5), 668–673. https://doi.org/10.1289/ehp.1104034

He, W., et al. (2020). Endophyte-assisted phytoremediation: mechanisms and current application strategies for soil mixed pollutants. *Critical Reviews in Biotechnology*, 40(1), 31–45. https://doi.org/10.1080/07388551.2019.1675582

Higgins, C. P., et al. (2005). Quantitative determination of perfluorochemicals in sediments and domestic sludge. *Environmental Science and Technology*, 39(11), 3946–3956. https://doi.org/10.1021/es048245p

Holly, M., et al. (2024). Evaluation of per- and polyfluoroalkyl substances leaching from biosolids and mitigation potential of biochar through undisturbed soil columns. *Environmental Science & Technology Water*, 4(2), 413–426. https://doi.org/10.1021/acsestwater.3c00414

Hoskins, T. D., et al. (2023). Are biosolid-impacted, stocked farm ponds understudied PFAS hotspots? SETAC North America 44th Annual Meeting, Nov 12–16, 2023, Louisville, KY.

Huang, S. & Jaffé, P. R. (2019). Defluorination of perfluorooctanoic acid (PFOA) and perfluorooctane sulfonate (PFOS) by Acidimicrobium sp. Strain A6. *Environmental Science & Technology*, 53(19), 11410–11419.

Huang, S., et al. (2022). Anaerobic degradation of perfluorooctanoic acid (PFOA) in biosolids by Acidimicrobium sp. strain A6. *Journal of Hazardous Materials*, 424, 127699. https://doi.org/10.1016/j.jhazmat.2021.127699

Huff, D. K., et al. (2020). Accumulation of six PFAS compounds by woody and herbaceous plants: potential for phytoextraction. International Journal of Phytoremediation, 22(14), 1538–1550. https://doi.org/10.1080/15226514.2020.1786004

Hughes, S. G. (2023). *PFAS in Biosolids: A Review of State Efforts & Opportunities for Action*. The Environmental Council of the States. https://www.ecos.org/wp-content/uploads/2023/01/PFAS-in-Biosolids-A-Review-of-State-Efforts-and-Opportunities-for-Action.pdf

Ibrahim, M. M., et al. (2023). Biochar rate-dependent regulation of extended nitrogen supply by modifying stable aggregates-N and microbial responses. *Carbon Research*, 2, 22. https://doi.org/10.1007/s44246-023-00056-2

Illango, A. K., et al. (2024). Uptake of per- and polyfluoroalkyl substances by Conservation Reserve Program's seed mix in biosolids-amended soil. *Environmental Pollution*, 363(2), 125235. https://doi.org/10.1016/j.envpol.2024.125235

Ingelido, A. M., et al. (2020) Serum concentrations of perfluorinated alkyl substances in farmers living in areas affected by water contamination in the Veneto Region (Northern Italy). *Environmental International*, 136, 105435. https://doi.org/10.1016/j.envint.2019.105435

Ippolito, J. A., et al. (2009). Water treatment residuals and biosolids long-term co-applications effects to semi-arid grassland soils and vegetation. *Soil Science Society of America Journal*, 73(6), 1880–1889. https://doi.org/10.2136/sssaj2008.0352

Ippolito, J. A., et al. (2011). Drinking water treatment residuals: a review of recent uses. *Journal of Environmental Quality*, 40(1), 1–12. https://doi.org/10.2134/jeq2010.0242

Islam, M. A., et al. (2025). Per- and polyfluoroalkyl substances (PFAS) contamination in agriculture and its potential conflict with circular economy. *Environmental Pollution*, 385, 127036. https://doi.org/10.1016/j.envpol.2025.127036

Jacob, P., et al. (2021). Target and nontarget analysis of per- and polyfluoralkyl substances in wastewater from electronics fabrication facilities. *Environmental Science and Technology*, 55(4), 2346–2356. https://doi.org/10.1021/acs.est.0c06690

Jha, G., et al. (2021). Per- and polyfluoroalkyl substances (PFAS) in integrated crop–livestock systems: environmental exposure and human health risks. *International Journal of Environmental Research and Public Health*, 18, 12550. https://doi.org/10.3390/ijerph182312550

Jian, J. M., et al. (2018). A short review on human exposure to and tissue distribution of per- and polyfluoroalkyl substances (PFASs). *Science of the Total Environment*, 636, 1058–1069. https://doi.org/10.1016/j.scitotenv.2018.04.380

Johnson, G. R. (2022). PFAS in soil and groundwater following historical land application of biosolids. *Water Research*, 211(August 2021). https://doi.org/10.1016/j.watres.2021.118035

Karagunduz, A., et al. (2015). Influence of surfactants on unsaturated water flow and solute transport. *Water Resources Research*, 51(4), 1977–1988. https://doi.org/10.1002/2014WR015845

Kavusi, E., et al. (2023). Critical review on phytoremediation of polyfluoroalkyl substances from environmental matrices: Need for global concern. *Environmental Research*, 217, 114844. https://doi.org/10.1016/j.envres.2022.114844

Khair Biek, S., et al. (2024). Assessing the impact of immobilisation on the bioavailability of PFAS to plants in contaminated Australian soils. *Environmental Science and Pollution Research*, 31, 20330–20342. https://doi.org/10.1007/s11356-024-32496-7

Khan, B., et al. (2023). Occurrence and bioaccumulation patterns of per- and polyfluoroalkyl substances (PFAS) in the marine environment. *ACS ES and T Water*, 3(5), 1243–1259. https://doi.org/10.1021/acsestwater.2c00296

Kibbey, T. C. G., et al. (2021). Source allocation of per- and polyfluoroalkyl substances (PFAS) with supervised machine learning: classification performance and the role of feature selection in an expanded dataset. *Chemosphere*, 275, 130124. https://doi.org/10.1016/j.chemosphere.2021.130124

Kim, H., et al. (2019). Science of the total environment field-scale evaluation of the uptake of Per fluoroalkyl substances from soil by rice in paddy fi elds in South Korea. *Science of the Total Environment*, 671, 714–721. https://doi.org/10.1016/j.scitotenv.2019.03.240

Kim, M. H., et al. (2012). Biodefluorination and biotransformation of fluorotelomer alcohols by two alkane-degrading Pseudomonas strains. *Biotechnology and Bioengineering*, 109(12), 3041–3048. https://doi.org/10.1002/bit.24561

Klamerus, J., et al. (2025). Field measurement of PFAS leaching at a long-term land-applied biosolids site. *Environmental Science & Technology*. https://pubs.acs.org/10.1021/acs.edt.5c08025

Kleinstreuer, N. & Hartung, T. (2024). Artificial intelligence (AI)–it's the end of the tox as we know it (and I feel fine). *Archives of Toxicology*, 98, 735–754. https://doi.org/10.1007/s00204-023-03666-2

Kolanczyk, R. C., et al. (2020). Transfer of per- and polyfluoroalkyl substances (PFAS) from feed into the eggs of laying hens. Part 2: toxicokinetic results including the role of precursors. *Journal of Agricultural and Food Chemistry*, 68(45), 12539–12548. https://doi.org/10.1021/acs.jafc.0c04485

Kookana, R. S., et al. (2022). Key properties governing sorption–desorption behaviour of poly- and perfluoroalkyl substances in saturated and unsaturated soils: a review. *Soil Research*, 61(2), 107–125. https://doi.org/10.1071/sr22183

Kornher, K., et al. (2025). Associations of PFAS and pesticides with lung function changes from adolescence to young adulthood in the ESPINA study. *International Journal of Hygiene and Environmental Health*, 265, 114526.

Kotthoff, M., et al. (2015). Perfluoroalkyl and polyfluoroalkyl substances in consumer products. *Environmental Science and Pollution Research*, 22(19), 14546–14559. https://doi.org/10.1007/s11356-015-4202-7

Kumar, R., et al. (2023). Microbial and thermal treatment techniques for degradation of PFAS in biosolids: a focus on degradation mechanisms and pathways. Journal of Hazardous Materials, 452, 131212 https://doi.org/10.1016/j.jhazmat.2023.131212

Kwiatkowski, C. F., et al. (2020). Scienti fic basis for managing PFAS as a chemical class. Environ. Sci. Technol. Lett. 7(8), 532–543 https://doi.org/10.1021/acs.estlett.0c00255

Langenback, B. & Wilson, M. (2021). Per- and polyfluoroalkyl substances (PFAS): significance and considerations within the regulatory framework of the USA. *International Journal of Research and Public Health*, 18, 11142. https://doi.org/10.3390/ijerph182111142

Lasee, S., et al. (2022). Targeted analysis and total oxidizable precursor assay of several insecticides for PFAS. *Journal of Hazardous Materials Letters*, 3(June), 100067. https://doi.org/10.1016/j.hazl.2022.100067

Lasters, R., et al. (2022). Home-produced eggs: an important human exposure pathway of perfluoroalkylated substances (PFAS) Chemosphere. https://doi.org/10.1016/j.chemosphere.2022.136283

Lazo, A. J. & Lee, L. S. (2024). Plant uptake of PFAS in soybeans. Paper presented at the ASA/SSSA Annual Meeting, San Antonio, TX.

Le, S., et al. (2022). Science of the total environment predicting the impact of salt mixtures on the air-water interfacial behavior of PFAS. *Science of the Total Environment*, 819, 151987. https://doi.org/10.1016/j.scitotenv.2021.151987

Lee, H., et al. (2010). Biodegradation of polyfluoroalkyl phosphates as a source of perfluorinated acids to the environment. *Environmental Science and Technology*, 44(9), 3305–3310. https://doi.org/10.1021/es9028183

Lee, H., et al. (2014a). Fate of polyfluoroalkyl phosphate diesters and their metabolites in biosolids-applied soil: biodegradation and plant uptake in greenhouse and field experiments. *Environmental Science and Technology*, 48(1), 340–349.

Lee, H., et al. (2014b). Fate of polyfluoroalkyl phosphate diesters and their metabolites in biosolids-applied soil: biodegradation and plant uptake in greenhouse and field experiments. *Environmental Science and Technology*, 48(1), 340–349. https://doi.org/10.1021/es403949z

Lee, J. W., et al. (2020). Adverse effects of perfluoroalkyl acids on fish and other aquatic organisms: a review. *Science of the Total Environment*, 707, 135334. https://doi.org/10.1016/j.scitotenv.2019.135334

Li, F., et al. (2019). Adsorption of perfluorinated acids onto soils: kinetics, isotherms, and influences of soil properties. *Science of the Total Environment*, 649, 504–514. https://doi.org/10.1016/j.scitotenv.2018.08.209

Li, H., et al. (2020). Transport behavior difference and transport model of long- and short-chain per- and polyfluoroalkyl substances in underground environmental media: a review. *Environmental Pollution*, 327, 121579.

Li, X., et al. (2024). A global dataset of biochar application effects on crop yield, soil properties, and greenhouse gas emissions. *Scientific Data*, 11, 57. https://doi.org/10.1038/s41597-023-02867-9

Li, Y., et al. (2018). A critical analysis of published data to discern the role of soil and sediment properties in determining sorption of per and polyfluoroalkyl substances (PFASs). *Science of the Total Environment*, 628–629, 110–120. https://doi.org/10.1016/j.scitotenv.2018.01.167

Liao, S., et al. (2022). In fluence of residual nonaqueous-phase liquids (NAPLs) on the transport and retention of per fluoroalkyl substances. https://doi.org/10.1021/acs.est.2c00858

Liao, S., et al. (2024). Simulating PFAS transport in effluent-irrigated farmland using modified PRZM5, LEACHM and Hydrus-1D models. Journal of Environmental Quality, 54(1), 54–65. http://doi.org/10.1002/jeq2.20639

Limbu, S., et al. (2024). A machine-learning-driven pathophysiology-based new approach method for the dose-dependent assessment of hazardous chemical mixtures and experimental validations. *Toxics*, 12(7), 481. https://doi.org/10.3390/toxics12070481

Lin, A. M., et al. (2024). Landfill gas: a major pathway for neutral per- and polyfluoroalkyl substance (PFAS) release. *Environmental Science and Technology Letters*, 11(7), 730–737. https://doi.org/10.1021/acs.estlett.4c00364

Lin, Y., et al. (2017). Identification of novel hydrogen-substituted polyfluoroalkyl ether sulfonates in environmental matrices near metal-plating facilities. *Environmental Science and Technology*, 51(20), 11588–11596. https://doi.org/10.1021/acs.est.7b02961

Liu, J., et al. (2023a). Correspondence on defluorination of perfluorooctanoic acid (PFOA) and perfluorooctane sulfonate (PFOS) by acidimicrobium sp. Strain A6. *Environmental Science & Technology*, 53(19), 11410–11419.

Liu, S., et al. (2023b). Concentrations of per- and polyfluoroalkyl substances in vegetables from Sydney and Newcastle, Australia. *Journal of the Science of Food and Agriculture*, 104, 6667–6675. https://doi.org/10.1002/jsfa.13491

Liu, X., et al. (2013). Biochar's effect on crop productivity and the dependence on experimental conditions-a meta-analysis of literature data. *Plant and Soil*, 373(1–2), 583–594. https://doi.org/10.1007/s11104-013-1806-x

Liu, Y., et al. (2019a). High-resolution mass spectrometry (HRMS) methods for nontarget discovery and characterization of poly- and per-fluoroalkyl substances (PFASs) in environmental and human samples. *TrAC - Trends in Analytical Chemistry*, 121, 115420. https://doi.org/10.1016/j.trac.2019.02.021

Liu, Z., et al. (2019b). Multiple crop bioaccumulation and human exposure of perfluoroalkyl substances around a mega fluorochemical industrial park, China: Implication for planting optimization and food safety. *Environment International*, 127(December 2018), 671–684. https://doi.org/10.1016/j.envint.2019.04.008

Longstaffe, J. G., et al. (2012). In-situ molecular-level elucidation of organofluorine binding sites in a whole peat soil. *Environmental Science & Technology*, 46(19), 10508–10513.

Lopez-Espinosa, M.-J., et al. (2012) Thyroid function and perfluoroalkyl acids in children living near a chemical plant. *Environmental Health Perspectives*, 120(7), 1036–1041. https://doi.org/10.1289/ehp.1104370

Lu, D., et al. (2020) Treatment train approaches for the remediation of per- and polyfluoroalkyl substances (PFAS): a critical review. *Journal of Hazardous Materials*, 386, 121963. https://doi.org/10.1016/j.jhazmat.2019.121963

Lu, X., et al. (2016). Adsorption behavior and mechanism of perfluorooctane sulfonate on nanosized inorganic oxides. *Journal of Colloid and Interface Science*, 474, 199–205. https://doi.org/10.1016/j.jcis.2016.04.032

Lyu, Y. & Brusseau, M. L. (2020). The influence of solution chemistry on air-water interfacial adsorption and transport of PFOA in unsaturated porous media. *Science of the Total Environment*, 713, 136744. https://doi.org/10.1016/j.scitotenv.2020.136744

Lyu, Y., et al. (2018). Adsorption of PFOA at the air-water interface during transport in unsaturated porous media. *Environmental Science and Technology*, 52(14), 7745–7753. https://doi.org/10.1021/acs.est.8b02348

Ma, T., et al. (2023). Toxicity of per- and polyfluoroalkyl substances to aquatic vertebrates. *Frontiers in Environmental Science*, 11, 1101100. https://doi.org/10.3389/fenvs.2023.1101100

Published by Burleigh Dodds Science Publishing Limited, 2025.

Mahinroosta, R. & Senevirathna, L. (2020). A review of the emerging treatment technologies for PFAS contaminated soils. *Journal of Environmental Management*, 255, 109896. https://doi.org/10.1016/j.jenvman.2019.109896

Maine CDC (2017). https://www.maine.gov/dep/spills/topics/pfas/Derivation-of-Action-Levels-for-PFOS-in-Cows-Milk-03.28.17.pdf

Maine CDC (2020). https://www.maine.gov/dep/spills/topics/pfas/PFOS-Action-Levels-for-Beef-Derivation-Memo-08.04.20.pdf

Mattsson, K., et al. (2015). Levels of perfluoroalkyl substances and risk of coronary heart disease: findings from a population-based longitudinal study. *Environmental Research*, 142, 148–154. https://doi.org/10.1016/j.envres.2015.06.033

Mayakaduwage, S., et al. (2022). Phytoremediation prospects of per- and polyfluoroalkyl substances: a review. *Environmental Research*, 212(Part B), 113311. https://doi.org/10.1016/j.envres.2022.113311

MDEP (2022). Background levels of PFAS and PAHHs in maine shallow soil study report (2022) prepared for the maine department of environmental protection, file No. 5060.00. https://www.maine.gov/dep/spills/topics/pfas/Maine_Background_PFAS_Study_Report.pdf

Mei, W., et al. (2021). Per- and polyfluoroalkyl substances (PFASs) in the soil–plant system: sorption, root uptake, and translocation. Environment International, 156, 106774. https://doi.org/10.1016/j.envint.2021.106642

Mejia Avendaño, S. & Liu, J. (2015). Production of PFOS from aerobic soil biotransformation of two perfluoroalkyl sulfonamide derivatives. *Chemosphere*, 119, 1084–1090. https://doi.org/10.1016/j.chemosphere.2014.09.059

Mejia-Avendaño, S., et al. (2020). Sorption of poly fluoroalkyl surfactants on surface soils: effect of molecular structures, soil properties, and solution chemistry. https://doi.org/10.1021/acs.est.9b04989

Merino, N., et al. (2018). Fungal biotransformation of 6:2 fluorotelomer alcohol. *Remediation*, 28(2), 59–70. https://doi.org/10.1002/rem.21550

Mikkonen, A. T., et al. (2023). Dynamic exposure and body burden models for per- and polyfluoroalkyl substances (PFAS) enable management of food safety risks in cattle. *Environment International*, 180, 108218. https://doi.org/10.1016/j.envint.2023.108218

Moodie, D., et al. (2021). Legacy and emerging per- and polyfluoroalkyl substances (PFASs) in Australian biosolids. *Chemosphere*, 270, 129143. https://doi.org/10.1016/j.chemosphere.2020.129143

Mroczko, O., et al. (2022). Spatiotemporal patterns of PFAS in water and crop tissue at a beneficial wastewater reuse site in central Pennsylvania. *Journal of Environmental Quality*, 51(6), 1282–1297. https://doi.org/10.1002/jeq2.20408.

Munoz, G., et al. (2020). Bioaccumulation of zwitterionic polyfluoroalkyl substances in earthworms exposed to aqueous film-forming foam impacted soils. *Environmental Science and Technology*, 54(3), 1687–1697. https://doi.org/10.1021/acs.est.9b05102

Munoz, G., et al. (2022). Target and nontarget screening of PFAS in biosolids, composts, and other organic waste products for land application in France. *Environmental Science & Technology*, 56, 6056–6068. https://doi.org/10.1021/acs.est.1c03697

Nascimento, R. A., et al. (2018). Sulfluramid use in Brazilian agriculture: a source of per- and polyfluoroalkyl substances (PFASs) to the environment. *Environmental Pollution*, 242, 1436–1443. https://doi.org/10.1016/j.envpol.2018.07.122

NASEM (National Academies of Sciences, Engineering, and Medicine). (2022). *Guidance on PFAS Exposure, Testing, and Clinical Follow-Up*. The National Academies Press.

Nason, S. L., et al. (2024). A comprehensive trial on PFAS remediation: hemp phytoextraction and PFAS degradation in harvested plants. *Environmental Science: Advances*, 3(2), 304–313. https://doi.org/10.1039/d3va00340j

Nickerson, A., et al. (2023). Effect of geochemical conditions on PFAS release from AFFF-impacted saturated soil columns. *Environmental Science: Processes & Impacts*, 25(3), 405–414. https://doi.org/10.1039/D2EM00367H

Ogawa, Y., et al. (2020). Current contributions of organofluorine compounds to the agrochemical industry. *IScience*, 23(9), 101467. https://doi.org/10.1016/j.isci.2020.101467

Oliver, D. P., et al. (2019). The role of surface charge and pH changes in tropical soils on sorption behaviour of per- and polyfluoroalkyl substances (PFASs). *Science of the Total Environment*, 673, 197–206. https://doi.org/10.1016/j.scitotenv.2019.04.055

Oliver, D. P., et al. (2020). Sorption behaviour of per- and polyfluoroalkyl substances (PFASs) as affected by the properties of coastal estuarine sediments. *Science of the Total Environment*, 720, 137263. https://doi.org/10.1016/j.scitotenv.2020.137263

Openiyi, E. & Lee, L. S. (2024). Biochar for maximizing soil quality and pollution prevention. Paper presented at the ASA/SSSA Annual Meeting, San Antonio, TX.

Openiyi, E. O., et al. (2024). Evaluating sorbents for reducing per- and polyfluoroalkyl substance mobility in biosolids-amended soil columns. Journal of Environmental Quality, 54(1), 118–131. https://doi.org/10.1002/jeq2.20658

Openiyi, E., et al. (2025). High carbon wood ash impact on grass uptake of per- and polyfluoroalkyl substances from contaminated agricultural soils. *Journal of Agricultural and Environmental Chemistry*, jf-2025-08985t.

Pandelides, Z,. et al. (2023). Critical review of Amphibian per- and polyfluoroalkyl substance ecotoxicity research studies: identification of screening levels in water and other useful resources for site-specific ecological risk assessments. *Environmental Toxicology and Chemistry*, 42(10), 2078–2090. https://doi.org/10.1002/etc.5695.

Panieri, E., et al. (2022). PFAS molecules: a major concern for the human health and the environment. *Toxics*, 10(2), 44. https://doi.org/10.3390/toxics10020044.

Pepper, I. L., et al. (2021). Incidence of PFAS in soil following long-term application of class B biosolids. *Science of The Total Environment*, 793, 148449. https://doi.org/https://doi.org/10.1016/j.scitotenv.2021.148449

Pepper, I. L., et al. (2023). Is PFAS from land applied municipal biosolids a significant source of human exposure via groundwater? *Science of the Total Environment*, 864, 160933. https://doi.org/10.1016/j.scitotenv.2022.161154

Peter, L. (2025). *Occurrence, fate and transport of per- and polyfluoroalkyl substances (PFAS) in agricultural systems* (PhD dissertation). Purdue University, West Lafayette, IN.

Peter, L. & Lee, L. S. (2025). Sources and pathways of PFAS occurrence in water sources: relative contribution of land-applied biosolids in an agricultural dominated watershed. Environmental Science & Technology, 59(1), 1–12. https://doi.org/10.1021/acs.est.4c09490

Peter, L. & Lee, L. S. (2024a, May 19–21). *Fate of per- and polyfluoroalkyl substances (PFAS) in the vadose zone following land-application of biosolids: impact of soil properties*. IUSS, Florence, Italy.

Peter, L. & Lee, L. S. (2024b, April 16–17). *Per-and polyfluoroalkyl substances (PFAS) in groundwater following long-term land application of biosolids.* NGWA, Tucson, Arizon.

Peter, L., et al. (2024). PFAS fate using lysimeters during degraded soil reclamation using biosolids. *Journal of Environmental Quality*, 54, 1–13. https://doi.org/10.1002/jeq2.20576

Peter, L., et al. (2025). PFAS retention and distribution in the vadose zone of three soil types impacted by biosolids application. *Journal of Environmental Management*, JEMA-D-25-27159.

Phelps, D. W., et al. (2024). Per- and polyfluoroalkyl substances in food packaging: migration, toxicity, and management strategies. *Environmental Science and Technology*, 58, 5670–5684. https://doi.org/10.1021/acs.est.3c03702

Place, B. J. & Field, J. A. (2012). Identification of novel fluorochemicals in aqueous film-forming foams used by the US military. *Environmental Science and Technology*, 46(13), 7120–7127. https://doi.org/10.1021/es301465n

Poothong, S., et al. (2020). Multiple pathways of human exposure to poly- and perfluoroalkyl substances (PFASs): from external exposure to human blood. *Environment International*, 134(October 2019), 105244. https://doi.org/10.1016/j.envint.2019.105244

Post, G. B., et al. (2017). Key scientific issues in developing drinking water guidelines for perfluoroalkyl acids: Contaminants of emerging concern. *PLoS Biology*, 15(12), e2002855. https://doi.org/10.1371/journal.pbio.2002855

Prevedouros, K., et al. (2006). Sources, fate and transport of perfluorocarboxylates. *Environmental Science and Technology*, 40(1), 32–44. https://doi.org/10.1021/es0512475

Qi, Y., et al. (2016). Spatial distribution and source apportionment of PFASs in surface sediments from five lake regions, China. *Scientific Reports*, 6(November 2015), 1–11. https://doi.org/10.1038/srep22674

Qiao, W., et al. (2018). Perfluoroalkyl substances (PFASs) influence the structure and function of soil bacterial community: Greenhouse experiment. *Science of the Total Environment*, 642, 1118–1126. https://doi.org/10.1016/j.scitotenv.2018.06.113

Qiu, Y., et al. (2025). Advances in water and nitrogen management for intercropping systems: crop growth and soil environment. *Agronomy*, 15(8), 2000. https://doi.org/10.3390/agronomy15082000

Quinnan, J., et al. (2022). Ex situ soil washing to remove PFAS adsorbed to soils from source zones. *Remediation*, 32(2), 151–166. https://doi.org/10.1002/rem.21727

Rankin, K., et al. (2016). A North American and global survey of perfluoroalkyl substances in surface soils: distribution patterns and mode of occurrence. *Chemosphere*, 161, 333–341. https://doi.org/10.1016/j.chemosphere.2016.06.109

Rappazzo, K. M., et al. (2017). Exposure to perfluorinated alkyl substances and health outcomes in children: a systematic review of the epidemiologic literature. *International Journal Research Public Health*, 14(7), 691. https://doi.org/10.3390/ijerph14070691

Rayne, S. & Forest, K. (2010). Theoretical studies on the pKa values of perfluoroalkyl carboxylic acids. *Journal of Molecular Structure: THEOCHEM*, 949(1–3), 60–69. https://doi.org/10.1016/j.theochem.2010.03.003

Royer, L., et al. (2015). Microbial transformation of 8:2 fluorotelomer acrylate and methacrylate in aerobic soils. *Chemosphere*, 129, 54–61.

Ruan, T., et al. (2013). Aerobic soil biotransformation of 6:2 fluorotelomer iodide. *Environmental Science and Technology*, 47(20), 11504–11511. https://doi.org/10.1021/es4018128

Ruyle, B. J., et al. (2025). High organofluorine concentrations in municipal wastewater affect downstream drinking water supplies for millions of Americans. Proceedings of the National Academy of Sciences, 122(3), e2417156122. https://doi.org/10.1073/pnas.2417156122

Salim, I., et al. (2019). Comparison of two receptor models PCA-MLR and PMF for source identification and apportionment of pollution carried by runoff from catchment and sub-watershed areas with mixed land cover in South Korea. *Science of the Total Environment*, 663, 764–775. https://doi.org/10.1016/j.scitotenv.2019.01.377

Scearce, A. E., et al. (2024). Uptake of per- and polyfluoroalkyl substances (PFAS) into lettuce (Lactuca sativa), tall fescue (Schedonorus arundinaceus) and tomato (Solanum lycopersicum): a greenhouse experiment evaluating bioconcentration factors and testing the effect of intercropping. *Environmental Advances*, 20, 100629. https://doi.org/10.1016/j.envpol.2024.125235

Schaefer, C. E., et al. (2021). Desorption isotherms for poly- and perfluoroalkyl substances in soil collected from an Aqueous Film-Forming Foam Source Area, Journal of Environmental Engineering, 148(1), 04021152. https://doi.org/10.1061/(ASCE)EE.1943-7870.0001952

Schaefer, C. E., et al. (2022). Release of poly- and perfluoroalkyl substances from finished biosolids in soil mesocosms. *Water Research*, 217, 118405. https://doi.org/10.1016/j.watres.2022.118405

Schymanski, E. L., et al. (2023). Per- and polyfluoroalkyl substances (PFAS) in PubChem: 7 million and growing. *Environmental Science and Technology*, 57(44), 16918–16928. https://doi.org/10.1021/acs.est.3c04855

Sepulvado, J. G., et al. (2011). Occurrence and fate of perfluorochemicals in soil following the land application of municipal biosolids. *Environmental Science and Technology*, 45(19), 8106–8112. https://doi.org/10.1021/es103903d

Shahsavari, E., et al. (2021). Challenges and current status of the biological treatment of PFAS-contaminated soils. Frontiers in Bioengineering and Biotechnology, 8, 602040. https://doi.org/10.3389/fbioe.2020.602040

Sharifan, H., et al. (2021). Fate and transport of per- and polyfluoroalkyl substances (PFASs) in the vadose zone. *Science of the Total Environment*, 771. https://doi.org/10.1016/j.scitotenv.2021.145427

Shaw, D. M. J., et al. (2019). Degradation and defluorination of 6:2 fluorotelomer sulfonamidoalkyl betaine and 6:2 fluorotelomer sulfonate by Gordonia sp. strain NB4-1Y under sulfur-limiting conditions. *Science of the Total Environment*, 647, 690–698. https://doi.org/10.1016/j.scitotenv.2018.08.012

Shi, Y., et al. (2018). Probing the differential tissue distribution and bioaccumulation behavior of per- and polyfluoroalkyl substances of varying chain-lengths, isomeric structures and functional groups in Crucian Carp. *Environmental Science and Technology*, 52(8), 4592–4600. https://doi.org/10.1021/acs.est.7b06128

Shigei, M., et al. (2020). Per- and polyfluoroalkyl substances in water and soil in wastewater-irrigated farmland in Jordan. *Science of the Total Environment*, 716, 137057. https://doi.org/10.1016/j.scitotenv.2020.137057

Silva, J. A. K., et al. (2019). Evaluating air-water and NAPL-water interfacial adsorption and retention of perfluorocarboxylic acids within the Vadose zone. *Journal of Contaminant Hydrology*, 223(April), 103472. https://doi.org/10.1016/j.jconhyd.2019.03.004

Silva, J. A. K., et al. (2022). Comparison of methods to estimate air-water interfacial areas for evaluating PFAS transport in the vadose zone. *Journal of Contaminant Hydrology*, 247, 103984. https://doi.org/10.1016/j.jconhyd.2022.103984

Silva, J. A. K., et al. (2021). Air-water interfacial adsorption coefficients for PFAS when present as a multi-component mixture. *Journal of Contaminant Hydrology*, 236, 103731. https://doi.org/10.1016/j.jconhyd.2020.103731

Simones, T. L., et al. (2024). Uptake of per- and polyfluoroalkyl substances in mixed forages on biosolid-amended farm fields. *Journal of Agricultural and Food Chemistry*, 72(42), 23108–23117. https://doi.org/10.1021/acs.jafc.4c02078

Sinclair, G. M., et al. (2020). What are the effects of PFAS exposure at environmentally relevant concentrations? *Chemosphere*, 258, 127340. https://doi.org/10.1016/j.chemosphere.2020.127340

Sørmo, E., et al. (2024). Stabilization of PFAS-contaminated soil with sewage sludge- and wood-based biochar sorbents. *Science of the Total Environment*, 922, 170971. https://doi.org/10.1016/j.scitotenv.2024.170971.

Sørmo, E., et al. (2021). Stabilization of PFAS-contaminated soil with activated biochar. *Science of the Total Environment*, 763, 144034. https://doi.org/10.1016/j.scitotenv.2020.144034

Stahl, T., et al. (2009). Carryover of perfluorooctanoic acid (PFOA) and perfluorooctane sulfonate (PFOS) from soil to plants. *Archives of Environmental Contamination and Toxicology*, 57(2), 289–298. https://doi.org/10.1007/s00244-008-9272-9

Steenland, K., et al. (2013). Ulcerative colitis and perfluorooctanoic acid (PFOA) in a highly exposed population of community residents and workers in the Mid-Ohio Valley. *Environmental Health Perspectives*, 121(8), 900–905. https://doi.org/10.1289/ehp.1206449

Sun, H., et al. (2011). Long-chain perfluorinated chemicals in digested sewage sludges in Switzerland. *Environmental Pollution*, 159(2), 654–662. https://doi.org/10.1016/j.envpol.2010.09.020

Sunderland, E. M., et al. (2019). A review of the pathways of human exposure to poly- and perfluoroalkyl substances (PFASs) and present understanding of health effects. *Journal of Exposure Science & Environmental Epidemiology*, 29(2), 131–147. https://doi.org/10.1038/s41370-018-0094-1

Thomas, T., et al. (2023). Global regulations around PFAS: the past, the present and the future. *ICRL*, 1, 3–17. https://icrl.lexxion.eu/data/article/18898/pdf/icrl_2023_01-005.pdf

Thompson, J. T., et al. (2023). Underestimation of per- and polyfluoroalkyl substances in biosolids: precursor transformation during conventional treatment. *Environmental Science and Technology*, 57(9), 3825–3832. https://doi.org/10.1021/acs.est.2c06189

Tow, E. W., et al. (2021). Managing and treating per- and polyfluoroalkyl substances (PFAS) in membrane concentrates. *AWWA Water Science*, 3, e1233. https://doi.org/10.1002/aws2.1233

Tseng, N., et al. (2014). Biotransformation of 6:2 fluorotelomer alcohol (6:2 FTOH) by a wood-rotting fungus. *Environmental Science and Technology*, 48(7), 4012–4020. https://doi.org/10.1021/es4057483

Turner, T., et al. (2019). Potential alternative reuse pathways for water treatment residuals: remaining barriers and questions–a review. *Water, Air, and Soil Pollution*, 230(9). https://doi.org/10.1007/s11270-019-4272-0

Uhme, A. C., et al. (2021). Sorption of PFOS in 114 well-characterized tropical and temperate soils: application of multivariate and artificial neural network analyses. *Environmental Science and Technology*, 55, 1779–1789. https://dx.doi.org/10.1021/acs.est.0c07202

Uhme, A. C., et al. (2024). A systematic investigation of single solute, binary and ternary PFAS transport in water-saturated soil using batch and 1-dimensional column studies: focus on mixture effects. *Journal of Hazardous Materials*, 462, 132688. https://doi.org/10.1016/j.jhazmat.2023.132688

Ulrich, H., et al. (2016) Getting on with persistent pollutants: decreasing trends of perfluoroalkyl acids (PFAAs) in sewage sludge. *Chemosphere*, 161, 527–535. https://doi.org/10.1016/j.chemosphere.2016.07.048

U.S.EPA (2020). 537.1 - 1. 1(2009), 1–50.

U.S.EPA (2024a). Biden-harris administration finalizes first-ever national drinking water standard to protect 100M people from PFAS pollution, press release April 10, 2024.

U.S.EPA (2024b). Basic information about sewage sludge and biosolids. https://www.epa.gov/biosolids/basic-information-about-biosolids

U.S.EPA (2024c). Method 1633 - Analysis of per- and polyfluoroalkyl substances (PFAS) in aqueous, solid, biosolids, and tissue samples by LC-MS/MS. EPA 821-R-24-001, January, 69 pp. www.epa.gov

U.S.EPA (2024d). Research grants for understanding PFAS uptake and bioaccumulation in plants and animals in agricultural, rural, and tribal communities. https://www.epa.gov/research-grants/research-grants-understanding-pfas-uptake-and-bioaccumulation-plants-and-animals

Valalamontes, A. & Adamopoulos, I. (2025). Safety of future food supply in Europe: challenges and sustainable solutions, academisa, Paper #2570, Rev 2.53. https://doi.org/10.13140/RG.2.2.29643.71202

Van Hamme, J. D., et al. (2013). Genomic and proteomic characterization of Gordonia sp. NB4-1Y in relation to 6 : 2 fluorotelomer sulfonate biodegradation. *Microbiology (United Kingdom)*, 159(8), 1618–1628. https://doi.org/10.1099/mic.0.068932-0

Vierke, L., et al. (2013). Estimation of the acid dissociation constant of perfluoroalkyl carboxylic acids through an experimental investigation of their water-to-air transport. *Environmental Science and Technology*, 47(19), 11032–11039. https://doi.org/10.1021/es402691z

Wang, F. & Shih, K. (2011). Adsorption of perfluorooctanesulfonate (PFOS) and perfluorooctanoate (PFOA) on alumina: influence of solution pH and cations. *Water Research*, 45(9), 2925–2930. https://doi.org/10.1016/j.watres.2011.03.007

Wang, F., et al. (2019). Protein-specific distribution patterns of perfluoroalkyl acids in egg yolk and albumen samples around a fluorochemical facility. *Science of the Total Environment*, 650(Part 2), 2697–2704. https://doi.org/10.1016/j.scitotenv.2018.10.006

Wang, L. K., et al. (2009). *Handbook of Environmental Engineering: Biosolids Engineering and Management*. Human Press. https://doi.org/10.1108/meq.2009.08320bae.001

Wang, N., et al. (2012). 5:3 Polyfluorinated acid aerobic biotransformation in activated sludge via novel "one-carbon removal pathways." *Chemosphere*, 87(5), 527–534. https://doi.org/10.1016/j.chemosphere.2011.12.056

Wang, W., et al. (2020). Uptake and accumulation of per-and polyfluoroalkyl substances in plants. *Chemosphere*, 261, 127584.

Wang, Z., et al. (2015). Fluorinated alternatives to long-chain per fl uoroalkyl carboxylic acids (PFCAs), per fluoroalkane sulfonic acids (PFSAs) and their potential precursors. *Environmental Science & Technology*, 60(2013), 242–248.

Washington, J. W., et al. (2010). Concentrations, distribution, and persistence of perfluoroalkylates in sludge-applied soils near Decatur, Alabama, USA. *Environmental Science and Technology*, 44(22), 8390–8396. https://doi.org/10.1021/es1003846

Whitehead, H. D., et al. (2021). Fluorinated compounds in North American cosmetics. *Environmental Science and Technology Letters*, 8(7), 538–544. https://doi.org/10.1021/acs.estlett.1c00240

Xie, M.-Y., et al. (2023). Glioma is associated with exposure to legacy and alternative per- and polyfluoroalkyl substances. *Journal of Hazardous Materials*, 441, 129819. https://doi.org/10.1016/j.jhazmat.2022.129819

Xie, M. Y., et al. (2024). Per- and polyfluoroalkyl substances (PFAS) exposure in plasma and their blood–brain barrier transmission efficiency–A pilot study. *Environment International*, 187(May), 108719. https://doi.org/10.1016/j.envint.2024.108719

Xiao, F., et al. (2019). Sorption and Desorption Mechanisms of Cationic and Zwitterionic Per- and Polyfluoroalkyl Substances in Natural Soils: Thermodynamics and Hysteresis, *Environmental Science and Technology*, 53, 11818–11827. http://dx.doi.org/10.1021/acs.est.9b05379

Xu, B., et al. (2023). Effects of perfluoroalkyl and polyfluoroalkyl substances (PFAS) on soil structure and function. *Soil Ecology Letters*, 5, 108–117. https://doi.org/10.1007/s42832-022-0143-5

Yao, X., et al. (2020). Amino acid transporters in plants: identification and function. *Plants*, 9(8), 972. htps://doi.org/10.3390/plants9080972

Yao, Y., et al. (2014). Distribution and primary source analysis of per- and poly-fluoroalkyl substances with different chain lengths in surface and groundwater in two cities, North China. *Ecotoxicology and Environmental Safety*, 108, 318–328.

Zabaleta, I., et al. (2018). Biodegradation and uptake of the pesticide sulfluramid in a soil-carrot mesocosm. *Environmental Science and Technology*, 52(5), 2603–2611. https://doi.org/10.1021/acs.est.7b03876

Zhang, D. Q., et al. (2020). Distribution of perfluoroalkyl substances (PFASs) in aquatic plant-based systems: from soil adsorption and plant uptake to effects on microbial community. *Environmental Pollution*, 257, 113575. https://doi.org/10.1016/j.envpol.2019.113575

Zhang, S., et al. (2013). 6:2 and 8:2 fluorotelomer alcohol anaerobic biotransformation in digester sludge from a WWTP under methanogenic conditions. *Environmental Science and Technology*, 47(9), 4227–4235. https://doi.org/10.1021/es4000824

Zhang, S., et al. (2016a). Biotransformation potential of 6:2 fluorotelomer sulfonate (6:2 FTSA) in aerobic and anaerobic sediment. *Chemosphere*, 154, 224–230. https://doi.org/10.1016/j.chemosphere.2016.03.062

Zhang, X., et al. (2016b). Source attribution of poly- and perfluoroalkyl substances (PFASs) in surface waters from Rhode Island and the New York Metropolitan Area.

Environmental Science and Technology Letters, 3(9), 316–321. https://doi.org/10.1021/acs.estlett.6b00255

Zhao, S., et al. (2017). Interaction effects on uptake and toxicity of perfluoroalkyl substances and cadmium in wheat (Triticum aestivum L.) and rapeseed (Brassica campestris L.) from co-contaminated soil. *Ecotoxicology and Environmental Safety*, 137(July 2016), 194–201. https://doi.org/10.1016/j.ecoenv.2016.12.007

Zhang, Z., et al. (2021). Adsorption of perfluorooctanoic acid (PFOA) and perfluorooctanesulfonic acid (PFOS) by aluminum-based drinking water treatment residuals. *Journal of Hazardous Materials Letters*, 2, 100034. https://doi.org/10.1016/j.hazl.2021.100034

Zhu, W., et al. (2022). Vermont-wide assessment of anthropogenic background concentrations of perfluoroalkyl substances in surface soils. *Journal of Hazardous Materials*, 438, 129479. https://doi.org/10.1016/j.jhazmat.2022.129479.

Published by Burleigh Dodds Science Publishing Limited, 2025.

Chapter 4

Pesticide residues as contaminants in agricultural soils

Vera Silva, Wageningen University and Research, The Netherlands; Judith Riedo, Freie Universität Berlin, Germany; Diana Vieira, European Commission, Joint Research Centre (JRC), Italy; Sergejus Ustinov, Food and Agriculture Organization of the United Nations, Italy; Pia Kotschik, German Environment Agency, Division IV Chemical Safety – Section IV 1.3. Plant Protection Products, Germany; Violette Geissen, Wageningen University and Research, The Netherlands; Antonio Bispo and Claire Froger, INRAE, UR Info&ols, France; and Shiva Sabzevari, Masaryk University, Czech Republic

1 Introduction

A pesticide is any chemical or biological substance that prevents, controls, or eliminates pests. There are 425 active substances currently approved for pest control in the European Union (EU, 2024), and around 600 are approved worldwide (McDougall, 2018). Pesticides are commonly classified by their function (e.g. herbicides, fungicides, insecticides), chemical composition (e.g. organophosphates, organochlorines), or mode of action (e.g. acetylcholinesterase inhibitors, sodium channel modulators), though other

http://dx.doi.org/10.19103/AS.2025.0151.07

classification systems exist for this diverse group of substances (Akashe et al., 2018). Approximately 90% of pesticide sales are related to the agricultural sector, where they are used as plant protection products. These are applied to soil or weeds to combat plants competing with crops for resources, or directly to crops to fight organisms that damage agricultural commodities. It is estimated that 26–40% of global crop production is lost to pests annually; without pesticides, these losses could potentially double (Oerke, 2006). Ensuring high crop yields is becoming increasingly difficult as we face degradation and depletion of arable land, more frequent extreme climate events, higher pest incidences, and increasing pest resistance (Jung et al., 2019; Lykogianni et al., 2021; Prăvălie et al., 2021).

Pesticides became increasingly important in agriculture in the 1940s, following the discovery of the insecticidal properties of dichlorodiphenyltrichloroethane, commonly known as DDT, and other organochlorine compounds. Pesticide sales grew 20–30-fold between the 1960s and 1990s, largely driven by the Green Revolution (Carvalho, 2017; Oerke, 2006). While sales have stabilized in most developed countries, they continue to rise in many developing nations (FAOSTAT, 2024). Global pesticide use now surpasses 4 million tonnes/year, with an average of 2.63 kg applied per hectare of cropland (FAO, 2020b). Their continued, intensive, and widespread use raises serious environmental and health concerns. The biologist Rachel Carson was one of the first to sound the alarm on the side effects of pesticides, exposing the dangers of DDT in her landmark book *Silent Spring* (Carson, 1962). Today, numerous studies address the environmental and health risks posed by pesticides (e.g. Ali et al., 2021; Buckwell et al., 2020; Geiger et al., 2010; IPBES, 2018; Maggi et al., 2021). Increasing attention is being paid to the hidden and external costs of pesticide use (Bourguet and Guillemaud, 2016), and the research agenda surrounding pesticides is gaining more public awareness and legal scrutiny.

Concerns about pesticides stem from two key issues: i) part of applied pesticides is released into the environment – with the amount varying greatly among crops and application methods (Carvalho, 2017) and ii) most pesticides – and/or their degradation products – are toxic to non-target-species (Colin et al., 2019; Francisco, 2011; Ullah et al., 2018), mobile and/or persistent in the environment (Masiá et al., 2013; Silva et al., 2019), and/or bio-accumulate/bioconcentrate in organisms and food chains (Goutner et al., 2012; Wang et al., 2011). A study published in *Nature Geoscience* (Tang et al., 2021) highlights the urgency of addressing pesticide-related challenges, emphasizing the need for more sustainable food production practices. The study reports that 64% of global agricultural land is at risk of pesticide pollution, with 31% considered at high risk – including regions with high biodiversity, water scarcity, and low- to lower-middle-income nations.

Published by Burleigh Dodds Science Publishing Limited, 2025.

The real extent of soil contamination by pesticide residues remains unknown. Most of the studies are conducted on this topic focus on agricultural land, on experimental sites or parcels, and pesticide, and spatial coverage is rather variable, and so are the underlying field and laboratory details (Sabzevari and Hofman, 2022). This prevents proper evaluation of the threats caused by pesticides to terrestrial and aquatic ecosystems and human health at a national, continental, or global level. Furthermore, many of the existing studies focused on long-banned toxic pesticides such as the organochlorine pesticides DDT or atrazine (Aichner et al., 2013; Wenzel et al., 2002), and until recently, little was known about the distribution of currently used pesticides (CUPs). More recently, research on the topic has accelerated, revealing a widespread occurrence of pesticides, banned and CUPs, in soils under all land uses (Froger et al., 2023; Hvězdová et al., 2018; Silva et al., 2019; Vieira et al., 2023; see Fig. 1 on the European case).

Aggregation of findings from different studies is not straightforward.. Pesticide figures in soil fluctuate according to an input-output dynamic model, as a function of the number and amount of pesticides reaching soil and the rates at which they degrade or are exported. The fate of pesticides is governed by physical, chemical, and biological processes, sensitive to regional factors (Alekseeva et al., 2014; Arias-Estévez et al., 2008; Gentil et al., 2019; Fig. 2). Main factors affecting the fate of pesticides in soil: i) soil types/properties (soil texture, soil permeability, organic matter content, soil pH), ii) pesticide properties (water solubility, degradation time, adsorption coefficient, etc), iii) management practices (application method, pesticide rate, and timing), and iv) site conditions (depth to groundwater, geologic conditions, climate; Pérez-Lucas et al., 2019). Adsorption-desorption dynamics play a key role in the fate and impacts of pesticides. Most pesticides adsorb strongly to soil particles, in particular organic matter, clays, oxi-hydroxides of iron, and manganese (García-Delgado et al., 2020; Weber et al., 2004), forming long-term bound residues. Adsorbed pesticides (as other adsorbed substances) are generally less mobile than those in solution (Arias-Estévez et al., 2008) but can still be transported to other areas and environmental compartments, through macropores, or via wind, water, or harvest erosion. Adsorbed substances tend to be also less bioavailable than those in solution, and consequently linked to lower or delayed toxicity (van Gestel, 2012). On the other hand, adsorbed pesticides are also less available to micro-organisms, and consequently degraded more slowly than those in solution (Guo et al., 2000). Microbial degradation is the most relevant degradation pathway for most pesticides and often involves the formation of metabolites, breakdown, or reaction products, which may have similar or even higher toxicity than the parent compounds (Karas et al., 2018; Vasileiadis et al., 2018). Pesticide impacts, in both source and receiving areas, depend on (i) the hazardous characteristics of the pesticides, (ii) exposure context, namely dose,

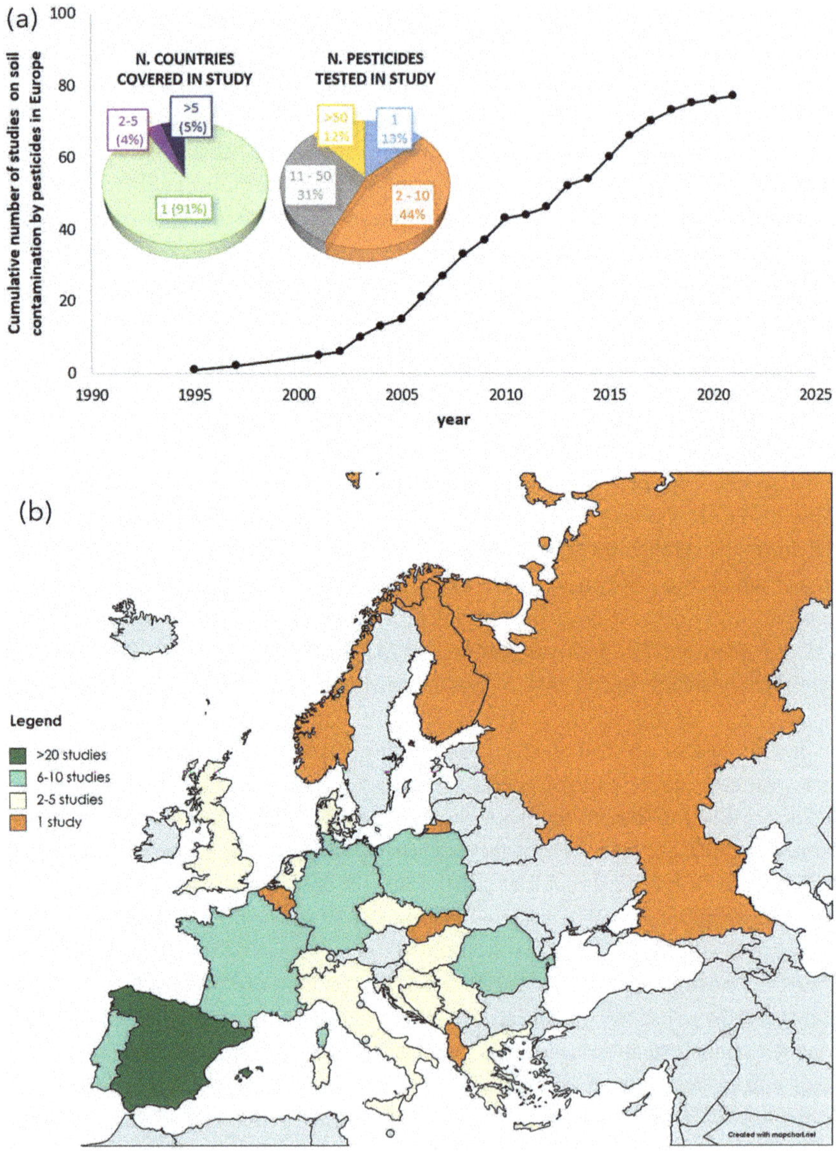

Figure 1 Overview of the European studies on soil contamination by pesticide residues. (a) number of studies over the years, number of compounds and countries covered; (b) map on the number of soil-pesticide studies across EU countries. Reprinted from Silva et al., 2022.

length, and/or frequency of exposure, and (iii) organisms' vulnerability (Arias-Estévez et al., 2008; Hayes and Laws, 1991). Interactions among pesticides and between pesticides and other contaminants may lead to different outcomes

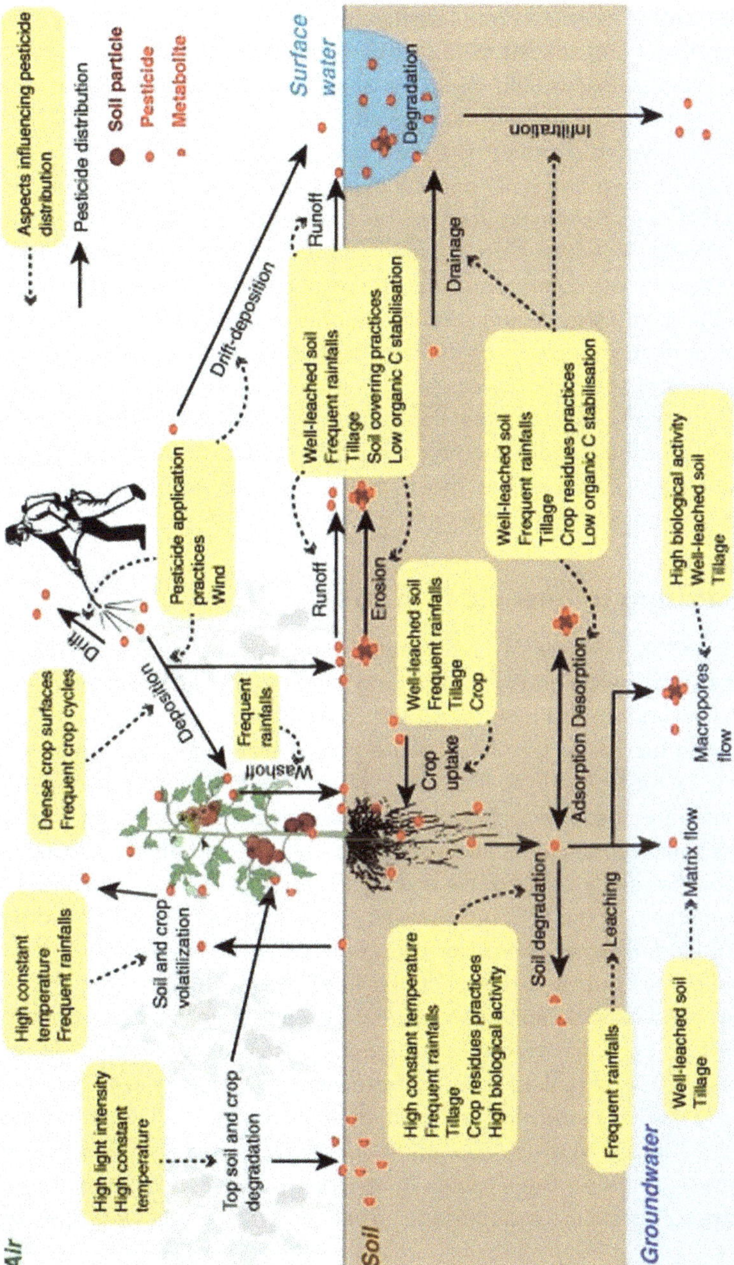

Figure 2 Key processes governing pesticide fate. Normal arrows represent the pesticide distribution, and dotted arrows represent aspects influencing pesticide distribution under tropical conditions. Reprinted from Gentil et al. (2019).

than expected based on an individual compound assessment (Ünlü Endirlik et al., 2023; Coors and De Meester, 2008).

In this chapter, we focus on soil-pesticide monitoring data, emphasizing key findings and challenges such as legacy contamination and the interpretation of results. We also discuss the dual role of soils, as both a sink and source of pesticide residues, underlining the need for regular, harmonized, and holistic monitoring programs. Additionally, we provide recommendations for better assessing and preventing soil contamination by pesticide residues, including an initial attempt at defining relative benchmarks for pesticide residues in agricultural land. Given the high fragmentation of existing monitoring data, we chose not to conduct compound or crop-specific analyses, instead aiming for an overarching, and whenever possible, global perspective. At the same time, it must be noticed that much of the data highlighted is from well-developed countries, in particular from European countries. This is because Europe is the source of most soil-pesticide data, has the strictest pesticide regulatory system in the world, and has legally binding reduction targets. Our focus is primarily on synthetic pesticides, used on most agricultural land, and most relevant for regulatory and pesticide reduction efforts.

2 Monitoring of pesticides in soil

As mentioned previously, it is estimated that 64% of global agricultural land is at risk of pesticide pollution (Tang et al., 2021). Unravelling effective contaminated figures and monitoring soil contamination by pesticide residues is urgently needed to assess and secure soil quality and health (Bach et al., 2020). Data on pesticides in soils are also valuable for the verification/validation of environmental fate models, to assess consequences/efficacy of pesticide use and regulatory measures (including restrictions and bans of substances), and support decisions on land use and agricultural practices (Hvězdová et al., 2018). At the same time, it is not realistic to monitor all pesticide residues in all units of land. Existing soil-pesticide monitoring studies and programs are usually policy- or risk-driven (see sections 4 and 5).

Most of the existing soil-pesticide data originate from individual studies (see Table 1 on European overview – the richest dataset worldwide). Sabzevari and Hofman (2022) reviewed soil data from currently/recently used pesticides, reporting 72 studies worldwide with numerical data available, 33 are European studies, covering a total of 306 compounds (200 active substances, 106 metabolites) – more on the findings in section 2.1. Overall, European or not, most of the soil-pesticide monitoring studies often cover a limited number of pesticides and/or a relatively small area, and altogether they vary greatly on sampling time (year and season), sampling strategy (soil depth), and analytical methods (scope, extraction, and limits of detection and quantification),

Published by Burleigh Dodds Science Publishing Limited, 2025.

Table 1 Recent studies on soil contamination by currently used pesticides in Europe. *TP = Transformation Product

Study	Local	Land cover	Sampling		Pesticides	Major findings
Chiaia-Hernandez et al. (2017)	Switzerland (NABO)	Agricultural soils	14 sites	1995–2008	80 pesticides 87 TPs*	**Occurrence**: All sites with at least a residue **Most frequent**: simazine (97%), atrazine (86%), terbuthylazine (62%)
Silva et al. (2018)	Europe	Agricultural soils	317 sites	2015	Glyphosate + AMPA	**Occurrence**: 45% of the sites with glyphosate or AMPA (18% with both) **Spatial disparities**: lower glyphosate degradation in dryer conditions
Hvězdová et al. (2018)	Czech Republic	Arable lands	75 sites	2015 (Feb–March)	53 pesticides 15 TPs	**Occurrence**: 99% of the sites with at least 1 residue; 64% with >3 pesticides **Most frequent**: triazines (90% of the sites) and conazoles (73%).
Silva et al. (2019)	Europe	Agricultural lands	317 sites	2015	76 pesticides	**Occurrence:** 80% of the sites with at least 1 residue, 58% >2 residues. **Most frequent:** glyphosate + AMPA, boscalid, epoxiconazole, tebuconazole
Kosubová et al. (2020)	Czech Republic	Arable lands	34 sites	2014–2017	60 pesticides 4 TPs	**Occurrence**: all sites with at least 1 residue; 19 pesticides non-detected **Most frequent**: hydroxyatrazine, epoxiconazole, carbendazime
Pelosi et al., (2021)	France	Arable lands, grasslands and hedgerows	180 sites	2016	31 pesticides	**Occurrence:** all soils with at least 1 residue **Most frequent**: diflufenican, imidaclopride, boscalid, epoxiconazole
Riedo et al., (2021)	Switzerland	Agricultural soils	100 sites	2016	38 pesticides 8 TPs	**Occurrence:** All sites with at least 1 residue **Most frequent:** 2-hydroxyatrazine, atrazine, chloridazon, linuron, terbutylazine, epoxiconazole

(*Continued*)

Table 1 (*Continued*)

Study	Local	Land cover	Sampling		Pesticides	Major findings
Geissen et al. (2021)	Spain, Portugal, Netherlands	Arable lands	4 sites, 340 samples	2015–2018	151 pesticides 15 TP	**Occurrence:** all the sites with at least 1 residue **Most frequent**: glyphosate, AMPA, metalaxyl, dimetomorph
Froger et al. (2023)	France	All land covers	47 sites	2019–2020	110 pesticides 1 TP	**Occurrence:** 98% of sites with at least 1 residue **Most frequent:** glyphosate, AMPA, fluxapyroxad, epoxiconazole,
Vieira et al. (2023)	Europe	All land covers	3473	2018	118 pesticides	**Occurrence:** 74.5% of sites with at least 1 residue, 57.1% 2 or more. **Most frequent:** AMPA, epoxiconazole, tebuconazole
Knuth et al. (2024)	Europe (10 countries)	Agricultural soils	10 study sites, 201 samples	2021	192 pesticides	**Occurrence:** 99% and 95% with >1 residue for conv. and organic fields **Most frequent:** p,p′-DDE, hexachlorobenzene

Published by Burleigh Dodds Science Publishing Limited, 2025.

hampering an overview of occurrence and levels of the residues in agricultural soils. While monitoring programs for pesticides in ground and surface waters are well established (e.g. Cooke et al., 2024; Baran et al., 2021), at least in multiple European countries, monitoring programs for soil are still being set up or have only recently been initiated (see Mason et al 2025 for overview of European monitoring systems, analyzed the upcoming European Soil Monitoring Law). Large-scale soil-pesticide assessments are especially scarce, with only three pan-European works so far (see Table 2 for their main characteristics and findings).

So far, the trend analysis performed within two LUCAS (Land Use and Coverage Area frame Survey) soil campaigns, the most comprehensive ones worldwide, indicates a higher incidence of pesticide residues (Vieira et al., 2023) and a higher toxicity for soil organisms in 2018 when compared to 2015 (Franco et al., 2024). Nonetheless, this comparison was performed within a limited number of sites (*n* = 73). As the most recent LUCAS 2022 campaign allowed performing sample collection in the same locations as in 2015, more

Table 2 Overview of European-wide soil - pesticides assessments: main characteristics and findings in agricultural soils

	Silva et al., 2019	Vieira et al., 2023	Knuth et al., 2024
Source of samples	LUCAS 2015 survey	LUCAS 2018 survey	SPRINT 2021 survey
Number of samples	317 (11 countries, 6 crops)	3473	201 (10 countries, 8 crops)
Land uses covered	Agricultural	Mostly agricultural	Agricultural (Conventional and Organic)
n pesticides tested	76	118	192
n samples 1 or + residues	83%	75%	97% (C: 99%; O:95%)
n samples with mixtures	58%	57%	88% (C: 96%; O: 79%)
Total pesticide level (max)	2.87 mg/kg	2.4 mg/kg	C: 28.7 mg/kg; O: 5.46mg/kg
Most frequent compounds	Glyphosate, AMPA, DDE, boscalid, epoxiconazole and tebuconazole, phthalimide	AMPA, Epoxiconazole, Tebuconazole	DDE, AMPA, HCB, chlorpyrifos, glyphosate
Main outputs:	MAP on Number residues in soil (NUTS2) MAP on total pesticide content in soil (NUTS2) Mixtures composition	Mixture risk indicator (RQ = MEC vs NOEC for soil organisms) (Franco et al., 2024)	Detection vs. application records Measured vs. predicted concentrations

Published by Burleigh Dodds Science Publishing Limited, 2025.

insights on trends are expected once those pesticide results are out . In the future, the expected integration of LUCAS soil with the Soil Monitoring Law framework will allow the set up of a monitoring system for the EU, and a more systematic assessment of soil pollution in general, and of pesticide residues in particular. SoilSTAT[1], CAP Strategic Plans, and Mission Soil projects will also promote better assessments and healthier soils. Examples of such projects include PREPSOIL, which, among others, will show how different approaches to soil monitoring may support the transition to sustainable land use; ARAGORN, on a mission to combat soil contamination in Europe; EDAPHOS, whose objectives involve implementing cutting-edge technologies to monitor and rejuvenate polluted soils while accelerating their recovery; or ISLANDR, on Information-Based Strategies for Land Remediation.

2.1 Focus and findings from soil-pesticide monitoring

The monitoring of pesticides in agricultural soils has been overlooked, and most likely underestimated worldwide (Cornu et al., 2023, Hvězdová et al., 2018). Although most of the recent studies cover >100 residues, and corroborate widespread contamination (often with more than 80% of the soils containing at least one residue, and in many, none of the soils were free of pesticides), there is still a significant gap in terms of existence of a harmonized, comprehensive, and most importantly regular soil monitoring programs (Sabzevari and Hofman, 2022). In this section, we will address and highlight this gap for Europe and globally by comparing the aggregated data obtained from 72 monitoring studies published in the world through 51 years (1969–2020) with the emphasis on currently used pesticides – CUPs, where regulatory/use restriction actions are possible (Sabzevari and Hofman, 2022). Residues of recently or long-banned compounds, which are often part of the mixtures in soil, are rather addressed by remediation actions.

In Europe, the first published study on soil monitoring by CUPs goes back to 1989, which was conducted in Spain (de Barreda et al., 1991). The first published soil-CUPs dataset ever dates to an American study performed in 1969: Pesticide residue levels in soils, FY 1969-National Soils Monitoring Program (Wiersma et al., 1972). The most frequently studied pesticides in European studies are triazines, followed by organophosphates, triazoles, and pyrethroids. This prioritization from the worldwide perspective was different: organophosphates were first, following triazines, pyrethroids, and then triazoles. This variation is likely attributable to the disparities in pesticide markets and approval across different regions and countries.

1 https://www.fao.org/land-water/databases-and-software/soilstat/en/

Published by Burleigh Dodds Science Publishing Limited, 2025.

Soil is a highly heterogeneous medium, where chemical concentrations can vary significantly over small spatial scales due to factors such as soil type, land use, and topography. For analyzing a chemical or other soil health indicator in a heterogeneous matrix like soil, the sampling strategy plays a significant role in the accuracy of the monitoring results. Besides factors such as sampling pattern, season, depth, and tools, the number of samples is of great importance. For example, soil depth, most studies focus on topsoil since most pesticide residues accumulate on the soil surface. The levels of glyphosate and of its main metabolite aminomethylphosphonic acid (AMPA) can be up to 2 to 3 times higher in the top 1–2 cm of the soil surface layer than deeper in the profile (Laitinen et al., 2006; Yang et al., 2015). Presenting concentrations as average concentrations of the topsoil layer (0–15/20/30 cm) may lead to underestimations of pesticide content on the soil surface, which in turn leads to underestimations of the potential export of pesticide residues to the surrounding environment by erosion processes (Silva et al., 2018).

A higher number of samples gives, in principle, more accurate characterization of contamination status and a more reliable understanding of pesticide distribution and temporal dynamics. In Fig. 3, two graphs present

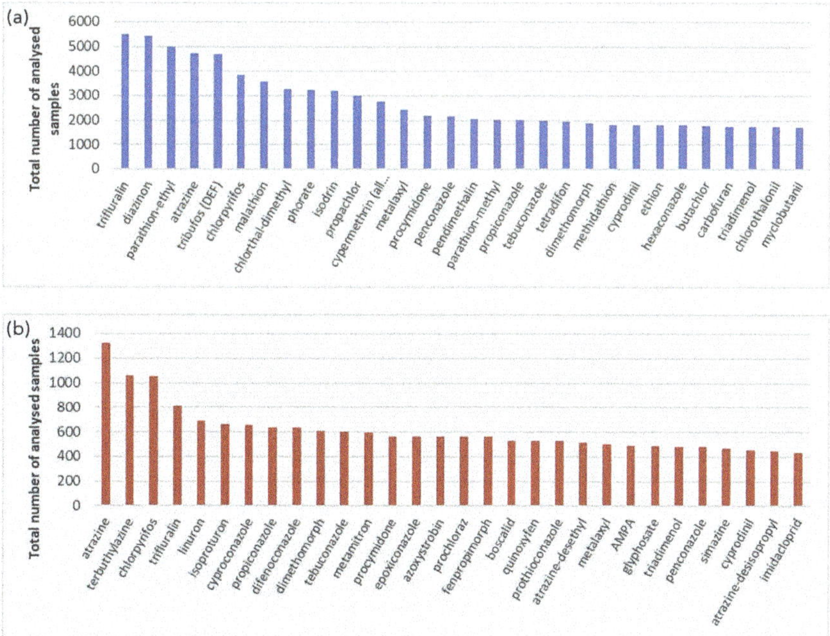

Figure 3 Top 30 pesticides analysed in a higher number of soil samples in the world (blue bars) and Europe (red bars). Original figure.

pesticides analyzed in an overall higher number of samples after aggregation of data from both worldwide and European monitoring studies. The graphs compare the focus of European studies and the studies carried out worldwide to show which pesticides drew more attention and were analyzed in a greater number of samples at both European and worldwide levels. It is worth noting that European data had a considerable share in the worldwide aggregated data due to the higher number of European studies involved in this aggregation. In Fig. 4, the total number of samples aggregated from all the studies carried out in each country is presented on a map. The highest number of samples was related to the United States, followed by South Korea, Spain, and China. Articles chosen for this aggregation were those particularly focusing on the monitoring of CUPs in agricultural soils from 1969 to 2020 (Sabzevari and Hofman, 2022). Since the primary data for aggregation were taken from separate studies done under different sampling and analysis protocols, the overall aggregated data gives a general point of view around soil-pesticide investigations/monitoring efforts, useful for finding the gaps and highlighting the necessity of performing more frequent and harmonized soil monitoring programs. Other crucial information, such as pesticide use, properties, degradation, and fate, is needed for other type of inferences

The aggregation on the detection frequency of the pesticides' occurrence in soil samples and the average concentration of the pesticide residue also gives useful information to estimate how critical the soil-pesticide concerns are . Note that none of these indicators translate automatically into risk to

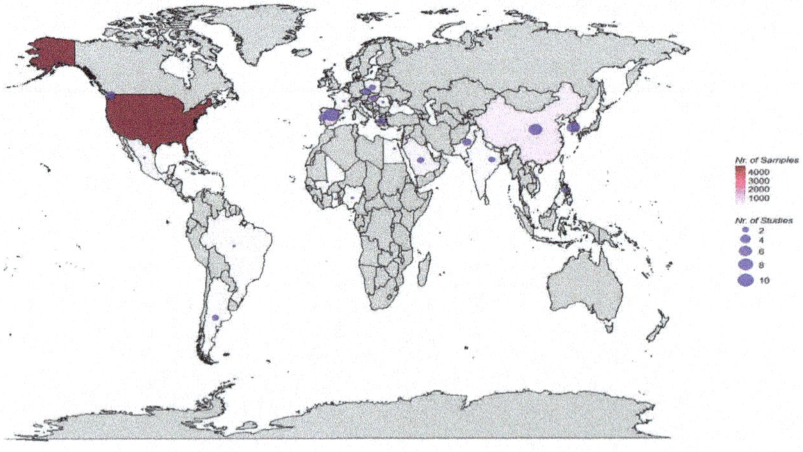

Figure 4 Overview of the number of samples and number of studies covering agricultural soil monitoring of CUPs (currently used pesticides). Covered studies are presented in Sabzevari and Hofman (2022). Original figure.

Published by Burleigh Dodds Science Publishing Limited, 2025.

soil biota and soil ecosystem.. Authorities define science-based safety standards and thresholds for different chemicals in different matrices. However, tresholds/limits for pesticides in soil are scarce and confined to a few countries. Thus, the numbers and results related to soil monitoring are useful to give a general comparative point of view to the readers rather than determining the related risks. More on the interpretation of monitoring results in section 6.3. Figures 5 and 6 present the graphs showing the pesticides with the highest overall detection frequency and overall average concentration in agricultural soil samples reported in European studies, and comparing them to their worldwide situation, respectively. Note also that the monitoring of CUP in soils points out the occurrence of mixtures of up to several dozen residues in the same sample. Specifically, the studies reported the same frequent molecules, including conazole fungicides (e.g. epoxiconazole, tebuconazole), neonicotinoids such as imidacloprid, and herbicides such as glyphosate. Some of them were recently banned due to their adverse health effects. Finally, to highlight the low number of transformation products targeted in most of the studies, corroborating a knowledge gap on their distribution and risks.

Finally, besides the previously-explored CUPs, soil also tends to present recently and long-banned compounds. The banned pesticides most frequently detected in soils are organochlorine pesticides such as triazine compounds (e.g. atrazine, simazine, and metabolites), DDT and its metabolites, lindane, or chlordecone (in tropical soils, particularly). The presence of these legacy compounds, often highly hazardous chemicals in soil, is still of much concern.

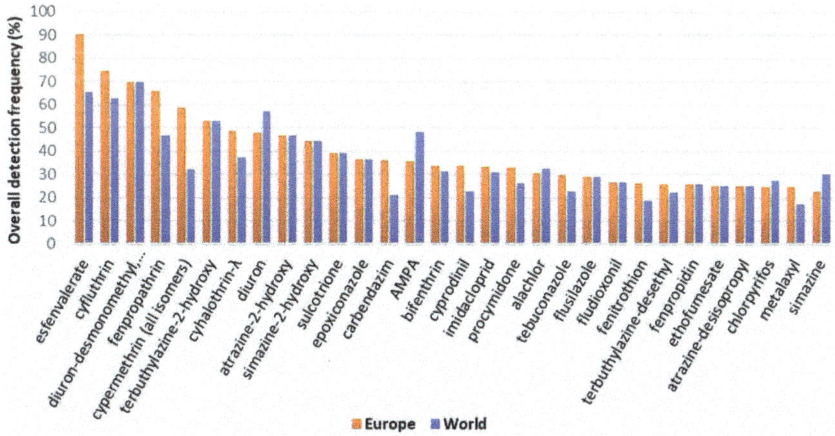

Figure 5 Top 30 pesticides showing the highest overall frequency of detection in European vs all the monitoring studies. The overall detection frequency exhibits which pesticides were detected more frequently on average in the soils, and was calculated as the average of all reported detection frequencies for each pesticide. Original figure.

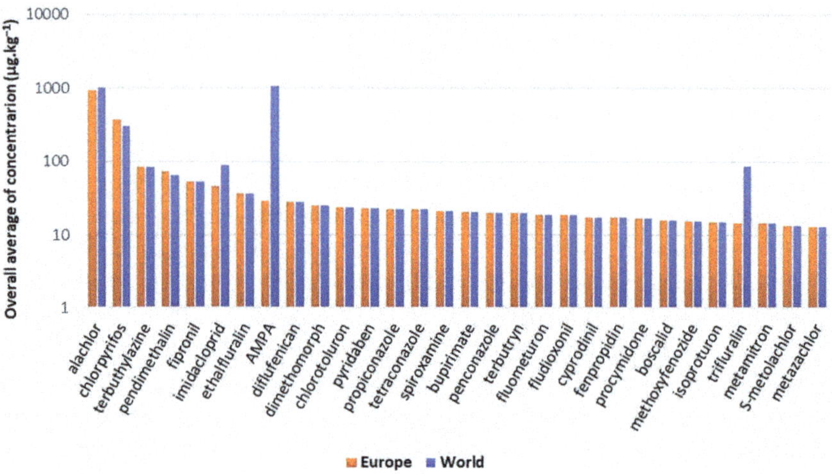

Figure 6 Top 30 pesticides showing the highest overall average of concentration in European vs all the monitoring studies. The overall average concentration was obtained by calculating the average of the concentrations reported in the studies for each pesticide. This parameter is of great value because it represents a more realistic view of pesticide residue concentration. Original figure.

This is because several studies point out their remanence and impacts on the environment, which can be enhanced by agricultural practices. Ploughing and the intensive use of herbicides such as glyphosate, leading to no soil cover, have been proven to lead to soil erosion and the release of DDT and chlordecone into rivers and lakes in France and the French West Indies (Sabatier et al., 2021, 2014).

2.2 Soil as a sink of pesticide residues

The occurrence of pesticides in agricultural soils/fields that are treated with pesticides is somewhat expected. Silva et al. (2019), the first European scale assessment on the topic, reported the presence of one or more pesticide residues in 83% of the European agricultural soils tested. Other, more recent studies corroborate these figures and the widespread contamination aspect. On the other hand, it is also expected that such pesticide residues are (fully) degraded after some time (in line with their half-life time and other pesticide degradation kinetics). Such an assumption is sometimes refuted by monitoring data, or at least the point of the real persistence of these chemicals, compared to their theoretical degradation times accounted at the homologation steps. These concerns are supported by, for example, : the presence of pesticides considered non-persistent (NP) and moderately persistent (MP) in soil samples collected 'before growing season', or by the presence of NP and MP compounds in soil under organic farming, long after conversion to organic, or in fallow lands no longer used (Froger et al., 2023). See Table 3 for an example of the differences in pesticide findings among soil collected before, during, and after the growing season. Moreover, the comparison between measurement and application-based prediction show that, rather often, there is a mismatch between these values(Froger et al., 2023; Knuth et al., 2024). A measured>predicted situation is most likely linked to a higher-than-expected persistence in soil, and possible accumulation over applications and sometimes carry-over from one year to the other. Additional off-site sources may also play a role. A measured<predicted situation is most likely linked with faster-than-expected degradation and/or transport off-site. On top of all these CUP-related scenarios and considerations, the frequent detection of recently and long-banned compounds in soil also corroborates its sink role. Their presence, sometimes long after the pesticide ban, stress (long-lasting) impacts of the placement ofhighly hazardous compounds in the market (candidates for substitution should be investigated, also accounting for this, as the impact does not stop at the time of the restriction/ban).

2.2.1 Case study – pesticide in soils managed under organic farming

In line with pesticide fate and persistence considerations mentioned previously, it is not completely surprising that pesticide residues are also found in soils where they are not used, or no longer used. Corroboration is however, limited; for instance, only a few recent studies have explored the occurrence of synthetic pesticides in soils managed under organic farming (Table 4). There are two main pathways explaining the presence of synthetic pesticides in 'organic soils' (here as synonym of soils managed under organic farming): 1) legacy effects – residues in 'organic soils' relate to previous conventional applications; 2) diffuse contamination – pesticides originatefrom adjacent conventional fields (spray drift, infiltration, runoff) or from farther locations (volatilization at source,long-range transport with subsequent atmospheric deposition (Bernasconi et al., 2021; Riedo et al., 2021). These pathways and flows bring practical and regulatory challenges, as the presence and levels of such compounds in soil and plants can be noncompliant with organic farming certification standards. Legacy contamination, especially by obsolete and very persistent pesticides, is hard to manage and has already led to critical situations for farmers where their products have been destroyed or declassified due to contamination (e.g. in 2023, the French Agency for Food, Environmental and Occupational Health&Safety installed a working group to review the environmental fate of lindane and identify concentrations in exposure media). Diffuse contamination can, however, be attenuated to a certain extent by spatially separating fields

Table 3 Overview of the pesticide residues found in European agricultural soils. Compounds are listed per class of soil persistence and marked in red when not approved at sampling time, in blue when approved at sampling time but not approved now, and green in case of approved substances. S-V and S-O = vegetable and orange production in Spain, respectively; P-G = grape production in Portugal; N-P = potato production in the Netherlands. I = insecticide, F = fungicide, H = herbicide, m=metabolite, - = not tested. Reprinted from Silva et al., 2022

		Before growing season	During growing season	After growing season		
		Geissen et al., 2021	Silva et al., 2019	Geissen et al., 2021		
		N-P, *n* = 34 (28 conv./6 org)	EU, *n* = 317 (unknown farm type)	P-G, *n* = 108 (108 conv.)	S-V, *n* = 108 (54 conv./54 org)	S-O, *n* = 90 (54 conv./36 org)
	Compound	Freq. (%)	Freq. (%)	Freq. (%)	Freq. (%)	Freq. (%)
Non-persistent	3-Hydroxycarbofuran (Im)	–	3	0	–	–
	Chloridazon (H)	36/0	–	–	–	–
	Chlorpyrifos (I)	–	1	0	6/0	7/0
	Cymoxanil (F)	0/0	2	–	0/0	–
	Cypermethrin (I)	–	–	–	22/0	–
	Fenpropimorph (F)	6/0	2	–	–	–
	Folpet (F)	–	7	4	–	0/0
	Glyphosate (H)	93/0	21	78	–	22/6
	Isoproturon (H)	–	1	–	–	0/0
	Kresoxim-methyl (F)	–	–	–	7/0	–
	Metamitron (H)	46/0	5	–	–	–
	Metolachlor (H)	18/0	–	–	–	–
	Metribuzin (H)	11/0	–	–	0/0	–
	Phthalimide (Fm)	–	19	–	–	0/0
	Propamocarb (F)	0/0	–	–	2/0	–

	Prosulfocarb (H)	100/100	–	–	–	–
	Prothioconazole (F)	7/0	7	–	–	–
	Terbuthylazine (H)	–	3	0	–	–
Moderately persistent	Atrazine (H)	–	<1	0	–	0/0
	Azoxystrobin (F)	86/0	7	4	72/0	–
	Cyflufenamid (F)	–	–	–	4/0	–
	Cyfluthrin (I)	–	–	–	2/0	–
	Cyhalothrin (I)	–	–	–	0/2	–
	Cyprodinil (F)	–	2	0	–	0/0
	Deltamethrin (I)	–	–	–	4/2	–
	Difenoconazole (F)	–	3	–	65/17	–
	Dimethomorph (F)	–	4	100	24/0	–
	Endosulfan alpha (I)	–	1	0	–	0/0
	Ethion (I)	–	0	–	–	4/0
	Linuron (H)	61/0	7	–	4/0	0/0
	Metalaxyl (F)	–	1	51	0/0	0/0
	Myclobutanil (F)	–	2	0	–	–
	Oxyfluorfen (H)	–	–	–	70/0	37/0
	Penconazole (F)	–	1	17	–	–
	Pirimicarb (I)	–	–	–	2/0	–
	Prochloraz (F)	–	2	–	–	33/17
	Propiconazole (F)	–	3	0	–	–

(*Continued*)

Table 3 (*Continued*)

		Before growing season	During growing season	After growing season		
		Geissen et al., 2021	Silva et al., 2019	Geissen et al., 2021		
		N-P, n = 34 (28 conv./6 org)	EU, n = 317 (unknown farm type)	P-G, n = 108 (108 conv.)	S-V, n = 108 (54 conv./54 org)	S-O, n = 90 (54 conv./36 org)
	Compound	Freq. (%)	Freq. (%)	Freq. (%)	Freq. (%)	Freq. (%)
	Propyzamide (H)	–	–	–	6/0	–
	Pyraclostrobin (F)	–	1	31	56/0	–
	Pyrimethanil (F)	–	–	4	–	–
	Quinoxyfen (F)	–	2	0	–	–
	Tebuconazole (F)	32/17	12	16	–	–
	Triadimenol (F)	–	<1	–	–	–
Persistent	Boscalid (F)	61/67	27	–	85/0	0/0
	Bixafen (F)	71/83	–	–	–	–
	Chlorantraniliprole (I)	–	–	–	100/7	–
	Cyproconazole (F)	–	2	–	–	–
	Epoxiconazole (F)	–	24	–	–	–
	Fluopicolide (F)	79/0	–	–	48/0	–
	Heptachlor (I)	–	<1	0	–	–
	Imidacloprid (I)	–	7	–	93/30	–
	Indoxacarb (I)	–	–	–	33/0	–
	Metrafenone (F)	–	–	–	56/0	–
	Pendimethalin (H)	43/0	–	–	63/0	–

	Procymidone (F)	–	<1	0	–	–
Very persistent	AMPA (Hm)	96/83	42	83	–	87/17
	Chlordane (I)	–	<1	0	–	–
	DDD op (Im)	0/0	7	0	–	0/3
	DDD pp (Im)	32/17	3	0	–	13/44
	DDE op (Im)	0/0	<1	0	–	0/0
	DDE pp (Im)	68/83	23	0	–	78/89
	DDT op (I)	0/0	<1	0	–	0/8
	DDT pp (I)	14/0	7	0	–	19/56
	Dieldrin (I)	–	5	11	–	–
	Hexachlorobenzene (F)	–	<1	0	–	–

Table 4 Overview of studies assessing synthetic pesticides in organic soils

Reference	Number of samples	Land use	Country	Concentration [ug/kg]	Years since organics	Number/type of pesticides tested
Knuth et al. 2024	10	Fruits	Switzerland	0.8–198	>=5	192 pesticide, 39 metabolites, 1 synergist
	11	Oilseeds	Czechia	51.6–5458		
	10	Cereals	Denmark	0–302.9		
	10	Broccoli	Spain	2.8–412		
	7	Grapes	France	3.6–79.7		
	10	Olives	Croatia	0–59.2		
	10	Vegetables	Italy	4–1377		
	10	Potatoes	Netherlands	6.9–205.7		
	12	Grapes	Portugal	0–218.2		
	10	Maize	Slovenia	0–116.3		
Geissen et al. 2021	6	Potatoes	Netherlands	<200	>10	36 pesticides
	54	Vegetable farms	Spain	<200	>5	38 pesticides
	36	Orange orchards	Spain	Max. 600	>5	75 pesticides
Panis et al. 2022	2	Farmland	France	0.6–6.5	Not available	31 pesticides
Froger et al. 2023	2	Farmland	France	12.4–14.5	1 and 9 years	110 pesticides and one metabolite (fungicides, herbicides, insecticides, acaricides, safeners)
Riedo et al. 2021	40	Farmland	Switzerland	0.05–90.31	3–29	38 pesticides and 8 metabolites
Humann-guilleminot et al. 2019	27 + 24	Farmland and ecological focus areas	Switzerland	Max 0.17	>10	five neonicotinoids (imidacloprid, clothianidin, thiamethoxam, thiacloprid, acetamiprid)

Published by Burleigh Dodds Science Publishing Limited, 2025.

Karanasios et al., 2018	13 + 22	Olive farms and non-treated sites	Greece	13–54	Not available	glyphosate and AMPA
De Geronimo et al. 2015	8	Experimental fields with no agricultural activity over past years	Argentina	20–140	Few years	18 pesticides
Ghabbour et al. 2012	2	Strawberry farms	Egypt	4–48	Not available	86 pesticides
Muendo et al. 2012	7	Experimental field plots left fallow for various periods	Kenya	0.05–8.25	3 months to 8 years	glyphosate, metribuzin, 2, 4-D, lindane, hexazinone, atrazine, diuron, alachlor, permethrin, DDT, aldrin, endosulfan, dieldrin, endrin

managed under organic farming from fields of conventional/integrated pest management farming and clustering the organically managed fields at the landscape scale (Larsen et al., 2024).

On the 'organic soils' data, Riedo et al. (2021) found several residues in organic soils, most of which could be explained by past conventional management and the persistence of the compounds much longer than expected. Other studies have found similar results, with contamination of AMPA, imidacloprid, clothianidin, thiacloprid, and acetamiprid in fields that had not been applied for more than ten years (Geissen et al., 2021, Humann-Guilleminot et al., 2019). These results suggest that the persistence of pesticides under field conditions, including potential cumulative contamination from repeated applications, has been underestimated. Moreover, organic and broader soil monitoring data corroboratediffuse contamination challenges. Not only have pesticide residues been found in areas such as forests and grasslands where no pesticides have been applied (Froger et al., 2023, Riedo et al., 2021) ascalculations have shown that the presence of some pesticides in soil, such as metalaxyl, S-metolachlor, terbuthylazine, and pirimicarb, could potentially be explained by atmospheric inputs (Riedo et al., 2021). The dissemination of dust from the sowing of insecticide-coated seeds (Humann-Guilleminot et al., 2019) and runoff from the washing of spray equipment used on nearby fields (Karanasios et al., 2018) have also being suggested as possible causes for the presence of CUPs in organic soils. Off-site contamination can be diminished through alterations in agricultural methodologies, such as the implementation of precision agriculture technologies or the establishment of buffer zones (Khan et al., 2024).

The number and concentration of pesticide residues in organically managed soils are typically (much) lower than in conventionally managed soils. Riedo et al. (2021) found that 'organic soils' had 50% fewer pesticide residues and a median concentration that was 85% lower than in conventional soils. However, they still found up to 32 different residues in 'organic soils'. Similarly, Geissen et al. (2021) observed 30% fewer pesticide residues in 'organic soils', with 70–90% lower concentrations than in conventional soils. Panis et al. (2022) also reported 50% fewer pesticide residues at lower concentrations (95% less) in 'organic soils' compared to conventional soils. Knuth et al. (2024) found mixtures of up to 12 residues in 'organic soil' samples, 60% less than in conventional samples. Froger et al. (2023) also found 60% fewer residues and concentrations that were as much as 90% lower than the conventional sites. The number and type of substances found in soil, regardless of farming system, depend strongly on the scope of the studies and the analytical method used. Nevertheless, all these studies stress that 'organic soils' are often not pesticide-free. Two remarks on the samples/sites that come up 'clean': (i) all studies with data on 'organic soils' (like most of on conventional managed soils) used a targeted list of analytes, so although unlikely other non-tested synthetic compounds may exist in the soil; and (ii) the presence and accumulation of copper in the soil as a result of copper-based fungicides (and of other pesticides allowed in organic farming) is highly probable. However, this topic is beyond the scope of this book chapter and would require further investigation, considering the physiological implications of the level of this micronutrient on crop growth.

2.3 Soils as a source of pesticide residues in the environment

Pesticide residues are released from soil into the atmosphere, surface water, groundwater, and biota via multiple pathways (Karasali et al., 2016), potentially reaching other fields, rural, urban, natural, and protected areas (Brühl et al., 2024). The most recent global land budget of pesticide use indicates that about 82% of agricultural pesticides break down into a cascade of by-products, about 10% stay in the topsoil (being prone to mobilization by erosion and transport to surface waters and oceans), and about 7% leaches to deep soil, causing aquifer contamination (Maggi et al., 2023). As highlighted by this equation, waterbodies, especially those located in intensive agricultural landscapes, are particularly vulnerable to pesticide contamination. Monitoring data corroborates the concerns and a decrease in water quality (Wolfram et al., 2021). Regulations have been adapted, and not only must the risk to waterbodies be assessed before approval in the market, as in the 1990s several highly water-soluble pesticides, like atrazine and simazine, were forbidden in the European market. Water contamination by pesticides is regularly monitored and tabulated against official thresholds. The Water Framework Directive (WFD; EC, 2000) sets environmental quality standards for some pesticides in surface water. The drinking water directive (EU, 2020) sets precautionary quality standards of 0.1 μg/L and 0.5 ug/L for individual and total pesticide content, respectively. In the upcoming sub-sections, we explore the role of water and wind erosion, and leaching – the main drivers of pesticide off-site contamination, alongside drift. Drift is not covered in this chapter, as drift reduction is more related to technological developments than land management and policies. See, for instance, de Jong et al. (2008) on the impacts of drift on terrestrial habitats, Khan et al. (2024) on drift reduction measures, and Fornasiero et al. (2017) on the impact of these reduction measures on pests, closing the cycle.

2.4 Transport of pesticides by water erosion

Discussion around the implications of frequent detection of pesticide residues in ground and surface water bodies started in the 80s (e.g. Holden 1986), culminating in the ban of several highly water-soluble pesticides from the European market in the 90s. Over a long time, pesticide transport from soils to water bodies was only considered to take place in the solute phase, and pesticides adsorbed to soil particles were assumed to stay in the agricultural fields. However, recent monitoring data suggest a high relevance of particulate-phase transport. Commelin et al. (2022), for instance, observed that, in a hilly Loess region in the Netherlands, 55% of the total off-site transport of pesticides is related to particulate transport, and that eroded sediments are enriched in pesticide residues compared to soil. This is because most pesticide residues adsorb to organic matter and the soil clay fraction – both highly prone to erosion. Several other studies corroborate high concentrations of pesticide residues in runoff water and eroded sediment after rainfall and runoff peaks, either via flume experiments (Yang et al., 2015), field erosion plots, or watershed monitoring (Peruzzo et al., 2008; Climent et al., 2019). Whereas the transport of pesticides in the solute phase is accounted for in FOCUS (Forum for Coordination of pesticide fate models and their Use; e.g. Pereira et al., 2017), the particulate transport process is not fully covered in the pre-market regulatory assessment. Therefore, the prediction of total off-site transport, namely in theeroded sediments reaching and accumulating at the bottom of of surface water bodies might have been underestimated by regulatory authorities. It becomes even more concerning when considering the slow degradation of pesticide residues in such sediments, under anaerobic and dark conditions, leading to their accumulation and potential later release (sediments are also a sink and a source of contaminants). Khurshid et al. (2024) describe the omnipresence of pesticide residue in European waterbed sediments, collected and analysed under the SPRINT project umbrella (on Sustainable Plant Protection Transition; Sprint-h2020.eu). These add to the findings of Chiaia-Hernandez et al. (2020), or to those of Gardes et al (2021), who detected post-ban releases of organochlorine pesticides from sediments of French water bodies. Sediment transport and sediment quality are of increasing interest in the WFD, as shown in the Common Implementation Strategy on 'Integrated sediment management guidelines and good practices in the context of the Water Framework Directive'.

2.5 Transport of pesticides by wind erosion

The fact that pesticide residues adsorbed to soil particles are transported into the atmosphere by wind erosion has been pointed out in the 80s, for DDTs. Nowadays, there is multiple evidence on this process, and airborne transport of pesticides is recognized as an important off-site pathway (Boonupara et al., 2023), and a major challenge for transboundary regulations. Bento et al. (2017), for instance, observed, in a wind tunnel experiment, high glyphosate and AMPA contents in the smallest soil eroded fractions (PM10 and less), which are easily inhaled and, therefore, may contribute to animal and human exposure to pesticides. Aparicio et al. (2018) also showed the enrichment of glyphosate and AMPA in wind-eroded material in Argentina and alerted to the potential of long-distance transport. Csányi and Farsang (2022) showed high enrichment factors for other pesticides (pendimethalin and chlorpyrifos) in the sediments transported by wind erosion compared to the content in soils.

Recent studies conducted in Germany (Kruse-Plaß et al., 2021) and across Europe, using passive air samplers (Santos et al. in prep.) show the presence of multiple pesticide residues in the atmosphere, including DDTs and other obsolete compounds that cannot be explained by drift, and that are of low volatility. A targeted study (Debler et al., 2024), conducted in Portugal and the Netherlands, which used high-volume active samplers, brings additional insights into this transport pathway. The authors reported a higher number of substances in the particulate than gaseous phase, and acknowledged detections outside the pesticide application period and the presence of non-approved pesticides which are likely resulting from emissions from soil/ erosion of contaminated soil particles. Additional evidence originates from the recent studies that detected a high number of (non-approved) pesticides in indoor dust of farmhouses, but also in houses of organic farmers and consumers living > 250 m from the fields where pesticides are applied (Navarro et al., 2023; Aparicio et al. in prep). Humans may be exposed to these pesticides through dust inhalation or ingestion, but the relative contribution to the overall exposure is unknown, and further research is needed on this.

2.6 Transport of pesticides by leaching

Numerous studies reveal the leaching of pesticides through the soil profile, often reaching groundwater. In Finland, for example, about 15% of groundwater samples situated in intensive agricultural areas contained pesticides and exceeded quality standards for pesticide content (FAO and UNEP, 2021). A sampling campaign in Switzerland showed that groundwater samples contained between 4 and 44 pesticides/metabolites. The metabolite chlorothalonil was the most frequently detected and was present in all groundwater samples, even in aquifers with low anthropogenic impact (Kiefer et al., 2019). These findings are aligned with European ones, where pesticides have been detected above the quality thresholds at 4–11% of groundwater monitoring sites (2013–2021 period). Exceedances were mostly caused by atrazine and its metabolites and bentazone (EEA, 2024). Note also that karst aquifers and springs are highly susceptible to contamination due to high permeability. Schorr et al. (2024), for instance, detected several pesticide residues in karst groundwater (in two agriculturally impacted Swiss karst springs), including compounds of low leachability, after rain events. Evidence of leaching and concerns go beyond Europe. Bexfield et al. (2020), for instance, investigated the presence of pesticide residues in groundwater used for public supply across the United States, and their human health implications. Panis et al. (2022) did something similar for Brazil, with a focus on cancer risk. Grondona et al. (2023) report on pesticides in aquifers from Latin America and the Caribbean. Affum et al. (2018) and Sishu et al. (2022) provide some of the first data on sub-Saharan Africa, Lari et al. (2014) do it for India, Shakerkhatibi et al. (2014) in Iran, etc.

To accurately assess the risks and impacts of pesticide pollution in the groundwater, it is important to conduct leaching tests under conditions that closely resemble those in the actual field or the intended end-use environment. A few leaching tests, including column experiments, batch tests, tank tests, and extractions, are available worldwide to assess the potential for pesticides to migrate from soil to groundwater. Two of the most used ones, including in Germany (DIN EN 12457–4, 2003), the United Arab Emirates (Environment Agency–Abu Dhabi, 2018), Australia (National Environmental Protection, 2013), and New Zealand (Contaminated Land Management Guidelines, 2024), are batch leaching tests (also known as static extraction tests), which are mostly used for inorganic contaminants, and dynamic leaching tests (such as column and diffusion-based tests), which are used for organic contaminants. The dynamic tests can simulate the continuous flow of water through soil, providing a more realistic assessment of contaminant leaching over time.

Comparing the results of different leaching tests can be challenging due to varying equilibrium conditions and test limitations. The main limitations are: (i) leaching tests often fail to reflect natural conditions because of lower liquid-to-solid ratios in the field. These tests are typically conducted under saturated conditions because chemical analysis requires a considerable volume of liquid, ranging from several to hundreds of milliliters; (ii) in unsaturated field conditions, materials are subject to neutralization by CO_2 and oxidation by O_2, which can influence leaching behavior; and (iii) many protocols do not require pH to be documented. Since pH is a major factor determining the extent of leaching, failing to report it makes it difficult to conduct a reliable assessment of contaminant release (van der Sloot et al., 2003).

Jurisdictions in the European Union and the United States of America have recognized the limitations of single-condition leaching tests and have progressively replaced these tests with several interlinked test methods that provide additional information about how leaching takes place from a solid material under a range of environmental conditions. New tests to determine how leaching varies with the pH and the liquid-to-solid ratio were developed in the EU as part of a harmonization process of test methods across member countries. U.S. Environmental Protection Agency adopted and updated these tests to form part of a new leaching environmental assessment framework consisting of four tests:

- Test Method 1313, which determines how liquid-solid partitioning varies with the pH of the leaching solution using a parallel batch extraction procedure;
- Test Method 1314, which determines how liquid-solid portioning varies with varying liquid-to-solid ratios using an up-flow percolation column procedure;
- Test Method 1315, which determines mass transfer rates of chemical constituents in leachate from monolithic and compacted granular materials (e.g. construction materials) using a semi-dynamic tank leaching procedure; and
- Test Method 1316, which determines how liquid-solid partitioning varies with the liquid-to-solid ratio using a parallel batch extraction procedure.

3 Trends and developments in analyzing pesticides in soil

Alongside the growing need for environmental protection, the methods of analysing pesticides in soil have evolved over recent years. These developments have led to more accurate, sensitive, and comprehensive monitoring of pesticide residues. Such technical breakthroughs in chemical analysis allowed the quantification of hundreds of pesticides at a more affordable price (Chiaia-Hernández et al., 2020; Ottenbros et al., 2023), and most of the most recent studies on soil contamination by pesticide residues included > 100 molecules (Table 1). Such developments are somehow (expected to be) reflected in the figures reported at the 'State of Soils in Europe Report 2024' (Akca et al., 2024), launched during the 4th EU Soil Observatory (EUSO) Stakeholder Forum on 22 October 2024, and in the upcoming FAO & ITPS (Food and Agriculture Organization & Intergovernmental Technical Panel on Soils) report on the status of the world's soil resources. FAO & ITPS (2015) provided the first global assessment of the status of the world's soil resources in 2015, and they are expected to deliver an update in 2025, providing a comprehensive assessment of how the world's soils have changed over the past 10 years.

3.1 Pesticide extraction

One of the most common methods used for the extraction of pesticides from soil is the solid-liquid extraction. This typically involves mixing the soil with a solvent, followed by agitation to facilitate the transfer of pesticide residues from the soil matrix into the liquid phase. Its adaptability to different types of pesticides is achieved by selecting appropriate solvents and fine-tuning the extraction conditions. Methanol is a commonly chosen solvent because of its ability to dissolve a wide range of organic compounds, including both polar and non-polar pesticides. Following the extraction process, the concentrated extract is analyzed using methods such as chromatography or spectrometry. This extraction method is specifically used when high-performance liquid chromatography (HPLC) with Diode Array Detector is used for the determination. Despite its effectiveness, this method has several drawbacks. It can be time-consuming, labor-intensive, and costly, which may limit its efficiency in large-scale or high-throughput analyses with a sort of subsequent shaking process (Andreu and Picó, 2004; Brinco et al., 2023).

Another common and widely accepted extraction method is the QuEChERS (Quick, Easy, Cheap, Effective, Rugged and Safe; Perestrelo et al., 2019). It was originally developed for the extraction of pesticides from fruits and vegetables, which has been widely accepted for scientific and environmental research, including for soils. QuEChERS is widely popular due to its low cost, low equipment requirements, rapid and efficient functionality, as well as its robust performance in extracting a broad spectrum of pesticides. It involves a single-step extraction using acetonitrile, combined with salting-out liquid-liquid partitioning using magnesium sulfate ($MgSO_4$) to separate the aqueous phase from the sample. A dispersive solid-phase extraction clean-up is then performed to remove impurities such as organic acids, excess water, and other interfering substances, utilizing a combination of primary and secondary amine sorbent and $MgSO_4$. The resulting extracts are then subjected to mass spectrometry and chromatographic separation for precise analysis of pesticide residues. Studies indicate that QuEChERS outperforms other methods for the analysis of multi-residue pesticides in soil samples (Perestrelo et al., 2019; Calatayud-Vernich, et al., 2016). The main disadvantage of the QuEChERS method is its lower efficiency for polar pesticides (Brinco et al., 2023; Bruzzoniti et al., 2014).

3.2 Pesticide detection

The traditional methods for the detection of pesticides, which have been used for about a century, are chromatographic-based methods. These methods, such as gas chromatography (GC) and HPLC coupled with mass spectrometry (MS), remain the most used techniques for pesticide analysis in environmental and food studies. Their continued popularity stems from their acceptable sensitivity and the accuracy, precision, and reliability of the results produced. When the compound being analyzed is non-polar and highly volatile, GC is the preferred method of analysis. For pesticides that are polar, non-volatile, or heat-sensitive, liquid chromatography (LC) is the more appropriate choice. As detectors, numerous alternatives to MS are available, such as electron capture detectors, or nitrogen and phosphorus detectors. While these detectors are highly capable of detecting specific pesticides, MS, with its ongoing development and advancements, stands out as a more versatile technology, offering greater sensitivity and selectivity for pesticide analysis. In addition, the development of high-resolution mass spectrometry (HRMS) techniques, such as Time-of-Flight and Orbitrap, has opened new avenues for more advanced analytical approaches in non-targeted screening, as HRMS techniques have become increasingly prevalent today (Li et al., 2021; Samsidar et al., 2018).

4 Pesticide regulation and monitoring in the European Union (EU)

In the European Union, Directive 2009/128/EC sets a framework for the sustainable use of pesticides (EU, 2009a), as the main regulation to control pesticide emission to the environment. This directive paved the way for the sales decline of the most hazardous substances (Eurostat, 2024a), but also allowed to assess an increasing number of derogations of non-authorized substances during the same period (Eurostat, 2024b). However, according to the Court of Auditors, both the pesticide statistics and the risk indicators present in such legislation require improvement to better assess policy progress (ECA, 2020). Silva et al. (2023) recently proposed a pesticide prioritization indicator (PPI) that follows a rationale similar to the European Comission indicators but that accounts for the occurrence and levels of pesticides in environmental matrices. PPI is also applicable to pesticide transformation products', which are a substantial part of mixtures found in soil, and other environmental and biological matrices, and accounts for individual pesticide characteristics. Hazardous characteristics vary strongly among pesticides of the same aproval status , corroborating some existing concerns around the EC indicators, that using a single group weighting factor can lead to misleading assessments. Other pesticide prioritization indicators have been developed over the last few years, that differ in the prioritization strategies and applications – see Vryzas et al. (2020) for an overview.

The recent proposal for a Soil Monitoring and Resilience Directive (SML) formalized the ambition of the European Commission towards a harmonized soil information system in the EU, given the need to identify and assess the impacts of soil pollution, and ultimately plan soil restoration. To do so, the SML provides a set of criteria for healthy soils and sets the ambition towards healthy soils in the EU by 2050. The SML approaches point source pollution and diffuse pollution differently, while on contaminated sites, a risk-based approach is applied; the descriptors for diffuse pollution are in line with the remaining descriptors for soil health (Vieira et al., 2024). The SML trialogue negotiations are ongoing, with a decision to be made whether (certain) pesticides should be monitored. The EC proposal does not explicitly require Member States to monitor pesticides in soil, and no threshold or listing of target substances is provided. Nevertheless, the need for their monitoring and for such a list is acknowledged by the Parliament position, and a connection between the SML and the WFD is mentioned for the soil contamination descriptor, as the WFD includes a watchlist mechanism listing substances under the priority list and watchlist.

The SML adds and follows up on existing soil and pesticide-related legal instruments. See SoilLEX[2] for an overview of soil instruments, worldwide (covering national, subnational, and local initiatives), Smith et al. (2024) on the policy frameworks and initiatives to improve the status of the world's soils, and finally Table 5 on pesticide-related conventions, codes, and legal frameworks. These all together support broader United Nations and International Union for Conservation of Nature goals, especially those around Land degradation neutrality (Sustainable Development Goal 15.3), and ecosystem restoration. At the same time, and more pesticide-focused, it is important to be aware of the existence of lock-ins and barriers towards the adoption of more sustainable plant protection practices (Frelih-Larsen et al., 2023), and of the factors limiting the effectiveness of policies aimed at reducing chemical inputs in agriculture (Brunelle et al., 2024). Brunelle et al. (2024) propose an approach for reducing chemical inputs in agriculture (Fig. 7), and stress that the 'combination of different policy instruments, such as standards, taxes, and subsidies, in a simplified and coherent way is required to increase effectiveness and ensure better coordination in the adoption of sustainable practices'.

5 Managing risks from pesticides in agricultural soils

There are multiple regulations in force to reduce the potential risks of pesticides to humans and the environment. As an example, this section summarizes the strict regulations in force in Europe. There are two pillars in Europe aiming to reduce the risks and impacts of pesticide use on human health and the environment. One pillar – the 2009/128/ECSUD – focuses on the sustainable use of pesticides by reducing the risks linked to their use, promoting alternative agricultural practices, encouraging the minimization of pesticide use, and ensuring their responsible use by obliging Member States to adopt National Action Plans including measurable reduction targets (EU, 2009a). Based on the reduction targets for pesticide use set in the Farm to Fork strategy (EC, 2020a), the SUD should have been updated to the Sustainable Use Regulation (EC, 2020b). However, its implementation failed within the European Parliament due to a combination of economic, food security, and political concerns, particularly around the perceived risks to farmers, food production, and agricultural competitiveness.

2 https://www.fao.org/soils-portal/soilex/soil-keywords/soil-monitoring/en/&https://www.fao.org/soils-portal/soilex/soil-keywords/soil-pollution/en/

Table 5 Main pesticide-related conventions, codes, and legal instruments. Reprinted from Silva et al., 2022

International conventions and agreements	
Basel Convention on the Control of Transboundary Movements of Hazardous Wastes and Their Disposals (adopted in 1989)	**Purpose**: reduce of hazardous waste generation and the promotion of environmentally sound management of hazardous wastes; restriction of transboundary movements of hazardous wastes.
Rotterdam Convention on Prior Informed Consent Procedure for Certain Hazardous Chemicals and Pesticides in International Trade (adopted in 1998)	**Purpose**: promote shared responsibility and cooperative efforts in the international trade of certain hazardous chemicals; facilitate information exchange about characteristics of such chemicals, for the decision-making process on import and export decisions.
Stockholm Convention on Persistent Organic Pollutants (adopted in 2001)	**Purpose**: restrict or eliminate the production and use of persistent organic pollutants (POPs); stockpiles and wastes consisting of or contaminated by POPs are managed in an environmentally sound manner.
International Code of Conduct on the Distribution and Use of Pesticides (adopted in 1985)	**Purpose**: guidance on pesticide management for public and private entities linked with the distribution and use of pesticides.
Intergovernmental Forum on Chemical Safety (established in 1994)	**Purpose**: promotion of chemical risk assessment and the environmentally sound management of chemicals.
International Code of Conduct on Pesticide Management (adopted in 2013)	**Purpose**: regulate and control the quality and suitability of pesticide products; ensure that pesticides are used effectively and efficiently in a sustainable manner to minimize adverse effects on human health and the environment while contributing to the sustainable improvement of agriculture.
Codex Maximum Residue Limits (adopted in 2017)	**Purpose**: establish internationally agreed food standards covering pesticide residues in or on food and feed.
EU legal frameworks	
Regulation (EC) No 1107/2009 of the European Parliament and of the Council of 21 October 2009 concerning the placing of plant protection products on the market. (consolidated version: 27/03/2021)	**Purpose**: ensure a high level of protection of both human and animal health and the environment, and at the same time to safeguard the competitiveness of Community agriculture.
Regulation (EC) No 396/2005 of the European Parliament and of the Council of 23 February 2005 on maximum residue levels of pesticides in or on food and feed of plant and animal origin (consolidated version: 10/10/2021).	**Purpose**: ensure a high level of consumer protection and harmonized Community provisions relating to maximum levels of pesticide residues in or on food and feed of plant and animal origin.

(*Continued*)

Table 5 (*Continued*)

EU legal frameworks	
Directive 2009/128/EC of the European Parliament and of the Council of 21 October 2009 establishing a framework for Community action to achieve the sustainable use of pesticides (SUD) (consolidated version: 26/07/2019)	**Purpose**: reduce the risks and impacts of pesticide use on human health and the environment, and promote the use of Integrated Pest Management (IPM) and of alternative approaches or techniques such as non-chemical alternatives to pesticides.
Regulation (EU) No 528/2012 of the European Parliament and of the Council of 22 May 2012 concerning the making available on the market and use of biocidal products (consolidated version: 10/06/2021)	**Purpose**: improve the functioning of the internal market through the harmonization of the rules on the making available on the market and the use of biocidal products, whilst ensuring a high level of protection of both human and animal health and the environment.
Regulation (EC) No 1185/2009 of the European Parliament and of the Council of 25 November 2009 Concerning Statistics on Pesticides (consolidated version: 09/03/2017)	**Purpose**: establish a common framework for the systematic production of Community statistics on the placing on the market and use of those plant protection products.
Directive 2009/127/EC of the European Parliament and of the Council of 21 October 2009 amending Directive 2006/42/EC with regard to machinery for pesticide application.	**Purpose**: introduce requirements for the inspection and maintenance to be carried out on machinery for pesticide application.
Regulation (EC) No 1272/2008 of the European Parliament and of the Council of 16 December 2008 on classification, labelling and packaging of substances and mixtures (consolidated version: 01/10/2021)	**Purpose**: ensure a high level of protection of human health and the environment, as well as the free movement of substances, mixtures, and articles.
Regulation (EU) 2018/848 of the European Parliament and of the Council of 30 May 2018 on organic production and labelling of organic products (consolidated version: 14/11/2020).	**Purpose**: establishes the principles of organic production and lays down the rules concerning organic production, related certification, and the use of indications referring to organic production in labelling and advertising, as well as rules on controls additional to those laid down in Regulation (EU) 2017/625.
EU Action Plan: 'Towards Zero Pollution for Air, Water and Soil' (adopted in 2021)	**Purpose**: reduce air, water, and soil pollution to not harmful levels, thus creating a toxic-free environment, by 2050.
A Farm to Fork Strategy for a fair, healthy and environmentally-friendly food system (COM/2020/381 final)	**Purpose**: reduce pollution from pesticides in air, water, and soil by cutting by 50% their overall use and risk, including the most hazardous ones, by 2030. Other goals focus on the excess of nutrients, antimicrobial resistance, and organic farming area.
Chemicals Strategy for Sustainability (COM(2020) 667 final)	**Purpose**: better protect citizens and the environment, and boost innovation for safe and sustainable chemicals.

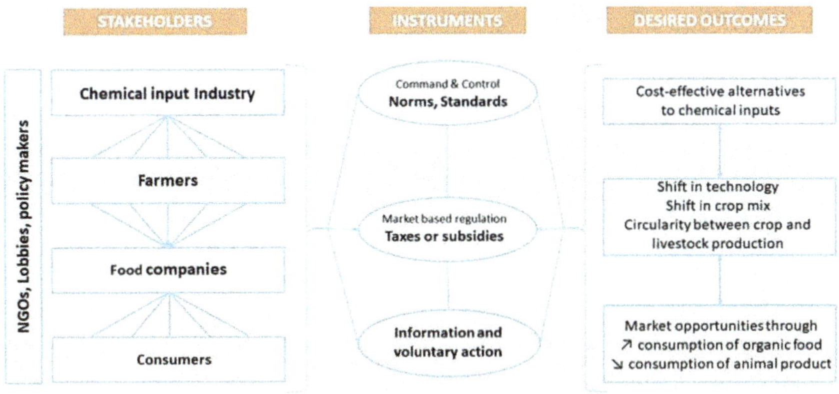

Figure 7 This diagram, reprinted from Brunelle et. al (2024), provides a summary of the approach proposed by these authors for reducing chemical inputs in agriculture. Their approach involves all stakeholders in the chain value, combining different public policy instruments to enhance their effectiveness, and sets precise targets for each stakeholder in the supply chain (desired outcome).

The second pillar – the EC Regulation No 1107/2009 – focuses on the approval process of pesticides. Before pesticides are placed on the European Union market, a comprehensive risk assessment is conducted, which aims to exclude unacceptable effects on humans as well as the environment (EC, 2011). The risk assessment within the authorization process for active substances as well as comercial products follows a pragmatic approach, aiming to characterize effects, exposure, and risks as simply as possible but as detailed as necessary, whereby effects on humans or the environment are assessed for the applied use of an active substance or a plant protection product. The evaluation process includes a large set of studies and risk assessment frameworks, which are used to predict the expected effects on humans and the environment by the application of pesticides in the environment (EU, 2009b). At the same time, several studies show an alarming declining trend for non-target organisms (FAO, 2020a; IPBES, 2018; UNEP, 2022), and exposure to pesticides is mentioned as one of the main reasons. Recent targeted studies confirm that pesticide residues occurring in the soil can cause harm to in-soil organisms (Froger et al., 2023; Cao et al., 2024; Franco et al., 2024; Meidl et al., 2024), and may result in long-term effects on communities (Beaumelle et al., 2023). Note that soils host 59% of the global biodiversity (Anthony et al., 2023) and provide numerous ecosystem services, such as carbon storage or water regulation (Adhikari and Hartemink, 2016). Deleterious effects and concerns go beyond soil. Indeed, pesticides, along with climate change, have been identified as a leading cause of global biodiversity loss, responsible for the decline of insects and pollinators (Janousek et al., 2023; Sánchez-Bayo and

Wyckhuys, 2019), birds (Stanton et al., 2018), or aquatic organisms (Navarro et al., 2024).

The question arises as to why, despite a comprehensive risk assessment framework being in place, multiple pesticide threats to in-soil and other non-target organisms are identified in Europe. The answer to this question is simple: the European current risk assessment framework needs to be updated. For in-soil scenarios, for example, only three standard species -earthworms, collembolans, and mites - are tested (whose representativeness is questioned), in artificial soils (which might not be suitable to address effects on wild living species, as well as communities of in-soil organisms occurring in agricultural soils). Several studies show differences in toxicity towards test organisms in the test outcome if natural soils are tested instead of artificial standard soils (e.g. Römbke and Sousa et al., 2017; Kotschik et al., 2018). Moreover, the current risk assessment for in-soil organisms neglects the occurrence of mixtures and interactions, expected to occur in soil by tank mixtures and spray series, and supported by monitoring and ecotoxicological data; and misses the calibration of data produced in the laboratory on standard test organisms and standard soils compared to occurring effects on in-soil communities in natural soils. The analysis of the current framework has not only identified some gaps and shortcomings but also stressed the need for a shift towards a system-based approach (Sousa et al., 2022). Responsible authorities have acknowledged the need to update the guidance document for in-soil organisms exposed to pesticides and initiated an update process of the current guidance document (EFSA, 2017, 2024; EC, 2024). PERA (Advancing the ERA of Plant Protection Products towards a system-based approach) and PARC (Partnership for the Assessment of Risks from Chemicals) are two outstanding initiatives in this regard. Research projects like H2020 SPRINT are already applying some of the improvement recommendations, testing mixtures, standard and native species, at mono and multi-species setups, and looking into a broader set of effect indicators. Preliminary results indicate a higher risk when using this more comprehensive and realistic conditions approach compared to the risk derived from the application of the current regulatory procedure. This raises alarming concerns around the implications of possible and likely recurrent underestimation of individual and combined risks of pesticides.

6 Key challenges in monitoring and assessing risks from pesticide residues in agricultural soils

6.1 Fragmented assessments

Most of the existing soil-pesticide studies advocate for a large, harmonized monitoring program, ideally covering different land uses and a greater number

Published by Burleigh Dodds Science Publishing Limited, 2025.

of residues. Such a program would be a step forward in the post-registration control (Streissl et al., 2018) and would allow the timely identification of potential risks caused by these substances. Such a program could be associated with existing soil monitoring efforts, taking advantage of the sampling but also other meaningful information available, such as agricultural practices and pesticide application records. Monitoring of pesticides based on existing soil monitoring programs has been partly done in Switzerland (Chiaia-Hernandez et al., 2017), France (Froger et al., 2023), with the first national maps of lindane in Orton et al. (2013), and in a subset of LUCAS Soil in Europe (Silva et al., 2019; Vieira et al., 2023). Doing so saves time and money, but sampling, sample preparation and conservation protocols may need to be adapted or expanded for this prupose, e.g. by sampling less in-depth to avoid the dilution of the pesticides in the soil matrix, and freeze-drying (part) of the samples to reduce volatilization and microbial degradation. Each of those programs had a specific sampling design, set based on the main campaign purpose, and a specific target list, often based on regional challenges and interests. The ongoing construction of the future Directive on Soil Monitoring and Resilience might be an ideal situation to work on a common list of pesticides to target, including long-banned pesticides as well as CUP and their metabolites, set based on current and past use, (eco)toxicity information, persistence in soils, but also contamination of other media (air, water). This is because recent studies show that pesticide residues with hazard classifications relevant to non-target species, including humans, are omnipresent in the environment and farmer residences (Silva et al., 2023), stressing the need for holistic assessments, or at least the alignment of different matrices programs towards holistic assessments, to fully assess pesticide contamination and impacts. Moreover, since a few soil monitoring networks are also being extended to the measurement of soil biodiversity (Imbert et al., 2021), and the Soil Monitoring and Resilience law also covers biological descriptors, another possible way forward would be to monitor also the body burden of pesticides in soil organisms and food webs (Pelosi et al., 2021).

6.2 No standardized operating procedure for determination of pesticide residues in soil

The absence of a globally harmonized SOP (Standardized Operating Procedure), on field and laboratory steps, leads to inconsistencies in how different teams/laboratories assess soil-pesticide contamination, hampering comparisons and straightforward aggregation of findings. Harmonized sampling methods and analytical procedures, including standardized Quality Assurance/Quality Control plans, are essential for addressing these challenges. Such standards would enable the consistent and accurate assessment of soil pollution, potentially facilitating better-informed decisions on land management and pesticide use. The International Organization for Standardization (ISO) Technical Committee 190 on Soil Quality has developed a comprehensive catalog of standard methods for the collection and analysis of soil contaminants, including the ISO 23646: 2022 – on the determination of organochlorine pesticides in soil. However, there is often a fee to access ISO documents, which builds on the unavoidable high cost of the analytical determinations and creates a barrier to their widespread adoption. LUCAS (Land Use and Coverage Area frame Survey) soil sampling procedures stand as a recent, open-access reference for fieldwork. The Global Soil Partnership is working in collaboration with experts worldwide to promote universal access to internationally developed standards. Through global networks such as the Global Soil Laboratory Network (GLOSOLAN) and the International Network on Soil Pollution(INSOP), efforts are underway to develop and disseminate an SOP specifically for determining pesticide residues in soil. This SOP will focus on the QuEChERS EN 15662 method (Quick, Easy, Cheap, Effective, Rugged, and Safe), solid-liquid extraction methods, and the application of LC-MS/MS, GC-MS/MS, and HPLC techniques. By creating and promoting these harmonized procedures, the goal is to ensure that all laboratories, regardless of location, have the tools needed to accurately measure and monitor pesticide residues in soil.

6.3 The interpretation of monitoring data

Currently, there are neither threshold values nor quality/health standards for pesticides in soil. Some countries have soil screening values (SSV) for some pesticides, but the type of soil screening values varies greatly among countries, as well as the values within the same SSV type (Carlon, 2007; German BBSchVo[3]). Most of the soil-pesticide SSV available relate to long-banned pesticides (especially DDT and atrazine), with only a few countries including values for 'other pesticides'. Significant efforts and developments have been made recently in this regard, which will hopefully help the development of a harmonized interpretation system on soil health, that should, among others, facilitate communication with stakeholders and regulators. This does not mean having one set of global/European soil-pesticide thresholds. Various conditioning factors play a role here and justify differences in risk assessment tools and approaches, and SSV among regions: geographical factors, climatic factors, cultural factors, regulatory and political factors, and even scientific factors (EEA, 2022).

3 BBodSchV.pdf (gesetze-im-internet.de)

Published by Burleigh Dodds Science Publishing Limited, 2025.

The European Environmental Agency picked up on JRC/Carlon et al. (2007) work, and the FAOLEX database underlying values, producing a landmark document on the topic: 'Soil monitoring in Europe – indicators and thresholds for soil health assessments' (EEA, 2022). The authors suggest using agricultural critical limits (e.g. in food products, for animal health, and in nearby surface water) to derive SSV. Soil quality would be considered suitable/acceptable if measured levels of pesticides in soil are below such risk-based SSV. If the pesticide level in the soil exceeds the SSV, then further actions are required. These cover further investigation, site-specific risk assessment (and often bioavailability assessment), remediation, and/or risk reduction measures. The EJP-SOIL proposed four different methods to set targets and thresholds for soil health indicators (fixed value, reference – relative to natural land uses, relative changes, and distribution), each with pros and cons, and a framework for soil health characterization (Fig. 8; Matson et al., 2024). We applied the distribution method to pesticide monitoring data, targeting relative benchmarking. This first attempt on pesticides was focused on the number of pesticides in soil, and the total pesticide content, and considered the most comprehensive datasets available for the purpose: Vieira et al. (2023) for overall, and crop-specific segregation; Knuth et al. (2024) for management type analyses, and Silva et al. (2022) for timing of sampling. The distribution-percentile values summarized in Table 6 can help with the positioning of individual assessments into current knowledge of soil-pesticide contamination but should be critically looked at, due to the possibility of bias in the percentiles, related to the contamination degree of the datasets used to derive the percentiles and the fact that there is not a linear relationship between pesticide figures in soil and risk to ecosystem and human health. For pesticides, an additional step may be required to the EJP-SOIL health framework, namely between the post method and the pre-threshold evaluation phase, where at least the hazardous characteristics of the pesticide in soils are accounted for, maybe by the addition of different weighting factors to different hazard classes. Finally, the outputs of recently initiated projects like BENCHMARKS and AI4SOIL are much anticipated and may lead to updates on the indicators of the EU Soil Law and the Global Soil Partnership program.

Defining chemical soil condition, in classes of slightly to seriously contaminated, or even soil quality/health (high, good, moderate, bad, poor), would be rather useful, however hard to do, as contamination relates to the number of pesticides in soil, the type of pesticides, the levels of the individual pesticides, etc. Accounting for a chemical and biological dimension, as done in the Water Framework Directive, seems a reasonable compromise between complexity and representativity in a quality/health assessment. In the absence of thresholds, individual pesticide levels can be compared against toxicological indicators for soil organisms, usually like EC_{50} (median effect concentration)

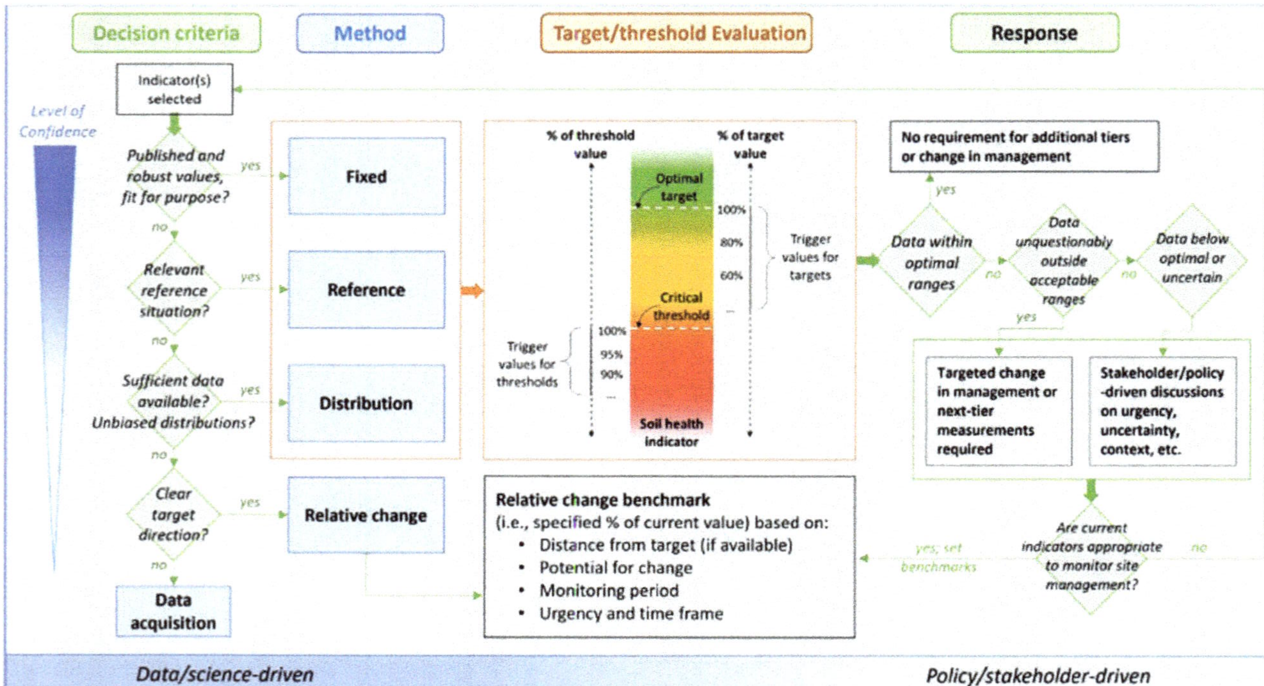

Figure 8 A framework for the selection/use of targets and thresholds for soil health indicators. The decision flowchart starts with the selection of a method, followed by an evaluation of data using trigger values (percentiles against which data is compared to assess how far it is from the target or threshold). Triggered responses may be changes in management or further data collection. Reprinted from Matson et al., 2024.

Table 6 Reference values for the number of pesticide residues in soil and total pesticide content. Relative contamination benchmark is proposed against basic distribution percentiles, of the most complete datasets for overall soil (Vieira et al., 2023), agricultural soil, segregated by main crop types (Vieira et al., 2023), agricultural soils, segregated by management type (Knuth et al., 2024), and agricultural soils, segregated by soil sampling timing (Silva et al., 2022)

	Number of pesticide residues					
		Min	25th	50th	75th	Max
Vieira et al., 2023	Overall, n = 3473	0	0	2	7	31
Vieira et al., 2023	Common wheat, n = 527	0	3	7	11	31
	Barley, n = 352	0	1	3	7	22
	Maize, n = 238	0	3	5	8	18
	Olive groves, n = 158	0	0	2	3	6
	Rape and turnip rape, n = 127	0	7	9	12	25
	Sunflower, n = 127	0	2	4	7	14
	Vineyards, n = 91	0	2	6	8	13
Knuth et al., 2024	Conventional, n = 100	0	5	7	13	21
	Organic, n = 101	0	2	3	4	12
Silva et al., 2023	Before growing season, n = 34	3	6	9	10	13
	During growing season, n = 317	0	1	2	4	13
	After growing season, n = 306	0	2	3	5	13
Total pesticide concentration (mg kg^{-1})						
		Min	**25th**	**50th**	**75th**	**Max**
Vieira et al., 2023	Overall, n = 3473	0	0.0024	0.0076	0.025	2.4
Vieira et al., 2023	Common wheat, n = 527	0	0.0026	0.0077	0.023	1.9
	Barley, n = 352	0	0.0024	0.0073	0.025	1.3
	Maize, n = 238	0	0.0021	0.0056	0.021	2.3
	Olive groves, n = 158	0	0.0083	0.0370	0.110	1.9
	Rape and turnip rape, n = 127	0	0.0022	0.0064	0.019	2.2
	Sunflower, n = 127	0	0.0023	0.0071	0.026	0.8
	Vineyards, n = 91	0	0.0029	0.0085	0.028	1.6
Knuth et al., 2024	Conventional, n = 100	0	0.0642	0.2508	0.9526	28.673
	Organic, n = 101	0	0.0046	0.0310	0.1163	5.458
Silva et al., 2023	Before growing season, n = 34	0.015	0.066	1.222	0.202	1.939
	During growing season, n =317	0	0.012	0.086	0.290	2.886
	After growing season, n = 306	0	0.043	0.215	0.712	11.603

and NOEC (No-Observed-Effect Concentration), but considering that almost all agricultural soils present mixtures of pesticide residues, which may interact among themselves, it would most likely underestimate the real risk to organisms and ecosystems. Risk quotients can be done based on site-specific mixtures, but validation is missing, including for the 5 and 10 assessment factors used on regulatory risk assessment. Ecotoxicological data on complex mixtures are rarely available, especially for concentrations and ratios as in the environment, and non-standard toxicity endpoints (Martin et al., 2021). Hopefully, such a knowledge gap is overcome soon via research like the one conducted in the SPRINT project, and/or the adoption of new pre-market risk assessment requirements.

6.4 Access to data

Access to pesticide distribution datasets is increasing as open science is becoming the rule. Platforms like IPCHEM, NORMAM, and ECHA are centralizing available information, making it easier to find and connect datasets, including with other contaminants, and steering data potentialities (see for instance NORMAN – Network of reference laboratories, research centres and related organisations for monitoring of emerging environmental substances – Prioritization of emerging substances framework, or EUSO SOIL pollution dashboard and watchlist). On the other hand, considering soil data, it may be more complicated to fully open all information collected as for several countries, the point georeferenced soil data are considered as personal data, therefore falling under the exception for environmental data sharing because of General Data Protection Regulation, which is foreseen in the Directive 2007/2/EC (Infrastructure for Spatial Information in Europe, known as the INSPIRE Directive) and Directive 2003/04/EC on public access to environmental information (Cornu et al., 2023). Opening the data will require the consent of all the owners of the land or will need special licenses or contracts stating what can be done and not done with that data (generally for research purposes). Another possible option would also be to anonymize the dataset by deleting or simplifying (with less precision) the coordinates of the sampling points or aggregating at higher scales so that the landowner cannot be identified.

Getting the consent from landowners for the open disclosure of point georeferenced soil data is, therefore, a prerequisite to overcome these sharing constraints, but owners may then fear the results, or the reporting and follow-ups, and refuse the sampling. Note also that some farmers may be quite interested in getting such information on their land for free that is generally quite expensive to get. There is then a sort of atrade-off between more difficulties in getting (open data) samples and having more information from the samples and monitoring program. Depending on the aim and scale of the study (e.g. monitoring soil health at the national level) the exact coordinates may not be needed since the objective is to follow with time the trends in soils. This is not the case when for instance measuring transfers in a river basin, where the exact locations become essential. Thus, when starting

Published by Burleigh Dodds Science Publishing Limited, 2025.

any monitoring activities, depending on the scale and the way all results will be presented and shared, it is essential to think from the beginning if the exact coordinates will be needed, and if so, to ask for the respective consent of landowners.

7 Conclusion and recommendations

- There is fragmented yet abundant data corroborating the presence of pesticide residues in agricultural soils. Considering that most studies have a targeted list of analytes, pesticide occurrence and levels in soil (and related environments) might be even higher. A monitoring strategy focused on the assessment of large groups of residues, ideally full screening, is expected to provide more realistic assessments. Repeated assessments, as happening in Europe with the regular LUCAS pesticide monitoring module, can inform on contamination trends, and the impact of soil and pesticide-related policies, at the point and large scale. Sequential assessments during the year, at least in a subset of samples, would help to identify variations of mixtures in soil, which depend on the grown crop and the used products.
- The presence of multiple pesticide residues in soil is the rule rather than the exception. Most samples contain mixtures of currently used and obsolete pesticides. The mixture composition is quite variable among crops and regions, highlighting different access, knowledge, and regulations regarding pesticide use, but also the important role of pedoclimatic features on pesticides' persistence and fate.
- Soil is a reservoir of pesticide residues. This is corroborated by the detection of compounds past degradation times, and the accumulation of CUPs within and over years. This ultimately increases the complexity of mixtures in soil and may prolong the pesticide in situ impacts. It also represents a latent thread for connected environments since soil can also act as a source of pesticide residues.
- Soils managed under organic farming present fewer residues and lower pesticide levels than soils from conventional fields. The presence of CUPs in these soils stresses legacy and diffuse pollution problems and challenges. Conversion to organic farming at the landscape level can reduce off-site contamination and even pesticide use in nearby conventional fields.
- The frequent detection of obsolete pesticides in soil relates to past pesticide use. Such legacy contamination stresses the long-lasting implications of pesticides once allowed in the market, including towards the achievement of legislative goals in Europe (McGinley et al., 2023), and should be a serious warning against continuing to have candidates of substitution and compounds with knowledge gaps in the market.

Further implementation of the precautionary principle should reduce such risks.

- Transport with water, a major contributor to offsite contamination, is highly investigated, including links and implications to environmental and human health. Transport via eroded sediments is, however, not fully addressed, although monitoring data from waterbed sediments and indoor dust highlight the relevance of such pathways. They also highlight the importance of holistic/*one health assessments* and the need for even more responsible and integrated land and water management.
- There is growing evidence of pesticide negative effects on the soil ecosystem, and overall non-target species. An update of the current framework for the risk assessment of pesticides for in-soil organisms, non-target plants, and non-target arthropods is urgently needed. This is for a more realistic characterization of pesticide risk, but also as a risk reduction strategy. A reduction of the pool of products in the market, which should be dominated by low-risk and more specific pesticides, should reduce pesticides' negative impacts and footprint significantly.
- Main challenges exist around the monitoring of pesticide residues in soil, the main ones being the lack of a harmonized monitoring program, the lack of field/lab SOP, the lack of threshold/quality/health values for interpretation of data, and limited access to monitoring data. Harmonization of procedures and monitoring programs, and centralization and open access to data, potentially via policy-related instruments (e.g. the upcoming Soil Monitoring and Resilience Law), may fix some of these. The definition of thresholds seems to be the most pressing issue, considering the ambitious pesticide, soil, and pollution targets, at the European and global scale. Considering all soil interlinkages, soil-pesticide thresholds should account not only for soil figures and context but also for the potential transport and risk to receiving compartments. It would be likely more effective to act at the beginning of the cascade (and trigger further investigations and reductions right at the soil/field scale), than act upon all knowledge gaps (in terms of transport pathways, establishing thresholds for sediments, air, indoor dust, etc).
- A transformative food production concept, supported by policy instruments, and further involvement of different food chain actors is required to break pesticide reliance and strongly reduce pesticide use. More support to farmers on the selection and application of pesticides should be facilitated when chemical control is unavoidable. Low-risk substances and precision applications should be the default in these cases.

Published by Burleigh Dodds Science Publishing Limited, 2025.

8 Acknowledgments

This work was financed supported by the H2020 grant agreement number of EJPSOIL, H2020 grant agreement no 862695, and by the H2020 SPRINT project (on Sustainable Plant Protection Transition), grant agreement no 862568. part of the text is an excerpt from the PhD Thesis (Silva, 2022).

9 Where to look for further information

Below is a short list of papers and reviews (mix of methodological, fate/process, large-scale monitoring and ecological impact) that together give a strong grounding for pesticide monitoring and associated challenges. Arias-Estévez et al. (2008) *provides a comprehensive overview of* sorption, mobility and degradation processes in soil. Schwarz et al. (2022) – on Microbial Pesticide Degradation - provides interesting insights around hotspots and why residues vary spatially. Pathak et al. (2022) and Kariyanna et al. (2024) cover environmental fate and effects of pesticides on human/ecosystem. See Swaine et al. (2025) for a systematic review/meta-analysis on pesticide impacts on soil organisms and soil health. Vryzas (2018) and Silva et al. (2023) provide evidence on cross-media movement, and Tang et al (2025) inform on transboundary impacts. Brinco (2023) covers analytical techniques and multiclass methods for soils. Franco et al., (2024) present ecological risk insights using the large LUCAS soil pesticide dataset. See Panagos et al. (2024) on more European efforts around soil, Alblas et al. (2025) on how to bridge Law and Soil Science to Promote Soil Health, and Finger (2024) for broader pesticide policy discussions.

10 References

Adhikari, K. & Hartemink, A.E. 2016. Linking soils to ecosystem services—a global review. *Geoderma* 262: 101–111.

Affum, A.O., et al. 2018. Distribution and risk assessment of banned and other current-use pesticides in surface and groundwaters consumed in an agricultural catchment dominated by cocoa crops in the Ankobra Basin, Ghana. *Science of the Total Environment* 633: 630–640.

Aichner, B., et al. 2013. Levels and spatial distribution of persistent organic pollutants in the environment: a case study of German forest soils. *Environmental Science & Technology* 47: 12703–12714. https://doi.org/10.1021/es4019833

Akashe, M.M., et al. 2018. Classification of pesticides: a review. *International Journal of Research in Ayurveda and Pharmacy* 9: 144–150.

Akca, E., et al. (eds) (2024). Publications Office of the European Union, Luxembourg. https://doi.org/10.2760/7007291

Alblas, E., et al. 2025. Bridging law and soil science to promote soil health. *European Journal of Soil Science* 76(4).

Alekseeva, T., et al. 2014. Effect of soil properties on pure and formulated mesotrione adsorption onto vertisol (Limagne plane, Puy-de-Dôme, France). *Chemosphere* 111: 177–183.

Ali, S., et al. 2021. Environmental and health effects of pesticide residues. *Sustainable Agriculture Reviews* 48: 311–336.

Andreu, V. & Picó, Y. 2004. Determination of pesticides and their degradation products in soil: critical review and comparison of methods. *Trends in Analytical Chemistry* 23: 772–789. https://doi.org/10.1016/j.trac.2004.07.008.

Anthony, M.A., et al. 2023. Enumerating soil biodiversity. *Proceedings of the National Academy of Sciences* 120: e2304663120. https://doi.org/10.1073/pnas.2304663120.

Aparicio, V.C., et al. 2018. Glyphosate and AMPA concentrations in wind-blown material under field conditions. *Land Degradation&Development* 29(5): 1317–1326.

Arias-Estévez, M., et al. 2008. The mobility and degradation of pesticides in soils and the pollution of groundwater resources. *Agriculture, Ecosystems & Environment* 123(4): 247–260.

Bach, E.M., et al. 2020. Soil biodiversity integrates solutions for a sustainable future. *Sustainability* 12: 266.

Baran, N., et al. 2021. Pesticides in groundwater at a national scale (France): impact of regulations, molecular properties, uses, hydrogeology and climatic conditions. *Science of The Total Environment* 791: 148061.

Beaumelle, L., et al. 2023. Pesticide effects on soil fauna communities–a meta-analysis. Journal of Applied Ecology 60(7): 1239–1253. https://doi.org/10.1111/1365-2664.14437

Bento, C.P.M., et al. 2017. Glyphosate and AMPA distribution in wind-eroded sediment derived from loess soil. *Environmental Pollution* 220: 1079–1089.

Bernasconi, C., et al. 2021. Evidence for soil pesticide contamination of an agroecological farm from a neighboring chemical-based production system. *Agriculture, Ecosystems & Environment* 313: 107341.

Bexfield, L.M., et al. 2020. Pesticides and pesticide degradates in groundwater used for public supply across the United States: occurrence and human-health context. *Environmental Science & Technology* 55(1): 362–372.

Boonupara, T., et al. 2023. Airborne pesticides from agricultural practices: a critical review of pathways, influencing factors, and human health implications. *Toxics* 11(10): 858.

Bourguet, D. and Guillemaud, T. 2016. The hidden and external costs of pesticide use. *Sustainable Agriculture Reviews* 19: 35–120.

Brinco, J., et al. 2023. Analysis of pesticide residues in soil: a review and comparison of methodologies. *Microchemical Journal* 195.

Brühl, C.A., et al. 2024. Widespread contamination of soils and vegetation with current use pesticide residues along altitudinal gradients in a European Alpine valley. *Communications Earth & Environment* 5(1): 72.

Brunelle, T., et al. 2024. Reducing chemical inputs in agriculture requires a system change. *Communications Earth & Environment* 5(1): 369.

Bruzzoniti, M. C., et al. 2014. QuEChERS sample preparation for the determination of pesticides and other organic residues in environmental matrices: a critical review. *Analytical and Bioanalytical Chemistry* 406(17): 4089–4116.

Buckwell, A., et al. 2020. *Crop protection & the EU food system. Where are they going?* RISE Foundation, Brussels.

Calatayud-Vernich, P., et al. 2016. Efficiency of QuEChERS approach for determining 52 pesticide residues in honey and honey bees. *MethodsX* 3: 452–458.

Cao, Y., et al. 2024. Effects of neonicotinoid residues on non-target soil animals: a case study of meta-analysis. *Journal of Hazardous Materials* 476: 135022. https://doi.org/10.1016/j.jhazmat.2024.135022.

Carlon, C. 2007. *Derivation methods of soil screening values in Europe. A review and evaluation of national procedures towards harmonization*. European Commission, Joint Research Centre, Ispra, EUR 22805-EN, 306 pp.

Carson, R. 1962. *Silent spring*. Houghton Mifflin, Boston, 368 p.

Carvalho, F.P. 2017. Pesticides, environment, and food safety. *Food and Energy Security* 6: 48–60.

Climent, M.J., et al. 2019. Residues of pesticides and some metabolites in dissolved and particulate phase in surface stream water of Cachapoal River Basin, Central Chile. *Environmental Pollution* 251: 90–101. https://doi.org/10.1016/j.envpol.2019.04.117

Chiaia-Hernandez, A.C., et al. 2017. Long-term persistence of pesticides and TPs in archived agricultural soil samples and comparison with pesticide application. *Environmental Science & Technology* 51: 10642–10651. https://doi.org/10.1021/acs.est.7b02529

Chiaia-Hernández, A.C., et al. 2020. Target and suspect screening analysis reveals persistent emerging organic contaminants in soils and sediments. *Science of the Total Environment* 740: 140181. https://doi.org/10.1016/j.scitotenv.2020.140181

Colin, T., et al. 2019. Traces of a neonicotinoid induce precocious foraging and reduce foraging performance in Honey Bees. *Environmental Science & Technology* 53: 8252–8261.

Commelin, M.C., et al. 2022. Pesticides are substantially transported in particulate phase, driven by land use, rainfall event and pesticide characteristics–a runoff and erosion study in a small agricultural catchment. *Frontiers in Environmental Science* 10: 830589.

Contaminated Land Management Guidelines. 2024. https://www.wasteminz.org.nz/Member%20Portal/2024%20Conference%20and%20Expo/Day%202%20-%20Tuesday/Arena%20Lounge/Session%204%20Te%20Matai%20-%20Chemistry/1610_Te%20Matai_Humphrey%20Girvin.pdf

Cooke, A.-K., et al. 2024. Analysing agricultural plant protection product concentrations in groundwater in Germany: Nationwide database with site and compound insights. *Environmental Research* 248: 118231.

Coors, A. and De Meester, L. 2008. Synergistic, antagonistic and additive effects of multiple stressors: predation threat, parasitism and pesticide exposure in Daphnia magna. *Journal of Applied Ecology* 45(6): 1820–1828.

Cornu, S., et al. 2023. National soil data in EU countries, where do we stand?. *European Journal of Soil Science* 74(4): e13398.

Csányi, K. and Farsang, A. 2022. Evaluation of off-site effects of wind-eroded sediments especially the content of pesticides. *Geographica Pannonica* 26(3): 273–283.

de Barreda, D.G., et al. 1991. Survey of herbicide residues in soil and wells in three citrus orchards in Valencia, Spain. *Weed Research* 31(3): 143–151.

de Gerónimo, E., et al. 2015. A simple and rapid analytical methodology based on liquid chromatography-tandem mass spectrometry for monitoring pesticide residues in soils from Argentina. *Analytical Methods* 7(22): 9504–9512.

de Jong, F.M.W., et al. 2008. Estimated nationwide effects of pesticide spray drift on terrestrial habitats in the Netherlands. *Journal of Environmental Management* 86(4): 721–730.

Debler, F., et al. 2024. Occurrence and distribution of pesticides and transformation products in ambient air in two European agricultural areas. *Science of The Total Environment*, 173705. https://doi.org/10.1016/j.scitotenv.2024.173705

DIN EN 12457-4. 2003. Charakterisierung von Abfällen – Auslaugung; Übereinstimungsuntersuchung für die Auslaugung von körnigen Abfällen und Schlämmen – Teil 4: Einstufiges Schüttelverfahren mit einem Flüssigkeits-/Feststoffverhältnis von 10 L/kg und einer Korngröße unter 10 mm (ohne oder mit Korngrößenreduzierung (Characterization of Waste – Leaching; Compliance Test for Leaching of Granular and Sludges – Part 4: One Stage Batch Test at a Liquid to Solid Ratio of 10 L/kg With Particle Size Below 10 mm (Without or with size reduction)). German Standardisation Organisation.

EC. 2000. Directive 2000/60/EC of the European Parliament and of the Council of 23 October 2000 establishing a framework for community action in the field of water policy. OJ L 327, 22.12.2000, pp. 1–73.

EC. 2011. Implementing Regulation (EC) No 1107/2009 of the European Parliament and of the Council as regards uniform principles for evaluation and authorisation of plant protection products (Text with EEA relevance(EU) Nr. 546/2011.

EC. 2020a. Communication from The Commission To The European Parliament, The Council, The European Economic and Social Committee and the Committee of the Regions - A farm to fork strategy for a fair, healthy and environmentally-friendly food system. Brussels, 20.5.2020 - COM(2020) 381 final.

EC. 2020b. Proposal for a Regulation of the European Parliament and of the Council on the sustainable use of plant protection products and amending Regulation (EU) 2021/2115. COM(2022) 305 final 2022/0196 (COD).

EC. 2024. https://ec.europa.eu/food/plant/pesticides/eu-pesticides-database/start/screen/active-substances

ECA. European Court of Auditors (ECA) 2020. *Special report on Sustainable use of plant protection products: limited progress in measuring and reducing risks.* Publications Office of the European Union, Luxembourg. https://www.eca.europa.eu/Lists/ECADocuments/SR20_05/SR_Pesticides_EN.pdf

EEA. 2022. Soil monitoring in Europe - Indicators and thresholds for soil health assessments. https://www.eea.europa.eu/publications/soil-monitoring-in-europe

EEA. 2024. https://www.eea.europa.eu/en/analysis/indicators/pesticides-in-rivers-lakes-and?activeAccordion=ecdb3bcf-bbe9-4978-b5cf-0b136399d9f8

EFSA. 2024. EFSA call for expressions of interest for external experts in working groups - Environmental risk assessment for terrestrial non-target organisms.

EFSA Panel on Plant Protection Products and their Residues (PPR), et al. 2017. Scientific opinion addressing the state of the science on risk assessment of plant protection products for in-soil organisms. *EFSA Journal* 15(2): e04690. https://doi.org/10.2903/j.efsa.2017.4690

Environment Agency - Abu Dhabi. 2018. https://www.wkcgroup.com/wpcontent/uploads/2023/04/Soil-Contamination-User-Guide-for-Abu-Dhabi-Emirate.pdf

EU. 2009a. DIRECTIVE 2009/128/EC of the European Parliament and of the Council of 21 October 2009 establishing a framework for Community action to achieve the sustainable use of pesticides.

EU. 2009b. Regulation 1107/2009 of the European Parliament and of the Council of 21 October 2009 concerning the placing of plant protection products on the market and repealing Council Directives 79/117/EEC and 91/414/EEC.

EU. 2020. DIRECTIVE (EU) 2020/2184 of the European Parliament and of the Council of 16 December 2020 on the quality of water intended for human consumption (recast). https://eur-lex.europa.eu/eli/dir/2020/2184/oj

EU. 2024. EU Pesticide database. https://ec.europa.eu/food/plant/pesticides/eu-pesticides-database/active-substances/?event=search.as

Eurostat. 2024a. https://food.ec.europa.eu/plants/pesticides/sustainable-use-pesticides/farm-fork-targets-progress/eu-trends_en#eu-trends-in-the-use-of-more-hazardous-pesticides

Eurostat. 2024b. https://food.ec.europa.eu/plants/pesticides/sustainable-use-pesticides/harmonised-risk-indicators/trends-eu_en#fig2

FAO. 2020a. *State of knowledge of soil biodiversity -Status, challenges and potentialities*. FAO, Rome, p. 618.

FAO. 2020b. *World food and agriculture - statistical yearbook*. FAO, Rome. https://doi.org/10.4060/cb1329en

FAO and ITPS. 2015. *Status of the world's soil resources (SWSR) – Technical summary*. Food and Agriculture Organization of the United Nations and Intergovernmental Technical Panel on Soils, Rome, Italy.

FAO and UNEP. 2021. *Global assessment of soil pollution - Summary for policy makers*. FAO, Rome. https://doi.org/10.4060/cb4827en

FAOSTAT. 2024. Pesticides use. https://www.fao.org/faostat/en/#data/RP/visualize

Finger, R. 2024. Europe's ambitious pesticide policy and its impact on agriculture and food systems. *Agricultural Economics* 55(2): 265–269.

Fornasiero, D., et al. 2017. Effect of spray drift reduction techniques on pests and predatory mites in orchards and vineyards. *Crop Protection* 98: 283–292.

Franco, A., et al. 2024. Evaluation of the ecological risk of pesticide residues from the European LUCAS Soil monitoring 2018 survey. *Integrated Environmental Assessment and Management* 20(5): 1639–1653. https://doi.org/10.1002/ieam.4917

Frelih-Larsen, M., et al., 2023. https://sprint-h2020.eu/index.php/my-sprint/project-documents/deliverables-and-reports/413-d7-1-final/file

Froger, C., et al. 2023. Pesticide residues in french soils: occurrence, risks, and persistence. *Environmental Science & Technology* 57: 7818–7827. https://doi.org/10.1021/acs.est.2c09591

García-Delgado, C., et al. 2020. Organic carbon nature determines the capacity of organic amendments to adsorb pesticides in soil. *Journal of Hazardous Materials* 390: 122162.

Gardes, T., et al. 2021. Historical and post-ban releases of organochlorine pesticides recorded in sediment deposits in an agricultural watershed, France. *Environmental Pollution* 288: 117769.

Geiger, F., et al. 2010. Persistent negative effects of pesticides on biodiversity and biological control potential on European farmland. *Basic and Applied Ecology* 11: 97–105.

Geissen, V., et al. 2021. Cocktails of pesticide residues in conventional and organic farming systems in Europe – Legacy of the past and turning point for the future. *Environmental Pollution* 278: 116827. https://doi.org/10.1016/j.envpol.2021.116827

Gentil, C., et al. 2019. Challenges and ways forward in pesticide emission and toxicity characterization modeling for tropical conditions. *The International Journal of Life Cycle Assessment* 25(7): 1290–1306.

Ghabbour, S. I. & Sobhy, H. M. 2012. Monitoring of pesticide residues in strawberry and soil from different farming systems in Egypt. https://www.semanticscholar.org/paper/Monitoring-of-Pesticide-Residues-in-Strawberry-and-Ghabbour-Sobhy/12d511aff0d555623475b0c7966738026c1c6653

Goutner, V., et al. 2012. Organochlorine pesticide residue concentrations and accumulation patterns in waterbirds and in their prey at Lake Kerkini, a Ramsar wetland, Greece. *Journal of Biological Research-Thessaloniki* 17: 154–168.

Grondona, S.I., et al. 2023. Pesticides in aquifers from Latin America and the Caribbean. *Science of The Total Environment* 901: 165992.

Guo, L., et al. 2000. Dependence of pesticide degradation on sorption: nonequilibrium model and application to soil reactors. *Journal of Contaminant Hydrology* 43: 45–62.

Hayes, W. and Laws, E. 1991. *Handbook of pesticide toxicology - Volume 1 - General principles*. Academic Press, Inc, San Diego.

Holden, P. W. (1986). *Pesticides and groundwater quality issues and problems in four states*. Academy Press, Washington.

Humann-Guilleminot, S., et al. 2019. A nation-wide survey of neonicotinoid insecticides in agricultural land with implications for agri-environment schemes. *Journal of Applied Ecology* 56: 1502–1514.

Hvězdová, M., et al. 2018. Currently and recently used pesticides in Central European arable soils. *Science of the Total Environment* 613–614: 361–370. https://doi.org/10.1016/j.scitotenv.2017.09.049

Imbert, C., et al. 2021, April. A soil biodiversity survey coupled with the national soil quality monitoring network?. In *Global Symposium on Soil Biodiversity*, pp. 46–52.

IPBES, Intergovernmental Science-Policy Platform on Biodiversity and Ecosystem Services. 2018. The IPBES assessment report on land degradation and restoration. In Montanarella, L., et al. (eds.). *Secretariat of the Intergovernmental Science-Policy Platform on Biodiversity and Ecosystem Services*, Bonn, Germany. 744 pp.

Janousek, W.M., et al. 2023. Recent and future declines of a historically widespread pollinator linked to climate, land cover, and pesticides. Proceedings of the National Academy of Sciences 120(5): e2211223120. https://doi.org/10.1073/pnas.2211223120

Jung, Y.H., et al. 2019. Climate change has likely already affected global food production. *PLoS One* 14: e0217148.

Karanasios, E., et al. 2018. Monitoring of glyphosate and AMPA in soil samples from two olive cultivation areas in Greece: aspects related to spray operators activities. *Environmental Monitoring and Assessment* 190: 361.

Karas, P.A., et al. 2018. Assessment of the impact of three pesticides on microbial dynamics and functions in a lab-to-field experimental approach. *Science of The Total Environment* 637–638: 636–646.

Karasali, H., et al. 2016. Pesticide residue concentration in soil following conventional and Low-Input Crop Management in a Mediterranean agro-ecosystem, in Central Greece. *Science of The Total Environment* 541: 130–142.

Kariyanna, B., et al. 2024. Comprehensive insights into pesticide residue dynamics: unraveling impact and management. *Chemical and Biological Technologies in Agriculture* 11(1).

Khan, F.A., et al. 2024. Spray drift reduction management in agriculture: a review. *Progress in Agricultural Engineering Sciences* 20(1): 1–36.

Khurshid, C., et al. 2024. Pesticide residues in European sediments: a significant concern for the aquatic systems?. *Environmental Research* 261: 119754. https://doi.org/10.1016/j.envres.2024.119754

Kiefer, K., et al. 2019. New relevant pesticide transformation products in groundwater detected using target and suspect screening for agricultural and urban micropollutants with LC-HRMS. *Water Research* 165: 114972.

Knuth, D., et al. 2024. Pesticide residues in organic and conventional agricultural soils across Europe: measured and predicted concentrations. *Environmental Science & Technology* 58(15): 6744–6752.

Kosubová, P., et al. 2020. Spatial and temporal distribution of the currently-used and recently-banned pesticides in arable soils of the Czech Republic. *Chemosphere* 254: 126902. https://doi.org/10.1016/j.chemosphere.2020.126902

Kotschik, P., et al. 2018. Which are the key soil parameters driving the toxicity of chemicals to soil organisms? Talk at 27th SETAC Europe Annual Meeting, 7–11 May 2017, Brussels.

Kruse-Plaß, M., et al. 2021. Pesticides and pesticide-related products in ambient air in Germany. *Environmental Sciences Europe* 33(1): 114.

Laitinen, P., et al. 2006. Fate of the herbicides glyphosate, glufosinate-ammonium, phenmedipham, ethofumesate and metamitron in two Finnish arable soils. *Pest Management Science* 62: 473–491.

Lari, S.Z., et al. 2014. Comparison of pesticide residues in surface water and ground water of agriculture intensive areas. *Journal of Environmental Health Science and Engineering* 12: 11.

Larsen, A.E., et al. 2024. Spillover effects of organic agriculture on pesticide use on nearby fields. *Science* 383(6689): 1308–1312.

Li, R., et al. 2021. Analysis of main components in Jujube and Mulberry extracts by high-sensitive HPLC-ESI-Q-TOF-MS/MS. *Journal of Chromatographic Science* 59(9): 806–812.

Lykogianni, M., et al. 2021. Do pesticides promote or hinder sustainability in agriculture? The challenge of sustainable use of pesticides in modern agriculture. *Science of The Total Environment* 795: 148625.

Maggi, F., et al. 2021. The pesticide health risk index: an application to the world's countries. *Science of The Total Environment* 801: 149731.

Maggi, F., et al. 2023. Agricultural pesticide land budget and river discharge to oceans. *Nature* 620: 10137–1017.

Martin, O., et al. 2021. Ten years of research on synergisms and antagonisms in chemical mixtures: a systematic review and quantitative reappraisal of mixture studies. *Environment International* 146.

Masiá, A., et al. 2013. Screening of currently used pesticides in water, sediments and biota of the Guadalquivir River Basin (Spain). *Journal of Hazardous Materials* 263: 95–104.

Matson, A., et al. 2024. Four approaches to setting soil health targets and thresholds in agricultural soils. *Journal of Environmental Management* 371: 123141.

Mason, E., et al. 2025. Monitoring systems of agricultural soils across Europe regarding the upcoming European soil monitoring law. European *Journal of Soil Science* 76(4). https://doi.org/10.1111/ejss.70163

McGinley, J., et al. 2023. Impact of historical legacy pesticides on achieving legislative goals in Europe. *Science of The Total Environment* 873: 162312.

McDougall, P. 2018. Evolution of the crop protection industry since 1960. https://croplife.org/wp-content/uploads/2018/11/Phillips-McDougall-Evolution-ofthe-Crop-Protection-Industry-since-1960-FINAL.pdf

Meidl, P., et al. 2024. Combined application of up to ten pesticides decreases key soil processes. *Environmental Science and Pollution Research* 31(8): 11995–12004. https://doi.org/10.1007/s11356-024-31836-x

Muendo, B. M., et al. 2012. Behavior of pesticide residues in agricultural soil and adjacent River Kuywa sediment and water samples from Nzoia sugarcane belt in Kenya. *The Environmentalist* 32(4): 433–444.

Navarro, I., et al. 2023. Occurrence of pesticide residues in indoor dust of farmworker households across Europe and Argentina. *Science of the Total Environment* 905: 167797. https://doi.org/10.1016/j.scitotenv.2023.167797

Navarro, I., et al. 2024. Assessing pesticide residues occurrence and risks in water systems: a Pan-European and Argentina perspective. Water Research 254: 119999. https://doi.org/10.1016/j.watres.2024.121419

National Environmental Protection. 2013. https://www8.austlii.edu.au/au/other/cth/AUEPI/2013/4.pdf

Oerke, E.C. 2006. Crop losses to pests. *The Journal of Agricultural Science* 144: 31–43.

Orton, T.G., et al. 2013. Spatial distribution of Lindane concentration in topsoil across France. *Science of the Total Environment* 443: 338–350. https://doi.org/10.1016/j.scitotenv.2012.10.103

Ottenbros, I., et al. 2023. Assessment of exposure to pesticide mixtures in five European countries by a harmonized urinary suspect screening approach. International Journal of Hygiene and Environmental Health 248: 114105. https://doi.org/10.1016/j.ijheh.2022.114105

Panagos, P., et al. 2024. A 1 billion euro mission: a soil deal for Europe. *European Journal of Soil Science* 75(1).

Panis, C., et al. 2022. Widespread pesticide contamination of drinking water and impact on cancer risk in Brazil. *Environment International* 165: 107321.

Pathak, V. M., et al. 2022. Current status of pesticide effects on environment, human health and it's eco-friendly management as bioremediation: a comprehensive review. *Frontiers in Microbiology* 13.

Pelosi, C., et al. 2021. Residues of currently used pesticides in soils and earthworms: a silent threat? Agriculture, Ecosystems & Environment 305: 107167. https://doi.org/10.1016/j.agee.2020.107167.

Pereira, A.S., et al. 2017. Evaluation of FOCUS surface water pesticide concentration predictions and risk assessment of field-measured pesticide mixtures—a crop-based approach under Mediterranean conditions. *Environmental Science and Pollution Research* 24(21): 17394–17406.

Perestrelo, R., et al. 2019. QuEChERS - Fundamentals, relevant improvements, applications and future trends. *Analytica Chimica Acta* 1070: 1–28.

Pérez-Lucas, G., et al. 2019. Environmental risk of groundwater pollution by pesticide leaching through the soil profile [Internet]. Pesticides - Use and Misuse and Their

Impact in the Environment. IntechOpen. http://dx.doi.org/10.5772/intechopen.82418.
Peruzzo, P.J., et al. 2008. Levels of glyphosate in surface waters, sediments and soils associated with direct sowing soybean cultivation in North Pampasic Region of Argentina. *Environmental Pollution* 156(1): 61–66. https://doi.org/10.1016/j.envpol.2008.01.015
Prăvălie, R., et al. 2021. Arable lands under the pressure of multiple land degradation processes: a global perspective. *Environmental Research* 194: 110697.
Riedo, J., et al. 2021. Widespread occurrence of pesticides in organically managed agricultural soils–the ghost of a conventional agricultural past? *Environmental Science & Technology* 55(5): 2919–2928. https://doi.org/10.1021/acs.est.0c06405
Römbke, J., et al. 2017. Final report. Performance of laboratory tests with soil organisms: influence of different test soils on the results of ecotoxicological tests. UBA Project No 77696.
Sabatier, P., et al. 2014. Long-term relationships among pesticide applications,mobility, and soil erosion in a vineyard watershed. *Proceedings of the National Academy of Sciences of the United States of America* 111: 15647–15652. https://doi.org/10.1073/pnas.1411512111
Sabatier, P., et al. 2021. Evidence of chlordecone resurrection by glyphosate in French West Indies. *Environmental Science & Technology* 55: 2296–2306. https://doi.org/10.1021/acs.est.0c05207
Sabzevari, S. and Hofman, J. 2022. A worldwide review of currently used pesticides' monitoring in agricultural soils. *Science of The Total Environment* 812: 152344.
Samsidar, A., et al. 2018. A review of extraction, analytical and advanced methods for determination of pesticides in environment and foodstuffs. *Trends in Food Science & Technology* 71: 188–201.
Sánchez-Bayo, F. and Wyckhuys, K.A.G. 2019. Worldwide decline of the entomofauna: a review of its drivers. *Biological Conservation* 232: 8–27. https://doi.org/10.1016/j.biocon.2019.01.020
Sanchez-Bayo, F. et al. 2011. Impacts of agricultural pesticides on terrestrial ecosystems. In: *Ecological Impacts of Toxic Chemicals (Open Access)* 63–87. https://library.oapen.org/bitstream/id/95397caf-f3b6-460b-bfb7-ada4b8412299/9781608051212.pdf
Schorr, J., et al. 2024. Tracing pesticide dynamics: high resolution offers new insights to karst groundwater quality. *Water Research* 267: 122412.
Schwarz, E., et al. 2022. Spatial control of microbial pesticide degradation in soil: a model-based scenario analysis. *Environmental Science & Technology* 56(20): 14427–14438.
Shakerkhatibi, M., et al. 2014. Pesticides residue in drinking groundwater resources of rural areas in the northwest of iran. *Health Promotion Perspectives* 4: 195–205.
Silva, V., et al. 2018. Distribution of glyphosate and aminomethylphosphonic acid (AMPA) in agricultural topsoils of the European Union. *Science of the Total Environment* 621: 1352–1359. https://doi.org/10.1016/j.scitotenv.2017.10.093
Silva, V., et al. 2019. Pesticide residues in European agricultural soils: a hidden reality unfolded. *Science of the Total Environment* 653: 1532–1545. https://doi.org/10.1016/j.scitotenv.2018.10.441
Sishu, F.K., et al. 2022. Pesticide contamination of surface and groundwater in an Ethiopian Highlands' Watershed. *Water* 14(21): 3446.

Silva, V., et al. 2023. Pesticide residues with hazard classifications relevant to non-target species including humans are omnipresent in the environment and farmer residences. *Environment International* 181: 107999.

Sishu, F.K., et al. 2022. Pesticide contamination of surface and groundwater in an Ethiopian Highlands' Watershed. *Water* 14(21): 3446.

Smith, P., et al. 2024. Status of the world's soils. *Annual Review of Environment and Resources* 49(1): 73–104.

Sousa, J.P., et al. 2022. Building a European partnership for next generation, systems-based environmental risk assessment (PERA): final roadmap report. *EFSA Supporting Publications* 19(8): 7546E. https://doi.org/10.2903/sp.efsa.2022.EN-7546

Stanton, R.L., et al. 2018. Analysis of trends and agricultural drivers of farmland bird declines in North America: a review. *Agriculture, Ecosystems & Environment* 254: 244–254. https://doi.org/10.1016/j.agee.2017.11.028

Streissl, F., et al. 2018. Linking pesticide marketing authorisations with environmental impact assessments through realistic landscape risk assessment paradigms. *Ecotoxicology* 27: 980–991. https://doi.org/10.1007/s10646-018-1962-0

Swaine, M., et al. 2025. Impact of pesticides on soil health: identification of key soil microbial indicators for ecotoxicological assessment strategies through meta-analysis. *FEMS Microbiology Ecology* 101(6).

Tang, F.H.M., et al. 2021. Risk of pesticide pollution at the global scale. *Nature Geoscience* 14: 206–210.

Tang, F. H. M., et al. 2025. Transboundary impacts of pesticide use in food production. *Nature Reviews Earth & Environment* 6(6): 383–400.

Ullah, S., et al. 2018. Cypermethrin induced toxicities in fish and adverse health outcomes: its prevention and control measure adaptation. *Journal of Environmental Management* 206: 863–871.

UNEP, United Nations Environment Programme. 2022. *The closing window -climate crisis calls for rapid transformation of societies. Emissions Gap Report 2022*: Nairobi: United Nations Environment Programme. https://www.unep.org/emissions-gap-report-2022.

Ünlü Endirlik, B., et al. 2023. Non-additive mixture effects of benzo[a]pyrene and pesticides in vitro and in vivo: Role of AhR signaling. *Environmental Pollution* 316: 120679.

van der Sloot, H.A., et al. 2003. *Leaching methods for soil, sludge and treated biowaste. Horizontal desk study 23: Leaching tests*. European Commission Joint Research Centre. https://www.leachingtests.com/app/uploads/sites/18/2021/08/hor_desk_23_leaching2.pdf

van Gestel, C.A.M. 2012. Soil ecotoxicology: state of the art and future directions. *ZooKeys* 176: 275–296.

Vasileiadis, S., et al. 2018. Blame it on the metabolite: 3,5-Dichloroaniline rather than the parent compound is responsible for the decreasing diversity and function of soil microorganisms. *Applied and Environmental Microbiology* 84: e01092-18.

Vieira, D.C.S., et al. 2023. Pesticides residues in European agricultural soils - Results from LUCAS 2018 soil module. https://doi.org/10.2760/86566

Vieira, D.C.S., et al. 2024. Soil pollution in the European Union – An outlook. *Environmental Science & Policy* 161: 103876.

Vryzas, Z. 2018. Pesticide fate in soil-sediment-water environment in relation to contamination preventing actions. *Current Opinion in Environmental Science & Health* 4: 5–9.

Vryzas, Z., et al. 2020. Pesticide prioritization approaches and limitations in environmental monitoring studies: from Europe to Latin America and the Caribbean. *Environment International* 143: 105958.

Wang, N., et al. 2011. Accumulation levels and characteristics of some pesticides in human adipose tissue samples from Southeast China. *Chemosphere* 84: 964–971.

Weber, J.B., et al. 2004. Calculating pesticide sorption coefficients (Kd) using selected soil properties. *Chemosphere* 55(2): 157–166.

Wenzel, K.D., et al. 2002. Fate of POPs (DDX, HCHs, PCBs) in upper soil layers of pine forests. *Science of the Total Environment* 286: 143–154. https://doi.org/10.1016/S0048-9697(01)00972-X

Wiersma, G.B., et al. 1972. Pesticide-residue levels in soils, FY 1969 - national soils monitoring program. *Pesticide Monitoring Journal* 6: 194–228.

Wolfram, J., et al. 2021. Water quality and ecological risks in European surface waters – Monitoring improves while water quality decreases. *Environment International* 152: 106479.

Yang, X., et al. 2015. Short-term transport of glyphosate with erosion in Chinese loess soil – A flume experiment. *Science of The Total Environment* 512–513: 406–414.

Chapter 5

Antibiotic residues as contaminants in agricultural soils

Nikola Rakonjac, Wageningen University, The Netherlands[1]; Joost Lahr, National Institute for Public Health and the Environment (RIVM), The Netherlands; Sören Thiele-Bruhn, Trier University, Germany; Bjorn J. A. Berendsen, Wageningen Food Safety Research, The Netherlands; Carlos A. Faúndez Urbina, Pontificia Universidad Católica de Chile, Chile; and Coen J. Ritsema, Wageningen University, The Netherlands

1 Introduction

Soil contamination research has increased over the past decades, reflecting a growing awareness of this issue globally (Guo et al., 2014). The diversity of man-made contaminants in the soil is constantly evolving due to advancements in agrochemicals and industrial processes. Moreover, the escalated focus on ensuring high-quality food, particularly with the rising global population, has led to increased attention to the health of the soil. Notably, over the last three decades in particular, the intensified practices in livestock farming and activities like wastewater reuse in agriculture have underscored the existence

1 *Corresponding author:* nikola.rakonjac@wur.nl.

http://dx.doi.org/10.19103/AS.2025.0151.08

of antibiotics in soils, categorizing them as an emerging concern. Figure 1 summarizes the number of scientific journal articles published each year from 1935 to 2023, sourced from the Scopus database using the search terms "antibiotics" and "soil".

Antibiotics are substances used, among other purposes, for the treatment or prevention of microbial diseases in both animals and humans. Their usage is global and is regulated, necessitating market authorization prior to application (EMEA, 2008; European Commission, 2021). The selection and quantity of these compounds used vary, influenced by factors such as the specific disease being addressed, geographical location (availability of certain antibiotics), and individual characteristics of the human or animal recipient (e.g. weight). Hutchings et al. (2019) provide a historical perspective on antibiotics, detailing their discovery, clinical introduction, and targeting principles.

A primary route through which veterinary antibiotics, specifically those administered to livestock animals, are introduced into the environment involves the excretion of urine and feces by treated animals, followed by the application of contaminated manure to agricultural soils (Boxall et al., 2004). In contrast, human antibiotics, typically reach agricultural soils through irrigation with treated wastewater (Negreanu et al., 2012). These pathways, along with other less prominent ones, are further discussed later in this chapter.

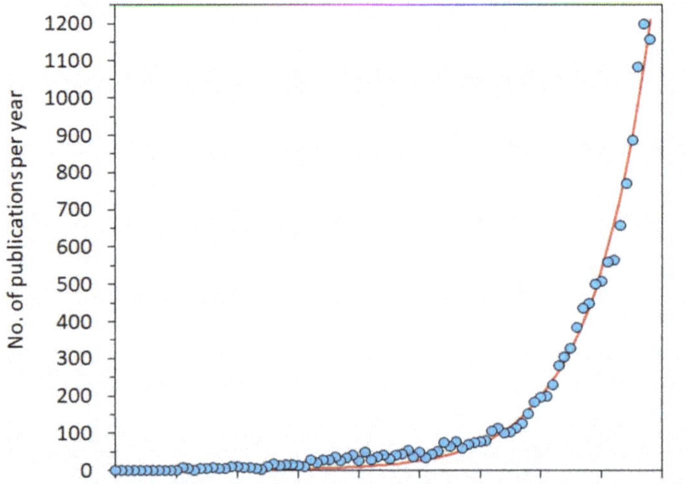

Figure 1 Number of scientific journal publications released annually from 1935 to 2023: results of a literature search conducted in the Scopus database, retrieved on February 8, 2024, using the search terms "antibiotics" and "'soil".

The two extensive reviews (Thiele-Bruhn, 2003b; Cycoń et al., 2019), published nearly two decades apart, and focusing on antibiotic residues in soils, indicate that the prevalent antibiotic groups detected in soil have remained consistent. These include Tetracyclines, Sulfonamides, β-lactams, Macrolides, Fluoroquinolones, and Aminoglycosides. Some antibiotics within the mentioned groups potentially have adverse effects on soil processes, organisms, or, eventually, human health through the food chain (Katsivelou et al., 2023; Nguyen et al., 2023). However, a comprehensive understanding of the potential extent of these effects remains challenging, partly attributed to the widely varying behavior of antibiotics within the soil, as documented in the global literature. This variability is influenced by the physico-chemical properties of the substances, soil structure and characteristics, environmental conditions, or a combination of these factors. Hence, there is a need for a shift towards standardization in studying the behavior of antibiotics in soil. Straightforward laboratory experiments and fit-for-purpose analytical methodologies would allow us to study all these aspects in an efficient, yet effective way. This would facilitate a more profound grasp of the antibiotics' fate in soil, offering valuable insights for accurate risk and effect assessments.

The fate of antibiotics in soil can also be explored through modeling approaches. Modeling offers certain advantages over experimental efforts by being applicable to a broad range of conditions, scenarios, and various types of compounds. Additionally, modeling can assist authorities and policymakers in identifying antibiotics potentially present in the soil compartment and assessing conditions that may increase the exposure to them. However, the accuracy and reliability of modeling results heavily rely on the availability and (un)certainty of input parameters. Here, monitoring and experimental data are essential. Therefore, a combination of all aspects is crucial in tiered methodologies that comprehensively address the environmental fate of antibiotics.

Given that the two mentioned reviews (Thiele-Bruhn, 2003b; Cycoń et al., 2019) already furnish a summary of concentrations found in soil worldwide over the past 40 years and their effects on soil microbial communities, these aspects are not extensively examined in this chapter but are rather briefly addressed. Therefore, the aim of this chapter is to:

- provide an overview of the potential environmental routes by which human and veterinary antibiotics may reach the agricultural soil;
- address processes affecting antibiotic fate in the soil;
- discuss modeling perspectives for predicting the residues and fate of antibiotics within soils;
- offer the latest insights into both the detection techniques for antibiotics in soil and the experimental efforts aimed at standardizing investigations into antibiotic behavior in soil;

- highlight key findings from the Dutch case study, where antibiotics were introduced into the soil through slurry manure injection;
- explore the issue of mixed antibiotic contaminations; and
- discuss potential future trends in this research area.

2 Antibiotic pathways to agricultural soils

There are various pathways through which antibiotic residues can reach the soil. These routes may involve variations in the types of antibiotics present in the soil, variances in the concentrations introduced, and differences in the duration of exposure over time periods.

2.1 Via manure (indirect application – veterinary antibiotics)

After administration to livestock animals, a significant proportion of veterinary antibiotics may be excreted unchanged through urine and feces. The amount excreted is influenced by various factors, such as the livestock sector, animal characteristics, antibiotic properties and type, and the method of administration (Kim et al., 2011). Depending on the combinations of these factors, the excreted portions of parent compounds may range from 3% to 90% (Rakonjac et al., 2022). However, certain antibiotics may undergo extensive metabolism post-administration. The degree to which an antibiotic is metabolized is an important

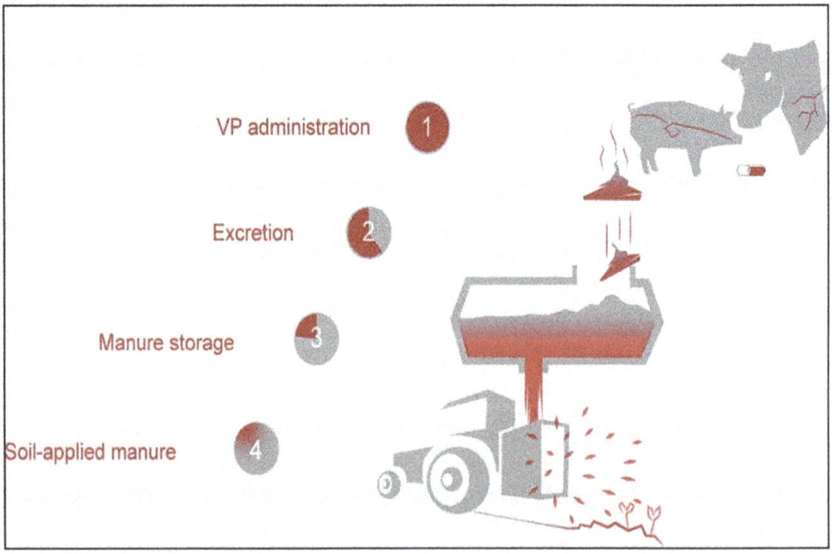

Figure 2 Visualization of the entry pathway of veterinary antibiotics into the environment through the application of post-storage manure to soil. Source: Reprinted from Rakonjac (2023).

Published by Burleigh Dodds Science Publishing Limited, 2025.

factor in environmental exposure, as the metabolites of some compounds exhibit reduced or no antibacterial activity. Also, some metabolites may be more environmentally relevant than the parent compound, especially when the metabolite is more toxic.

Following excretion, antibiotic residues usually end up in manure storage, where they might undergo further dissipation. The degree of dissipation is influenced by factors like storage conditions, duration, temperature, physico-chemical properties of the antibiotics, and practices related to manure handling and management. After the storage period, which for instance lasts around 6 months in the Netherlands (Lagerwerf et al., 2019), any remaining antibiotic fractions are introduced into the soil through the application of manure. This pathway is depicted in Fig. 2.

2.2 Via grazing manure (direct application – veterinary antibiotics)

In addition to residues from stored manure, veterinary antibiotics can also directly reach the soil through manure or urine deposited by animals during pasture seasons (e.g. dairy cows). This pathway is noteworthy because it bypasses the storage period and associated dissipation processes, potentially leading to higher concentrations of antibiotics compared to those in stored manure. However, on a larger scale, this route is generally less significant due to lower quantities of produced manure and a limited number of agricultural fields used for this purpose (Rakonjac et al., 2023).

2.3 Products from manure processing facilities (veterinary antibiotics)

The processing of animal manure is considered, alongside feeding measures and manure export, as a potential solution to alleviate pressure on the fertilizer market and mitigate associated pollution. One approach involves the separation of manure into solid and liquid fractions. The solid fraction can be utilized as an energy source, while the liquid (thin) fraction serves as a substitute for fertilizers. Similar to raw manure, this liquid fraction may contain elevated concentrations of antibiotics, as observed in storage tanks of processing installations by Lahr et al. (2014). Therefore, these residues could potentially be deposited onto the soil when using this substitute fertilizer.

2.4 Via wastewater irrigation and sludge disposal (human antibiotics)

Worldwide, continuous efforts are underway to ensure a reliable and sustainable freshwater supply, with a critical emphasis on optimizing the utilization of

Published by Burleigh Dodds Science Publishing Limited, 2025.

current water resources. One notable initiative in this regard involves the reuse of (treated) wastewater for irrigation. This practice has been implemented for numerous years in arid and semi-arid regions, but it has also garnered increased attention in temperate climates (Bixio et al., 2006). While wastewater irrigation serves as a beneficial water and nutrient source, it also presents risks. The abundance of contaminants, antibiotics included, present in wastewater, can pose threats to public health and the environment.

Human antibiotics are usually released into the sewer, either after excretion from treated patients or through improper disposal via toilets (Kusturica et al., 2017). Consequently, in urban settings, hospitals are often considered hotspots for antibiotic emissions (Herrmann et al., 2015). Once these residues enter the sewer system, they are typically conveyed to wastewater treatment plants, where incomplete elimination may occur, leading to their potential transfer into surface water (Kümmerer, 2003). Subsequently, they may further infiltrate soils when this water is used for irrigation.

The diversity of antibiotics in wastewater used for agricultural soil irrigation depends on its source and the treatment technologies applied before reuse. Existing literature consistently notes significant variations in antibiotic concentrations in such water sources (Michael et al., 2013; Kulkarni et al., 2017), with the possibility of soil contamination by complex mixtures (Kostich et al., 2014). It is important to highlight that the treatment of wastewater before irrigation is not always feasible in low- and middle-income countries due to economic constraints, potentially leading to more severe pollution in such cases.

Agricultural soil pollution can also occur through waste disposal practices. The sludge generated in wastewater treatment plants is sometimes utilized as a soil fertilizer in agricultural fields. While this practice positively influences soil fertility and remains essential in many countries globally (Schmidt et al., 2006; Coors et al., 2016), it might introduce unwanted contaminants (e.g. human antibiotics) to agricultural soils. It is worth noting that this practice is not prevalent in some countries, such as the Netherlands, where the sludge is predominantly incinerated, processed, and exported rather than being applied locally.

2.5 Combination of veterinary and human antibiotics

While the individual pathways of these two groups are well-understood, a knowledge gap exists when considering the combination of these scenarios. For instance, when the same agricultural soils are affected by both manure application and treated wastewater irrigation, they are exposed to both human and veterinary antibiotics (Pan and Chu, 2017). This implies potential exposure to entirely different antibiotics during various time intervals, as well as to the same antibiotics over extended periods, encompassing both the manuring season and irrigation cycles. Therefore, it is crucial to consider this aspect when developing strategies for environmental risk assessment of these compounds.

3 Fate of (veterinary) antibiotics in the soil

Once mixtures of antibiotics and excreta penetrate into the soil, the antibiotics tend to remain in the soil solution, where they are exposed to various processes. The relevant processes, as well as biotic effects caused by antibiotics, are schematically depicted in Fig. 3. While points 1 (medication) and 2 (input) from this figure have been previously discussed and are primarily associated with veterinary antibiotics, this section will concentrate on points 3 and beyond.

In the case of application onto the soil surface, antibiotics may undergo photodegradation by sunlight (4.1, Fig. 3), a process that can be quantitatively significant depending on the substance and substrate properties (Boreen et al., 2005; Wolters and Steffens, 2005; Thiele-Bruhn and Peters, 2007). Yet, photodegradation is fully inhibited as soon as the antibiotics are translocated into soil depths deeper than 0.5 mm (Miller and Donaldson, 1994). Once in the soil, degradation rates of antibiotics can vary depending on the compound, soil type (Berendsen et al., 2021), and depth (Rakonjac et al., 2023). In contrast, volatilization from the soil into the atmosphere (4.2, Fig. 3) is negligible because all these polar chemicals are characterized by very low vapor pressure and Henry coefficients (Thiele-Bruhn, 2003a). Transport of antibiotics by erosion and surface run-off (4.3, Fig. 3) occurs on sloping surfaces, especially grassland,

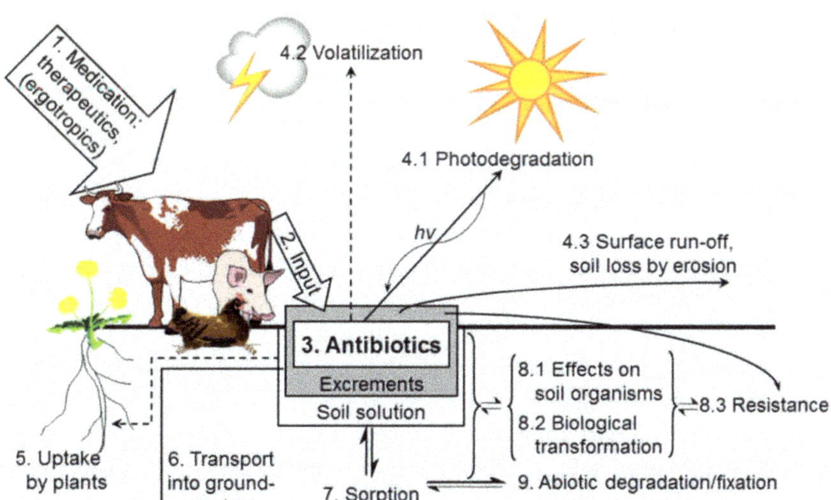

Figure 3 Input, distribution, degradation, and effects of veterinary antibiotics in the soil. Source: Reprinted from (Thiele-Bruhn, 2019).

Published by Burleigh Dodds Science Publishing Limited, 2025.

and is further accelerated by manure (Müller et al., 2003; Kreuzig et al., 2005). Thereby, run-off from grassland is increased due to the temporary surface sealing and hydrophobic effect of manure crusts (Stamm et al., 2003; Burkhardt et al., 2005). These transport processes are relevant especially in areas with dynamic relief. Yet, they are not specific for antibiotics but affect all chemicals and pollutants that have been introduced into soil by manure or other excreta.

Uptake by plants (5, Fig. 3) was confirmed for several plant types and antibiotics from different structural classes (Boxall et al., 2006; Dolliver et al., 2007; Liu et al., 2009). On one hand, the portions taken up rarely exceed a few percent of the total amount of antibiotics in soil (Langhammer et al., 1990; Grote et al., 2007; Freitag et al., 2008), on the other hand, adverse effects on plant growth and yield (Michelini, 2012; Michelini et al., 2012; Michelini et al., 2013), as well as on endophytes (Michelini et al., 2015), have been determined.

Transport and leaching into larger soil depths and groundwater (6, Fig. 3), respectively, occur (Kay et al., 2004; Kay et al., 2005; Höper et al., 2007; Weiß et al., 2007; Rakonjac et al., 2023). Yet, only smaller portions of the antibiotic contamination are dislocated (Aust et al., 2008b; Aust et al., 2010), while large fractions will remain in the soil (Förster et al., 2009; Rosendahl et al., 2011). The transport is largely bound to cotransport with particles and colloids or fast preferential flow (Hamscher, 2007; Höper et al., 2007; Xing et al., 2015; Zhou et al., 2016). Concentrations in groundwater of antibiotics that have been leached from manure treatment of soil are typically below 0.1 μg/L for each single compound (Hamscher, 2007; Hannappel et al., 2017), although also higher concentrations have been determined. Yet, analyses of groundwater samples showed that mixtures of several antibiotics frequently occur so that the summed concentration might even exceed 0.1 μg/L, a concentration that has been proposed as threshold value by the German UBA (Hannappel et al., 2017; Aust et al., 2010).

Sorption (7, Fig. 3) to the soil exchange sites, i.e. not only soil organic matter but also clay minerals and pedogenic oxides (Figueroa et al., 2004; Figueroa and Mackay, 2005; Gao and Pedersen, 2005; Schwarz et al., 2012), governs the mobility and bioavailability of antibiotics. Amphoteric and polar antibiotics are sorbed to soil components by specific and unspecific mechanisms (Tolls, 2001; Thiele-Bruhn et al., 2004). Sorption varies depending on the physico-chemical properties of the antibiotics, content and composition of soil organic matter as well as soil minerals, pH, and also soil moisture (Thiele, 2000; ter Laak et al., 2006a; ter Laak et al., 2006b). Furthermore, sorption and mobility, respectively, of antibiotics are significantly altered in the presence of organic waste material such as manure (Thiele-Bruhn and Aust, 2004; Kahle and Stamm, 2007; Bourdat-Deschamps et al., 2017). Comprehensive review articles on soil sorption of antibiotics were published by Wang and Wang (2015) and Wegst-Uhrich et al. (2014). Antibiotics are differently retarded in soils, which can be

Published by Burleigh Dodds Science Publishing Limited, 2025.

read from sorption coefficients. Differences largely depend on the structural class with smaller variation among individual compounds within an antibiotic class (Wegst-Uhrich et al., 2014; Wang and Wang, 2015). Antibiotics from some structural classes exhibit rather low sorption coefficients. Consequently, there is a higher risk that these antibiotics are leached from soil and may contaminate groundwater resources (Hamscher, 2007; Höper et al., 2007; Aust et al., 2010; Balzer et al., 2016). Sorption further varies with the pH-dependent speciation of the antibiotics. Compared to the sorption of the neutral species, it typically increases with the proportion of cationic species. On the other hand, sorption slightly declines with the increasing formation of anionic species (ter Laak et al., 2006b; Vasudevan et al., 2009; Figueroa-Diva et al., 2010). The formation of charged species depends on the pH of the soil and the acid dissociation constant (pK_a) of the specific antibiotic, resulting in the increased formation of cationic species at lower pH and of anionic species at higher pH. Among the different mechanisms responsible for the sorption of organic chemicals in soil, especially specific sorption mechanisms contribute to the sorption of antibiotics (Thiele-Bruhn et al., 2004; Figueroa-Diva et al., 2010; Schwarz et al., 2012; Wang and Wang, 2015). Consequently, it must be assumed that different individual antibiotic compounds compete for such specific sorption sites.

Antibiotics exert specific effects on biota even at very low dosage, hence, adverse effects on soil organisms (8.1, Fig. 3) and especially on microorganisms must be expected. In response, microorganisms react to contamination with the biological transformation (8.2, Fig. 3) of organic compounds and in this case of antibiotics (Schwarz et al., 2010; Girardi et al., 2011; Grossberger et al., 2014). However, this, as well as immobilization, in soil does not necessarily has a direct and/or immediate feedback on the – once initiated – negative effects of antibiotics on soil biota. It was shown that even soil-bound antibiotics can exert adverse effects and that effects can proceed even beyond the dissipation of the antibiotics (Chander et al., 2002; Chander et al., 2005; Hammesfahr, 2011; Peng et al., 2015). Effects on and reactions of microorganisms lead to an altered and increased resistance (8.3, Fig. 3) as a natural reaction (Schmitt et al., 2006; Heuer and Smalla, 2007; Heuer et al., 2011; Marti et al., 2013), which is triggered by the specific action of the antibiotics or through the input of resistant microorganisms with contaminated excrements. A reduction in the spread of resistance and in the resistance level subsequently occurs after the addition of an antibiotic to soil ended (Sengeløv et al., 2003; Heuer and Smalla, 2007). However, it is supposed that the restoration of a total population to its former status, including antibiotic sensitivity, is unlikely. Both types of pollution, the addition of antibiotics and of antibiotic-resistant microorganisms to soil, can affect the structure and function of environmental microbial populations. Regarding resistance genes, the situation is more complex, since genes are not "degradable pollutants" but auto-replicating elements (Grenni et al., 2018). Comprehensive information on the effect of antibiotics on resistance levels in the environment can be found in a literature study by Schmitt et al. (2017).

Furthermore, abiotic degradation and fixation/sequestration processes (9, Fig. 3) contribute to the reduction of antibiotic contents in soil (Schwarz et al., 2015; Hurtado et al., 2017). Thereby, it is assumed that the subsequent release of previously sequestered antibiotics is low (Rosendahl et al., 2011).

4 Modeling approaches

Sources, pathways, fate, and behavior of antibiotics in the soil environment can be evaluated through dedicated experiments and monitoring studies or by using modeling approaches. The advantage of a modeling approach is its applicability to a broad range of environmental conditions and different types of antibiotics, whereas experimental efforts provide results for specific situations only. Conducting scenario analyses is also a significant benefit of modeling. However, to ensure accuracy and reliability of modeling result, monitoring and experimental data are essential.

The European Medicines Agency provides a guideline on environmental impact assessment for veterinary medicinal products (EMEA, 2008). This guideline addresses different modeling perspectives per environmental compartment. In principle, it is recommended that when estimating medicine residues in the environment a stepwise (tiered) approach should be followed, using simple equations to provide an initial standard assessment and moving on to more complex modeling approach when a more refined estimate of exposure is required. Regarding the latter, a series of mechanistic environmental models and accompanying scenarios has been created by working groups in Europe known as FOCUS (Forum for the Co-ordination of Pesticide Fate Models and Their Use) to simulate the fate and transport of agrochemicals in the environment. Since FOCUS models are primarily designed for the exposure assessment of pesticides, before they can be used for modeling the exposure of antibiotics they have to be tailored likewise.

FOCUS soil calculations are relatively straightforward, focusing on the degradation rate of chemicals applied to a fixed soil depth while ignoring potential losses from volatilization, runoff, and leaching. This guidance recommends using "best fit" degradation kinetics instead of relying solely on conventional first-order fits to experimental data. In addition to determining the appropriate degradation kinetics of the compound in soil, these calculations require the associated application rates to soil (i.e. mass of chemical per land area). The EMEA guidelines (EMEA, 2008) offer further recommendations on incorporating antibiotics instead of pesticides in the calculations mentioned earlier and interpreting the modeling results.

Published by Burleigh Dodds Science Publishing Limited, 2025.

One of the major uncertainties in modeling approaches is how to address compound sorption. The default assumption is that sorption behavior is primarily influenced by sorption to organic carbon. However, since veterinary antibiotics could be large molecules with several functional groups and a tendency to speciate into ionic species at certain environmental pH values, other soil components with polar and/or charged surfaces might also act as sorbents. Sorption behavior is likely to be highly pH dependent. These factors may necessitate updates to existing models. For more information on models addressing sorption specifically related to veterinary antibiotics in soil-manure systems, we refer to Wang et al. (2015). The primary shortcoming of current modeling approaches is their focus on individual environmental compartments rather than the entire chain of processes. This compartmentalization overlooks how the outcomes in one compartment might significantly impact the subsequent phases. For instance, neglecting soil leaching and the role of macropore flow could be critical, depending on soil type and compound chemical characteristics, as these processes can contribute to groundwater contamination. Models that accurately conceptualize and simulate macropore flow are possibly essential for antibiotic risk assessments at regional, national, or global scales. However, these models often require numerous physical parameters, many of which are challenging to estimate independently under field conditions. Improving parameter estimation would enhance model accuracy and our understanding of antibiotic transport. The SWAP-PEARL model (Berg et al., 2016; Kroes et al., 2017) used for pesticide leaching (Faúndez Urbina et al., 2020) could serve as a valuable reference for similar studies on antibiotics. In general, to model the fate of antibiotics in soil, an interdisciplinary approach is essential. In addition to studying soil–compound interactions, there is a need for a better understanding of agricultural practices like manure spreading, which can vary widely across countries and regions. These practices, coupled with spatially variable environmental conditions such as hydrology, can greatly influence the interactions mentioned earlier and hinder accurate risk assessment of these compounds.

5 Controlled laboratory studies

As indicated earlier, the fate of chemical substances in soil depends on two important parameters: persistence and mobility. The persistence, usually expressed as DT50, refers to the kinetic degradation of the native antibiotic. The mobility, usually expressed as K_d, is related to the transport of the antibiotic from soil to other reservoirs, e.g. surface or groundwater.

The persistence of a chemical compound is contingent upon both its inherent physico-chemical characteristics and on the properties of the soil in which it resides, encompassing biotic and abiotic elements such as the fraction of organic matter, pH and the active microbiome. Some studies reported the fate (and environmental risks) of selected antibiotics in manure (Berendsen et al., 2018), or soil after applying manure containing antibiotics (Hamscher et al., 2002; Jacobsen et al., 2004; Brambilla et al., 2007; Martínez-Carballo et al., 2007; Stoob et al., 2007; Uslu et al., 2008; Domínguez et al., 2014; Spielmeyer et al., 2017; Rietra et al., 2024). The methodologies applied in these studies are partially derived from the standardized approach method 307 of the Organization for Economic Co-operation and Development (OECD, 2002). The OECD guidelines describe procedures for aerobic and anaerobic degradation experiments and also mention experiments with sterilized soils to determine the contribution of abiotic transformation. The guidelines suggest to study at least four different soils to determine degradation rates and rely on relatively large portions of soil (50–200 g) introducing potential errors related to inhomogeneity. Furthermore, many studies focus on the persistence of the native substance only. However, it is recognized that degradation products can still exert a negative impact on the environment. Therefore, degradation studies followed by a study to annotate potential degradation products should be included in strategies to assess the fate of chemicals in the agricultural environment.

The mobility of a chemical compound is related to the soil–water partitioning coefficient (K_d), representing the likelihood of the chemical being adsorbed to soil particles. As adsorption primarily takes place through the partition into soil organic matter, a K_d normalized to the organic carbon content of the soil is usually reported, which is expressed as an organic carbon–water partition coefficient (K_{OC}) (Wegst-Uhrich et al., 2014). Mechanisms other than hydrophobic partitioning, e.g. ionic interactions, complex formation, and hydrogen bonding, are not accurately accounted for using this approach which results in underestimation of K_{OC} (Tolls, 2001). Several methods have been used for the estimation of K_{OC}, of which most are derived from OECD method 106 (batch equilibrium method) (OECD, 2000) or method 312 (soil column leaching method) (OECD, 2004). The column leaching test is described for native and aged residues, the latter referring to the test substance and its transformation products present in soil after prolonged exposure to the soil. The test relies on a 30 cm soil column and requires careful uniform packing of this column to obtain reproducible results. Artificial rain is applied to the soil column at a rate of 200 mm over a period of 48 hours, which is equal to a 251 mL volume for a soil column with a 4 cm inner diameter. Both the soil and the leachate volume are relatively high. The soil has to be adequately homogenized before analysis to reduce additional uncertainty and the large leachate volume leads to a highly diluted aquatic fraction which demands high sensitivity of the analytical method for determination of the chemical's concentration.

As the persistence and mobility of antibiotics clearly depend on soil composition, there is a need for good modeling approaches to allow estimation of the DT50 and K_d for a large range of different soils. Recently, a study was performed to predict these parameters for eight antibiotics based on selected soil properties (Rietra et al., 2024). It was concluded that both the range in DT50 and K_d can be predicted reasonably well using regression models that account for the variation in soil type and sampling depth. Organic matter, iron oxides, pH, and clay content appear to be the most important soil parameters impacting these parameters. Furthermore, simple yet effective laboratory procedures to determine DT50 and K_d in a large number of different soils and for a large number of different chemicals are required to obtain data to feed the models. Such procedures allow the determination of the persistence and mobility of specific substance–soil(-substrate) combinations and to understand the mechanistic principles of the interactions of the antibiotic in their environment.

5.1 Analytical detection techniques

Fit-for-purpose analytical techniques are crucial to study the fate of any chemical compound in the soil–water nexus. Several analytical challenges need to be overcome. First of all, the methods need to include a wide range of antibiotics with highly diverse structures and physico-chemical properties (Thiele-Bruhn, 2003b). Second, the matrix itself is a challenge due to its high variation: soil can be a rather simple sandy soil with relatively low organic matter, but it can also be a loamy soil or a soil with high clay content which makes effective extraction of the chemicals in it extremely challenging. Furthermore, the addition of highly diverse substrates to soil can alter the soil composition and structure and complicate the analysis. Preferably, the analytical methods are applicable to all types of soils. Third, methods should be able to achieve detection of antibiotic residues at extremely low concentrations, preferably in the low microgram per kilograom range. Only then the soil interactions can be studied at environmentally relevant concentrations. Last, the methods have to be accurate, precise, and highly selective to prevent false positives from occurring and to prevent natural soil constituents from interfering in the detection of the relevant chemicals and their transformation products.

To achieve such a high standard, the detection method of choice is tandem mass spectrometry. Currently, triple-quadrupole mass spectrometry hyphenated to liquid chromatography is the method of choice for antibiotic analysis. A review of methodologies applied for environmental and food matrices was presented by Thiele-Bruhn (2003b) in 2003. In the early 2010s methods for soil analysis or related complex matrices (e.g. sludge and manure) were scarce and included 1–5 different antibiotic groups. More recently multi-methods were developed (Blackwell et al., 2004; Janusch et al., 2014; Wallace and Aga, 2016) and in 2019 a study was published especially focusing on the extraction of also the bound fraction of antibiotics in manure and soil (Jansen et al., 2019). An extraction solvent of 0.125% trifluoroacetic acid in acetonitrile with the addition of McIlvain-EDTA buffer proved to be the most effective. In the latter study, it was also demonstrated that by changing the extraction solvent either the total residual concentration can be determined or only the biologically available fraction; the fraction of the antibiotic that is freely present in the soil environment and has the potential to move among reservoirs or be transported into a cell.

To achieve the low detection limits and the high selectivity required, methods usually rely on solid phase extraction (SPE) as a clean-up and concentration technique. The extract is transferred to an activated SPE cartridge which is designed to retain the antibiotics of interest. After washing the cartridge, the antibiotics are eluted by an organic solvent that is evaporated to minimize the final extract volume. The concentrated extract is then injected onto the LC-MS/MS system.

Even though not often reported in the literature, these methods are not applicable to all types of soils and substrates. Thorough validation studies are usually lacking or are only focused on relatively simple sandy soils. In general, absolute recoveries tend to be extremely low, especially for soils with high clay or peat content hampering the accurate and precise analysis of antibiotic content in such soils.

5.2 Standardized fate studies

The OECD procedures aim for standardization of the determination of persistence and mobility of chemicals in soil. They rely on the availability of relatively large portions of soil and for both fate parameters the difficulty in homogenizing these soils after the addition of the chemicals tested, adds to the uncertainty of the determined parameters. Aiming for an effective and high-throughput approach, a strategy to determine the fate of chemicals in soil was reported in 2021 (Berendsen et al., 2021). This strategy includes an easy-to-apply persistence and mobility study and also considers the degradation of antibiotics to antimicrobial active transformation products (Fig. 4).

First, the strategy aims to categorize the antibiotics into one of three groups: (1) persistent, (2) degradation to inactive transformation products, and (3) degradation to active transformation product(s). Note that from an environmental point of view, the second category is preferred. To achieve this, a simple degradation study is proposed based on a miniaturization of OECD guidelines 307 (OECD, 2002). The advantage of this miniaturized setup is that the complete soil aliquot is extracted and thus no errors due to inhomogeneity are introduced. Also this allows to perform a study without the need for kilograms of soil material.

Published by Burleigh Dodds Science Publishing Limited, 2025.

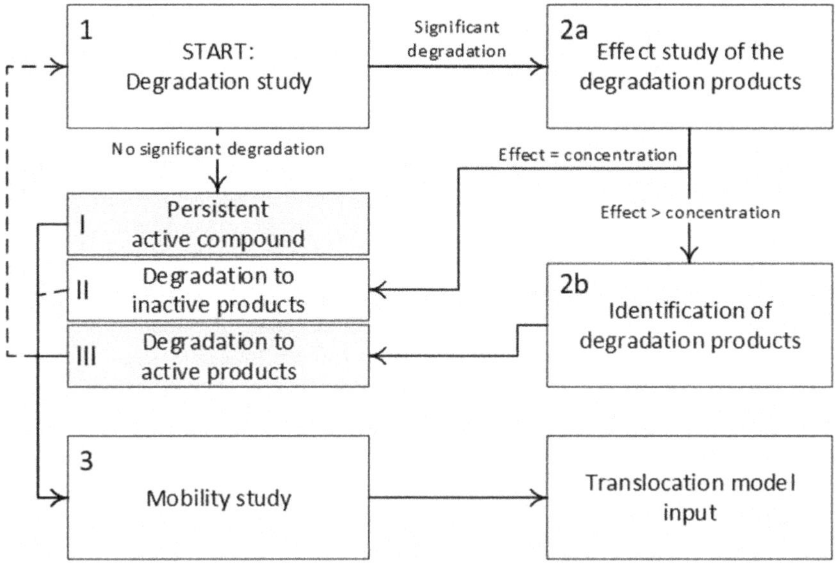

Figure 4 Schematic presentation of the reported strategy to determine the fate of chemicals in the soil water nexus. Source: Reprinted from Berendsen et al. (2021).

In case chemical degradation of the antibiotic was observed, the degradation was also examined using an effect-based test which recorded the antimicrobial activity instead of the concentration of the native antibiotic. In case chemical degradation was significantly larger than the decrease in antimicrobial activity over the same period of time, it was concluded that an antimicrobial active transformation product was formed during soil incubation. From the eight antibiotics included in that specific study, this was clearly the case for tylosin A. Further investigation using high-resolution mass spectrometry is required to (tentatively) identify the transformation product(s). For tylosin A, it was found that, during soil incubation, the native drug is transformed to tylosin A acid, a related substance that is still an antimicrobial (yet somewhat less potent) and more persistent.

Especially for the persistent chemicals, the persistence study should be followed by a mobility study. The use of a column leaching setup proved to be most suitable. Again, a miniaturized setup based on the OECD guidelines is most applicable, especially in case many different soils are included in the study. Soil columns, each containing 5 g of soil, were prepared using an SPE manifold. A 10 mL volume of artificial rain proved to yield K_{OC} values that were comparable to the values previously reported in the literature. The smaller volumes again prevent additional errors due to homogeneity, as the complete soil aliquot and leachate volume are analyzed to determine the antibiotic concentration in each of them. Furthermore, instead of 48 hours, this experiment lasted 1 hour to prevent any significant degradation of the native antibiotic inside the soil column. As a result, the throughput was increased allowing to study a larger number of different soils and soil-substrate combinations, and different types of chemicals.

6 Dutch case study

In the Netherlands, manure from intensive livestock farming is spread onto arable land and grassland in considerable amounts. Consequently, together with Belgium, the Netherlands represents one of the highest nitrogen input regions in Europe (de Vries et al., 2021). The majority of this manure, over 95%, is applied in a slurry form, with the common practice of injecting slurry into the soil to mitigate ammonia emissions into the air. Besides, applied manure may contain a broad range of veterinary antibiotics of different quantities, as detailed in Rakonjac et al. (2022), and could serve as a significant source for introducing these compounds into surface water (Rakonjac et al., 2024).

In order to study the environmental fate of various compounds injected with slurry, a field survey was conducted in the province of Gelderland in the Netherlands. Five pig-rearing and five veal calve-rearing farms situated on sandy soils were selected. In the spring of 2017, the slurry collected over the winter season at these facilities was injected into arable fields (9 locations) and grassland (1 location). During four months following injection, concentrations were monitored in soil (around 25 cm below the surface), in the shallow groundwater of the fields (max. 2.5 m) and in the surface water and sediment of adjacent ditches. The results were reported by (Lahr et al., 2018). Here, a summary of the findings for antibiotics is presented.

Most antibiotics reported to have been used at the farms were detected in slurry from the pits before land application. These included oxytetracycline, doxycycline, and sulfadiazine at both types of farms and flumequine and tilmicosin at veal farms only (Table 1). Concentrations were highest in slurry from veal farms. Other compounds were detected occasionally. Not all antibiotics used at the farms were represented in the analysis package (e.g. penicillin).

Several of the more hydrophobic antibiotics were found frequently in the soil of the fields after application and persisted to various degrees, especially

Table 1 Concentration ranges (μg/kg fresh weight) of the most frequently detected veterinary antibiotics in slurry from intensive pig and veal rearing facilities before application on land

Substance	Veal calve slurry (μg/kg f.w.)	Pig slurry (μg/kg f.w.)
Oxytetracycline	700–5000	<4–71
Doxycycline	350–2000	<5–700
Sulfadiazine	<1–200	<1–100
Flumequine	<3–1700	<7
Tilmicosin	24-300	<1

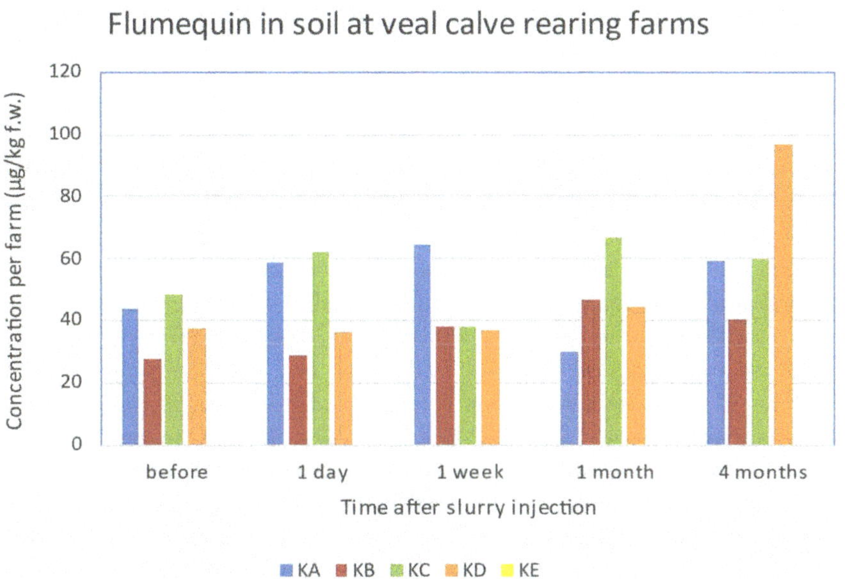

Figure 5 Concentrations (µg/kg fresh weight) of the quinolone antibiotic flumequine in the soil of five veal calve rearing farms (KA to KE) before and after slurry injection. Farm KE confirmed that flumequine was not used in their facility.

oxytetracycline, flumequine and tilmicosin. The latter two are typical of veal calve-rearing and may persist for over a year in soil because they were already found in the soils before the yearly slurry injection (see Fig. 5, for flumequine). Despite not being used at the farms, chlortetracycline was also detected in the soil at most farms. Its exact origin remains unclear.

As expected, instead of in soil, the more soluble antibiotics tended to be found more often in groundwater and/or surface water of adjacent ditches at the fields injected with slurry, most notably sulfadimidine (synonym of sulfamethazine) and sulfadiazine. The use of sulfadimidine as an antibiotic was not reported by the farms in the study. It is not sure where it stems from. Oxytetracycline, doxycycline, chlortetracycline, and flumequine were occasionally detected in the sediments of the ditches near the fields. In five regional streams in the same region as the farms, sampled once in July of the same year, only sulfadiazine was detected in the surface water and at only one occasion.

The results clearly demonstrate that slurry injection leads to contamination of the soil of agricultural fields by antibiotics and potentially also to some contamination of the groundwater and surface water around the fields of farms with intensive livestock rearing.

The findings of this limited study warrant a more in-depth assessment of the risks of substances injected with slurry to soil life, aquatic organisms, and the production of safe drinking water from groundwater.

7 Mixed contaminations with antibiotics in excreta and soil

Generally, it must be expected that contaminated soils not only contain one contaminant but mixtures of several pollutants (Posthuma et al., 2008; Backhaus, 2014; Jahnke et al., 2016). This also applies to antibiotics that have been recognized as widespread soil pollutants (Thiele-Bruhn, 2003b; Boxall et al., 2004; Sarmah et al., 2006). The fact that mixtures of veterinary antibiotics end up in soil is due to three major reasons. First, combinations of antibiotics are frequently administered to livestock (Fig. 6). Second, different antibiotics are given to different livestock animals and to different live stages of the animals because different pathogens and diseases typically occur and have to be treated. As a result, antibiotic mixtures are created when the excrements of different animals are collected in the same manure tank or lagoon. Third, excreta with mixed contamination are subsequently and often repeatedly spread onto agricultural fields when they are reused as fertilizer. The assumed accumulation of antibiotics after spreading to soil with organic fertilizer was previously confirmed in various studies (Hamscher et al., 2002; Höper et al., 2002; Hamscher, 2007; Aust et al., 2008a). This is also due to a

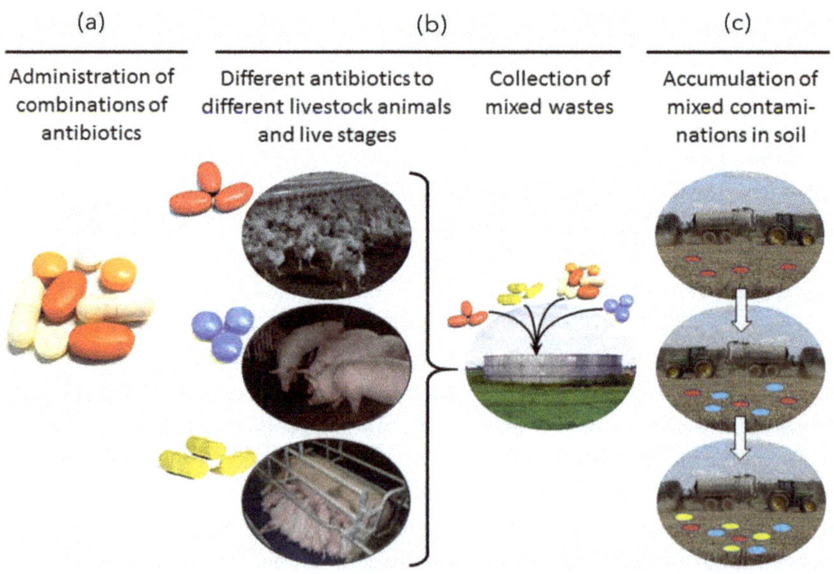

Figure 6 Formation of mixed contamination in the environment through (a) medication by using combinations of antibiotics, (b) on the farm by collecting wastes from different livestock animals and live stages, (c) by input and accumulation of residual contaminations from repeated manure application. Source: Own figure from S. Thiele-Bruhn (Univ. Trier). Single photographs with permission from M. Arenz-Leufen, I. Rönnefahrt, and UBA and from Christoph-Schulz (2018), doi: 10.1007/s00003-017-1144-7, open access.

retarded degradation in soil (Braschi et al., 2013), even of otherwise rather unstable antibiotics such as penicillins (Kotzerke et al., 2010). The repeated application of manure contaminated with antibiotics can lead to a seemingly continuous contamination level that is termed as apparent persistence or pseudo-persistence (Hamscher et al., 2005; Hamscher, 2007; Bottoni and Caroli, 2018). Not least, mixtures of antibiotic parent compounds and their metabolites, some of which exert an (altered) antibiotic activity as well, develop in soil upon biotransformation of the compounds (Pollard and Morra, 2018; Schwarz et al., 2010). A similar scenario for the accumulation of antibiotics exists for antibiotics for human use. There, also (i) combinations of antibiotics are often used for medication, (ii) mixtures accumulate over time in wastewater canals and wastewater treatment plants, and (iii) accumulate in soil with the repeated use of wastewater for irrigation and sewage sludge for fertilization, respectively (Monteiro and Boxall, 2010; Verlicchi et al., 2012; Verlicchi and Zambello, 2015).

Sorption competition between the components of a mixed contamination can occur in soil and increase the mobility and bioavailability of the contaminants. Sorption competition occurs in particular with mixtures of antibiotics of the same compound class, which compete for the same sorption sites in the soil. In addition, sorption competition increases with increasing concentrations. The resulting sorption coefficients are about 1/3 to 2/3 lower than for the individual compounds (Conkle et al., 2010; Ahmed et al., 2017). The sorption kinetics are considerably delayed, as has been shown for a single sulfonamide compared to a mixture of three sulfonamides (Thiele-Bruhn, 2019).

Regulatory limits of chemicals and ecological risk assessment are usually based on the effect of single compounds, but not taking into account mixture effects (Natal-da-Luz et al., 2011). Two characteristics in particular make the combined toxic effect of a mixture of antibiotics an important issue for hazard and risk assessment: (i) the ecotoxicity of a mixture of antibiotics is usually higher than the effect of the individual components, and consequently (ii) such a mixture may exhibit significant ecotoxicity even if all the individual contaminants are present in such low concentrations that they would not cause toxic effects on their own (Backhaus, 2014). Furthermore, it is expected that the environmental fate and effects of mixtures of antibiotics will differ from that of single compounds, because antibiotics from different structural classes have different modes of action leading to synergistically increased or antagonistically decreased effects, while in other cases additive effects are assumed. These different possibilities of a purely additive effect of mixed contaminations or antagonistic or synergistic interactions between the components are described by the two models of concentration additivity (CA) and independent action (IA) (Grimme et al., 2000). Although there are new developments, e.g. with thermodynamically based models (Gobas et al., 2018), these two models

Published by Burleigh Dodds Science Publishing Limited, 2025.

are widely used. This makes it important to note that the effects of antibiotic mixtures on the biological effects and chemical fate in the soil environment are still largely unknown (Boxall et al., 2012).

Using not single compounds but combinations of different antibiotics and/or further pharmaceuticals and components such as metals (e.g. silver Ag, copper Cu, zinc Zn) is common practice (Takahashi et al., 1987; Russell and Hugo, 1994; Haeili et al., 2014; Poole, 2017). Combinations of antibiotics and other pharmaceuticals are meant to synergistically increase the antibiotic effect and/or to reduce the formation of resistance (Riviere, 2011; Bollenbach, 2015). The use of antibiotic combinations and/or of antibiotics mixed with other synergistic compounds is actually increasing steadily and will probably continue to increase in the future (Keith et al., 2005; Caminero et al., 2010). Various studies showed that even old pharmaceuticals, sorted out in the past due to widespread resistance formation, can be revived in new combinations (Malik et al., 2014; Baym et al., 2016). The frequent use of mixtures of antibiotics in livestock production is exemplarily shown on veterinary prescriptions for medicated feedstuff during a one-year period on data from Mecklenburg-Western Pomerania, Germany (Table 2). It is shown that in 47% of all prescriptions mostly two but up to five antibiotics and other pharmaceuticals had been mixed (Thiele-Bruhn et al., 2003).

Since the use of different antibiotics in parallel and in close succession is common and significant amounts of antibiotics as well as other drugs and elements with antibiotic properties, such as copper and zinc, are excreted by the drug-treated organisms (Sarmah et al., 2006; Du and Liu, 2012), the mixed contamination of wastes is widespread. Published studies cover excreta and manure, wastewater and sewage sludge. They consistently report that manure, digestate, wastewater, sewage sludge, and compost are often contaminated

Table 2 Number of antibiotic compounds used and mixed in medicated feedstuff; data base: 2097 veterinary manufacturing orders for pharmaceuticals' admixture to feedstuff in the state of Mecklenburg-Western Pomerania, Germany (period October 2000 until September 2001; data from Thiele-Bruhn (2003a))

No. of components	5	4	3	2	2-5	1	0
	Including non-antibiotic compounds[a]						
No. of prescriptions	3	92	346	546	987	1110	
% of prescriptions	0.1	4.4	16.5	26.0	47.1	52.9	
	Only antibiotic pharmaceuticals						
No. of prescriptions	2	54	246	504	806	1290	1[b]
% of prescriptions	0.1	2.6	11.7	24.0	38.5	61.5	0.05

[a]Non-antibiotic compounds: Zn oxide, bromhexine hydrochloride, citric acid, vitamins A and D3, Mn carbonate.
[b]Feed contained only a vitamin mix.

Published by Burleigh Dodds Science Publishing Limited, 2025.

with two or more (>20) individual antibiotic compounds (Sattelberger, 1999; Winckler et al., 2004; McClellan and Halden, 2010; Arenz-Leufen, 2012; Clara et al., 2013; Li et al., 2013; Ratsak et al., 2013; An et al., 2015; Hannappel et al., 2017). As the waste materials are usually used as organic fertilizer and applied to agricultural soils, the corresponding environmental samples are consequently often also contaminated with pharmaceuticals and antibiotics in particular (Monteiro and Boxall, 2010; Li et al., 2013; Łukaszewicz et al., 2017). The studies in Table 3 show that not only single antibiotic compounds but mixtures of several compounds frequently occur in soil, so that the combined concentrations clearly exceed natural background levels. Natural soil contents of antibiotics, which are formed by the secondary metabolism of autochthonous soil microorganisms, are in the range of a few μg/kg and are mainly found in the rhizosphere of soil (Soulides, 1965; Lumsden et al., 1992). The introduced anthropogenic antibiotics belong to different structural classes and exhibit different modes of action (Hu et al., 2008; Karci and Balcioğlu, 2009; Ok et al., 2011). Thereby, antibiotics from some structural classes are more abundant (e.g. tetracyclines) while others have been determined in much lower concentrations (e.g. sulfonamides) (Thiele-Bruhn, 2003b; Hembrock-Heger et al., 2011; Hamscher and Mohring, 2012). This is due to (i) a different usage and thus release into the environment, and (ii) a different persistence and mobility in soil. Based on these consistent findings, it can therefore be concluded that the occurrence of mixed contamination in the terrestrial environment is the rule rather than the exception.

In Table 3, data from eight studies were combined, presenting concentrations of two and up to 14 different antibiotics in soils. Often, however, the number of substances found in environmental samples seems to be limited less by the contamination status of the soil than by the number of substances investigated.

Interactions between antibiotics and also further pharmaceuticals are known from clinical and pharmacological research. Similar mixture effects are expected on non-target organisms in the (soil) environment. Interactions of antibiotic pharmaceutical were reviewed by Bollenbach (2015). Synergistic effects of pharmaceutical combinations can be due to

- effects on the uptake: the first pharmaceutical increases the permeability of the cell membrane for the second pharmaceutical,
- direct physical interaction: pharmaceuticals reciprocally stabilize the binding to the target site,
- targeting of sequential metabolic steps: e.g., sulfonamides and trimethoprim both affect the folic acid biosynthesis pathway with different mode of action (inhibit dihydrofolate reductase and dihydropteroate synthetase, respectively).

Table 3 Contamination level of soils with antibiotics. Numbers of samples (No. sample), numbers of antibiotics analyzed in the study (No. antibiotics), and highest numbers of antibiotics detected in one individual sample (max No. antib) are indicated

Study	Sattelberger (1999)			Hamscher et al. (2002)			Kues et al. (2002)			Clara et al. (2013)			An et al. (2015)			Gildemeister et al. (2011)			Hannappel et al. (2017)			Zhou et al. (2012)		
No. sample	61			14			42			8			70			2			104			5		
No. antibiotics	11			4			2			7			8			9			11			13		
max No. antib.	4			2			2			2			>4			7			2			14		
	#[a]	**av.**[b]	**max**[c]	**#**[a]	**av.**[b]	**max**[c]	**#**[a]	**av.**[b]	**max**[c]	**#**[a]	**av.**[b]	**max**[c]	**#**[a]	**av.**[b]	**max**[c]	**#**[a]	**av.**[b]	**max**[c]	**#**[a]	**av.**[b]	**max**[c]	**#**[a]	**av.**[b]	**max**[c]
		µg/kg	µg/kg		µg/kg	µg/kg		µg/kg	µg/kg		µg/kg	µg/kg		µg/kg	µg/kg		µg/kg	µg/kg		µg/kg	µg/kg		µg/kg	µg/kg
Tetracycline	0			14	31.7	191.3	30	21.5	85.0				27	241	976	2	168	332				3	350	1010
Oxytetracycline	0			0									34	609	1399	2	0.1	0.2				4	368	1410
Chlortetracycline	3	87	160	14	61.2	747.5	32	10.2	38.2				45	718	1590							4	3402	12900
Doxycycline													19	120	871							2	259	499
Trimethoprim	2	50	100							2	6.2	6.5				1	118	118	1	7.0	7.0	2	1.60	3.20
Sulfadimidine	2	50	100							0			4	3.5	11.5	1	589	589	0			2	1.85	3.69
Sulfadiazine	0									0			20	71.5	760				0			1	4.95	4.95
Sulfadoxine	0																		0					
Sulfathiazole	0																		0					
Sulfamethoxaz.	2	50	100							1	3.0	3.0	13	19.4	672				1	2.0	2.0			
Sulfamerazine													12	65.8	311				0					
Sulfamethoxypy.																			0					
Sulfachloropy.																			0			1	5.11	5.11
Sulfadimethox.																			0					
Sulfaethoxypy.																			0					

Sulfaguanidine													1	[d]<LOQ	[d]<LOQ
Sulfamonomethox.													2	2.69	5.37
Sulfaquinoxaline													1	2.02	2.02
Ciprofloxacin	5	200	370							2	2.6	3.4	3	8.10	14.0
Enrofloxacin	4	103	200										4	28.4	95.8
Norfloxacin													4	25.4	58.8
Ofloxacin													3	74.3	113
Pefloxacin													2	20.5	31.0
Difloxacin													1	4.01	4.01
Tylosin	0			0											
Erythromycin							5	23.6	35.0	1	0.3	0.3	1	[d]<LOQ	[d]<LOQ
Roxythromycin							0								
Clarythromycin							0								
Lincomycin													1	92.3	92.3
Polypeptide										1	33.1	33.1			

#[a] number of samples in which the specific antibiotic was detected.
av.[b] average concentration in all samples. Average was calculated only for data above the limit of detection.
max[c] highest concentration determined in one of the investigated samples.
[d]<LOQ detected below limit of quantification.

Altered effects of antibiotic mixtures vs. the single compounds have been reported for various combinations of antibiotics. These individual case studies showed deviating effects, e.g. on the soil bacterial community composition, microbial enzyme activities and abundance of functional genes in particular of those related to the nitrogen cycle, community level physiological profiles (Ma et al., 2014; Cleary et al., 2016; David et al., 2016). Additionally, deviating effects of mixtures of antibiotics with metals or other organic agents such as pesticides have been reported (Kong et al., 2006; Yan et al., 2011; Fang et al., 2016; Liu et al., 2016). Furthermore effects on faunal species such as earthworm *Eisenia fetida* or the soil nematode *Caenorhabditis elegans* have been reported (Jonker et al., 2004; Dong et al., 2011).

However, in the case of environmental contamination, it seems difficult to predict whether the effects of the mixture are additive or independent of each other. Typically, it is assumed that pharmaceuticals having the same mode of action cause additive toxic effects, when they are combined (Vasquez et al., 2014). However, experimental findings show that binary mixtures of antibiotics from the same chemical class exhibit not only additive (sulfonamides) but also synergistic effects (e.g. macrolides, tetracyclines, and fluoroquinolones) (Yang et al., 2008). Thiele-Bruhn (2015) reported synergistic effects also for a binary mixture of sulfonamides. On the other hand, interactions between bactericidal and bacteriostatic antibiotics are largely antagonistic (Ocampo et al., 2014) as was shown by a study on the effects of binary mixtures of 21 antibiotics on *E. coli* as test organism. However, the general conclusion by Ocampo et al. (2014) was somewhat disproved by the findings of Christensen et al. (2006). In the presence of sludge bacteria numerous synergistic effects and in less cases concentration additivity were found even for combinations of bacteriostatic and bactericidal antibiotics (Christensen et al., 2006). Results from Yang et al. (2008) point to the fact that combinations of bacteriostatic antibiotics (within one or between different compound classes) exert additive or synergistic effects, while combinations of bactericidal antibiotics are additive (within compound class) or are antagonistic (between different compound classes). Furthermore, effects of mixtures frequently change with the concentrations of the mixed compounds and combinatory effects may shift, e.g. from antagonism at a low concentration level to concentration additivity at a medium level and to a synergistic effect at high concentrations (González-Pleiter et al., 2013).

One particular aspect of the undesirable effects of antibiotics in the environment is the increase in antibiotic resistance. Synergistic combinations can lead to stronger and also faster resistance development compared to that in the presence of single antibiotics (Hegreness et al., 2008). A combination of tetracycline, sulfamonomethoxine and ciprofloxacin added to soil led within 7 days and 20 days to 1.7 to 11-fold higher abundance of sulfamonomethoxine and ciprofloxacin resistant bacteria but not to increased tetracycline-resistant bacteria compared to control soils contaminated with the single compounds (Ma et al., 2014). In contrast, antagonistic pharmaceutical combinations lead to slower resistance evolution than synergistic ones (Hegreness et al., 2008; Singh and Yeh, 2017). Also metals such as Cu, Zn, Cd, and Hg exert antimicrobial effects and can promote a co-selection for antibiotic resistance (Baker-Austin et al., 2006; Seiler and Berendonk, 2012). The release of such metals combined with antibiotics into the (soil) environment in the course of agricultural activities may result in a combined selection and co-selection of antimicrobial resistant bacteria, which even makes agricultural soils hot-spots of the formation of antibiotic resistance in the environment (Seiler and Berendonk, 2012). However, the state of knowledge about adverse effects of antibiotics and, in particular, antibiotic mixtures in the environment is still incomplete and contradictory, so that it is not actually possible to derive predicted no-effect concentrations (PNEC) that trigger an increased abundance of microbial resistance genes in the soil environment (Schönfeld et al., 2017).

8 Conclusion and future trends

The intensified practices in livestock farming and the adoption of waste (water) reuse in agriculture have highlighted concerns about the presence of antibiotics in soils. Various pathways contribute to the introduction of antibiotic residues into the soil compartment, leading to variations in the types of antibiotics, concentrations, and exposure durations. Once antibiotics enter the soil, they undergo diverse processes, with persistence and mobility being crucial parameters influencing their fate.

Soils are commonly exposed to a mixture of antibiotics rather than a single antibiotic, and this phenomenon can be attributed to three primary factors. First, livestock often receives combinations of antibiotics. Second, diverse antibiotics are administered to different livestock and at various life stages to address specific pathogens and diseases. Consequently, excrements from different animals are collected in the same manure tank or lagoon, resulting in the creation of antibiotic mixtures. Third, excreta with mixed contamination is subsequently and frequently spread onto agricultural fields during reuse as fertilizer. Consistent findings indicate that the occurrence of mixed antibiotic contamination in the terrestrial environment is the norm rather than the exception. This is a crucial consideration in the risk assessment of these compounds.

Furthermore, addressing the complex behavior of both individual antibiotics and mixtures in soil requires a shift toward standardization in studying their behavior. Implementing straightforward laboratory experiments and employing fit-for-purpose analytical methodologies would facilitate the investigation of these aspects efficiently and effectively. This approach would improve our understanding of antibiotics' fate in soil, offering valuable insights for accurate risk and effect assessments. This chapter offers some suggestions to move in that direction.

Here are a few challenges that remain to be addressed in the future:

- insufficient publicly disclosed data on the administration rates of veterinary antibiotics per animal sector. Ongoing initiative: Big Data strategy for veterinary medicines in the EU | European Medicines Agency (europa.eu);
- uncertainty about the excretion rates of antibiotics;
- clear definition of persistent antimicrobials and preference to use antibiotics that degrade within the system (animal/manure/soil);
- lack of established environmental threshold values for pharmaceuticals; and
- interaction between pharmaceuticals and other soil contaminants, particularly in relation to antibiotic resistance. For instance, recent research has explored soil plastispheres (Zhu et al., 2022) and Quaternary ammonium compounds (Mulder et al., 2018).

9 Acknowledgments

The research of the Dutch case study was performed by Wageningen Environmental Research, financed by the Province of Gelderland (The Netherlands) and co-financed by the waterboards 'Vallei & Veluwe' and 'Rijn & IJssel', the Dutch Ministry of Infrastructure and Water Management and the Dutch Ministry of Agriculture, Nature and Food Quality (project no. BO 20.004 127). Drinking water company Vitens contributed in kind. Furthermore, part of the text (in Sections 3 and 7) is an excerpt from the Report No. (UBA-FB) FB000071 32/2019 (UBA Project No. 92954) published as UBA Texte 32/2019 (Thiele-Bruhn, 2019). Likewise, part of the text (in Sections 2 and 4) is an excerpt from the PhD Thesis (Rakonjac, 2023).

10 Where to look for further information

A standard introduction to the topic of veterinary antibiotics in the environment can be found in Boxall et al. (2004) and Sarmah et al. (2006). If the focus is specifically on soil, a comprehensive overview is provided by Thiele-Bruhn (2003b). For more details on the uptake of veterinary medicines from soil into plants, Boxall et al. (2006) offers one of the foundational reviews. For further insight into experimental and testing procedures, the OECD guidelines, as mentioned earlier in this chapter, can be consulted.

In addition to the studies referenced in this chapter, valuable information on the (environmental) characteristics and behavior of specific antibiotic compounds can be found in the following sources:

- European Medicines Agency (https://www.ema.europa.eu/en/homepage);
- European Chemicals Agency (https://echa.europa.eu);
- Veterinary Substances DataBase (https://sitem.herts.ac.uk/aeru/vsdb/index.htm).

For a broader perspective on medicines in the environment and the interplay between scientific and regulatory issues, as well as how these have evolved over the years, one might refer to Boxall et al. (2012, 2024). For an even wider scope, Van den Brink et al. (2018) offers valuable insights.

11 References

Ahmed, M. B., Zhou, J. L., Ngo, H. H., et al. (2017) Single and competitive sorption properties and mechanism of functionalized biochar for removing sulfonamide antibiotics from water. *Chemical Engineering Journal* 311: 348–358.

An, J., Chen, H., Wei, S., et al. (2015) Antibiotic contamination in animal manure, soil, and sewage sludge in Shenyang, Northeast China. *Environmental Earth Sciences* 74: 5077–5086.

Arenz-Leufen, M. (2012) *Manure composition and its influence on the input and transfer of veterinary antibiotics in soil.* Trier: Universitäts-Druckerei.

Aust, M. O., Godlinski, F., Travis, G. R., et al. (2008a) Distribution of sulfamethazine, chlortetracycline and tylosin in manure and soil of Canadian feedlots after subtherapeutic use in cattle. *Environmental Pollution* 156: 1243–1251.

Aust, M. O., Thiele-Bruhn, S., Jandl, G., et al. (2008b) Determination of six membered ring sulfonamides used in veterinary medicine in manure, soil and soil leachates using LC-ESI mass spectrometry. *Fresenius Environmental Bulletin* 17: 331–340.

Aust, M. O., Thiele-Bruhn, S., Seeger, J., et al. (2010) Sulfonamides leach from sandy loam soils under common agricultural practice. *Water, Air, and Soil Pollution* 211: 143–156.

Backhaus, T. (2014) Medicines, shaken and stirred: a critical review on the ecotoxicology of pharmaceutical mixtures. *Philosophical Transactions of the Royal Society of London. Series B, Biological Sciences* 369. https://doi.org/10.1098/rstb.2013.0585

Baker-Austin, C., Wright, M. S., Stepanauskas, R., et al. (2006) Co-selection of antibiotic and metal resistance. *Trends in Microbiology* 14: 176–182.

Balzer, F., Zöhlke, S. and Hannappel, S. (2016) Antibiotics in groundwater under locations with high livestock density in Germany. *Water Science and Technology: Water Supply* 16: 1361–1369.

Baym, M., Stone, L. K. and Kishony, R. (2016) Multidrug evolutionary strategies to reverse antibiotic resistance. *Science* 351. https://doi.org/10.1126/science.aad3292

Published by Burleigh Dodds Science Publishing Limited, 2025.

Berendsen, B. J. A., Lahr, J., Nibbeling, C., et al. (2018) The persistence of a broad range of antibiotics during calve, pig and broiler manure storage. *Chemosphere* 204: 267–276.

Berendsen, B. J. A., Roelofs, G., van Zanten, B., et al. (2021) A strategy to determine the fate of active chemical compounds in soil; applied to antimicrobially active substances. *Chemosphere* 279: 130495.

Berg, F. v. d., Tiktak, A., Boesten, J. J. T. I., et al. (2016) *PEARL model for pesticide behaviour and emissions in soil-plant systems*. Wageningen: WOT Natuur & Milieu.

Bixio, D., Thoeye, C., De Koning, J., et al. (2006) Wastewater reuse in Europe. *Desalination* 187: 89–101.

Blackwell, P. A., Holten Lützhøft, H.-C., Ma, H.-P., et al. (2004) Ultrasonic extraction of veterinary antibiotics from soils and pig slurry with SPE clean-up and LC–UV and fluorescence detection. *Talanta* 64: 1058–1064.

Bollenbach, T. (2015) Antimicrobial interactions: mechanisms and implications for drug discovery and resistance evolution. *Current Opinion in Microbiology* 27: 1–9.

Boreen, A. L., Arnold, W. A. and McNeill, K. (2005) Triplet-sensitized photodegradation of sulfa drugs containing six-membered heterocyclic groups: identification of an SO2 extrusion photoproduct. *Environmental Science and Technology* 39: 3630–3638.

Bottoni, P. and Caroli, S. (2018) Presence of residues and metabolites of pharmaceuticals in environmental compartments, food commodities and workplaces: a review spanning the three-year period 2014–2016. *Microchemical Journal* 136: 2–24.

Bourdat-Deschamps, M., Ferhi, S., Bernet, N., et al. (2017) Fate and impacts of pharmaceuticals and personal care products after repeated applications of organic waste products in long-term field experiments. *Science of the Total Environment* 607–608: 271–280.

Boxall, A. B., Fogg, L. A., Blackwell, P. A., et al. (2004) Veterinary medicines in the environment. *Reviews of Environmental Contamination Toxicology* 180: 1–91.

Boxall, A. B. A., Johnson, P., Smith, E. J., et al. (2006) Uptake of veterinary medicines from soils into plants. *Journal of Agricultural and Food Chemistry* 54: 2288–2297.

Boxall, A. B. A., Rudd, M. A. , Brooks, B. W., et al. (2012) Pharmaceuticals and personal care products in the environment: what are the big questions? *Environmental Health Perspectives* 120: 1221–1229.

Boxall, A. B. A. and Brooks, B. W. (2024) Pharmaceuticals and Personal Care Products in the Environment: What Progress Has Been Made in Addressing the Big Research Questions?. *Environ Toxicol Chem* 43: 481–487. https://doi.org/10.1002/etc.5827

Brambilla, G., Patrizii, M., De Filippis, S. P., et al. (2007) Oxytetracycline as environmental contaminant in arable lands. *Analytica Chimica Acta* 586: 326–329.

Braschi, I., Blasioli, S., Fellet, C., et al. (2013) Persistence and degradation of new β-lactam antibiotics in the soil and water environment. *Chemosphere* 93: 152–159.

Burkhardt, M., Stamm, C., Waul, C., et al. (2005) Surface runoff and transport of sulfonamide antibiotics and tracers on manured grassland. *Journal of Environmental Quality* 34: 1363–1371.

Caminero, J. A., Sotgiu, G., Zumla, A., et al. (2010) Best drug treatment for multidrug-resistant and extensively drug-resistant tuberculosis. *The Lancet Infectious Diseases* 10: 621–629.

Chander, K., Ulrich, E., Joergensen, R. G., et al. (2002) Decomposition of 14C labelled wheat straw in repeatedly fumigated and non-fumigated soils with different levels of heavy metal contamination. *Biology and Fertility of Soils* 35: 86–91.

Chander, Y., Kumar, K., Goyal, S. M., et al. (2005) Antibacterial activity of soil-bound antibiotics. *Journal of Environmental Quality* 34: 1952–1957.

Christensen, A. M., Ingerslev, F. and Baun, A. (2006) Ecotoxicity of mixtures of antibiotics used in aquacultures. *Environmental Toxicology and Chemistry* 25: 2208.

Christoph-Schulz, I. (2018) SocialLab – Nutztierhaltung im Spiegel der Gesellschaft. *Journal of Consumer Protection and Food Safety* 13: 145–236.

Clara, M., Scheffknecht, C. and Weiß, S. (2013) *Eintrag von Arzneimittelwirkstoffen in die Umwelt.* Bregenz: Umweltbundesamt, Umweltinstitut des Landes Vorarlberg.

Cleary, D. W., Bishop, A. H., Zhang, L., et al. (2016) Long-term antibiotic exposure in soil is associated with changes in microbial community structure and prevalence of class 1 integrons. *FEMS Microbiology Ecology* 92.

Conkle, J. L., Lattao, C., White, J. R., et al. (2010) Competitive sorption and desorption behavior for three fluoroquinolone antibiotics in a wastewater treatment wetland soil. *Chemosphere* 80: 1353–1359.

Coors, A., Edwards, M., Lorenz, P., et al. (2016) Biosolids applied to agricultural land: influence on structural and functional endpoints of soil fauna on a short-and long-term scale. *Science of the Total Environment* 562: 312–326.

Cycoń, M., Mrozik, A. and Piotrowska-Seget, Z. (2019) Antibiotics in the soil environment–degradation and their impact on microbial activity and diversity. *Frontiers in Microbiology* 10.

David, V., Roose-Amsaleg, C., Nelieu, S., et al. (2016) Mixture effects of antibiotics on soil microbial nitrogen processes. *Presentation at 1st International Conference on Risk Assessment of Pharmaceuticals in the Environment - ICRAPHE; Paris, France; 8–9 September 2016.* http://prodinra.inra.fr/record/382231

de Vries, W., Schulte-Uebbing, L., Kros, H., et al. (2021) Spatially explicit boundaries for agricultural nitrogen inputs in the European Union to meet air and water quality targets. *Science of the Total Environment* 786: 147283.

Dolliver, H., Kumar, K. and Gupta, S. (2007) Sulfamethazine uptake by plants from manure-amended soil. *Journal of Environmental Quality* 36: 1224–1230.

Domínguez, C., Flores, C., Caixach, J., et al. (2014) Evaluation of antibiotic mobility in soil associated with swine-slurry soil amendment under cropping conditions. *Environmental Science and Pollution Research* 21: 12336–12344.

Dong, L., Xie, X. and Zhou, Q. (2011) The DNA damage and joint effects of two typical veterinary drugs on the earthworm *Eisenia foetida*. *Huanjing Kexue Xuebao/Acta Scientiae Circumstantiae* 31: 2798–2803.

Du, L. and Liu, W. (2012) Occurrence, fate, and ecotoxicity of antibiotics in agro-ecosystems. A review. *Agronomy for Sustainable Development* 32: 309–327.

EMEA, R. (2008) Environmental impact assessment for veterinary medicinal products in support of the VICH guidelines GL6 and GL38. EMEA/CVMP/ERA/418282/2005-Rev. 1.

European Commission. (2021) Union Register of medicinal products for human use. Comirnaty [Internet]. Available from: https://ec.europa.eu/health/documents/community-register/html/h1528.htm

Fang, H., Han, L., Cui, Y., et al. (2016) Changes in soil microbial community structure and function associated with degradation and resistance of carbendazim and chlortetracycline during repeated treatments. *Science of the Total Environment* 572: 1203–1212.

Faúndez Urbina, C. A., van den Berg, F., van Dam, J. C., et al. (2020) Parameter sensitivity of SWAP–PEARL models for pesticide leaching in macroporous soils. *Vadose Zone Journal* 19: e20075.

Figueroa-Diva, R. A., Vasudevan, D. and Mackay, A. A. (2010) Trends in soil sorption coefficients within common antimicrobial families. *Chemosphere* 79: 786–793.

Figueroa, R. A., Leonard, A. and Mackay, A. A. (2004) Modeling tetracycline antibiotic sorption to clays. *Environmental Science & Technology* 38: 476–483.

Figueroa, R. A. and Mackay, A. A. (2005) Sorption of oxytetracycline to iron oxides and iron oxide-rich soils. *Environmental Science & Technology* 39: 6664–6671.

Förster, M., Laabs, V., Lamshöft, M., et al. (2009) Sequestration of manure-applied sulfadiazine residues in soils. *Environmental Science & Technology* 43: 1824–1830.

Freitag, M., Yolcu, D. H., Hayen, H., et al. (2008) Screening of antibiotic transfer from soil into cereals in areas with high livestock concentration in Northrhine Westphalia (Germany). *Screening zum antibiotika-transfer aus dem boden in getreide in regionen Nordrhein-Westfalens mit großen viehbeständen* 3: 174–184.

Gao, J. A. and Pedersen, J. A. (2005) Adsorption of sulfonamide antimicrobial agents to clay minerals. *Environmental Science & Technology* 39: 9509–9516.

Gildemeister, D., Tauchnitz, N. and Berkner, S. (2011) *Tierarzneimittelmischungen in landwirtschaftlich genutzten Böden*. Poster presentation at conference: Böden verstehen – Böden nutzen – Böden fit machen. DBG, 03.09.– 09.09.2011, Berlin. Mitteilungen der Deutschen Bodenkundlichen Gesellschaft.

Girardi, C., Greve, J., Lamshöft, M., et al. (2011) Biodegradation of ciprofloxacin in water and soil and its effects on the microbial communities. *Journal of Hazardous Materials* 198: 22–30.

Gobas, F. A. P. C., Mayer, P., Parkerton, T. F., et al. (2018) A chemical activity approach to exposure and risk assessment of chemicals: focus articles are part of a regular series intended to sharpen understanding of current and emerging topics of interest to the scientific community. *Environmental Toxicology and Chemistry* 37: 1235–1251.

González-Pleiter, M., Gonzalo, S., Rodea-Palomares, I., et al. (2013) Toxicity of five antibiotics and their mixtures towards photosynthetic aquatic organisms: implications for environmental risk assessment. *Water Research* 47: 2050–2064.

Grenni, P., Ancona, V. and Barra Caracciolo, A. (2018) Ecological effects of antibiotics on natural ecosystems: a review. *Microchemical Journal* 136: 25–39.

Grimme, L. H., Altenburger, R., Backhaus, T., et al. (2000) Combined effects of environmental pollutants in ecotoxicology: biometrical models as concepts for prediction and their experimental proof (Review). *Umweltwissenschaften und Schadstoff-Forschung* 12: 226–234.

Grossberger, A., Hadar, Y., Borch, T., et al. (2014) Biodegradability of pharmaceutical compounds in agricultural soils irrigated with treated wastewater. *Environmental Pollution* 185: 168–177.

Grote, M., Schwake-Anduschus, C., Michel, R., et al. (2007) Incorporation of veterinary antibiotics into crops from manured soil. *Landbauforschung Völkenrode* 57: 25–32.

Guo, K., Liu, Y. F., Zeng, C., et al. (2014) Global research on soil contamination from 1999 to 2012: a bibliometric analysis. *Acta Agriculturae Scandinavica, Section B–Soil & Plant Science* 64: 377–391.

Haeili, M., Moore, C., Davis, C. J. C., et al. (2014) Copper complexation screen reveals compounds with potent antibiotic properties against methicillin-resistant Staphylococcus aureus. *Antimicrobial Agents and Chemotherapy* 58: 3727–3736.

Hammesfahr, U. (2011) *Alterations in soil microbial structure and function due to different manure and antibiotic loads.* Trier: Universitäts-Druckerei.

Hamscher, G. (2007) Tierarzneimittel in Boden, Sicker- und Grundwasser: Langzeituntersuchungen an Dauerbeobachtungsflächen in Niedersachsen 2000–2006. In: Röder, R., Weiß, K. and Sengl, M. (eds). Münchener Beiträge zur Abwasser-, Fischerei- und Flussbiologie. Oldenbourg Industrieverlag, München, 61–82.

Hamscher, G. and Mohring, S. A. I. (2012) Tierarzneimittel in Böden und in der aquatischen Umwelt. *Chemie Ingenieur Technik* 84: 1052–1061.

Hamscher, G., Pawelzick, H. T., Höper, H., et al. (2005) Different behavior of tetracyclines and sulfonamides in sandy soils after repeated fertilization with liquid manure. *Environmental Toxicology and Chemistry* 24: 861–868.

Hamscher, G., Sczesny, S., Höper, H., et al. (2002) Determination of persistent tetracycline residues in soil fertilized with liquid manure by high-performance liquid chromatography with electrospray ionization tandem mass spectrometry. *Analytical Chemistry* 74: 1509–1518.

Hannappel, S., Köpp, C. and Rejman-Rasinska, E. (2017) *Regionaler Themenbericht: Ermittlung der Ursachen des Eintrages von Tierarzneimitteln in das oberflächennahe Grundwasser; Datenauswertung 2012 bis 2016.* Norden: Niedersächsischer Landesbetrieb für Wasserwirtschaft, Küsten- und Naturschutz (NLWKN).

Hegreness, M., Shoresh, N., Damian, D., et al. (2008) Accelerated evolution of resistance in multidrug environments. *Proceedings of the National Academy of Sciences of the United States of America* 105: 13977–13981.

Hembrock-Heger, A., Nießner, M. and Reupert, R. (2011) Tierarzneimittel in landwirtschaftlich genutzten Böden und oberflächennahem Grundwasser in Nordrhein-Westfalen. *Bodenschutz* 4: 109–113.

Herrmann, M., Olsson, O., Fiehn, R., et al. (2015) The significance of different health institutions and their respective contributions of active pharmaceutical ingredients to wastewater. *Environment International* 85: 61–76.

Heuer, H. and Smalla, K. (2007) Manure and sulfadiazine synergistically increased bacterial antibiotic resistance in soil over at least two months. *Environmental Microbiology* 9: 657–666.

Heuer, H., Solehati, Q., Zimmerling, U., et al. (2011) Accumulation of sulfonamide resistance genes in arable soils due to repeated application of manure containing sulfadiazine. *Applied and Environmental Microbiology* 77: 2527–2530.

Höper. H., Kues, J., Nau, H., et al. (2002) Eintrag und Verbleib von Tierarzneimittelwirkstoffen in Böden. *Bodenschutz* 4: 141–148.

Höper, H., Schäfer, W. and Hamscher, G. (2007) Zur Stoffverlagerung von Tierarzneimittelwirkstoffen mit dem Sickerwasser. In: Röder, R., Weiß, K. and Sengl, M. (eds) *Tierarzneimittel in der Umwelt.* München: Oldenbourg Industrieverlag, 83–104.

Hu, X. G., Luo, Y., Zhou, Q. X., et al. (2008) Determination of thirteen antibiotics residues in manure by solid phase extraction and high performance liquid chromatography. *Fenxi Huaxue/ Chinese Journal of Analytical Chemistry* 36: 1162–1166.

Hurtado, C., Montano-Chávez, Y. N., Domínguez, C., et al. (2017) Degradation of emerging organic contaminants in an agricultural soil: decoupling biotic and abiotic processes. *Water, Air, and Soil Pollution* 228. https://doi.org/10.1007/s11270-017-3402-9

Hutchings, M. I., Truman, A. W. and Wilkinson, B. (2019) Antibiotics: past, present and future. *Current Opinion in Microbiology* 51: 72–80.

Jacobsen, A. M., Halling-Sørensen, B., Ingerslev, F., et al. (2004) Simultaneous extraction of tetracycline, macrolide and sulfonamide antibiotics from agricultural soils using pressurised liquid extraction, followed by solid-phase extraction and liquid chromatography–tandem mass spectrometry. *Journal of Chromatography A* 1038: 157–170.

Jahnke, A., Mayer, P., Schäfer, S., et al. (2016) Strategies for transferring mixtures of organic contaminants from aquatic environments into bioassays. *Environmental Science & Technology* 50: 5424–5431.

Jansen, L. J. M., van de Schans, M. G. M., de Boer, D., et al. (2019) A new extraction procedure to abate the burden of non-extractable antibiotic residues in manure. *Chemosphere* 224: 544–553.

Janusch, F., Scherz, G., Mohring, S. A. I., et al. (2014) Determination of fluoroquinolones in chicken feces – A new liquid–liquid extraction method combined with LC–MS/MS. *Environmental Toxicology and Pharmacology* 38: 792–799.

Jonker, M. J., Sweijen, R. A. J. C. and Kammenga, J. E. (2004) Toxicity of simple mixtures to the nematode *Caenorhabditis elegans* in relation to soil sorption. *Environmental Toxicology and Chemistry* 23: 480.

Kahle, M. and Stamm, C. (2007) Sorption of the veterinary antimicrobial sulfathiazole to organic materials of different origin. *Environmental Science and Technology* 41: 132–138.

Karci, A. and Balcioğlu, I. A. (2009) Investigation of the tetracycline, sulfonamide, and fluoroquinolone antimicrobial compounds in animal manure and agricultural soils in Turkey. *Science of The Total Environment* 407: 4652–4664.

Katsivelou, E., Perruchon, C., Karas, P. A., et al. (2023) Accelerated dissipation, soil microbial toxicity and dispersal of antimicrobial resistance in soils repeatedly exposed to tiamulin, tilmicosin and sulfamethoxazole. *Science of the Total Environment* 893: 164817.

Kay, P., Blackwell, P. A. and Boxall, A. B. A. (2004) Fate of veterinary antibiotics in a macroporous tile drained clay soil. *Environmental Toxicology and Chemistry* 23: 1136–1144.

Kay, P., Blackwell, P. A. and Boxall, A. B. A. (2005) A lysimeter experiment to investigate the leaching of veterinary antibiotics through a clay soil and comparison with field data. *Environmental Pollution* 134: 333–341.

Keith, C. T., Borisy, A. A. and Stockwell, B. R. (2005) Multicomponent therapeutics for networked systems. *Nature Reviews Drug Discovery* 4: 71–78.

Kim, K.-R., Owens, G., Kwon, S.-I., et al. (2011) Occurrence and environmental fate of veterinary antibiotics in the terrestrial environment. *Water, Air, & Soil Pollution* 214: 163–174.

Kong, W. D., Zhu, Y. G., Fu, B. J., et al. (2006) The veterinary antibiotic oxytetracycline and Cu influence functional diversity of the soil microbial community. *Environmental Pollution* 143: 129–137.

Kostich, M. S., Batt, A. L. and Lazorchak, J. M. (2014) Concentrations of prioritized pharmaceuticals in effluents from 50 large wastewater treatment plants in the US and implications for risk estimation. *Environmental Pollution* 184: 354–359.

Published by Burleigh Dodds Science Publishing Limited, 2025.

Kotzerke, A., Hammesfahr, U., Kleineidam, K., et al. (2010) Influence of difloxacin-contaminated manure on microbial community structure and function in soils. *Biology and Fertility of Soils*. https://doi.org/10.1007/s00374-010-0517-1

Kreuzig, R., Höltge, S., Brunotte, J., et al. (2005) Test-plot studies on runoff of sulfonamides from manured soils after sprinkler irrigation. *Environmental Toxicology and Chemistry* 24: 777–781.

Kroes, J. G., Bartholomeus, R., et al. (2017) *Swap version 4: theory description and user manual*. Wageningen Environmental Research Place of Publication: Wageningen

Kues, J., Höper, H., Pawelzick, H. T., et al. (2002) Gehalte von Tierarzneimitteln in Wirtschaftsdüngern, Eintrag in Böden und Abbauverhalten. In: B. BMVEL (ed) *KTBL-Schrift: Landwirtschaftliche Verwertung von Klärschlamm, Gülle und anderen Düngern unter Berücksichtigung des Umwelt- und Verbraucherschutzes*. Darmstadt: KTBL, 317–322.

Kulkarni, P., Olson, N. D., Raspanti, G. A., et al. (2017) Antibiotic concentrations decrease during wastewater treatment but persist at low levels in reclaimed water. *International Journal of Environmental Research and Public Health* 14: 668.

Kümmerer, K. (2003) Significance of antibiotics in the environment. *Journal of Antimicrobial Chemotherapy* 52: 5–7.

Kusturica, M., Tomas, A. and Sabo, A. (2017) Disposal of unused drugs: knowledge and behavior among people around the world. *Reviews of Environmental Contamination and Toxicology* 240: 71–104.

Lagerwerf, L., Bannink, A., Van Bruggen, C., et al. (2019) Methodology for estimating emissions from agriculture in the Netherlands: calculations of CH4, NH3, N2O, NOx, NMVOC, PM10, PM2. 5 and CO2 with the National Emission Model for Agriculture (NEMA), Update 2019. Statutory Research Tasks Unit for Nature & the Environment.

Lahr, J., Derksen, A., Wipfler, L., et al. (2018) Diergeneesmiddelen & hormonen in het milieu door de toediening van drijfmest: Een verkennende studie in de Provincie Gelderland naar antibiotica, antiparasitaire middelen, coccidiostatica en natuurlijke hormonen in mest, (water)bodem, grondwater en oppervlaktewater. Wageningen: Wageningen Environmental Research. https://edepot.wur.nl/455340

Lahr, J., Laak, T. L. t. and Derksen, A. (2014) *Screening van hot spots van nieuwe verontreinigingen: een pilot studie in bodem, grondwater en oppervlaktewater*. Wageningen: Alterra.

Langhammer, J. P., Führ, F. and Büning-Pfaue, H. (1990) Verbleib von Sulfonamid-Rückständen aus der Gülle in Boden und Nutzpflanze. *Lebensmittelchemie* 44: 93.

Li, W., Shi, Y., Gao, L., et al. (2013) Occurrence, distribution and potential affecting factors of antibiotics in sewage sludge of wastewater treatment plants in China. *Science of the Total Environment* 445–446: 306–313.

Liu, A., Cao, H., Yang, Y., et al. (2016) Combinational effects of sulfomethoxazole and copper on soil microbial community and function. *Environmental Science and Pollution Research* 23: 4235–4241.

Liu, F., Ying, G. G., Tao, R., et al. (2009) Effects of six selected antibiotics on plant growth and soil microbial and enzymatic activities. *Environmental Pollution* 157: 1636–1642.

Łukaszewicz, P., Maszkowska, J., Mulkiewicz, E., et al. (2017) Impact of veterinary pharmaceuticals on the agricultural environment: a re-inspection. *Reviews of Environmental Contamination and Toxicology* 243:89–148. https://doi.org/10.1007/398_2016_16

Lumsden, R. D., Locke, J. C., Adkins, S. T., et al. (1992) Isolation and localization of the antibiotic gliotoxin produced by Gliocladium virens from alginate prill in soil and soilless media. *Phytopathol* 82: 230–235.

Ma, J., Lin, H., Sun, W., et al. (2014) Soil microbial systems respond differentially to tetracycline, sulfamonomethoxine, and ciprofloxacin entering soil under pot experimental conditions alone and in combination. *Environmental Science and Pollution Research* 21: 7436–7448.

Malik, M., Li, L., Zhao, X., et al. (2014) Lethal synergy involving bicyclomycin: an approach for reviving old antibiotics. *Journal of Antimicrobial Chemotherapy* 69: 3227–3235.

Marti, R., Scott, A., Tien, Y. C., et al. (2013) Impact of manure fertilization on the abundance of antibiotic-resistant bacteria and frequency of detection of antibiotic resistance genes in soil and on vegetables at harvest. *Applied and Environmental Microbiology* 79: 5701–5709.

Martínez-Carballo, E., González-Barreiro, C., Scharf, S., et al. (2007) Environmental monitoring study of selected veterinary antibiotics in animal manure and soils in Austria. *Environmental Pollution* 148: 570–579.

McClellan, K. and Halden, R. U. (2010) Pharmaceuticals and personal care products in archived U.S. biosolids from the 2001 EPA national sewage sludge survey. *Water Research* 44: 658–668.

Michael, I., Rizzo, L., McArdell, C., et al. (2013) Urban wastewater treatment plants as hotspots for the release of antibiotics in the environment: a review. *Water Research* 47: 957–995.

Michelini, L. (2012) *Sulfonamide accumulation and effects on herbaceous and woody plants and microorganisms.* Padua: Università degli Studi di Padova.

Michelini, L., La Rocca, N., Rascio, N., et al. (2013) Structural and functional alterations induced by two sulfonamide antibiotics on barley plants. *Plant Physiology and Biochemistry* 67: 55–62.

Michelini, L., Meggio, F., Reichel, R., et al. (2015) Sulfadiazine uptake and effects in common hazel (*Corylus avellana* L.). *Environmental Science and Pollution Research* 22: 13362–13371.

Michelini, L., Reichel, R., Werner, W., et al. (2012) Sulfadiazine uptake and effects on *Salix fragilis* L. and *Zea mays* L. plants. *Water, Air, and Soil Pollution* 223: 5243–5257.

Miller, G. C. and Donaldson, S. G. (1994) Factors affecting photolysis of organic compounds on soils. In: Helz, G. R., Zepp, R. G. and Currier, R. W. (eds) *Aquatic and Surface Photochemistry.* Boca Raton, FL: Lewis Publishers, 97–109.

Monteiro, S. C. and Boxall, A. B. A. (2010) Occurrence and fate of human pharmaceuticals in the environment. *Reviews of Environmental Contamination and Toxicology* 202: 53–154.

Mulder, I., Siemens, J., Sentek, V., et al. (2018) Quaternary ammonium compounds in soil: implications for antibiotic resistance development. *Reviews in Environmental Science and Bio/Technology* 17: 159–185.

Müller, S. R., Singer, H., Stoob, K., et al. (2003) Occurence and fate of antibiotics in manure, soil and water. *Mitteilungen aus Lebensmittelwissenschaften und Hygiene* 94: 574–578.

Natal-da-Luz, T., Ojeda, G., Pratas, J., et al. (2011) Toxicity to Eisenia andrei and Folsomia candida of a metal mixture applied to soil directly or via an organic matrix. *Ecotoxicology and Environmental Safety* 74: 1715–1720.

Published by Burleigh Dodds Science Publishing Limited, 2025.

Negreanu, Y., Pasternak, Z., Jurkevitch, E., et al. (2012) Impact of treated wastewater irrigation on antibiotic resistance in agricultural soils. *Environmental Science & Technology* 46: 4800–4808.

Nguyen, M-K., Lin, C., Nguyen, H-L., et al. (2023) Occurrence, fate, and potential risk of pharmaceutical pollutants in agriculture: challenges and environmentally friendly solutions. *Science of The Total Environment* 899: 165323.

Ocampo, P. S., Lázár, V., Papp, B., et al. (2014) Antagonism between bacteriostatic and bactericidal antibiotics is prevalent. *Antimicrobial Agents and Chemotherapy* 58: 4573–4582.

OECD. (2000) *Test No. 106: adsorption – Desorption Using a Batch Equilibrium Method.*

OECD. (2002) *Test No. 307: aerobic and Anaerobic Transformation in Soil.*

OECD. (2004) *Test No. 312: leaching in Soil Columns.*

Ok, Y. S., Kim, S. C., Kim, K. R., et al. (2011) Monitoring of selected veterinary antibiotics in environmental compartments near a composting facility in Gangwon Province, Korea. *Environmental Monitoring and Assessment* 174: 693–701.

Pan, M. and Chu, L. M. (2017) Transfer of antibiotics from wastewater or animal manure to soil and edible crops. *Environmental Pollution* 231: 829–836.

Peng, F. J., Ying, G. G., Liu, Y. S., et al. (2015) Joint antibacterial activity of soil-adsorbed antibiotics trimethoprim and sulfamethazine. *Science of The Total Environment* 506–507: 58–65.

Pollard, A. T. and Morra, M. J. (2018) Fate of tetracycline antibiotics in dairy manure-amended soils. *Environmental Reviews* 26: 102–112.

Poole, K. (2017) At the nexus of antibiotics and metals: the impact of Cu and Zn on antibiotic activity and resistance. *Trends in Microbiology* 25: 820–832.

Posthuma, L., Eijsackers, H. J. P., Koelmans, A. A., et al. (2008) Ecological effects of diffuse mixed pollution are site-specific and require higher-tier risk assessment to improve site management decisions: a discussion paper. *Science of the Total Environment* 406: 503–517.

Rakonjac, N. (2023) *Emissions of veterinary pharmaceuticals from livestock breeding in the Netherlands.* Wageningen: Wageningen University.

Rakonjac, N., Roex, E. and Beeltje, H. (2024) Surface water monitoring of chemicals associated with animal husbandry in an agricultural region in the Netherlands using passive sampling. *Environmental Monitoring and Assessment* 196: 670.

Rakonjac, N., van der Zee, S. E. A. T. M., Wipfler, L., et al. (2022) Emission estimation and prioritization of veterinary pharmaceuticals in manure slurries applied to soil. *Science of the Total Environment* 815: 152938.

Rakonjac, N., van der Zee, S. E. A. T. M., Wipfler, L., et al. (2023) An analytical framework on the leaching potential of veterinary pharmaceuticals: a case study for the Netherlands. *Science of the Total Environment* 859: 160310.

Ratsak, C., Guhl, B., Zühlke, S., et al. (2013) Veterinary antibiotic residues in manure and digestates in Northrhine-Westfalia. *Environmental Sciences Europe* 25: 7.

Rietra, R. P. J. J., Berendsen, B. J. A., Mi-Gegotek, Y., et al. (2024) Prediction of the mobility and persistence of eight antibiotics based on soil characteristics. *Heliyon* 10: e23718.

Riviere, J. E. (2011) *Comparative pharmacokinetics: principles, techniques and applications.* Hoboken, NJ: Wiley-Blackwell.

Rosendahl, I., Siemens, J., Groeneweg, J., et al. (2011) Dissipation and sequestration of the veterinary antibiotic sulfadiazine and its metabolites under field conditions. *Environmental Science and Technology* 45: 5216–5222.

Russell, A. D. and Hugo, W. B. (1994) Antimicrobial activity and action of silver. *Progress in Medicinal Chemistry*. 351–370.

Sarmah, A. K., Meyer, M. T. and Boxall, A. B. A. (2006) A global perspective on the use, sales, exposure pathways, occurrence, fate and effects of veterinary antibiotics (VAs) in the environment. *Chemosphere* 65: 725–759.

Sattelberger, R. (1999) *Drug residues in the environment*. Wien, Österreich: Berichte des Umweltbundesamtes, Umweltbundesamt GmbH.

Schmidt, J. E., Christensen, N., Batstone, D. J., et al. (2006) Safe recycling of sewage sludge on agricultural land–biowaste. *Process Safety and Environmental Protection* 84: 253–257.

Schmitt, H., Stoob, K., Hamscher, G., et al. (2006) Tetracyclines and tetracycline resistance in agricultural soils: microcosm and field studies. *Microbial Ecology*. 51: 267–276.

Schmitt, H., ter Laak, T. and Duis, K. (2017) *Development and dissemination of antibiotic resistance in the environment under environmentally relevant concentrations of antibiotics and its risk assessment*. Dessau-Roßlau: Umweltbundesamt.

Schönfeld, J., Konradi, S., Berkner, S., et al. (2017) Antimikrobielle Resistenzen in der Umwelt – Gibt es Neues zum bekannten Phänomen? *UMID – Umwelt + Mensch Informationsdienst* 2: 7–19.

Schwarz, J., Aust, M. O. and Thiele-Bruhn, S. (2010) Metabolites from fungal laccase-catalysed transformation of sulfonamides. *Chemosphere* 81: 1469–1476.

Schwarz, J., Knicker, H., Schaumann, G. E., et al. (2015) Enzymatic transformation and bonding of sulfonamide antibiotics to model humic substances. *Journal of Chemistry* 2015.

Schwarz, J., Thiele-Bruhn, S., Eckhardt, K.-U., et al. (2012) Sorption of sulfonamide antibiotics to soil organic sorbents: batch experiments with model compounds and computational chemistry. *ISRN Soil Science* 2012: 10.

Seiler, C. and Berendonk, T. U. (2012) Heavy metal driven co-selection of antibiotic resistance in soil and water bodies impacted by agriculture and aquaculture. *Frontiers in Microbiology* 3. https://doi.org/10.3389/fmicb.2012.00399

Sengeløv, G., Agerso, Y., Halling-Sørensen, B., et al. (2003) Bacterial antibiotic resistance levels in Danish farmland as a result of treatment with pig manure slurry. *Environment International* 28: 587–595.

Singh, N. and Yeh, P. J. (2017) Suppressive drug combinations and their potential to combat antibiotic resistance. *Journal of Antibiotics* 70: 1033–1042.

Soulides, D. A. (1965) Antibiotics in soils. VII. Production of streptomycin and tetracyclines in soils. *Soil Science* 100: 200–206.

Spielmeyer, A., Höper, H. and Hamscher, G. (2017) Long-term monitoring of sulfonamide leaching from manure amended soil into groundwater. *Chemosphere* 177: 232–238.

Stamm, C., Burkhardt, M., Vollmer, T., et al. (2003) Gülle-Einfluss auf Abschwemmungsprozesse von Grünland. *Posterbeitrag bei der Jahrestagung der Deutschen Bodenkundlichen Gesellschaft, 30.08.-07.09.2003, Frankfurt (Oder).*

Stoob, K., Singer, H. P, Mueller, S. R., et al. (2007) Dissipation and transport of veterinary sulfonamide antibiotics after manure application to grassland in a small catchment. *Environmental Science & Technology* 41: 7349–7355.

Takahashi, K., Takita, T. and Umezawa, H. (1987) Activation of bleomycin-Fe(III) by bleomycin-Cu(II) and cysteine. *The Journal of Antibiotics* 40: 542–546.

Published by Burleigh Dodds Science Publishing Limited, 2025.

ter Laak, T. T, Gebbink, W. and Tolls, J. (2006a) Estimation of soil sorption coefficients of veterinary pharmaceuticals from soil properties. *Environmental Toxicology and Chemistry* 25: 933–941.

ter Laak, T. T., Gebbink, W. A. and Tolls, J. (2006b) The effect of pH and ionic strength on the sorption of sulfachloropyridazine, tylosin, and oxytetracycline to soil. *Environmental Toxicology and Chemistry* 25: 904–911.

Thiele-Bruhn, S. (2003a) Antibiotika. In: Blume, H. P., Felix-Hennigsen, P., Fischer, W. R., et al. (eds) *Handbuch der Bodenkunde*. Landsberg/Lech: ecomed Verlagsgesellschaft, 1–19.

Thiele-Bruhn, S. (2003b) Pharmaceutical antibiotic compounds in soils – a review. *Journal of Plant Nutrition and Soil Science* 166: 145–167.

Thiele-Bruhn, S. (2015) *Das Umweltverhalten und Effekte pharmazeutischer Antibiotika in Böden unterscheiden sich zwischen Einzelsubstanzen und Gemischen*. Oral presentation. Jahrestagung der Deutschen Bodenkundlichen Gesellschaft, Munich, 7–9 September 2015.

Thiele-Bruhn, S. (2019) *Environmental risks from mixtures of antibiotic pharmaceuticals in soils - a literature review*. Dessau-Roßlau, Germany: Umweltbundesamt.

Thiele-Bruhn, S. and Aust, M. O. (2004) Effects of pig slurry on the sorption of sulfonamide antibiotics in soil. *Archives of Environmental Contamination and Toxicology* 47: 31–39.

Thiele-Bruhn, S., Mogk, A. and Freitag, D. (2003) Einsatz von Tierarzneimitteln zur Anwendung bei landwirtschaftlichen Nutztieren in Mecklenburg-Vorpommern. *Berichte* über *Landwirtschaft* 81: 374–391.

Thiele-Bruhn, S. and Peters, D. (2007) Photodegradation of pharmaceutical antibiotics on slurry and soil surfaces. *Landbauforschung Völkenrode* 57: 13–23.

Thiele-Bruhn, S., Seibicke, T., Schulten, H. R., et al. (2004) Sorption of sulfonamide pharmaceutical antibiotics on whole soils and particle-size fractions. *Journal of Environmental Quality* 33: 1331–1342.

Thiele, S. (2000) Adsorption of the antibiotic pharmaceutical compound sulfapyridine by a long-term differently fertilized loess Chernozem. *Journal of Plant Nutrition and Soil Science* 163: 589–594.

Tolls, J. (2001) Sorption of veterinary pharmaceuticals in soils: a review. *Environmental Science and Technology* 35: 3397–3406.

Uslu, M. Ö., Yediler, A., Balcıoğlu, I. A., et al. (2008) Analysis and sorption behavior of fluoroquinolones in solid matrices. *Water, Air, and Soil Pollution* 190: 55–63.

Van den Brink, P. J., Boxall, A. B. A., Maltby, L., Brooks, B. W., Rudd, M. A., Backhaus, T., Spurgeon, D., Verougstraete, V., Ajao, C., Ankley, G. T., Apitz, S. E., Arnold, K., Brodin, T., Cañedo-Argüelles, M., Chapman, J., Corrales, J., Coutellec, M.-A., Fernandes, T. F., Fick, J., Ford, A. T., Giménez Papiol, G., Groh, K. J., Hutchinson, T. H., Kruger, H., Kukkonen, J. V. K., Loutseti, S., Marshall, S., Muir, D., Ortiz-Santaliestra, M. E., Paul, K. B., Rico, A., Rodea-Palomares, I., Römbke, J., Rydberg, T., Segner, H., Smit, M., van Gestel, C. A. M., Vighi, M., Werner, I., Zimmer, E. I. and van Wensem, J. (2018), Toward sustainable environmental quality: Priority research questions for Europe. *Environ Toxicol Chem*, 37: 2281-2295. https://doi.org/10.1002/etc.4205

Vasquez, M. I., Lambrianides, A., Schneider, M., et al. (2014) Environmental side effects of pharmaceutical cocktails: what we know and what we should know. *Journal of Hazardous Materials* 279: 169–189.

Vasudevan, D., Bruland, G. L., Torrance, B. S., et al. (2009) pH-dependent ciprofloxacin sorption to soils: interaction mechanisms and soil factors influencing sorption. *Geoderma* 151: 68–76.

Verlicchi, P., Al Aukidy, M. and Zambello, E. (2012) Occurrence of pharmaceutical compounds in urban wastewater: removal, mass load and environmental risk after a secondary treatment: a review. *Science of the Total Environment* 429: 123–155.

Verlicchi, P. and Zambello, E. (2015) Pharmaceuticals and personal care products in untreated and treated sewage sludge: occurrence and environmental risk in the case of application on soil: a critical review. *Science of the Total Environment* 538: 750–767.

Wallace, J. S. and Aga, D. S. (2016) Enhancing extraction and detection of veterinary antibiotics in solid and liquid fractions of manure. *Journal of Environmental Quality* 45: 471–479.

Wang, N., Guo, X., Xu, J., et al. (2015) Sorption and transport of five sulfonamide antibiotics in agricultural soil and soil-manure systems. *Journal of Environmental Science and Health B* 50: 23–33.

Wang, S. and Wang, H. (2015) Adsorption behavior of antibiotic in soil environment: a critical review. *Frontiers of Environmental Science and Engineering* 9: 565–574.

Wegst-Uhrich, S. R., Navarro, D. A. G., Zimmerman, L., et al. (2014) Assessing antibiotic sorption in soil: a literature review and new case studies on sulfonamides and macrolides. *Chemistry Central Journal* 8.

Weiß, K., Schüssler, W. and Porzelt, M. (2007) Auswaschung von Sulfadimidin, Enrofloxazin und Flubendazol in das Sickerwasser. In: Röder, R., Weiß, K. and Sengl, M. (eds) *Münchener Beiträge zur Abwasser-, Fischerei- und Flussbiologie*. Munich: Bayerisches Landesamt für Umwelt, Oldenbourg Verlag, 37–60.

Winckler, C., Engels, H., Hund-Rinke, K., et al. (2004) *Verhalten von Tetrazyklinen und anderen Veterinärantibiotika in Wirtschaftsdünger und Boden*. Berlin: Umweltbundesamt.

Wolters, A. and Steffens, M. (2005) Photodegradation of antibiotics on soil surfaces: laboratory studies on sulfadiazine in an ozone-controlled environment. *Environmental Science and Technology* 39: 6071–6078.

Xing, Y., Chen, X., Zhuang, J., et al. (2015) What happens when pharmaceuticals meet colloids. *Ecotoxicology* 24: 2100–2114.

Yan, H., Wang, D., Dong, B., et al. (2011) Dissipation of carbendazim and chloramphenicol alone and in combination and their effects on soil fungal:bacterial ratios and soil enzyme activities. *Chemosphere* 84: 634–641.

Yang, L. H., Ying, G. G., Su, H. C., et al. (2008) Growth-inhibiting effects of 12 antibacterial agents and their mixtures on the freshwater microalga Pseudokirchneriella subcapitata. *Environmental Toxicology and Chemistry* 27: 1201–1208.

Zhou, D., Thiele-Bruhn, S., Arenz-Leufen, M. G., et al. (2016) Impact of manure-related DOM on sulfonamide transport in arable soils. *Journal of Contaminant Hydrology* 192: 118–128.

Zhou, L. J., Ying, G. G., Liu, S., et al. (2012) Simultaneous determination of human and veterinary antibiotics in various environmental matrices by rapid resolution liquid chromatography-electrospray ionization tandem mass spectrometry. *Journal of Chromatography A* 1244: 123–138.

Zhu, D., Ma, J., Li, G., et al. (2022) Soil plastispheres as hotspots of antibiotic resistance genes and potential pathogens. *The ISME Journal* 16: 521–532.

Chapter 6

Wastewater as a contamination risk for agricultural soils

Guy J. Levy, Asher Bar-Tal, Eddie Cytryn, and Mikhail Borisover, Agricultural Research Organization, Volcani Institute, Israel

1 Introduction

The intensification of agriculture over the past century was achieved by the introduction of crop varieties (breeding programs), fertilization and irrigation, three of the main pillars of the Green Revolution. In fact, agricultural production is currently accountable for 85% of global water consumption and is projected to double by 2050. The irrigated area is expected to rise by a factor of 1.9 by 2050 (Pfister et al., 2011). However, irrigated agriculture in many parts of the world, especially in semi-arid and arid regions, is under threat due to the current and expected shortage of fresh water, facilitated by the ever-increasing demand for fresh water for direct use (i.e. domestic and industrial), the rapidly growing global population and increasing demand for food production, and especially meat and dairy products. As climate change progresses, the prevalence and intensity of global water shortages is predicted to increase, thereby highlighting the need for alternative sources of water.

The constant rise in population leads to a growing volume of municipal sewage that needs to be disposed of. Water recycling and the use of treated domestic sewage effluents, herein referred to as treated wastewater (TWW) for

http://dx.doi.org/10.19103/AS.2025.0151.10

agriculture, industry and non-potable urban and environmental purposes, can be a highly effective strategy to exploit a sustainable water resource in water-scarce areas. Irrigation with TWW can also contribute significantly to nutrient conservation and environmental protection. However, TWW is considered to be of inferior quality compared with fresh water due to elevated salinity and sodicity, presence of toxic inorganic ions and anthropogenic organic pollutants susceptible to migration toward surface and groundwater; all of which may result in contamination of the irrigated lands. It is important, therefore, to recognize and elucidate the processes by which irrigation with TWW may lead to contamination of the soil ecosystem and develop efficient management practices that can mitigate the hazards involved in TWW irrigation.

2 Wastewater sources, composition, treatment systems and methods

The volume of municipal sewage is growing rapidly due to the following factors: the constant rise in population, higher standard of living leading to higher domestic consumption of water and global urbanization of the majority of the population that is increasingly growing, especially in developing countries (Jones et al., 2021). Domestic sewage water is the major source of municipal sewage water as the number of households that are connected to a central collective municipal sewage system is also growing rapidly and in many countries, sewage is collected from over 90% of the households and high proportion of it is treated to secondary or tertiary level (Fig. 1) (OECD Statistic, 22.1.2024).

The industrial sector also produces sewage that can be a source for TWW. However, the high level of organic and inorganic contaminants including toxic materials in industrial sewage complicates its treatment (Iannelli and Giraldi, 2011). Of special concern are various emerging contaminants like detergents, pharmaceutics and personal care products (PPCPs), including antibiotics and microplastics (Hashem and Qi, 2021; Yalin et al., 2023), and industries bearing heavy metalloids, such as Cd, Cr, Cu, Ni, As, Pb and Zn, that are most hazardous (Iannelli and Giraldi, 2011). Because of their high solubility in aquatic environments, heavy metalloids can be absorbed by living organisms present there. Moreover, once they enter the food chain, large concentrations of heavy metals may accumulate in the human body and cause serious health disorders (Barakat, 2011). Hence, in most developed countries, the collected industrial sewage is pretreated separately from the domestic sewage (Yalin et al., 2023).

Another possible source of water that can contribute to the volume of the municipal sewage water is stormwater run-off in urban areas (Iannelli and Giraldi, 2011). In developed countries, this source of wastewater is collected separately from the domestic sewage water to avoid events of overflooding of the sewage

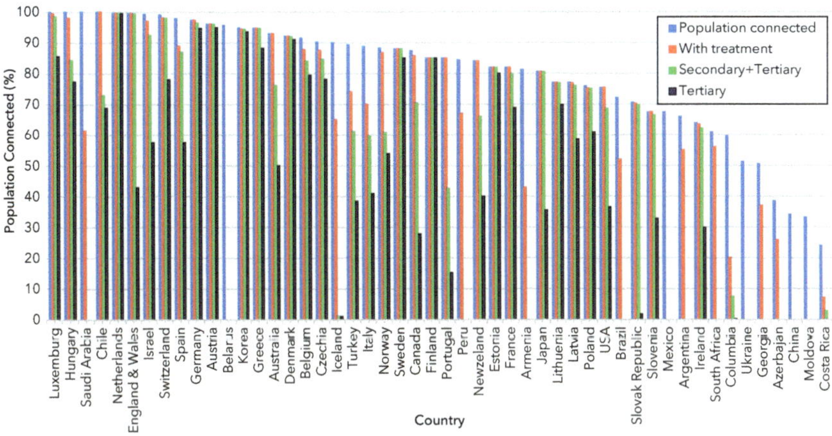

Figure 1 Percentage of total population connected to sewage water collecting systems, connected with treatments, secondary + tertiary and tertiary treatments by country. Data for most countries is for 2021 or 2022. Note the lack of information on (i) percentage of population connected with treatment (Belarus, China, Mexico, Moldova, Ukraine) and (ii) percentage of tertiary treatment (Argentina, Armenia, Belarus, Brazil, China, Mexico, Moldova, Peru, Saudi Arabia, South Africa, Ukraine). Data derived from the OECD Statistic (Organization for Economic Co-Operation and Development). https://stats.oecd.org/Index.aspx?QueryId=116818# (22.01.2024).

water treatment systems by extremely high volumes of sewage water blended with the storm water run-off (Shishegar et al., 2018). Consequently, the focus of this chapter is the domestic sewage water and the TWW produced from domestic municipal sewage water.

Worldwide, about 1.6–6.3% of the TWW is used for irrigation in agriculture, but the percentage is much higher in arid and semi-arid regions with a record value of 86% in Israel, followed by 63% in Kuwait, 46% in California (USA), 44% in Florida (USA), 44% in Pakistan, 34.6% in Saudi Arabia, 22% in Spain, 14% in Australia, 11.5% in Peru, and all the other countries below 10% (Table 1).

The major characteristics of domestic sewage water are the high concentrations of suspended solids, total dissolved solids (TDSs), dissolved organic matter (DOM), nutrients (mainly organic and inorganic nitrogen forms and phosphorus), salts and microbial mass including pathogens. The concentrations of these components are becoming lower as the standards of living are higher due to the high outflow of water from washing machines, dishwashers and bathing that dilute the domestic sewage water (Henze and Ledin, 2001). Secondary contaminants of high risk to the environment and human health are detergents, pharmaceuticals and personal care products (PPCPs), including antibiotics (Hashem and Qi, 2021; Yalin et al., 2023). Industrial sewage water may contain a wide range of hazardous materials, organic and inorganic and heavy metals (Iannelli and Giraldi, 2011; Hashem and Qi, 2021).

Table 1. Percentage of the produced wastewater used for agricultural irrigation in various countries around the world

Country	Percentage of the produced wasterwater used for agricultural irrigation	Reporting year
Algeria[a]	1.2	2012
Australia[b]	14.0	2010
Brazil[a]	0.1	2008
China[a]	2.7	2012
Egypt[a]	4.1	2011
France[c]	0.1	2014
Greece[b]	1.1	2010
Iran[a]	9.3	2010
Iraq[a]	0.9	2012
Israel[d]	86.0	2016
Japan[e]	7.0	2017
Kuwait[a]	63.0	2016
Lebanon[a]	1.0	2011
Morroco[a]	2.0	2010
Mexico[a]	5.4	2010
Peru[a]	11.5	2011
Pakistan[a]	44.0	2006
South Africa[a]	0.2	2009
Spain[f]	22.0	2018
Saudi Arabia[a]	34.6	2010
Senegal[a]	3.0	2010
Tunisial[g]	25.0	2008
United Kingdom[h]	0.1	2009
United States[i]		
California	46.0	2012
Florida	44.0	2012

[a]FAO (2019).
[b]United Nations Environment Program (2015).
[c]Ait-Mouheb et al. (2018).
[d]Tal (2016).
[e]Hara et al. (2016).
[f]Navarro (2018).
[g]Kellis et al. (2013).
[h]United Kingdom Environment Agency (2009).
[i]USEPA (2012).
Source: Adapted from: Leonel and Tonetti, 2021.

Table 2 presents data on the chemical and biological constituents and the physical properties of wastewater and their sources.

Sewage water treatment involves a series of processes aimed at removing contaminants from wastewater before it is reused or released into the

Table 2. Main chemical and biological characteristics of wastewater and their sources

Characteristic	Sources
Chemical constituents	
Organic	
Carbohydrates	Domestic, commercial and industrial wastes
Fats, oils, and grease	Domestic, commercial and industrial wastes
Pesticides	Agricultural and industrial wastes
Phenols	Industrial wastes
Proteins	Domestic, commercial and industrial wastes
Surfactants	Domestic, commercial and industrial wastes
Volatile organic compounds	Domestic, commercial and industrial wastes
Emerging concern pollutants	
Pharmaceutical compounds	Domestic, commercial and industrial wastes
Antibiotics	Domestic, commercial and industrial wastes
Personal care products	Domestic, commercial and industrial wastes
Flame retardants	Commercial and industrial wastes
Microplastics	Domestic, commercial and industrial wastes
Disinfectants	Domestic, commercial and industrial wastes
Inorganic	
Alkaline	Domestic wastes, domestic water supply and groundwater infiltration
Chlorides	Domestic wastes, domestic water supply, groundwater infiltration and water softeners
Heavy metals	Industrial wastes
Nitrogen	Domestic and agricultural wastes
Acids	Domestic, commercial and industrial wastes
Phosphorus	Domestic, commercial and industrial wastes natural runoff
Sulfur	Domestic water supply, domestic and industrial wastes
Toxic compounds	Industrial wastes
Biological constituents	
Bacteria	Domestic wastes, surface water infiltration and treatment plants
Archae	Domestic wastes, surface water infiltration and treatment plants
Protista	Domestic wastes, treatment plants
Viruses	Domestic wastes

Source: After Hashem and Qi (2021); Yalin et al. (2023).

environment. The specific methods used may vary depending on the scale of the treatment plant, the quality of the incoming sewage and the intended use of the treated water. A brief presentation of sewage treatment systems and methods is given below in relation to the hazard of contamination by the use of the resultant TWW for irrigation. More detailed and comprehensive descriptions of the principles of sewage water treatments are out of the scope of this chapter and the reader can find them in other publications (Iannelli and Giraldi, 2011). The common levels of sewage treatment are primary, secondary and tertiary treatments described below (Hashem and Qi, 2021).

Primary treatment: screening – the sewage passes through screens to remove large objects and debris, settling – suspended solids settle by gravity, and sludge accumulates at the bottom. This is a basic treatment that does not remove most of the soluble contaminants from the sewage.

Secondary treatment: this level of treatment comprises a number of alternative methods. (i) Activated sludge process: microorganisms are added to the sewage to digest organic matter (OM). The mixture is aerated to encourage microbial growth, and then the activated sludge is separated from the treated water. (ii) Trickling filters: sewage is sprayed over a bed of rocks or plastic media, promoting the growth of microorganisms that break down OM. (iii) Rotating biological contactors (RBCs): wastewater is passed over rotating disks with attached microorganisms that treat the sewage. (iv) Constructed wetlands: natural processes in wetland vegetation help break down organic pollutants. All the above-presented methods of secondary treatment reduce the concentrations of DOM, organic contaminants, heavy metals and pathogenic microorganisms in the treated water, thus enabling safer reuse of the water for irrigation under careful health and environmental guidelines.

Tertiary treatment: this level is the next stage of water treatment of pollutants removal for safer use with fewer limitations. It includes the following main methods: (i) filtration, using sand or membrane filters, for removing remaining suspended solids and (ii) chemical treatments using coagulants and disinfectants for removing remaining contaminants and pathogens. In addition, specific processes of phosphorus and nitrogen removal are employed at either or both the secondary and tertiary treatments to reduce environmental pollution by these nutrients.

Further advanced treatments can be used for removing most pollutants and reducing bacteria concentration below the minimal value enabling unrestricted reuse of the water for all purposes, including release to the environment, unrestricted irrigation and even domestic water supply. The advanced methods include membrane bioreactor, which is a combination of biological treatment and membrane filtration for finer particle removal. Ultrafiltration (UF) and reverse osmosis (RO) membranes can effectively remove dissolved salts, microorganisms and other contaminants but produce large volumes of residual contaminant water which requires further treatment (Finnerty et al., 2024). There are other alternative methods for further purification of the secondary TWW, including soil aquifer treatment (SAT) (Iannelli and Giraldi, 2011). The secondary treated water is applied to the soil surface or injected into the ground. The soil acts as a natural filter, removing remaining impurities and pathogens as the water percolates downward. This process mimics the natural recharge of groundwater. Further, post-treatments of the supplied TWW in reservoirs and in the irrigating systems will be described in Section 6.1.

3 Inorganic contaminants in wastewater

3.1 Salinity

Wastewater contains a higher load of dissolved salts compared to its fresh water of origin. The removal of organic contaminants during sewage water treatments does not affect the inorganic salts, while evaporation increases the concentration of the inorganic salts in the water that is used for the irrigated crops (Table 3). Hence, use of TWW may result in a significant soil salinity buildup in the root zone within a few years of application (Marcar et al., 2011). A recent meta-analysis by Gao et al. (2021) revealed that irrigation with TWW led to an average increase in soil salinity expressed in terms of electrical conductivity (EC) by 57% and resulted in an increase in the content of K (90.1%), Na (59.9%), Mg (25.2%) and Ca (17.3%).

The higher concentration of salts in the TWW-irrigation water adversely influences plant growth through (i) a rapid effect (hours to days) of osmotic pressure, which limits the ability of plants to absorb water from the soil solution (Munns, 1993) and (ii) slower effects of specific ion toxicity (mainly Na, Cl and B) and nutrients imbalance (Grattan and Grieve, 1992). Excluding the specific ion effects (see detailed discussion in section 3.2), crop growth retardation due to the reduction of water uptake is related to the osmotic potential, ψ_o (bar) which is a function of the salinity of the soil solution expressed in terms of EC (EC_e, dS m^{-1}) at 25°C as follows (Rhoades, 1996):

$$\psi_o = -0.39 x EC_e \quad [1]$$

Because decreasing ψ_o (increasing absolute value) has the net effect of reducing water availability (Munns, 1993), plants growing under saline conditions often appear to be suffering from drought. Threshold concentrations for classifying crop tolerance to salinity can be found in Maas (1990), while drop in crop yield (in relative terms) with the increase in EC_e for various crops is given in Maas and Hoffman (1977) using the salinity threshold (coefficient a) and the slope value (coefficient b) which describes the rate of the decline in yield with increasing salinity in the following linear equation:

Table 3 Typical chemical characteristics of sewage effluent water that contribute to salinity

Parameter	Units	Mean[a]	Min.[a]	Max.[a]	Increase[b,c]
Total nitrogen (N)	mg L^{-1}	15.2	2.8	39.0	–
Total phosphorus (P)	mg L^{-1}	5.9	0.0	12.0	–
pH	pH	7.9	6.2	9.8	–
Total dissolved salts (TDSs)	mg L^{-1}	675	145	1224	150–380
Electrical conductivity (EC)	dS m^{-1}	1.3	0.2	2.9	–
Sodium adsorption ratio (SAR)	$(mmol_c L^{-1})^{0.5}$	6	3	12.2	–
Sodium (Na)	mg L^{-1}	181	62.0	312.0	40–70
Calcium (Ca)	mg L^{-1}	35	10	74.0	6–16
Magnesium (Mg)	mg L^{-1}	19	6	40.0	4–10
Chloride (Cl)	mg L^{-1}	135	9.3	340.0	20–50
Potassium (K)	mg L^{-1}	–	–	–	7–15
Sulphate (SO_4)	mg L^{-1}	30	20	50	15–30
Bicarbonate (HCO_3)	mg L^{-1}	–	–	–	50–100

[a]NRMMC and EPHC (2006). Mean, minimum and maximum are from median values from a range of reports.
[b]Asano et al. (2007).
[c]Increases from domestic water values by effluent were stated, based on a medium strength wastewater with no commercial and industrial additions (Asano et al., 2007).
- = not available
Source: Adapted from: Marcar et al. (2011).

$$Yr = 100\text{-}bx(EC_e\text{-}a) \qquad [2]$$

Where Yr is the relative yield (%) of that below the threshold value and EC_e is the average effective EC in the root zone. Further research has presented more complicated functions that correlate yield with soil salinity by exponential ratios (Steppuhn et al., 1998) and a combination of a number of linear segments that describe different stages of crop response to salinity (Shani et al., 2001). Estimating EC_e from the EC of the irrigating water is challenging because it is dependent on the irrigation system, frequency of irrigation and soil properties, therefore, Letey et al. (2011) suggested the use of transient-state models that account for the impact of continually changing salinity on crop performance.

The magnitude of the adverse effects of TWW salinity on crop production depends, in addition to crop tolerance to salts, on the wastewater treatment grade, type of irrigation management (e.g. system and frequency of irrigation), soil properties, especially soil texture, affecting water and salt transport, and the soil water extraction pattern by the active roots distributed within the soil profile. Hence, relating crop performance solely to changes in salinity could be problematic. However, reports of some recent studies indicate that crop tolerance to salinity plays a crucial role in determining TWW irrigation impact on crop performance. Erel et al. (2019) reported that the higher salinity level in TWW compared with fresh water had no adverse effects on olive tree development and yield. In their meta-analysis, Gao et al. (2021) reported a decrease in crop yield of nearly 6.5%. Yet, other studies have reported a 20–40% decrease in the yield of avocado (Assouline and Narkis, 2013) and a decrease in grapefruit yield and tree performance (Paudel et al., 2018) after long-term irrigation with TWW, which was ascribed, at least in part, to the higher salinity in TWW compared with fresh water (Assouline et al., 2015; Paudel et al., 2019).

To avoid salt accumulation in the root zone during repeated irrigation with saline water, the common remedy is to apply water in volumes greater than evapotranspiration to deliberately cause a fraction of the applied water to flow downward through the root zone and flush away the excess salts. The required volume of excess water to leach the salt downward is termed the 'leaching requirement' (Corwin et al., 2007). Consequently, salinity hazards associated with TWW-irrigation due to the greater leaching fraction needed (compared with fresh water) include (i) reduced water-use efficiency because more water is needed for irrigation, (ii) an increase of net mass of salt that is added to the soil environment following the elevated volume of water used for irrigation, (iii) greater risk to salinization of groundwater, and (iv) rise of shallow groundwater containing the excess salts which further increases the risk of soil root zone salinization and reduction of yield.

3.2 Specific ion effects (phytotoxicity) (Na, Cl, B)

Specific ion effects occur due to the accumulation of specific ions in the plant, thus causing specific harm to metabolic processes and tissue (Grattana et al., 2015). These effects may occur in sensitive plants at relatively low salinity conditions below the threshold of osmotic pressure. The common ions in TWW that cause specific ion damages are Cl, Na and B (Grattan et al., 2015; Yalin et al., 2023). The ionic composition of TWW and the high concentration of dissolved organic carbon (DOC) may influence the effects of these ions. In the following sections, the specific ion effects of Cl, Na and B and possible interactions between these ions in TWW are presented.

3.2.1 Sodium

Sodium is a beneficial element for several higher plant species, including partial replacement of potassium, but it is mainly known as phytotoxic for plants when its concentration is above a critical value typical for each species (Kronzucker et al., 2013). The extent of sodium phytotoxicity depends on the plant species, soil conditions, climate and the concentration of sodium in the soil. Sodium toxicity is based mainly on its competition with K; therefore, the cytosolic K/Na ratio, rather than the Na concentrations alone, causes the deleterious effects of elevated plant Na uptake (George et al., 2012). Sodium displaces other ions from binding sites of enzymes and thus impairs cellular functioning. The competition of Na with K leads to impaired protein synthesis and photosynthesis. In some sensitive plants, Na accumulates in the roots resulting in retardation of roots growth (George et al., 2012). TWW are enriched with Na compared to fresh water by 40–70 mg l^{-1}, whereas the enrichment with Ca and Mg is much lower, 6–16 mg l^{-1} and 4–10 mg l^{-1}, respectively (Table 3). Sufficient Ca is essential for the sodium tolerance of sensitive crops. Irrigation with TWW may lead to Na toxicity due imbalance with magnesium and calcium, leading to retardation of growth and physiological disorders (Grattan et al., 2015). These authors suggested that, as a general guide for citrus, Ca and Mg concentrations in the soil solution above 1 mmol l^{-1} (40 mg l^{-1} and 24 mg l^{-1}, respectively) are nutritionally adequate in non-saline, sodic soils. With such levels of Ca^{2+} and Mg^{2+} in the soil solution, the threshold for sodium toxicity for citrus is 115 mg l^{-1}.

Albeit all of the above, the specific toxicity of Na to sensitive crops irrigated by TWW has been reported in numerous publications (Yalin et al., 2023). Suarez and Gonzalez-Rubio (2017) suggested the following mechanism for the effect of DOC in the wastewater on an elevation of soil solution sodium adsorption ratio (SAR). Decomposition of DOC over time leads to additional alkalinity of the soil solution that, in turn, enhances the precipitation of some Ca in the solution as calcite, thus raising the SAR of the soil solution. We suggest that the high concentration of bicarbonate and DOC in the TWW produces chelates with Ca (Metzger et al., 1983; Fotovat et al., 1996) and the bicarbonate produces ion pairs with Ca and Mg ions, reducing their activity relative to Na. Consequently, these processes lead to the enhancement of the uptake of Na ions by crops irrigated with TWW and to the accumulation of toxic levels of Na in the leaves (Paudel et al., 2018). Accumulation of Na in trunk and root tissues can also have detrimental effects (Netzer et al., 2014; Yalin et al., 2017). Paudel et al. (2019) suggested that the reduced oxygen availability in TWW interrupts the mechanism of Na exclusion by the roots, therefore, forced aeration can improve the performance of citrus irrigated with TWW or fresh water enriched with Na.

Sodium is a major contaminant for soils irrigated by TWW (Gao et al., 2021). The negative effects of high sodium concentrations in TWW on soil structure are presented and discussed separately in Section 3.4 Sodicity. Although Na is adsorbed by the charged surfaces of clay minerals, it is often displaced by the divalent cations Ca^{+2} and Mg^{+2}. Consequently, the content of Na in soils irrigated with TWW increases with depth (Lado et al., 2012; Levy et al., 2014) and it can be leached downward to the groundwater. Indeed, Na has been found as a major component of the elevated salinity of aquifers below fields irrigated with TWW (Mohanavelu et al., 2021).

3.2.2 Chloride

Chloride is an essential micronutrient for plants, but it can become phytotoxic at elevated concentrations. Like sodium, chloride is toxic to plants when it reaches a high concentration in the cytoplasm. Elevated chloride levels can interfere with the uptake of nitrate, leading to nitrate deficiency and negative impact on growth and development (Xu et al., 2000; Grattan et al., 2015; Yassuor et al., 2020). High concentrations of chloride ions in the soil can disrupt the balance of ions within plant cells, leading to ion toxicity. It displaces other ions from binding sites of enzymes and thus impairs cellular functioning (George et al., 2012). High chloride concentrations can disrupt protein synthesis, photosynthesis, enzyme activities, and other cellular functions (George et al., 2012). In sensitive plants, growth inhibition and injury of the foliage (chlorosis and necrosis on the margins of mature leaves) occur even at low salinity, while Cl^- accumulates in these tissues (George et al., 2012; Xu et al., 2000). The phytotoxic effects of chloride on plants depend on various factors, including plant species, soil type, and environmental conditions. The range of threshold concentrations for Cl^- toxicity in saturated soil is 10–80 mmol l^{-1} for sensitive to tolerant crops, respectively (Xu et al., 2000). The critical toxicity concentration in plant leaves is about 4–7 and 15–50 mg g^{-1} for Cl-sensitive and Cl-tolerant plant species, respectively. The threshold of Cl^- toxicity in saturated soil solution for citrus scions is in the range of 700–1700 mg l^{-1}. The concentration of Cl^- in the soil solution is estimated to be three times higher than that in the irrigating water (Letey et al., 2011). Consequently, the corresponding concentrations in the irrigation water are 233–567 mg l^{-1} (Grattan et al., 2015). Toxicity symptoms will appear on the margins and tips of older leaves first, because Cl^- is not mobile in most plants (Grattan et al., 2015; Xu et al., 2000).

TWW is enriched with Cl^- compared to fresh water, the average gap is 20–50 mg l^{-1} (Table 3). Long-term irrigation with TWW reduces the efficiency of Cl^- leaching due to the decrease in soil hydraulic conductivity and to the accumulation of Chloride ions in the root zone (Russo et al., 2020). Although the gap in Cl^- concentrations between TWW and FW is smaller than that of Na, high concentrations of Cl^- in leaves with symptoms of Chloride toxicity in crops irrigated with TWW were reported in many publications (Grattan et al., 2015; Yalin et al., 2023).

Chloride is a major component of the salinity of TWW and therefore it is a major contaminant of soils irrigated by TWW (Yalin et al., 2023). The common practice in irrigation with brackish water, including TWW is surplus irrigation for leaching the accumulated salts below the root zone (Assouline et al., 2015). Since chloride is not adsorbed by soil particles, it is easily leached downward below the root zone by excess irrigation dose and rainfalls (Russo et al., 2020; Erel et al., 2019). The mass of Cl^- leached below the root zone under irrigation with TWW is much greater than under fresh water because of the higher leaching fraction needed (Assouline et al., 2015; Russo et al., 2020). Consequently, contamination of groundwater by chloride is a major problem worldwide and especially in aquifers below fields irrigated with TWW. According to Elgallal et al. (2016), it is expected that irrigation with TWW will have a high salinity (Na and Cl) impact on soil and plants with the likelihood of groundwater salinization being of medium impact, whereas the likelihood of surface water salinization is unlikely and the expected impact is low.

3.2.3 Boron

Boric acid is moderately soluble in water and acts as a weak Lewis acid (Yermiyahu et al., 2011):

$$B(OH)_3 + 2H_2O \rightleftarrows B(OH)_4^- + H_3O^+ \qquad [3]$$

The reaction constant (K) of $B(OH)_3$ (Equation 3) is 5.8×10^{-10} at 20°C (Owen, 1934 in Yermiyahu et al., 2011). Further dissociation beyond borate ($B(OH)_4^-$) takes place at pH above 13, therefore, it can be ignored in soils. Equation [3] and the value of K indicate that the molecule $B(OH)_3$ is the dominant species of B in soil solution below pH 9.5. The implications of this property of B in water and soil solution will be discussed herein with regard to B uptake by plants, B adsorption and mobility in soil and B in RO-borne desalinized water.

Boron is an essential micronutrient for plants, but it can become phytotoxic at elevated concentrations. Unlike other nutrients, B is taken up as an uncharged molecule $B(OH)_3$ that passes through plant membranes easily. Consequently, uptake of B increases linearly as B concentration in the soil solution increases up to toxic values. While plants require boron for various physiological processes, such as cell wall formation, sugar transport and seed development, excessive boron levels can lead to toxicity (Brown et al., 2002). Boron toxicity is crop-specific and generally involves a number of metabolic processes (Reid et al., 2004), including reduced meristem expansion, development of necrotic areas in mature tissues (thus reducing photosynthetic capacity) and reduced supply of photosynthates to developing regions of the plant. It leads to chlorosis and necrosis in the tissues where B accumulates, beginning with the edges of mature leaves (Brown and Shelp, 1997). Plants vary in their tolerance to boron, and sensitive plants are more prone to showing symptoms of toxicity. The range of B concentrations between deficiency and toxicity is very narrow (Grieve et al., 2012; Yermiyahu et al., 2011), especially in sensitive plants like citrus, where deficiency and toxicity occur with leaf B concentrations at or below 15–25 mg kg^{-1} and above 250 mg kg^{-1}, respectively (Chapman 1968a, 1968b in Grattan et al., 2015). The threshold of B concentration in the soil solution for sensitive tree crops like citrus, avocado, grape, fig, walnut, pecan and stone fruits is 0.5–0.75 mg l^{-1}, while more than 2.0 mg l^{-1} is harmful to most crop species (Maas, 1990). Salinity and excessive B availability do not have interactive effects in most plant species investigated so far (George et al., 2012; Yermiyahu et al., 2011).

Boron concentration in the soil solution is determined by (i) soluble B entering the soil-solution system through groundwater or irrigation water or from the soil mineral fraction; and (ii) B adsorption/desorption reactions on the soil solid phase. The transport of B is mainly affected by the adsorption/desorption processes. As uncharged polar molecules, the $B(OH)_3$ is adsorbed by clay mineral surfaces, metal oxides, carbonates and OM by ligand exchange mechanism in contrast to Cl^- and NO_3^-, that are indifferent or rejected by these soil components (Yermiyahu et al., 2011). The capacity of soil to adsorb B varies greatly with the content of these constituents in the soil. Consequently, B in soil is much less mobile than Cl^{-1} and NO_3^- and it is not easily leached out of the irrigated root zone by rainfall and excess irrigation (Yermiyahu et al., 2011). It takes up to three times more water to reclaim B-affected soils than salt-affected soils (Ayers and Westcot, 1985 in Grattan et al., 2015). Consequently, irrigation with TWW containing high B concentration leads to accumulation of B in the soil, especially in clayey and calcareous soils (Lado et al., 2012; Yermiyahu et al., 2011). This B accumulation damages soil quality and limits the type of crops that can be grown in the affected soils. The risk of contamination of groundwater by B is low due to its low mobility in soil; however, there is a risk of groundwater contamination by B after long-term irrigation with TWW (Khatri et al., 2020; Nikolaou et al., 2020), after reaching saturation of the soil adsorption capacity for B.

3.3 Contamination by nutrients (nitrate, phosphate)

Irrigation with TWW is considered as a means to recycle nutrients. This benefit of irrigation with TWW is out of the scope of the current chapter and interested readers can find such information in recent reviews and book chapters (Bar-Tal et al., 2011; Chojnacka et al., 2020; Khatri et al., 2020; Ofori et al., 2021; Yalin et al., 2023). On the other hand, irrigation with TWW containing high concentrations of inorganic and organic forms of nutrients is considered as a risk to the environment (Ofori et al., 2021; Yalin et al., 2023). High N and P concentrations in surface water and coastal marine environments enhance the growth of algae, leading to eutrophication of the water and hypoxia (Sparks, 2003). High nitrate (NO_3^-) concentrations in drinking water originating from ground or surface water have been related to the 'blue baby' syndrome (methemoglobinemia), abortion in women and nitrate is suspected as a carcinogenic substance (Sparks, 2003).

The global N cycle is important for the whole biosphere and the environment. It includes several chemical reactions that have been investigated and described in detail in numerous publications (e.g. Bar-Tal et al., 2011). This chapter will briefly describe the N forms and transformations in sewage, in the production of TWW and in the soil. Sewage and soil contain organic and mineral N with the organic form being the major component of N in sewage. The production of mineral N from organic N and the opposite reaction are termed mineralization and immobilization, respectively. The direction of the reaction, namely net mineralization or immobilization depends on the C/N ratio in the OM. When this ratio is >20, immobilization takes place, whereas at lower C/N ratios organic N is mineralized. The major forms of nitrogen in the soil–water–air system are N_2, NH_3, NO, N_2O, NH_4^+, NO_2^-, NO_3^-. The transformations of these mineral N species are presented in Table 4. Most of these reactions are mediated by microorganisms except the dissolution of NH_3 (ammonia) in water that produces NH_4^+ (ammonium) and hydroxide and the reverse reaction of NH_4^+ with hydroxide to produce NH_3 (ammonia), and NH_3 emission from the water.

Ammonium in soil is partitioned between the soil solution and the exchange complex of the clay minerals. The cation exchange capacity (CEC) of the soil and its cationic composition govern the concentration of NH_4^+ in the soil solution. This process influences the mobility of NH_4^+ in soil, which is much slower than that of NO_3^- being an anion that is not adsorbed by the soil solid phase. Three-layered clay minerals in soil affect NH_4^+ and potassium concentration in soil solution by an additional mechanism called fixation. Fixed NH_4^+ is an important pool of mineral N in clay soils containing this type of clay that reduces the mobility of NH_4^+ in soil. The aqueous form of NH_3 is in equilibrium with the gas form. From the description of the production of NH_3 from NH_4^+, it is clear that evolution of NH_3 from the solution increases with the pH.

3.3.1 Nitrogen in treated wastewater

The global annual nitrogen in TWW in 2010 was 10.4 Tg and the projected quantity in 2050 will be 13.4–17.9 Tg, depending on the fraction of nutrient removal, technological and socio-economic progress in developing countries (van Puijenbroek et al., 2019). The total annual nitrogen produced worldwide by the Haber-Bosh process in 2010 was 120 Tg, of which 96 Tg was used for N fertilizers (Fowler et al., 2013). Raw sewage water contains high concentrations of organic and mineral N, from 20 mg l^{-1} to 100 mg l^{-1} (Feigin et al., 1991). The concentration of N in TWW decreases with the progress in the quality of the treatments. The fraction of N removal in primary, secondary, tertiary and quaternary is approximately 0.1, 0.35, 0.8 and 0.95, respectively (van Puijenbroek et al., 2019). Consequently, the mean total N in the TWW used

Table 4 Main transformations of mineral N in the environment

Reaction	Reactant	Product	Equation	Conditions
Nitrification* (step 1)	NH_4^+	NO_2^-	$NH_4^+{}_{(aq)}+1.5O_{2(g)}\rightarrow NO_{2(aq)}+H_2O_{(l)}+2H^+{}_{(aq)}$	Both steps consume oxygen and are mediated by microorganisms during the production of TWW and also in soil under aerobic conditions.
Nitrification (step 2)	NO_2^-	NO_3^-	$NO_2^-{}_{(aq)}+0.5O_{2(g)}\rightarrow NO_{3(aq)}$	
Denitrification	NO_3	NO, N_2O, N_2	$NO_3^-{}_{(aq)} \rightarrow NO_2^-{}_{(aq)} + 0.5O_{2(g)}NO_2^-{}_{(aq)} \rightarrow NO_{(g)} + 0.5O_{2(g)}NO^-{}_{(g)} \rightarrow 0.5N_{2(g)} + 0.5O_{2(g)}$	All three steps release oxygen and are mediated by microorganisms during the removal of nitrogen from secondary TWW and also in soil under anaerobic conditions and high organic matter content.
N fixation	N_2	NH_3	$N_{2(g)} + 3H_{2(g)} \rightarrow NH_{3(g)}$	Natural fixation of atmospheric N_2 is done mainly by Rhizobium microorganisms in association with the roots of legume plants. There is also minor fixation by free-living microorganisms.Industrial fixation is done by the Haber-Bosch process.
Hydrolysis of ammonia to ammonium	NH_3	NH_4^+	$NH_{3(g)}+H_2O_{(l)}\rightarrow NH_4^+{}_{(aq)}+OH^-{}_{(aq)}$	Chemical equilibrium without microorganisms. In the reverse reaction, ammonia is produced from ammonium.

*Over the past decade, many 'alternative' N-transformation pathways have been detected and identified, for example alternative pathways of ammonium oxidation by various bacteria (see Daims et al., 2016). https://www.cell.com/trends/microbiology/fulltext/S0966-842X(16)30045-2.

for irrigation is 15 mg l^{-1}, with a range of 3 mg l^{-1} to 39 mg l^{-1} (Table 3). For some crops, the amount of N added to soil through TWW irrigation can be similar to the amount commonly applied via fertilization with fresh water (Bar-Tal et al., 2011; Erel et al., 2019). An example of N load that matches the consumption is that of irrigating citrus orchard in a Mediterranean climate with 10 000 m^3 ha^{-1} TWW containing 20 mg l^{-1} N; the total N load is 200 kg ha^{-1}, that is very close to the recommended fertilization. However, when the total applied N with TWW exceeds the uptake by the crops, there is a high risk of environmental contamination. Using the aforementioned example of the citrus orchard in Mediterranean climate, if N concentration in the irrigation water is higher than 20 mg l^{-1} by 1 or 10 mg l^{-1}, the excess N that poses a risk to groundwater pollution is 10 kg ha^{-1} and 100 kg ha^{-1}, respectively. Additional risk for environmental pollution from irrigation with TWW is the mismatch of the timing of N supply by TWW irrigation with the consumption of N as a function of time. In conventional fertilization, growers can match the timing of N application with changes in the consumption of crops due to physiological stages and environmental conditions, whereas in TWW, the concentration of N is not controlled by the growers. An additional factor that contributes to excess N fertilization by growers using TWW is their lack of information on the N content in the TWW and/or because they do not trust the effectiveness of N supplied by the TWW.

The main forms of N in sewage and TWW are organic N and NH_4^+. The proportion of NH_4^+ in TWW increases with the improvement of the treatment process, and subsequently, it is the major form in secondary and tertiary TWW. Unlike alkaline soils, in acid soils, ammonium (NH_4) and organic N from effluent irrigation were found as the main forms of excess N transported below the root zone, whereas nitrate leaching was relatively small under a normal irrigation dose (Livesley et al., 2007 in Bar-Tal et al., 2011). Dissolved organic N (DON) has been found to be more mobile in soils than ammonium (Zhou et al., 2003), and high leaching of DON below the root zone was found in lysimeters and field studies applied with TWW containing high DON (Bar-Tal et al., 2011).

However, in the majority of observations in fields irrigated with TWW, the dominant form in the soil solution is NO_3^- and its dominance increases with vertical downward leaching deep into the soil profile. This is a result of the rapid nitrification of the applied NH_4^+ in semi-arid and arid climates and the low NH_4^+ concentration in the solution due to its adsorption to the exchangeable complex. Consequently, NO_3^- is the main possible N contaminant from irrigation with TWW. Contamination of groundwater by NO_3^- is a major problem worldwide (Spalding and Exner, 1993; Erisman et al., 2008). In Israel, it has led to the disqualification of a greater number of drinking water wells than any other environmental contaminant at the beginning of the twenty-first century (Elhanany, 2009 in Kurtzman et al., 2013). When nitrogen inputs by the applied

TWW exceed the amount of nitrogen needed by the plants, the excess nitrate is then easily leached by the irrigation and rainwater to deeper soil layers, finally reaching groundwater (Bar-Tal et al., 2011; Kurtzman et al., 2013; Libutti and Monteleone, 2017). Irrigation with TWW also contributes to the pollution of surface water and coastal marine by N through runoff and soil erosion. The estimated annual load of N by TWW to surface water is 10.4 Tg nitrogen (van Puijenbroek et al., 2019), which is approximately 10% of total nitrogen fertilizers (100 Tg N $year^{-1}$) produced by the Haber-Bocsh process (Tsunemi et al., 2023). The estimated annual load of N by TWW to coastal marine worldwide is 6.2 Tg nitrogen, which is approximately 40% of the total nitrogen load from agriculture (Tuholske et al., 2021). In a study of TWW irrigation in which 4.68 t N ha^{-1} was applied over 11 years (426 kg N ha^{-1} $year^{-1}$) to wetland and forest systems, the transport of effluent N via the stream was estimated as 263 t (29%) of the applied N (Tozer et al., 2005 in Bar-Tal et al., 2011). Based on a literature review of publications from semi-arid and arid regions, Elgallal et al. (2016) concluded that irrigation with TWW with high N concentration will have a high negative impact on groundwater due to its high mobility in soil and therefore lower impact on surface water.

An additional potential environmental concern from irrigation with TWW is the emission of ammonia (NH_3) and N oxides to the atmosphere. The ammonium (NH_4) concentration in the TWW, pH at the soil surface and temperature are the factors that directly govern NH_3 volatilization from soil. Smith et al. (1996) reported that the loss of NH_3 from the soil after TWW application under maximum evaporative conditions, high temperature and wind speed reached 24% of the TWW-applied ammonium in 2 days. However, they observed much higher NH_3 volatilization from an equivalent application of fertilizer N in urea form. Smith et al. (1996) further found a strong linear relationship between ammonia emission and water evaporation following irrigation with TWW. Direct volatilization of NH_3 from the applied irrigating water before reaching the soil has also been reported. Irrigation methods that increase the contact of water drops with the air accelerate volatilization, whereas drip irrigation decreases it. Consequently, it is expected that the order of potential N losses through ammonia volatilization is sprinklers > surface drip irrigation > sub-surface drip irrigation. Yet, data in the literature show that N Loss by NH_3 through ammonia volatilization from TWW irrigation is usually low and can be minimized by proper management (Bar-Tal et al., 2011).

In addition to ammonia volatilization, denitrification (the transformation of NO_3^- to N_2O) is a major route for N_2O emission from soil. Yet, by and large, the quantities of applied N losses as gases from TWW-irrigated soils are relatively small fraction of the applied N. The emission of N_2O is of great environmental importance because it provides a considerable contribution to the GHG emissions.

Low-quality TWW with a high OM content is expected to enhance denitrification. Higher denitrification is anticipated in clay soils than in coarse-textured, well-drained soils (Bar-Tal et al., 2011). The effect of irrigation dose and frequency of irrigation of TWW containing high concentration of N (32.9, 7.5 and 11.5 mg l^{-1} of NO_3^-, NH_4^+ and organic N, respectively) on denitrification was investigated in a column laboratory experiment (Monnett et al., 1995). As expected, increasing the effluent application rate from 1.25 cm $week^{-1}$ to 2.5 cm $week^{-1}$ elevated the average gaseous losses of N from 5.3% to 26.2% of the applied N, respectively, but the proportions of N_2O and N_2 in this study were unknown. In another laboratory study, Master et al. (2003) reported negligible N_2O emissions from a clay soil irrigated with TWW, whereas, when the soils were water saturated with TWW that was subjective to long evaporative conditions that resulted in a high salinity load, N losses by denitrification reached a high value of 3.1% of total applied N. Tozer et al. (2005) reported that denitrification accounted for 3% of the applied N by irrigation with municipal TWW to a system of forest and wetland for 11 years. Comparison of N_2O emissions from field plots irrigated with TWW and FW indicated that cumulative N_2O emissions were higher in irrigation with TWW by 2.1 times than in FW; yet, TWW and mix of TWW with FW treatments showed similar N_2O emissions (7.6 kg N ha^{-1}) (Gal 2020). Note that the mean monthly N_2O emission from TWW is higher than 1% of the applied N (the mean estimated value by the ICCP) with a maximum monthly rate of 2.5%. Further field investigations employing advanced methods and continuous measurements of N_2O emissions from TWW and FW plots are required.

3.3.2 Phosphorus in treated wasterwater

The global P cycle is an important element cycle for the biosphere and environment. It includes several chemical reactions which have been investigated and described in detail in numerous publications (e.g. Bar-Tal et al., 2011). In this chapter, we will briefly describe the P forms and transformations in sewage water, in the production of TWW and in the soil. The global annual P in TWW in 2010 was 1.5 Tg and the projected quantity in 2050 will be 1.6–2.4 Tg, depending on the fraction of nutrient removal, technological and socio-economic progress in developing countries (van Puijenbroek et al., 2019). The major component of P in sewage is organic P. Normally, wastewater contains low concentrations of P that may vary according to the level and methods of treatment. Mineral P, mainly orthophosphate (including different forms of H_3PO_4), is produced during the decomposition of OM in the processes of TWW production. Further decomposition of organic P occurs in the soil. According to current knowledge, ~65% of organic P added via annual TWW irrigation in arid zones is converted to mineral P (Bar-Tal et al., 2011). In plants' rhizosphere, the mineralization rate is much faster due to phosphatase excretion by roots. The pH of the soil solution governs the dissociation of H_3PO_4 and the distribution of the phosphate species. The $H_2PO_4^-$ is the dominant species in acid soils up to pH 7.0 and HPO_4^{-2} becomes the dominant species in basic soils above pH 7.0. The concentration of P in the soil solution is governed by two mechanisms (Lindsay, 1979). The first mechanism is a very fast (on a scale of seconds or minutes) adsorption onto the soil particles as a result of the high affinity of P ions to charged surfaces. Adsorption of mineral P is controlled by the pH, percentage and type of clays, SOM, Al and Fe oxides and carbonates. The distribution of the mineral P between the solid phase and the soil solution can be described by sorption models like Langmuir and Freundlich (Barrow, 2023). Then, the mineral P distribution in the soil profile and drainage can be estimated with models that incorporate sorption equations into solution transport. The second mechanism is slow (on a scale of hours or days) and involves precipitation and dissolution of metal-P compounds with Al, Fe and Mn under acidic pHs and with Ca and Mg under neutral or basic pHs (Lindsay, 1979). An additional long-term mechanism is the diffusion of P into and out of soil particles (Barrow, 2023).

After application of exogenous P to the soil, the concentration of P decreases gradually due to a change in the equilibrium with minerals with lower solubility. Consequently, P is stable in soils and is therefore considered beneficial with no negative impacts on the environment. The range of mineral P concentrations in soil solutions is 0.1 mg l^{-1} to 1.0 mg l^{-1}. This is the case even if TWW with high concentrations of phosphorus is applied over long periods of time (WHO, 2006).

The processes and factors affecting short- and long-term TWW-borne P behavior in soil involve processes influencing mineral and organic P sorption that affect P distribution in the soil profile between solid and solution phases and transport (runoff and drainage) in soils. Phytic acid, which is a major organic P form in TWW, was found to have stronger affinity to soil particles than the mineral form (Bar-Tal et al., 2011). Therefore, slower mobility of total P from irrigation with TWW could be expected. Irrigation with TWW may also slow down P transport in soil indirectly through reducing of soil hydraulic conductivity On the other hand, observed high correlation between soil P availability and DOC content can also affect P mobility in soil. A recent study showed that soil DOM properties that were affected by the type of applied organic residue determined soil P sorption capacity (Wang et al., 2024). It was found that high molecular weight (HMW)-fractionated DOM, which possesses unique characteristics such as strong aromaticity, had stronger inhibitory effects on P sorption than low-molecular-weight-fractionated DOM by 16–20%. The mechanism is primarily attributed to competition for soil P sorption sites and a decrease in soil surface charge, collectively suppressing P sorption in soil and thus enhancing its mobility. An additional possible mechanism is chelation of mineral P by HMW–DOM that is also expected to enhance the mobility of mineral P in soil. However, recent studies reported opposite effects of applied sewage sludge on P sorption to soil (Freiberg et al., 2022, 2023). OM, both in solid and dissolved forms, may enhance sorption of phosphates to soils that would depend on mineralogy and solution electrolytes (e.g. calcium or sodium). Data analysis of a long-term effluent irrigation survey in orchards grown on variable soil types in Israel (Tarchitzky et al., 2006 in Bar-Tal et al., 2011) indicated that in TWW-irrigated soils P adsorption was the main mechanism controlling P concentration in the soil solution, while under freshwater irrigation the effect of Ca–P precipitation could not be discounted. The lower Ca–P precipitation under irrigation with TWW was attributed to Ca–P crystallization inhibition by unknown OM molecules derived from TWW. Thus, it is difficult to predict the overall effect of DOM in TWW on P transport in soil as contrasting effects have been observed (Bar-Tal et al., 2011). Elgallal et al. (2016) concluded that it is unlikely that P in TWW will affect deep groundwater, while shallow groundwater may be affected. However, in sandy soils that contain small fractions of clays, metal oxides and SOM, P can be transported downward and accumulate in the groundwater (Sparks, 2003).

Generally, application of mineral and organic P to fields leads to its accumulation at shallow depth over time, where it could also reach surface water through soil erosion and runoff thus contributing to eutrophication (WHO, 2006; Qadir and Scott, 2010). Runoff P concentrations in most case study surveyed soils already exceeded the permitted eutrophication threshold concentration of 0.02–0.05 mg l^{-1} P (Tarchitzky et al., 2006 in Bar-Tal et al., 2011). The U.S. EPA set 0.10 mg l^{-1} as the maximum total P concentration in streams, and 0.05 mg l^{-1} in streams entering lakes or reservoirs.

Future research on P management optimization under TWW irrigation should focus on elucidating behavior of organic P components in soil and studying DOM–P interactions in soil. Due to the complexity of P–OM interactions, model-based decision support systems should be developed to predict the fate of inorganic and organic P in different soils. Such tools will be helpful in optimizing irrigation with TWW for minimizing environmental P pollution without compromising crop yields.

3.4 Sodicity

Sodium-affected (sodic) soils exhibit poor soil water and air ratios which adversely affect water movement in the soil, root growth and plant production and make the soil difficult to farm (Levy and Nachshon, 2023). Sodicity level of the irrigation water or soil solution is expressed by its SAR. Sodicity of the soil *per se* is defined by the exchangeable sodium percentage (ESP) which describes the fraction of adsorbed Na from the CEC of the soil. Following exchange reactions that take place between the soil solution (or applied water) and the exchangeable phase, both parameters are highly correlated. To date, there is no widely accepted definition of a sodic soil. Some studies suggested that an ESP of 15 should separate sodic from non-sodic soils (USSL Staff, 1954; McNeal and Coleman, 1966). More recent studies advocated that soils with ESP > 5 should be considered as sodic soils (e.g. Mace and Arnheim, 2001; Levy et al., 2005). It should be borne in mind that the sodicity level in a soil that would define it as a sodic soil is closely and inversely related to the ionic strength of the soil solution and/or that of the irrigation water (Sumner et al., 1998).

The larger increase in Na content in TWW compared with Ca and Mg (Gao et al., 2021) results in SAR values that are higher than those of fresh water (Ofori et al., 2021 and references cited therein). Yet, large variability exists in the SAR of TWW (Table 1). In regions where TWW originates from domestic streams and is subjected to advanced secondary and even tertiary treatment, as for instance, is the case in Israel, the SAR of the TWW is ~4. Such SAR is not expected to pose a threat to soil structure stability and its hydraulic properties. However, the presence of DOM in the TWW could adversely affect the soil solution SAR. It has been suggested that the effective SAR of the TWW is greater than that commonly calculated based on the total concentration of cations in the solution (Halliwell et al., 2001). The increase in effective SAR can occur through the formation of complexes between Ca^{2+} and Mg^{2+} and the organic ligands present in the TWW (Metzger et al., 1983) that would result in lower concentration of free forms (i.e. lower activity) of these cations in the soil solution and thus higher SAR values. The difference between the commonly calculated SAR and the effective one could reach 48% due to the great affinity of the divalent cations to organic ligands (Fotovat et al., 1996). DOM may also have a negative direct effect on soil structure stability and water flow in soil that arises from pH-dependent destabilization of clay suspensions, where negatively charged DOM components undergo interactions with clay edges, thus preventing edge-to-face clay interactions and flocculation (Tarchitzky et al., 1993). This direct effect of DOM was considered to be equivalent to an increase in SAR of 2–3 units (Suarez and Gonzalez-Rubio, 2017). Conversely, Assouline et al. (2016) reported a linear reduction of the Gapon constant with the increase in the ratio of DOC to calcite following irrigation with TWW. This finding implies that (i) with the increase in DOC (at a given calcite content), the affinity of the soil to Na decreases and that of Ca increases, and (ii) the SAR of the soil solution decreases. It is evident that the current understanding of the impact of DOC on Ca–Na exchange in soil requires further elucidation.

Studies worldwide have shown that long-term irrigation with TWW, mainly in orchards, leads to the buildup of soil sodicity, especially in subsurface soil layers where the ESP can exceed 10 (e.g. Levy et al., 2014; De las Heras and Manas, 2020; Hashem and Qi, 2021; Phogat et al., 2021). This sodicity buildup in orchards could be related to (i) the irrigation water being applied continuously on the same limited volume of soil ('wet bulb') for many years, and (ii) following the common practice of 'no-till' in orchards, the effects of tillage, that mixes sodium-affected soil with soil that has not been exposed to the irrigation water and thus mitigate the sodicity hazard on the field scale, are missing.

Both the aforementioned phenomena may explain reports on the deterioration of soil physical and hydraulic properties, in orchards grown in clayey soils, following long-term irrigation with TWW (e.g. Assouline and Narkis, 2011; Bardhan et al., 2016, Nemera et al., 2023). The impaired soil properties may also lead to aeration problems (Nemera et al., 2023) that subsequently lead to poor crop functioning and productivity (Paudel et al., 2019;, Nemera et al., 2020, 2021). It may be concluded that irrigation with TWW can pose a threat to the soil environment and its functioning due to imposing some sodic behavior, especially in clayey soils.

4 Contamination by the organic fraction in wastewater

4.1 *Origin, nature, characterization and selected properties of organic matter in treated wastewater*

OM in TWW is physically and chemically heterogeneous and includes dissolved, colloidal and particulate components such as natural OM, water-soluble microbial products, diverse anthropogenic chemicals (and the numerous and often poorly characterized products of their degradation), cellulose fibres and microorganisms (Drewes and Fox, 2000; Shon et al., 2006; Michael-Kordatou et al., 2015). Its composition depends on the specific origin of wastewater, the domestic, industrial and agricultural contributions to the sewage influent and on the wastewater treatment technology and operational conditions of the treatment facility (Guo et al., 2011; Michael-Kordatou et al., 2015).

DOM, operationally defined as passing the 0.45 μm filter, is considered generally a major part of TWW OM albeit its particular fraction in OM is also source- and treatment-dependent. Wang et al. (2014) found that 83–87% of the total organic carbon (TOC) of secondary TWW from a municipal wastewater plant is associated with DOM. Sánez et al. (2023) reported the DOC to be 89–98% of the TOC content in effluent following activated-sludge treatment and 75–100% of TOC in the final effluent following UV treatment.

TWW is a very complex system for chemical characterization. Its complexity and multiplicity of chemical constituents are revealed using ultra-high-resolution mass spectrometry. Zhang et al. (2021) indicated that in the secondary TWW samples, including those before and after ozonation, 9004 molecular formulae were found using Fourier transform ion cyclotron resonance mass spectrometry (FT-ICR-MS). At the same time, even such a powerful and insightful technique as FT-ICR-MS drastically underestimates the real number of different chemical components present in DOM. Thus, Ruan et al. (2023) summarized that FT-ICR-MS is limited in its ability to discriminate chemical structures and isomers associated with the same formulae and characterizing molecules >1000 Da. This suggests that a large number of formulae detected is an indication of the even greater and poorly accounted variety of organic chemical structures present in TWW. Ray et al. (2020) indicated that no established methods are presented currently that could allow the identification of all the components of DOM in TWW. The intrinsic (and variable) chemical complexity of TWW composition infers a great variety of TWW properties and, therefore, the variability of the potential effects on the soil environment. Hence, there is an importance of characterizing DOM in multiple terms, associated with its physico-chemical and biological properties, and various methods should be used together for this goal.

Characterization of TWW OM using different methods was summarized in Michael-Kordatou et al. (2015) and Tran et al. (2015). Among the multiple characterization methods, the most common are (i) the use of neutral and ion-exchange resins and polymers for DOM fractionation into hydrophobic/hydrophilic and acidic/basic/neutral fractions, (ii) characterization of aromaticity of OM based on specific UV absorbance, (iii) determination of molecular weight distribution, (iv) identification of different groups of fluorescent components in DOM, such as proteinaceous (containing certain amino acids emitting light in UV-visible region) and humic-like substances, and (v) biodegradability. Proportions between different hydrophobic/hydrophilic and acidic/basic/neutral fractions in TWW as well as their relations with molecular mass distribution are variable and not necessarily consistent, depending on different wastewater sources and treatment techniques (Wang and Chen, 2018). These authors proposed that the biodegradability increases with the content of hydrophilic and low molecular mass DOM components.

UV-visible spectra of TWW DOM are capable of yielding empirical indices based on specific absorbance, spectral slopes and absorbance ratios thus suggesting estimations for aromatic content, humification degree and molecular weight of DOM (Ray et al., 2020). Specific UV absorbance (SUVA) at 254 nm taken as an empirical measure of the aromatic content of DOM was indicated to increase through the biological treatment but decrease when oxidative treatment is involved, such that the range of the effluent SUVA values among different treatments varied between 0.65 and 2.9 (in L m^{-1} mg^{-1}; Wang and Chen, 2018).

Hu et al. (2016) showed that DOM in TWW appears to have higher hydrophilic and lower molecular mass fractions and be more degradable as compared with drinking water DOM. About 56–63% of organics in secondary TWW were found to have molecular masses less than 1 kDa (Wang et al., 2014).

On the other hand, Wang and Chen (2018) indicated that generally biological treatment led to the degradation of the low molecular mass fractions and the accumulation of resistant higher molecular mass substances. DOM present in TWW from different biologically treated sewage may show different molecular weight distributions (Guo et al., 2011).

Excitation-emission matrices of fluorescence coupled with multivariate analysis such as parallel factor (PARAFAC) analysis characterize the fluorescent composition of DOM in TWW, its biodegradability, and accumulation of certain fractions resistant to biological water treatment and degradation (Cohen et al., 2014; Carstea et al., 2016; Shi et al., 2021). In particular, proteinaceous components of fluorescent DOM revealed by tryptophan- and tyrosine-like fluorescence are expected to undergo biodegradation, unlike humic-like substances. The PARAFAC-isolated fluorescent components may correlate with chemical oxygen demand and total microbial activity in water (Cohen et al., 2014).

The scarcity of information on the chemical composition of TWW OM necessitates the use of more sophisticated methods (e.g. modern nuclear magnetic resonance techniques using multiple nuclei ^{13}C, ^{15}N, ^{19}F, ^{31}P, Anaraki et al., 2021).

4.2 Negative effects on soil due to influx of treated wastewater-associated organic matter

Use of TWW has numerous negative effects on the irrigated soil that are associated with the OM load in the TWW. The possible increase of the effective SAR of the soil solution due to preferential binding of divalent cations (Ca and Mg) by DOM and the direct DOM effects enhancing soil structure degradation are discussed in detail in Section 3.1 of this chapter.

Accumulation of TWW-derived organic compounds in soil induces soil hydrophobicity (Wallach et al., 2005; Nadav et al., 2013). This may alter water dynamics, reduce infiltration capacities, increase overland flow and cause preferential flow paths, making the spatial wetting pattern uneven (Leuther et al., 2018; 2019). Tolkin et al. (2020) mentioned that although TWW irrigation-induced soil hydrophobicity is addressed in multiple works, little is known about the specific TWW-derived chemical fractions leading to soil repellence. Examining the ability of humic acid, dextran (representing TWW polysaccharides) and valeric acid (representing TWW fatty acids) to induce soil repellence, Tolkin et al. (2020) concluded that humic acid was most effective in making soil hydrophobic. The recommendation is to remove elevated concentrations of humic molecules, polysaccharides and/or fatty acids from TWW to avoid inducing soil water repellency in fields irrigated with TWW (Tolkin et al., 2020).

To summarize, multiple studies have indicated that the negative effects of TWW irrigation on physico-chemical soil properties are associated with the input of OM. Yet more work and attention are needed to elucidate the role of specific chemical fractions and constituents within the TWW OM.

4.3 Wastewater-borne organic contaminants

Multiple anthropogenic organic contaminants and their degradation and transformation products are present in TWW, potentially leading to the transfer of pollutants to soil and plants and subsequently the food chain and to groundwater and surface water (Barbosa et al., 2016; Rogowska et al., 2020). Some important organic contaminants present in TWW and reviewed in this book include petroleum hydrocarbons, polycyclic aromatic hydrocarbons (PAHs), per- and poly-fluoroalkyl substances (PFASs), flame retardants, chlorinated compounds, micro/nanoplastics, pesticides and antibiotic residuals. In addition, anthropogenic organic contaminants may include surfactants (Becerra-Castro et al., 2015), hormones (Schröder et al., 2016), chemicals belonging to a broad family of pharmaceuticals and personal care products (PPCPs, Anand et al., 2022), quaternary ammonium organics (DeLeo et al., 2020), as well as organic carbon nanomaterials (e.g. fullerenes, Kunhikrishnan et al., 2015) and many other substances. The classification of anthropogenic organic contaminants is often mixed, using both their belonging to certain chemical groups (such as PAHs, PFAS) or certain uses (home pesticides, PPCPs). The appearance of new chemicals and their transformation and degradation products in TWW is expected to grow as a result of global industrialization and the requirement for new substances (Gao, 2024). A small fraction of organic contaminants in TWW is controlled by legal regulations (Rogowska et al., 2020; Helmecke et al., 2020; Parida et al., 2021), which may also vary between different regions. Therefore, micropollutants bearing a potential (or even proven) environmental and agricultural risk may not be monitored, and their presence in TWW can be overlooked. The concentrations of organic contaminants in TWW may widely vary, depending on their source, water treatment technologies and extent of removal associated also with the contaminant's nature and properties (solubility, volatility, red/ox properties, ability to undergo adsorption reactions and ion exchange, photo- and biodegradability).

The presence of some anthropogenic organic chemicals in TWW, their properties, toxicity and unwanted environmental and agricultural effects were discussed in the literature (e.g. Fatta-Kassinos et al., 2011; Ofori et al., 2021). Renai et al. (2020) reviewed the literature focused on pharmaceuticals and some other organic pollutants and discussed the impact of the TWW use for agricultural needs, behavior of organic micropollutants in soil, transfer to crops and related risks. Ben Mordechay et al. (2023) examined multiple

TWW-derived contaminants of emerging concern, including carbamazepine, 1H-benzotriazole, lamotrigine, venlafaxine and thiabendazole, in soil solutions and indicated that a single growing season using TWW irrigation is sufficient to introduce wastewater-derived contaminants to the arable soil system thus making them potentially bioavailable for plant uptake. More evidence for the accumulation of the contaminants of emerging concern in soil due to TWW irrigation and their presence in edible crops was also provided by others (e.g. Ben Mordechay et al., 2021; Ben Mordechay et al., 2022).

A separate group of the organic contaminants of agricultural and environmental concern appears in TWW when the latter undergoes disinfection, e.g., foremost, chlorination, thus producing disinfection byproducts (DBPs) that further may accumulate in soil (Albolafio et al., 2022; Xiao et al., 2023). In this case, DOM demonstrates an indirect toxic potential since it is an important precursor of DBP formation (Michael-Kordatou et al., 2015; Yan et al., 2023). The potential of toxic DBP formation due to water disinfection may be estimated using relations between DBP formation and the OM descriptors, such as concentration of total or dissolved carbon, UV absorbance and intensity of fluorescence emission (Manivannan and Borisover, 2020).

4.4 Contributions of wastewater-borne organic matter to the mobility/transport of contaminants in the soil profile

TWW-borne OM has the potential to affect the mobility of chemical contaminants in the soil profile. Mechanistic effects of TWW irrigation on soil sorption and transport of pesticides and other organic micro-contaminants in soils were reviewed by Gerstl and Graber (2011), who discussed some simplified mechanisms of contaminant distribution between the aqueous phase, soil and DOM. Depending on relationships between the soil sorption affinity of free pollutants and the tendency of pollutants to sorb as DOM-bound complexes, contrasting effects of DOM on soil sorption and subsurface transport may be expected. When water-soluble complexes formed between DOM and pollutants undergo weak soil sorption, the presence of OM in the soil solution reduces soil sorption and facilitates the transport of pollutants. In contrast, when such complexes are characterized by significant sorption capacity, DOM assists in soil sorption of contaminants and leads to retardation of their movement. This interplay of different interactions leading to possibly opposite environmental scenarios is dictated by the nature and concentrations of both the DOM and the contaminant.

The above-described concept provides a useful schematic framework for understanding the effect of DOM on the transport of contaminants. Its principles may apply to the effect of TWW OM on the transport of both organic pollutants and inorganic chemicals capable of forming water-soluble complexes with

DOM. However, multiple aspects may need to be accounted for to get a more realistic understanding of how DOM may influence the transport of pollutants in the soil environment.

- Long-term TWW irrigation may modify SOM composition, e.g., introducing hydrophobic components (Chen et al., 2010; Nadav et al., 2013), coating soil particles (Tolkin et al., 2020) and affecting sorption of chemicals (Wang & Chen, 2018). Hence, comparing the sorption of chemicals and their mobility in the same soil irrigated by water of different quality may not be limited by examination of the concentration (and composition) of DOM in soil solution.
- Soil–DOM interactions are generally irreversible (Oren & Chefetz, 2012; Avneri-Katz et al., 2017), which suggests that solution DOM concentration available for forming water-soluble complexes with chemicals can hardly be linked, using equilibrium isotherms, to DOM sorbed by soil.
- The competition between DOM and contaminants for sorption sites and even site blocking due to DOM–soil interactions, both factors contributing to chemical distribution from soil to solution phase and enhancing its transport, may occur (Peña et al., 2020).
- DOM in soils undergoes fractionation due to biological and physicochemical processes. In particular, adsorption on soils and mineral surfaces leads to DOM fractionation (Banaitis et al., 2006; Gjettermann et al., 2007; Avneri-Katz et al., 2017). Different DOM fractions are capable of different effects on soil sorption of organic chemicals (Wu et al., 2021) and differently interact with heavy metals in solutions (Fine et al., 2018; Wei et al., 2021). Therefore, generally, DOC concentration is not necessarily a sufficient descriptor of DOM presence in soil solutions, and care is needed to account for the dynamic DOM composition.
- DOM may be involved in chemical reactions with the participation of contaminants, e.g., by mediating the microbial reduction processes and mobilizing metal(loid)s, redox reactions and methylation (Polubesova & Chefetz, 2014; Abdelrady et al., 2020; Li et al., 2021). Some DOM components (e.g. humic substances) may enhance photolytic degradation of pesticides (Peña et al., 2020). Therefore, DOM may need to be accounted for when considering the role of chemical reactions occurring with contaminants in the soil environment in contaminant transport.

5 Microbial contaminants in wastewater irrigated soils

TWW is a source of various microbial contaminants that can accumulate in soil and edible crops, and subsequently pose a potential epidemiological threat to exposed humans and animals. Microbial pathogen levels in TWW are dictated by pathogen loads in untreated (raw) sewage and the level of wastewater treatment. Water quality regulations for TWW regularly involve enumeration of standardized indicator microbes such as fecal coliforms, thermotolerant coliforms or *Escherichia coli* (Blumenthal et al., 2000). The global increase in antibiotic resistance, which significantly complicates treatment of several bacterial pathogens, and the realization that wastewater is a reservoir of antibiotic-resistant bacteria and genes, has resulted in exploration of antibiotic resistance in TWW and downstream environments (Hong et al., 2018).

Several studies have investigated microbial pathogens and antibiotic resistance indicators in untreated wastewater and TWW and in TWW-irrigated crops, but relatively few have explored the fate of these pathogen indicators in soil (Ibekwe et al., 2018). While some studies suggest that antibiotic-resistant bacteria and associated resistance genes can persist in soil irrigated with TWW (Kampouris et al., 2021; Palacios et al., 2017), there is substantial evidence indicating that soil acts as an 'ecological barrier' that impedes colonization by fecal pathogens (Marano et al., 2019; Negreanu et al., 2012; Marano et al., 2019). This is supported by Campos et al. (2000), who showed rapid decline of fecal coliforms, somatic and F+ coliphages (viral fecal indicators) in wastewater irrigated soil (Campos et al., 2000), and by Marano et al. (2021), who demonstrated rapid decline of antibiotic-resistant bacteria in soil microcosms. Collectively, it was hypothesized that the phenomenon can be attributed to both abiotic stress (i.e. moisture content, OM concentration and pH) and biotic factors such as predation and native soil microbiomes that facilitate soil resilience (Frenk et al., 2014).

Albeit soil resilience, recent studies indicate that antibiotic resistance indicators (bacteria and associated resistance genes) can persist at below detection levels in TWW-irrigated soils, and proliferate following inoculation in selective media (Marano et al., 2021). This persistence was found to be inversely proportional to water quality, occurring in secondary effluent irrigated soils but not in tertiary effluent irrigated soils, suggesting that irrigation with high-quality water can circumvent pathogen accumulation in soil. In the future, quantitative microbial risk assessment models (Mara et al., 2007) can be applied to elucidate the balance between wastewater-associated pathogens entering irrigated soil and the natural capacity of soil to attenuate these pathogens.

Recent studies have suggested that conventional pathogen indicators used to estimate fecal contamination in wastewater and impacted environments (i.e. *E. coli*) do not always correlate with other microbial pathogens such as enteric parasites or viruses (Weller et al., 2020), and consequently, additional surveillance targets are required to establish health risks (Ofori et al., 2021). Attention should also be paid to emerging fecal pathogens such as the protozoan parasite *Cyclospora cayetanensis*, considering that it can be transmitted to humans through fecal-contaminated soil (Almeria et al., 2019). Recent cases in local produce in the United States (Rehme, 2023) have led the FDA to begin screening produce and soil for this parasite, especially in TWW-irrigated crops (Almeria, 2018, FDA, personal communication).

6 Management solutions to mitigate hazards involved in irrigation with wastewater

6.1 Management of wastewater quality at source

Hazards involved in irrigation with wastewater can be eliminated or minimized by a variety of regulations, management and wastewater treatments detailed below:

1. Separating industrial and highly polluted sources of sewage from the household source by separate collecting and treatment systems and continuous monitoring of hazardous chemicals (Yalin et al., 2023 and references cited therein).
2. Regulating the use of various chemicals and disposing hazardous chemicals that are not well removed through the sewage treatment plants. For example, reducing sources of elements that cause soil structure destruction like Na, K and Mg and toxic elements to plants like B, Na and Cl. Eliminating the use of Na-based ion exchange water softening devices to lower Na concentration and the SAR in TWW (Yalin et al., 2023). Replacing boron with other oxidizing chemicals eliminated excess B concentrations in TWW in Israel (Tarchitzky et al., 2004). Regulating B concentration in desalinated water to meet the critical value in irrigation water for sensitive crops like citrus led to low levels of B in the TWW in Israel (Shtull-Trauring et al., 2020).
3. Implementating physical, chemical and biological methods for removing biological and chemicals that are hazardous from the collected sewage. These methods are described above in Section 2. Note that it is possible to upscale the quality of TWW by UF and RO to remove all hazardous materials and organisms; however, these methods require high-energy consumption (Yalin et al., 2023 and references cited therein).
4. Improving the quality of the water supplied to the household may have an indirect effect on the quality of the TWW. The growing production of desalinized water for domestic consumption in Israel in the last decade led to an indirect positive effect on the reduction of the salinity of TWW (Shtull-Trauring et al., 2020), and a decrease in the mean concentrations

of Na and Cl in leaves of various crops in Israel (Yassour et al., 2020; Raveh and Ben-Gal, 2016).

6.2 Agronomic management solutions

The aforementioned hazards discussed negative effects that might be inflicted by intensive use of TWW for irrigation can be mitigated, at least in part, by adopting a variety of agronomic management practices. Such practices include the following:

1 Increase of the leaching fraction (LF) of the irrigation amount – following the greater salinity in TWW compared with that in fresh water of its origin, the LF needs to be increased such that the salinity in the root zone does not reach a level that is toxic to the crop grown. The drawback of this management practice is that the salt load added to the soil environment increases and with it increases the risk of salinization of groundwater. Segal et al. (2011) reported 1.75 times increase in Cl load in the water transported below the root zone under TWW irrigation compared with fresh water.

2 Adjustment and regulation of fertilization – because TWW contains some N, P and K (their concentration will depend on the level of treatment the TWW has been subjected to), it is advisable to consider the amount of each element when calculating the amount of fertilizers to be added to the irrigation water (Chojnacka et al., 2020; Pedrero et al., 2020; Yassour et al., 2020). Shtull-Trauring et al. (2022) suggested indices for evaluating the environmental risk of groundwater contamination by N, P and K from TWW. Those indices are based on N, P and K concentration in the TWW, crop water demand and N, P and K consumption by various crops and the sensitivity of aquifers to contamination by these elements. The latter is controlled by the soil properties and the aquifer depth. Erel et al. (2019) reported that no negative impact on olives was observed under irrigation with secondary TWW with no addition of fertilizers. This approach saves money for the farmers and mitigates contamination of the soil environment and groundwater by excess N (Kurtzman et al., 2013).

3 Change in irrigation management – One of the consequences of soil sodicity, especially in clayey soils, is impaired aeration. A possible solution is applying the irrigation water over a larger space and time intervals, thus allowing greater aeration periods (Yalin et al., 2021). This solution can be achieved by increasing the number of drip lines used along the tree compared with those used in fresh water, while maintaining the same amount of irrigation water. Friedman et al. (2010) noted that the use of three drip lines in TWW compared with two or one

line of fresh water in a citrus orchard improved aeration in the root zone. Yet, results regarding this approach have not been conclusive. Nemera et al. (2023) did not observe an increase in aeration in the root zone of an avocado plantation upon increasing the number of line drippers from two to three.

4 Use of trenches filled with coarse material – It has been suggested that creating a well-drained area for root growth using a coarse material, e.g. tuff that is known for its high porosity and high saturated hydraulic conductivity (Papadopoulos et al., 2008; Bar-Tal et al., 2019), may alleviate problems associated with diminished aeration in the root zone. Recent studies by Nemera et al. (2020, 2021, 2023) have shown that constructing tuff trenches (0.3 m wide and 0.3 m deep) along both sides of the tree rows were effective at improving soil aeration and subsequently several physiological parameters (e.g. sap flow, photosynthesis, stomatal conductance and canopy conductance) in an avocado orchard irrigated with TWW.
5 Change in water quality – In cases where a source of a better quality water (compared with TWW) is available for irrigation, a mixture based on mixing of the two different waters prior to their application to the soil, using a constant, predetermined, user-controlled mixing ratio will mitigate water salinity at the root zone. In addition, a periodic alteration of water quality between TWW and fresh water has also been proposed (Phogat et al., 2021). Based on a numerical analysis, Russo et al. (2015, 2020) proposed that once a threshold salinity level at a given soil depth is attained following irrigation with TWW, water of better quality is used until soil solution salinity reaches a predetermined lower salinity level that will then allow returning to the use of TWW. Use of this more sophisticated approach, compared with the predetermined mixing ratio or periodic altering of water quality, controls the salinity level at the root zone more accurately and may save water from a better quality source (Russo et al., 2015, 2020). Yet, in extreme cases, where the use of TWW for irrigating leads to a very high average salinity level, TWW should be replaced with potable water.
6 Use of surfactants – Repeated on-site application of surface-active chemicals may help in remediating the adverse effects of TWW irrigation that turns the soil surface water repellent. Spreading of the abovementioned chemicals onto the soil surface makes the soil wettable and subsequently the distribution of soil water and chemicals is more uniform (Ogunmokun and Wallach, 2021, 2024).
7 Use and breeding of saline-tolerant cultivars and crops – Selection of appropriate crops may reduce damages due to irrigation with TWW. Some crops may reduce potential heavy metal(loid)s health risk in

wastewater-contaminated agricultural soils (Cao et al., 2022). Use of specific rootstocks can minimize the negative effects of irrigation with TWW by alleviating the risk of accumulation of toxic elements such as B, Na, Cl and heavy metals in the leaves of fruit trees and vegetables (Edelstein and Ben-Hur, 2012).

8 Optimize the management of irrigation with TWW – Use of water–soil–crops modeling can lead to minimizing damage to the soil, crops and the environment (Assouline et al., 2015; Kurtzman et al., 2013; Russo et al., 2020; Styczen et al., 2010).

7 Conclusion

TWW is a significant resource for agricultural production worldwide. Use of TWW for irrigation of cultivated lands includes benefits (e.g. conservation of nutrients) and drawbacks such as risks to crop production, soil ecosystem, the environment and human health. The risks of using TWW arise from its inferior quality compared to its water of origin that leads to (i) soil water elevated osmotic pressure, (ii) addition of toxic ions to the bulk soil and its solution, (iii) deterioration of soil physical and hydraulic properties, (iv) introduction of TWW-borne OM contaminants to the soil, (v) enhanced mobility of chemical contaminants by the TWW-borne DOM in the soil profile that ultimately may reach surface and ground water, (vi) epidemiological threats to humans and animals, and (vii) increased abundance and diversity of antibiotic resistant bacteria and genes.

Long-term sustainability of TWW irrigation necessitates more comprehensive understanding of the processes by which irrigation with TWW leads to the abovementioned ill-effects to crops, soil and the environment. Additionally, further development of wastewater treatment processes to obtain TWW of better quality, adjustment of irrigation methods, and crop and soil management practices that mitigate the abovementioned hazards caused by TWW irrigation are also required.

8 References

Abdelrady, A., et al. (2020). Characterisation of the impact of dissolved organic matter on iron, manganese, and arsenic mobilisation during bank filtration. *Journal of Environmental Management, 258*. https://doi.org/10.1016/j.jenvman.2019.110003

Albolafio, S., et al. (2022). Strategies for mitigating chlorinated disinfection byproducts in wastewater treatment plants. *Chemosphere, 288*. https://doi.org/10.1016/j.chemosphere.2021.132583

Almeria, S., Cinar, H.N. and Dubey, J.P., 2019. Cyclospora cayetanensis and cyclosporiasis: an update. *Microorganisms*, *7*(9), 317.

Almeria, S., et al. (2018). Evaluation of the US Food and Drug Administration validated method for detection of *Cyclospora cayetanensis* in high-risk fresh produce matrices and a method modification for a prepared dish. *Food Microbiology, 76*, 497–503. https://doi.org/10.1016/j.fm.2018.07.013

Anand, U., et al. (2022). Occurrence, transformation, bioaccumulation, risk and analysis of pharmaceutical and personal care products from wastewater: a review. *Environmental Chemistry Letters*, *20*(6), 3883–3904. https://doi.org/10.1007/s10311-022-01498-7

Anaraki, M. T., et al. (2021). NMR spectroscopy of wastewater: a review, case study, and future potential. *Progress in Nuclear Magnetic Resonance Spectroscopy*, *126–127*, 121–180. https://doi.org/10.1016/j.pnmrs.2021.08.001

Asano, T., et al. (2007). *Water Reuse. Issues, Technologies and Applications. Metcalf and Eddy*. McGraw Hill, New York.

Assouline, S. and Narkis, K. (2011). Effects of long-term irrigation with treated wastewater on the hydraulic properties of a clayey soil. *Water Resources Research*, *47*, W08530. https://doi.org/10.1029/2011WR010498

Assouline, S. and Narkis, K. (2013). Effect of long term irrigation with treated wastewater on the root zone environment. *Vadose Zone Journal, 12*(2). https://doi.org/10.2136/vzj2012.0216

Assouline, S., et al. (2015). Balancing water scarcity and quality for sustainable irrigated agriculture, *Water Resources Research*, *51*, 3419–3436.

Assouline, S. K., et al. (2016). Combined effect of sodicity and organic matter on soil properties under long-term irrigation with treated wastewater. *Vadose Zone Journal*, *15*(4), 1–10.

Avneri-Katz, S., et al. (2017). Adsorptive fractionation of dissolved organic matter (DOM) by mineral soil: macroscale approach and molecular insight. *Organic Geochemistry*, *103*, 113–124. https://doi.org/http://dx.doi.org/10.1016/j.orggeochem.2016.11.004

Banaitis, M. R., et al. (2006). Investigating sorption-driven dissolved organic matter fractionation by multidimensional fluorescence spectroscopy and PARAFAC. *Journal of Colloid and Interface Science*, *304*(1), 271–276. isi:000241593100035

Barakat, M. A. (2011). New trends in removing heavy metals from industrial wastewater. *Arabian Journal of Chemistry*, *4*(4), 361–377.

Barbosa, M. O., et al. (2016). Occurrence and removal of organic micropollutants: an overview of the watch list of EU Decision 2015/495. *Water Research*, *94*, 257–279. https://doi.org/10.1016/j.watres.2016.02.047

Bardhan, G., et al. (2016). Changes in the hydraulic properties of a clay soil under long-term irrigation with treated wastewater. *Geoderma*, *264*, 1–9.

Barrow, N. J. (2023). Equations to describe the amount and rate of sorption. *European Journal of Soil Science*, *74*(2). https://doi.org/10.1111/ejss.13355

Bar-Tal, A., et al. (2011). Major minerals. In G. Levy, et al. (Eds.), *Use of Treated Waste Water in Agriculture: Impacts on Soil Environment and Crop* (pp. 129–215). Wiley-Blackwell, UK.

Bar-Tal, A., et al. (2019). Inorganic and synthetic organic components of soilless culture and potting mixes. In M. Raviv, et al. (Eds.), *Soilless Culture: Theory and Practice* (pp. 259–301). Academic Press, Elsevier, UK.

Becerra-Castro, C., et al. (2015). Wastewater reuse in irrigation: a microbiological perspective on implications in soil fertility and human and environmental health.

Environment International, *75*(0), 117–135. http://www.sciencedirect.com/science/article/pii/S0160412014003237

Ben Mordechay, E., et al. (2021). Pharmaceuticals in edible crops irrigated with reclaimed wastewater: evidence from a large survey in Israel. *Journal of Hazardous Materials*, *416*. https://doi.org/10.1016/j.jhazmat.2021.126184

Ben Mordechay, E., et al. (2022). Fate of contaminants of emerging concern in the reclaimed wastewater-soil-plant continuum. *Science of the Total Environment*, *822*. https://doi.org/10.1016/j.scitotenv.2022.153574

Ben Mordechay, E., et al. (2023). Wastewater-derived contaminants of emerging concern: concentrations in soil solution under simulated irrigation scenarios. *Soil & Environmental Health*, *1*(3), 100036. https://doi.org/10.1016/j.seh.2023.100036

Blumenthal, U. J., et al. (2000). Guidelines for the microbiological quality of treated wastewater used in agriculture: recommendations for revising WHO guidelines. *Bulletin of the World Health Organization*, *78*, 1104–1116.

Brown, P. H. and Shelp, V. (1997). Boron mobility in plants. *Plant and Soil*, *193*, 85–101.

Brown, P. H., et al. (2002). Boron in plant biology. *Plant Biology*, *4*, 205–223.

Campos, C., et al. (2000). Behaviour of the fecal pollution indicators in a soil irrigated with treated wastewater under onsurface and subsurface drip irrigation. *Water Science and Technology*, *42*(1–2), 75–79.

Cao, C., et al. (2022). Crop selection reduces potential heavy metal(loid)s health risk in wastewater contaminated agricultural soils. *Science of the Total Environment*, *819*, 152502.

Carstea, E. M., et al. (2016). Fluorescence spectroscopy for wastewater monitoring: a review. *Water Research*, *95*, 205–219. https://doi.org/10.1016/j.watres.2016.03.021

Chen, Y., et al. (2010). Organic matter in wastewater and treated wastewater-irrigated soils: properties and effects. In G. J. Levy, et al. (Eds.), *Treated Wastewater in Agriculture* (pp. 400–417). Wiley, London. https://doi.org/10.1002/9781444328561.ch13

Chojnacka, K., et al. (2020). A transition from conventional irrigation to fertigation with reclaimed wastewater: prospects and challenges. *Renewable and Sustainable Energy Reviews*, *130*, 109959.

Cohen, E., et al. (2014). Fluorescent components of organic matter in wastewater: efficacy and selectivity of the water treatment. *Water Research*, *55*. https://doi.org/10.1016/j.watres.2014.02.040

Corwin, D. L., et al. (2007). Leaching requirement for soil salinity control: Steady-state versus transient models. *Agricultural Water Management*, 90, 165–180.

Daims, H., et al. (2016). A new perspective on microbes formerly known as Nitrite-oxidizing bacteria. *Trends in Microbiology*, *24*, 699–712.

De las Heras, J. and Mañas, P. (2020). Reclaimed wastewater to irrigate olive groves and vineyards: effects on soil properties. *Agronomy*, *10*, 649. https://doi.org/10.3390/agronomy10050649

DeLeo, P. C., et al. (2020). Assessment of ecological hazards and environmental fate of disinfectant quaternary ammonium compounds. *Ecotoxicology and Environmental Safety*, *206*. https://doi.org/10.1016/j.ecoenv.2020.111116

Drewes, J. E. and Fox, P. (2000). Effect of drinking water sources on reclaimed water quality in water reuse systems. *Water Environment Research*, *72*(3), 353–362. https://doi.org/10.2175/106143000x137581

Edelstein, M. and Ben Hur, M. (2012). Use of grafting to mitigate chemical stresses in vegetables under arid and semiarid conditions. *Advances in Environmental Research*, *20*, 163–179.

Edirisooriya, E. M. N. T., et al. (2024). Economic feasibility of developing alternative water supplies for agricultural irrigation. *Current Opinion in Chemical Engineering*, *43*, 100987.

Elgallal, M., et al. (2016). Assessment of potential risks associated with chemicals in wastewater used for irrigation in arid and semiarid zones: a review. *Agricultural Water Management*, *177*, 419–431.

Erel, R., et al. (2019). Long-term irrigation with reclaimed wastewater: implications on nutrient management, soil chemistry and olive (Olea europaea L.) performance. *Agricultural Water Management*, *213*, 324–335. https://doi.org/10.1016/j.agwat.2018.10.033

Erisman, J. W., et al. (2008). How a century of ammonia synthesis changed the world. *Nature Geoscience*, *1*, 636–639.

Fatta-Kassinos, D., et al. (2011). The risks associated with wastewater reuse and xenobiotics in the agroecological environment. *Science of the Total Environment*, *409*(19), 3555–3563. https://doi.org/10.1016/j.scitotenv.2010.03.036

Feigin, A., et al. (1991). Irrigation with treated sewage effluent: management for environmental protection. In J. K. Matsushima (Ed.), *Advances Series In Agricultural Sciences 17*. Springer-Verlag Berlin, Heidelberg, New York.

Fine, P., et al. (2018). Properties of the DOM in soil irrigated with wastewater effluent and its interaction with copper ions. *Water, Air, and Soil Pollution*, *229*(1). https://doi.org/10.1007/s11270-017-3627-7

Finnerty, C. T. K. et al. (2024). The future of municipal wastewater reuse concetrate management: drivers, challenges, and opportunities. *Environmental Science & Technology*, *58*, 3–16.

Fowler, D., et al. (2013). The global nitrogen cycle in the twenty-first century. *Philosophical Transactions of the Royal Society B*, *368*, 20130164.

Fotovat, A., et al. (1996). Analysis of indigenous zinc in alkaline sodic soil solutions by graphite furnace atomic absorption spectrometry. *Communications in Soil Science and Plant Analysis*, *27*, 2997–3012.

Freiberg, Y., et al. (2022). Biosolids increase phosphate adsorption of semi-arid Mediterranean soils. *Journal of Environmental Management*, *305*, 114361.

Freiberg, Y., et al. (2023). Effect of biosolid-derived dissolved organic matter on orthophosphate sorption to soils depends on clay mineralogy and solution composition. *Environmental Science and Pollution Research*, *30*(53), 113649–113659.

Frenk, S., et al. (2014). Resilience of soil bacterial community to irrigation with water of different qualities under M editerranean climate. *Environmental Microbiology*, *16*(2), 559–569.

Friedman, S., et al. (2010). *Studying long term effects of wetted area and leaching management on trees and soil salinity response for optimizing drip irrigation in orchards with recycled effluent water*. Final report, submitted to the Chief Scientist, Ministry of Agriculture and Rural Development, State of Israel (in Hebrew).

Gal, A. (2020). *Effects of different agricultural managements and TWW irrigation regimes on greenhouse gas emissions from avocado orchard grown on a heavy clayey soil*. M.Sc. thesis, submitted to Bar-Ilan University (in Hebrew).

Gao, P. (2024). Chasing "emerging" contaminants: an endless journey toward environmental health. *Environmental Science and Technology*, *58*(4), 1790–1792. https://doi.org/10.1021/acs.est.3c10630

Gao, Y., et al. (2021). Changes in soil salinity under treated wastewater irrigation: a meta-analysis. *Agricultural Water Management*, *255*, 106986.

George, E., et al. (2012). Adaptation of plants to adverse chemical soil conditions. In P. Mrachschner (Ed.), *Mrachschner's Mineral Nutrition of Higher Plants*. Elsevier, London, UK.

Gerstl, Z. and Graber, E. (2011). Impact of irrigation with treated wastewater on pesticides and other organic microcontaminants in soils. In G. J. Levy, et al. (Eds.), *Treated Wastewater in Agriculture: Use and Impacts on the Soil Environment and Crops* (pp. 382–399). Blackwell Publishing Ltd, Hoboken. https://doi.org/10.1002/9781444328561

Gjettermann, B., et al. (2007). Sorption and fractionation of dissolved organic matter and associated phosphorus in agricultural soil. *Journal of Environmental Quality*, *36*(3), 753–763. https://doi.org/10.2134/jeq2006.0081

Grattan, S. R., et al. (2015). Assessing the suitability of saline wastewater for irrigation of *Citrus* spp.: emphasis on boron and specific-ion interactions. *Agricultural Water Management*, *157*, 48–58.

Grattan, S. R. and Grieve, C. M. (1992). Mineral element acquisition and growth response of plants grown in saline environments. *Agriculture, Ecosystems & Environment*, *38*, 275–300.

Grieve, C. M., et al. (2012). Plant salt tolerance. In W. W. Wallender and K. K. Tanji (Eds.), *ASCE Manual and Reports on Engineering Practice no. 71 Agricultural Salinity Assessment and Management* (Chapter 13, 2nd ed., 405–459). ASCE, Reston, VA.

Grattana, S. R., et al. (2015). Assessing the suitability of saline wastewaters for irrigation of Citrus spp.: emphasis on boron and specific-ion interactions. *Agricultural Water Management*, *157*, 48–58.

Guo, J., et al. (2011). Dissolved organic matter in biologically treated sewage effluent (BTSE): characteristics and comparison. *Desalination*, *278*(1–3), 365–372. https://doi.org/10.1016/j.desal.2011.05.057

Halliwell, D. J., et al. (2001). A review of the effects of wastewater sodium on soil physical properties and their implications for irrigation systems. *Australian Journal of Soil Research*, *39*, 1259–1267.

Hashem, S. M. and Qi, X. (2021). Treated wastewater irrigation – a review. *Water*, *13*, 1527. https://doi.org/10.3390/w13111527

Helmecke, M., et al. (2020). Regulating water reuse for agricultural irrigation: risks related to organic micro-contaminants. *Environmental Sciences Europe*, *32*(1). https://doi.org/10.1186/s12302-019-0283-0

Henze, M. and Ledin, A. (2001). Types, characteristics and quantities of classic, combined domestic wastewaters. In P. Lens, et al. (Eds.), *Decentralised Sanitation and Reuse: Concepts, Systems and Implementation* (Vol. Chapter 4, pp. 59–72). IWA Publishing, London.

Hong, P. Y., et al. (2018). Reusing treated wastewater: consideration of the safety aspects associated with antibiotic-resistant bacteria and antibiotic resistance genes. *Water*, *10*(3), 244.

Hu, H. Y., et al. (2016). Differences in dissolved organic matter between reclaimed water source and drinking water source. *Science of the Total Environment, 551–552*, 133–142. https://doi.org/10.1016/j.scitotenv.2015.12.111

Iannelli, R. and Giraldi, D. (2011). Sources and composition of sewage effluent; treatment systems and methods. In G. J. Levy (Eds.), *Treated Wastewater in Agriculture: Use and Impacts on the Soil Environment and Crops* (pp. 1–50). Blackwell Publishing Ltd, Oxford, UK.

Ibekwe, A. M., et al. (2018). Impact of treated wastewater for irrigation on soil microbial communities. *Science of the Total Environment*, *622*, 1603–1610.

Jones, E. R., et al. (2021). Country-level and gridded estimates of wastewater production, collection, treatment and reuse. *Earth System Science Data*, 13, 237–254.

Kampouris, I. D., et al. (2021). Treated wastewater irrigation promotes the spread of antibiotic resistance into subsoil pore-water. Environment International, 146, p. 106190.

Khatri, N., et al. (2020). Sewage water: from waste to resource – A review. *Environmental Claims Journal*, *33*(2), 108–135.

Kronzucker, H. J., et al. (2013). Sodium as nutriend and toxicant. *Plant and Soil*, *369*, 1–23.

Kunhikrishnan, A., et al. (2015). Sources, distribution, environmental fate, and ecological effects of nanomaterials in wastewater streams. *Critical Reviews in Environmental Science and Technology*, *45*(4), 277–318. https://doi.org/10.1080/10643389.2013.852407

Kurtzman, D., et al. (2013). Nitrate fluxes to groundwater under citrus orchards in a Mediterranean climate: observations, calibrated models, simulations and agro-hydrological conclusions. *Journal of Contaminant Hydrology*, *151*, 93–104.

Lado, M., et al. (2012). Changes in chemical properties of semiarid soils under long-term secondary-treated wastewater irrigation. *Soil Science Society of America Journal*, *76*, 1358–1369.

Leonel, L. P. and Tonetti, A. L. (2021). Wastewater reuse for crop irrigation: crop yield, soil and human health implications based on giardiasis epidemiology. *Science of the Total Environment*, *775*, 145833.

Letey, J., et al. (2011). Evaluation of soil salinity leaching requirementguidelines. *Agricultural Water Management*, *98*, 502–506.

Leuther, F., et al. (2018). Quantitative analysis of wetting front instabilities in soil caused by treated waste water irrigation. *Geoderma*, *319*, 132–141. https://doi.org/10.1016/j.geoderma.2018.01.004

Leuther, F., et al. (2019). Structure and hydraulic properties in soils under long-term irrigation with treated wastewater. *Geoderma*, *333*, 90–98. https://doi.org/10.1016/j.geoderma.2018.07.015

Levy, G. J. and Nachshon, U. (2023). Sodic soils. In M. Goss and M. Oliver (Eds.), *Encyclopedia of Soils in the Environment* (Vol. 1, pp. 368–382). Elsevier, Oxford. https://doi.org/10.1016/B978-0-12-822974-3.00030-6

Levy, G. J., et al. (2005). Saturated hydraulic conductivity of semi arid soils: combined effects of salinity, sodicity and rate of wetting. *Soil Science Society of America Journal*, *69*, 653–662.

Levy, G. J., et al. (2014). Long term irrigation with treated wastewater (TWW) and soil sodification. *Biosystems Engineering*, *128*, 4–10.

Li, Y., et al. (2021). Different dissolved organic matters regulate the bioavailability of heavy metals and rhizosphere microbial activity in a plant-wetland soil system. *Journal of Environmental Chemical Engineering*, *9*(6). https://doi.org/10.1016/j.jece.2021.106823

Libutti, A. and Monteleone, M. (2017). Soil vs. groundwater: the quality dilemma. Managing nitrogen leaching and salinity control under irrigated agriculture in Mediterranean conditions. *Agricultural Water Management*, *186*, 40–50. http://dx.doi.org/10.1016/j.agwat.2017.02.019

Lindsay, W. L. (1979). *Chemical Equilibrium in Soils*. Wiley, New York, USA.

Maas, E. V. (1990). Crop salt tolerance. In K. K. Tanji (Ed.), *Agricultural Salinity Assessment and Management*. ASCE Manuals and Reports on Engineering Practice no. 71 (pp. 262–304). ASCE, New York.

Maas, E. V. and Hoffman, G. J. (1977). Crop salt tolerance-current assessment. *Journal of Irrigation and Drainage Engineering, ASCE*, *103*, 115–134.

Mace, J. E. and Amrhein, C. (2001). Leaching and reclamation of a soil irrigated with moderate SAR waters. *Soil Science Society of America Journal*, *65*, 199–204.

Manivannan, B. and Borisover, M. (2020). Strengths of correlations with formation of chlorination disinfection byproducts: effects of predictor type and other factors. *Environmental Science and Pollution Research*, *27*(5). https://doi.org/10.1007/s11356-019-06976-0

Mara, D. D., et al. (2007). Health risks in wastewater irrigation: comparing estimates from quantitative microbial risk analyses and epidemiological studies. *Journal of Water and Health*, *5*(1), 39–50. https://doi.org/10.2166/wh.2006.055

Marano, R. B., et al. (2019). Antibiotic resistance and class 1 integron gene dynamics along effluent, reclaimed wastewater irrigated soil, crop continua: elucidating potential risks and ecological constraints. *Water Research*, *164*, 114906.

Marano, R. B., et al. (2021). Hidden resistome: enrichment reveals the presence of clinically relevant antibiotic resistance determinants in treated wastewater-irrigated soils. *Environmental Science & Technology*, *55*(10), 6814–6827.

Marcar, N. E., et al. (2011). Salinity. In G. L. Levy, et al. (Eds.), *Treated Wastewater in Agriculture: Use and Impacts on the Soil Environment and Crops* (pp. 286–305). Wiley-Blackwell, Oxford, UK.

Master, Y., et al. (2003). Gaseous nitrogen emissions and mineral nitrogen transformations as affected by reclaimed effluent application. *Journal of Environmental Quality*, *32*, 1204–1211.

McNeal, B. L. and Coleman, N. T. (1966). Effect of solution composition on soil hydraulic conductivity. *Soil Science Society of America, Proceedings*, *30*, 308–312.

Metzger, L., et al. (1983). Soil hydraulic conductivity as affected by physical and chemical-properties of effluents. *Agronomie*, *3*, 771–777.

Michael-Kordatou, I., et al. (2015). Dissolved effluent organic matter: characteristics and potential implications in wastewater treatment and reuse applications. *Water Research*, *77*(0), 213–248. http://www.sciencedirect.com/science/article/pii/S0043135415001931

Mohanavelu, A., et al. (2021). Irrigation induced salinity and sodicity hazards on soil and groundwater: an overview of its causes, impacts and mitigation strategies. *Agriculture*, *11*. https://doi.org/10.3390/agriculture11100983

Monnett, G. T., et al. (1995). Effects of domestic wastewater spray irrigatin on denitrification rates. *Journal of Environmental Quality*, *24*, 940–946.

Munns, R. (1993). Physiological processes limiting plant growth in saline soil: some dogmas and hypotheses. *Plant, Cell & Environment*, *16*, 15–24.

Nadav, I., et al. (2013). Induction of soil water repellency following irrigation with treated wastewater: effects of irrigation water quality and soil texture. *Irrigation Science*, *31*(3), 385–394. https://doi.org/10.1007/s00271-011-0316-y

Negreanu, Y., et al. (2012). Impact of treated wastewater irrigation on antibiotic resistance in agricultural soils. *Environmental Science & Technology*, *46*(9), 4800–4808.

Nemera, D. B., et al. (2020). Tuff trenches improve sap flow in avocado trees affected by long-term irrigation with treated wastewater. *Acta Horticulturae*, *1300*, 7–12. https://doi.org/10.17660/ActaHortic.2020.1300.2

Nemera, D. B., et al. (2021). Mitigating negative effects of long-term treated wastewater irrigation: leaf gas exchange and water use efficiency response of avocado trees (*Persea americana* Mill.). *Agricultural Water Management*, *257*, 107126. https://doi.org/10.1016/j.agwat.2021.107126

Nemera, D. B., et al. (2023). Remediation and mitigation measures to counteract orchard soil degradation by treated wastewater irrigation. *Soil and Tillage Research*, *234*, 105846. https://doi.org/10.1016/j.still.2023.105846

Netzer, Y., et al. (2014). Effects of irrigation using treated wastewater on table grape vineyards: dynamics of sodium accumulation in soil and plant. *Irrigation Science*, *32*, 283–294.

Nikolaou, G., et al. (2020). Irrigation groundwater quality characteristics: a case study of Cyprus. *Atmosphere*, *11*, 302.

NRMMC, EPHC. (2006). *Australian guidelines for water recycling: managing health and environmental risks*. Phase 1. Natl. Water Quality Management Strategy 21. Natural Resource Management Ministerial Council. Environmental Protection and Heritage Council Australia Health Ministers' Conference, Canberra, Australia.

Obayomi, O., et al. (2019). Importance of soil texture to the fate of pathogens introduced by irrigation with treated wastewater. *Science of the Total Environment*, *653*, 886–896.

Ofori, S., et al. (2021). Treated wastewater reuse for irrigation: pros and cons. *Science of The Total Environment*, *760*, 114026. https://doi.org/10.1016/j.scitotenv.2020.144026

Ogunmokun, F. A. and Wallach, R. (2021). Remediating the adverse effects of treated wastewater irrigation by repeated on-surface surfactant application. *Water Resources Research*, *57*(6). https://doi.org/10.1029/2020WR029429

Ogunmokun, F. A. and Wallach, R. (2024). Effect of surfactant surface and interfacial tension reduction on infiltration into hydrophobic porous media. *Geoderma*, *441*, 116735. https://doi.org/https://doi.org/10.1016/j.geoderma.2023.116735

Oren, A. and Chefetz, B. (2012). Sorptive and desorptive fractionation of dissolved organic matter by mineral soil matrices. *Journal of Environmental Quality*, *41*(2), 526–533. https://doi.org/10.2134/jeq2011.0362

Palacios, O. A., et al. (2017). Monitoring of indicator and multidrug resistant bacteria in agricultural soils under different irrigation patterns. *Agricultural Water Management*, *184*, 19–27.

Papadopoulos, A. P., et al. (2008). Inorganic and synthetic organic components of soilless culture and potting mixes. In M. Raviv and J. Heinrich Lieth (Eds.), *Soilless Culture: Theory and Practice* (pp. 505–544). Academic Press, San Diego.

Parida, V. K., et al. (2021). Emerging contaminants in wastewater: a critical review on occurrence, existing legislations, risk assessment, and sustainable treatment alternatives. *Journal of Environmental Chemical Engineering*, *9*(5). https://doi.org/10.1016/j.jece.2021.105966

Paudel, I., et al. (2018). Treated wastewater irrigation: soil variables and Grapefruit tree performance. *Agricultural Water Management*, *204*, 126–137. https://doi.org/10.1016/j.agwat.201.04.006

Paudel, I., et al. (2019). Tolerance of citrus rootstocks to poor water quality is improved by root zone aeration via selective uptake of ions, higher photosynthesis and carbon storage. *Scientia Horticulture*, *251*, 9–19.

Pedrero, F., et al. (2020). Opportunities for expanding the use of wastewaters for irrigation of olives. *Agricultural Water Management*, *241*, 106333.

Peña, A., et al. (2020). A review of the impact of wastewater on the fate of pesticides in soils: effect of some soil and solution properties. *Science of The Total Environment*, *718*, 134468. https://doi.org/https://doi.org/10.1016/j.scitotenv.2019.134468

Pfister, S., et al. (2011). Projected water consumption in future global agriculture: scenarios and related impacts. *Science of the Total Environment*, *409*, 4206–4216. https://doi.org/10.1016/j.scitotenv.2011.07.019

Phogat, V., et al. (2021). Modelling salinity and sodicity risks of long-term use of recycled water for irrigation of horticultural crops. *Soil Systems*, *5*, 49.

Polubesova, T. and Chefetz, B. (2014). DOM-affected transformation of contaminants on mineral surfaces: a review. *Critical Reviews in Environmental Science and Technology*, *44*(3), 223–254. https://doi.org/10.1080/10643389.2012.710455

Qadir, M. and Scott, C. A. (2010). Non-pathogenic tradeoffs of wastewater irrigation. In P. Drechsel, et al. (Eds.), *Wastewater Irrigation and Health: Assessing and Mitigating Risks in Low-Income Countries* (pp. 101–126). Earthscan-International Development Research Centre (IDRC)-International Water Management Institute (IWMI), UK.

Raveh, E. and Ben-Gal, A. (2016). Irrigation with water containing salts: evidence from a macro-data national case study in Israel. *Agricultural Water Management*, *170*, 176–179.

Ray, S. K., et al. (2020). Recent advances in the characterization and the treatment methods of effluent organic matter. *Membrane and Water Treatment*, *11*(4), 257–274. https://doi.org/10.12989/mwt.2020.11.4.257

Rehme, P. (2023). Notes from the field: doubling of cyclosporiasis cases partially attributable to a Salad Kit–Florida, 2021–2022. *MMWR. Morbidity and Mortality Weekly Report*, 72.

Reid, R. J., et al. (2004). A critical analysis of the causes of boron toxicity in plants. *Plant, Cell and Environment*, *27*, 1405–1414.

Renai, L., et al. (2020). Chapter four - impact of the use of treated wastewater for agricultural need: behavior of organic micropollutants in soil, transfer to crops, and related risks. In P. B. T.-A. in C. P. Verlicchi Environmental Management and Protection (Ed.), *Wastewater Treatment and Reuse – Lessons Learned in Technological Developments and Management Issues* (Vol. 6, pp. 103–135). Elsevier, London, UK. https://doi.org/https://doi.org/10.1016/bs.apmp.2020.07.008

Rhoades, J. D. (1996). Salinity: electrical conductivity and total dissolved solids. In D. Sparks, et al.(Eds.), *Methods of Soil Analysis*. https://doi.org/10.2136/sssabookser5.3.c14

Rogowska, J., et al. (2020). Micropollutants in treated wastewater. *Ambio*, *49*(2), 487–503. https://doi.org/10.1007/s13280-019-01219-5

Ruan, M., et al. (2023). Molecular-level exploration of properties of dissolved organic matter in natural and engineered water systems: a critical review of FTICR-MS application. *Critical Reviews in Environmental Science and Technology*, *53*(16), 1534–1562. https://doi.org/10.1080/10643389.2022.2157167

Russo, D., et al. (2015). Salinity control in a clay soil beneath an orchard irrigated with treated waste water in the presence of a high water table: a numerical study. *Journal of Hydrology*, *531*, 198–213.

Russo, D., et al. (2020). Improving water uptake by trees planted on a clayey soil and irrigated with low-quality water by various management means: a numerical study. *Agricultural Water Management*, *229*, 105891. https://doi.org/10.1016/j.agwat.2019.105891

Sánez, J. M., et al. (2023). Transformation of organic carbon through medium pressure (polychromatic) UV disinfection of wastewater effluent during wet weather events. *Science of the Total Environment*, *896*. https://doi.org/10.1016/j.scitotenv.2023.165149

Schröder, P., et al. (2016). Status of hormones and painkillers in wastewater effluents across several European states–considerations for the EU watch list concerning estradiols and diclofenac. *Environmental Science and Pollution Research*, *23*(13), 12835–12866. https://doi.org/10.1007/s11356-016-6503-x

Segal, E., et al. (2011). Olive orchard irrigation with reclaimed wastewater: agronomic and environmental considerations. *Agriculture, Ecosystems & Environment*, *140*, 454–461.

Shani, U. and Dudley, L. M. (2001). Field studies of crop response to water and salt stress. *Soil Science Society of America Journal*, *65*, 1522–1528.

Shi, W., et al. (2021). Monitoring dissolved organic matter in wastewater and drinking water treatments using spectroscopic analysis and ultra-high resolution mass spectrometry. *Water Research*, *188*. https://doi.org/10.1016/j.watres.2020.116406

Shishegar, S., et al. (2018). Optimization methods applied to stormwater management problems: a review. *Urban Water Journal*, *15*(3), 276–286.

Shon, H. K., et al. (2006). Effluent organic matter (EfOM) in wastewater: constituents, effects, and treatment. *Critical Reviews in Environmental Science and Technology*, *36*(4), 327–374. https://doi.org/10.1080/10643380600580011

Shtull-Trauring, E., et al. (2020). Reducing salinity of treated waste water with large scale desalination. *Water Research*, *186*, 116322.

Shtull-Trauring, E., et al. (2022). NPK in treated wastewater irrigation: regional scale indices to minimize environmental pollution and optimize crop nutritional supply. *Science of the Total Environment*, *806*, 150387.

Smith, C. J., et al. (1996). Ammonia volatilization from soil irrigated with urban sewage effluent. *Australian Journal of Soil Research*, *34*, 789–802.

Sparks, D. L. (2003). *Environmental Soil Chemistry*. Academic Press, Elsevier, San-Diego, CA, USA.

Spalding, R. F. and Exner, M. E. (1993). Occurrence of nitrate in groundwater: a review. *Journal of Environmental Quality*, *22*, 392–402. https://doi.org/10.2134/jeq1993.00472425002200030002x

Steppuhn, H., et al. (1998). Evaluating Russian wild ryegrass emergence from saline seedbeds using the Gompertz function. *Canadian Agricultural Engineering*, *40*, 241–247.

Styczen, M., et al. (2010). Management model for decision support when applying low quality water in irrigation. *Agricultural Water Management*, *98*, 478–481.

Suarez, D. L. and Gonzalez-Rubio, A. (2017). Effects of the dissolved organic carbon of treated municipal wastewater on soil infiltration as related to sodium adsorption ratio and pH. *Soil Science Society of America Journal*, *81*, 602–611.

Sumner, M. E., et al. (1998). Sodic soils: a reappraisal. In M. E. Sumner and R. Naidu (Eds.), *Sodic Soils: Distribution, Processes, Management and Environmental Consequences* (pp. 3–17). Oxford University Press, Oxford.

Tarchitzky, J., et al. (2004). The environmentally problematic bleaching agents. In M. J. Schick (Ed.), *Handbook of Detergents* (pp. 645–662). Marcel Decker, Inc., New York.

Tarchitzky, J., et al. (1993). Humic substances and pH effects on sodium and calcium montmorillonite flocculation and dispersion. *Soil Science Society of America Journal*, *52*, 1449–1452.

Tolkin, S., et al. (2020). Humic acid, dextran, and valeric acid representing macromolecules similar to compounds prevailing in treated wastewater induce soil hydrophobicity. *Journal of Soils and Sediments*, *20*(10), 3548–3556. https://doi.org/10.1007/s11368-020-02687-z

Tozer, W. C., et al. (2005). Using ^{15}N to determine a budget for effluent-derived nitrogen applied to forest. *Isotopes in Environmental and Health Studies*, *41*, 13–30.

Tsunemi, K., et al. (2023). Estimation of the potential global nitrogen flow in a nitrogen recycling system with industrial countermeasures. *Sustainability*, *15*, 6042.

Tuholske, C., et al. (2021). Mapping global inputs and impacts from of human sewage in coastal ecosystems. *PLoS One*, *16*(11), e0258898. https://doi.org/10.1371/journal.pone.0258898

Tran, N. H., et al. (2015). A critical review on characterization strategies of organic matter for wastewater and water treatment processes. In *Bioresource Technology* (Vol. 193, pp. 523–533). https://doi.org/10.1016/j.biortech.2015.06.091

USSL Staff. (1954). *Diagnosis and Improvement of Saline and Alkali Soils Agric Handb 60*. USDA, Washington, DC.

van Puijenbroek, P. J. T. M., et al. (2019). Global nitrogen and phosphate in urban wastewater based on the shared socio-economic pathways. *Journal of Environmental Management*, *231*, 446–456.

Wallach, R., et al. (2005). Soil water repellency induced by long-term irrigation with treated sewage effluent. *Journal of Environmental Quality*, *34*(5), 1910–1920. https://doi.org/10.2134/jeq2005.0073

Wang, H.-B., et al. (2024). High-molecular-weight dissolved organic matter enhanced phosphorus availability in paddy soils: Evidence from field and microcosm experiments. *Soil & Tillage Research*, *240*, 106099.

Wang, M. and Chen, Y. (2018). Generation and characterization of DOM in wastewater treatment processes. *Chemosphere*, *201*, 96–109. https://doi.org/https://doi.org/10.1016/j.chemosphere.2018.02.124

Wang, S., et al. (2014). In-depth characterization of secondary effluent from a municipal wastewater treatment plant located in Northern China for advanced treatment. *Water Science and Technology*, *69*(7), 1482–1488. https://doi.org/10.2166/wst.2014.040

Wei, L., et al. (2021). Forms and mobility of heavy metals/metalloids in sewage-irrigated soils in the North China Plain. *Journal of Soils and Sediments*, *21*(1), 215–234. https://doi.org/10.1007/s11368-020-02744-7

Weller, D., et al. (2020). Complex interactions between weather, and microbial and physicochemical water quality impact the likelihood of detecting foodborne pathogens in agricultural water. *Frontiers in Microbiology*, *11*. https://doi.org/10.3389/fmicb.2020.00134

WHO. (2006). *Guidelines for the Safe Use of Wastewater, Excreta and Grey Water*. Volume 2. Wastewater Use in Agriculture, Geneva.

Wu, D., et al. (2021). Mechanisms by which different polar fractions of dissolved organic matter affect sorption of the herbicide MCPA in ferralsol. *Journal of Hazardous Materials*, *416*. https://doi.org/10.1016/j.jhazmat.2021.125774

Xiao, R., et al. (2023). Disinfection by-products as environmental contaminants of emerging concern: a review on their occurrence, fate and removal in the urban water cycle. *Critical Reviews in Environmental Science and Technology*, *53*(1), 19–46. https://doi.org/10.1080/10643389.2022.2043101

Xu, G., et al. (2000). Advances in chloride nutrition of plants. *Advances in Agronomy*, *68*, 97–150.

Yalin, D., et al. (2017). Insights from "The hidden half": the impact of root-zone oxygen and redox dynamics on the response of avocado to long-term irrigation with treated wastewater in clayey soil. *Israel Journal of Plant Sciences*, *64*(1), 1–18.

Yalin, D., et al. (2021). Soil oxygen and water dynamics underlying hypoxic conditions in the root-zone of avocado irrigated with treated wastewater in clay soil. *Soil and Tillage Research*, *212*, 105039.

Yalin, D., et al. (2023). Mitigating risks and maximizing sustainability of treated wastewater reuse for irrigation. *Water Research X*, *21*, 100203.

Yan, P., et al. (2023). Fluorescence spectroscopic characterization of dissolved organic matter in the wastewater treatment plant and hybrid constructed wetlands coupling system in winter: a case study in eastern China. *Environmental Technology and Innovation*, *32*. https://doi.org/10.1016/j.eti.2023.103399

Yassour, H., et al. (2020). Consequences of irrigation and fertigation of vegetable crops with variable quality water: Israel as a case study. *Agricultural Water Management*, *242*, 106362.

Yermiyahu, U., et al. (2011). Boron. In G. J. Levy, et al. (Eds.), *Treated Wastewater in Agriculture, Use and Impacts on the Soil Environment and Crops* (pp. 216–233). Blackwell Publishing Ltd, Hoboken.

Zhang, B., et al. (2021). Unveiling the transformation of dissolved organic matter during ozonation of municipal secondary effluent based on FT-ICR-MS and spectral analysis. *Water Research*, *188*. https://doi.org/10.1016/j.watres.2020.116484

Zhou, J., et al. (2003). Mineralization of organic N originating in treated effluent used for irrigation. *Nutrient Cycling in Agroecosystems*, *67*, 205–213.

Part 2

Remediation techniques

Chapter 7

Advances in analytical techniques for detecting contaminants in soils

Liang Wang and Ravi Naidu, University of Newcastle, Australia and CRC for Contamination Assessment and Remediation of the Environment (crcCARE), Australia; and Ying Cheng, University of Newcastle, Australia

1 Introduction

The spectrum of contaminants is broad, encompassing both inorganic and organic substances. Soil contaminants refer to substances that, when present at levels exceeding environmental or human health threshold levels, can cause harm to the environment, plants, animals, and even humans. Soil contamination, a byproduct of both anthropogenic activities and certain natural processes, poses potential hazards to ecosystems and human health. These contaminants can be both naturally occurring or introduced through anthropogenic activities.

Within the realm of inorganic contaminants, common contaminants are heavy metal(loid)s such as lead (Pb), cadmium (Cd), mercury (Hg), and arsenic (As), which can originate from industrial processes, agricultural runoff, or atmospheric deposition. These contaminants, often sourced from industrial effluents, mining processes, and agricultural practices, become detrimental when concentrated in the environment [1]. Prolonged exposure to these metals can lead to a range of health issues, including neurological disorders and cancer. Other contaminants like hexavalent chromium (Cr) [2] and selenium (Se) [3] often used in industrial applications, can also be carcinogenic. Another significant challenge stems from salts, which, when introduced through

http://dx.doi.org/10.19103/AS.2025.0151.11

continuing or over-irrigation or inferior water quality, can hinder plant growth and disrupt the soil's intrinsic structure.

The domain of organic contaminants is equally intricate. Modern agriculture's reliance on pesticides, including compounds like dichlorodiphenyltrichloroethane and glyphosate, has implications for soil health and water quality, as residues can accumulate and leach [4]. The ramifications of the petroleum industry's byproducts, such as benzene and toluene, are similarly consequential for the soil's microbial community and its overall vitality [5]. Furthermore, persistent chemicals like polychlorinated biphenyls (PCBs) and dioxins, despite their dwindling industrial use, remain environmental concerns [6].

Emerging contaminants present another layer of complexity. Compounds like per- and polyfluoroalkyl substances (PFAS), due to their resistance to degradation, are becoming omnipresent. Simultaneously, novel challenges such as microplastics [7], nanomaterials [4], brominated flame retardants [8], and antibiotic resistance [9] are entering the spotlight, given their unclear long-term impacts and environmental trajectories. Recent attention has also been drawn to the residues from pharmaceuticals and personal care products, which, when entering the soil, can disrupt natural processes [10].

The introduction of radioactive elements, such as uranium (U) and plutonium (Pu), through mining activities or nuclear incidents, adds another layer of complexity to soil contamination. Addressing the issue of soil contamination requires detailed monitoring, remediation strategies, and preventive measures to safeguard environmental and human health.

Amidst the accelerating pace of industrialization and urbanization, there emerges an imperative need to employ advanced monitoring methodologies to effectively address potential soil contaminations. Early and precise detection of these contaminants is pivotal, as it serves not only to forestall land value depreciation but also to minimize the associated remediation expenses. Furthermore, with the introduction of novel chemicals into the environment, the requisition for refined detection modalities becomes increasingly significant.

A prominent category of soil contaminants is characterized by diffuse contaminants, alternatively termed as non-point source pollutants. Contrary to pollutants which emanate from distinct and easily locatable sources, diffuse contaminants have a more expansive and generalized origin though in low concentrations. For instance, agricultural activities might inadvertently introduce excessive fertilizers, pesticides, and herbicides into the soil matrix. Similarly, precipitation in urban areas has the potential to transport contaminants, encompassing oils and assorted chemicals, from urban infrastructures into the surrounding soil. The act of deforestation further exacerbates this contamination scenario by inducing soil erosion, thereby dispersing soil particles and concomitant pollutants over larger geographies. Another dimension of this contamination spectrum is atmospheric deposition, wherein airborne pollutants, over time, settle onto terrestrial and aquatic environments.

Considering the extensive and heterogeneous sources of these contaminants, devising effective strategies for their assessment, monitoring, and management becomes a convoluted task. It is pertinent to highlight that such diffuse contaminants often manifest in exceedingly low concentrations, occasionally reaching sub-ppb levels. This necessitates the deployment of highly sensitive analytical instruments to ensure their accurate detection.

In the vast tapestry of our environment contaminants, two analytical pillars, spectroscopy and mass spectrometry, have consistently risen to this challenge, offering unparalleled insights (Fig. 1). Spectroscopy, through techniques like atomic absorption spectroscopy (AAS), Fourier-transform infrared (FTIR), and nuclear magnetic resonance (NMR), delves deep into the molecular realm,

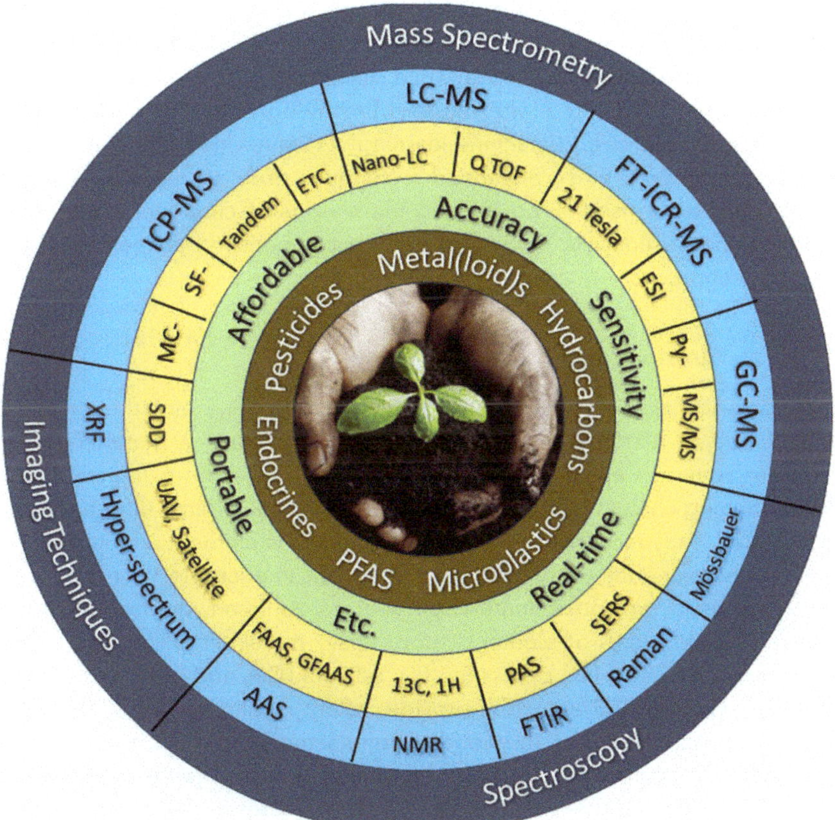

Figure 1 The range of spectroscopy, mass spectrometry, and imaging techniques for contaminant detection discussed in this chapter.

unveiling the intricacies of matter and its relationship with electromagnetic radiation. Mass spectrometry, on the other hand, is a masterclass in precision and depth, segregating ions and revealing elemental and structural narratives that define soils and the environment at large. Further augmenting our analytical prowess are imaging technologies such as X-ray fluorescence (XRF) and hyperspectral imaging (HSI), bridging micro-level detections with macro-scale assessments. This review synthesizes the capabilities and advancements of these tools, underscoring their instrumental role in decoding and preserving the delicate balance of our environment. As we venture further, these methodologies stand as our compass, guiding us toward a richer, more nuanced understanding of the world we inhabit.

2 Imaging technologies

At the forefront of environmental research and assessment, imaging technologies, particularly XRF and HSI, are pivotal. X-ray fluorescence offers a deep dive into elemental compositions, becoming an indispensable tool for detecting contaminants in soils, while HSI shines in environmental studies, especially when integrated with modern technologies like unmanned aerial vehicle (UAV) systems and machine learning. As we navigate the complexities of our environment, these advanced techniques promise not only accuracy but also real-time insights, positioning them as game-changers in the realm of environmental analysis. Table 1 presents a comparison between these two imaging techniques.

2.1 X-ray fluorescence

X-ray fluorescence is an analytical technique used to determine the elemental composition of materials. When bombarded with X-rays, a material's inner-shell electrons are ejected and replaced by electrons from higher energy levels, emitting characteristic X-ray radiation due to the energy difference [11]. The emitted X-ray radiation's energy is unique to the element and its electron transition, allowing the determination of elements in a sample.

When used for contaminated soil analysis, XRF can rapidly detect and quantify a wide range of elements, from heavy metal(loid)s like Pd, As, and Hg to nutrient metals like potassium (K), calcium (Ca), and magnesium (Mg). Modern handheld XRF analyzers provide instant results and allow for in-situ measurements, eliminating the need to send samples to a lab in many cases [12]. This non-destructive method can detect a broad range of elements simultaneously in a single measurement. In practical terms, for many metals of environmental concern like Pb, Cd, Zn, and Cu, the detection limits of modern portable XRF instruments are typically in the range of a few parts per million

Table 1 Comparison of imaging technologies

	XRF	Hyperspectral
Measurement principle	X-ray fluorescence (elemental analysis)	Light reflectance/absorption (molecular and mineralogical analysis)
Detection type	Discrete elements	Broad range of spectral signatures
Strengths	Rapid and quantitative analysis for most elements	Simultaneous detection of multiple organic and inorganic contaminants
Weaknesses	Difficulty in detecting light elements (e.g. C, H, N); Soil matrix can interfere with readings	Requires complex data analysis and interpretation Sensitive to environmental conditions (e.g. illumination) Quantitative analysis can be challenging
Applications	Screening and mapping for heavy metals and other inorganic contaminants	Identifying organic contaminants (e.g. PAHs, pesticides) Studying soil organic matter and mineralogy
Cost	Moderate to high (depending on instrument type)	Moderate to high (depending on instrument type and software)
Portability	Portable and handheld instruments available	UAV, satellites carried

(ppm) or even parts per billion (ppb) [11]. Ravansari et al. recently reviewed the application of portable XRF for environmental assessment of soils [13]. López-Núñez also reviewed portable XRF analysis of soil organic amendments [14].

Silicon drift detectors (SDDs) represent a significant advancement in the field of XRF analysis. Unlike traditional silicon PIN diode detectors or Si(Li) detectors, SDDs have a specially designed structure that uses a series of ringed electrodes to 'drift' the generated charge (from incident X-rays) to a small collection electrode [15]. This design minimizes the capacitance and thus allows for faster signal processing and better energy resolution. Traditional Si(Li) detectors required cooling using liquid nitrogen to operate efficiently. In contrast, SDDs can be cooled electronically, typically using Peltier coolers, making them more user-friendly and portable.

X-ray fluorescence imaging, also known as XRF microscopy, takes the principles of XRF spectroscopy and extends them to produce spatially resolved images based on the elemental composition of a sample [16]. In XRF imaging, similar to other XRF instruments in terms of sensitivity (ppm to ppb), a precisely focused X-ray beam is methodically moved across a soil sample in a raster or scanning pattern. The recent application of XRF microscopy is focusing on soil properties and nutrients. For example, van der Bom et al. applied XRF imaging to assess the distribution, speciation, and bioavailability of phosphorus in soils [17]. Synchrotron XRF microscopy was applied to study the root-induced soil deformation by rhizosphere soil chemistry [18].

3 Hyperspectral imaging

Visible-near infrared (vis-NIR) spectroscopy covers the visible (400–700 nm) and near-infrared (700–2500 nm) regions. These spectral ranges, which are primarily utilized in HSI, are valuable tools in environmental studies. Visible-near infrared provides information based on overtone and combination vibrations of fundamental molecular vibrations. It is more sensitive to bulk properties and less to specific molecular bonds. The technology is commonly used for analyzing factors such as water quality, forest health, and soil properties among other applications. Visible-near infrared can rapidly analyze soil samples with little to no sample preparation. With advancements in technology, portable vis-NIR devices can be used in the field for real-time analysis. There is potential to integrate vis-NIR with remote sensing techniques (such as satellite or drone imagery) to predict soil properties at larger scales [19, 20]. Hyperspectral sensors measure the reflected sunlight across many contiguous spectral bands [21]. While a colour camera might capture red, green, and blue bands, a hyperspectral sensor can capture tens to hundreds of narrow bands, spanning from the visible to the near-infrared, and sometimes even the shortwave infrared. The specific pattern of reflection and absorption gives every material a unique 'spectral signature', as demonstrated in Fig. 2. Stuart et al. has reviewed the application of HSI integration UAV systems [22]. Lassalle et al. [21] reviewed the UAV-embedded HSI for monitoring oil contamination in vegetated areas. Satellite-based HSI was widely applied for atmospheric pollutions [23] and marine pollutions [24].

To predict soil contaminants using HSI, it is necessary to develop a calibration model using methods such as machine learning or partial least squares regression. For example, Soni et al. [25] reviewed the use of machine learning techniques in conjunction with HSI to identify bacterial, fungal, and viral contaminants. Their review also highlighted that pairing HSI with microscopy allows for the detection of distinctive signals from microorganisms, irrespective of their cell size. Huang et al. [26] reviewed the applications of HSI combined with machine learning models for the online detection of microplastics.

Nevertheless, a model calibrated on a specific soil dataset might underperform when introduced to diverse soil types. Overfitting during machine learning training is a concern, leading to models that excel with familiar data but falter with novel, unseen data, given that soils from distinct regions or subjected to varied land uses can exhibit unique spectral patterns. As more researchers use vis-NIR, there is potential for building larger shared databases, which can be used to train more robust machine learning models.

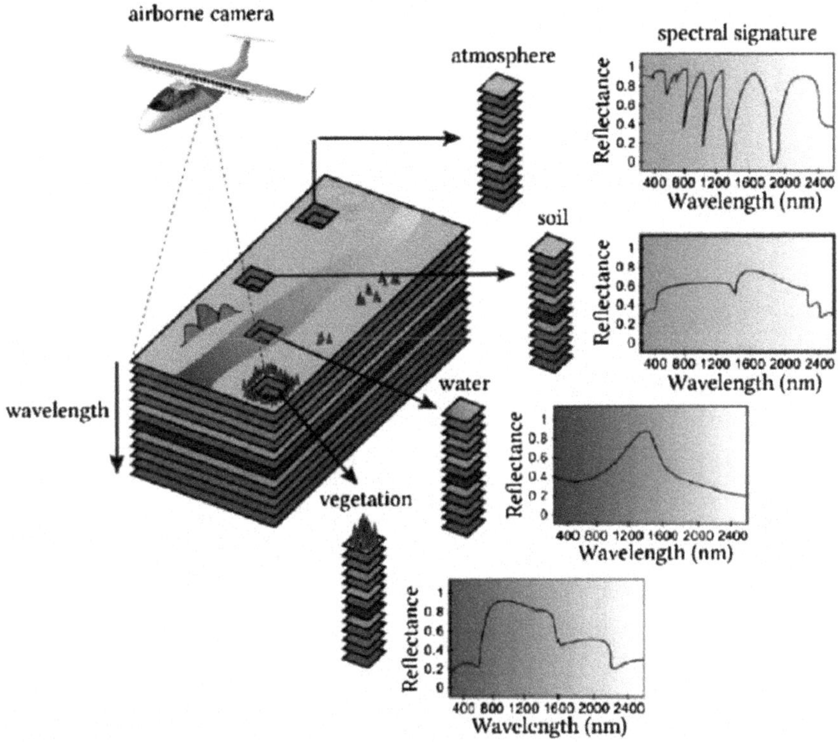

Figure 2 Principle of passive hyperspectral imagery [21].

4 Spectroscopy

Spectroscopy, with its deep-seated roots in analytical science, illuminates the interaction between matter and electromagnetic radiation, thereby offering profound insights into the molecular and atomic dimensions of materials. This review has covered an array of spectroscopic techniques, each honed for specific analytical challenges. Atomic absorption spectroscopy emerges as a trusted ally in the realm of elemental detection, especially concerning soils contaminated with heavy metals. Shifting our lens to Mössbauer spectroscopy, we recognize its invaluable contribution to discerning the magnetic characteristics and oxidation states, particularly highlighting the crucial role of iron in soils. Fourier-transform infrared spectroscopy unveils the intricate spectral 'fingerprints' of organic entities, enabling the identification of a spectrum of contaminants, from ubiquitous hydrocarbons to emerging microplastics. While Raman spectroscopy traverses a parallel path with FTIR in the vibrational domain, its offshoot, surface-enhanced Raman spectroscopy (SERS), amplifies its capabilities, especially in the ultra-sensitive detection of

Table 2 Comparison of different spectroscopy techniques

Spectroscopy	AAS	Mössbauer	FTIR	Raman	NMR
Working principle	Atomic absorption of UV/visible light	Gamma-ray interaction with specific atomic nuclei	Absorption of infrared radiation by functional groups	Inelastic scattering of laser light by molecules	Measurement of magnetic properties of atomic nuclei
LoD	ppb for heavy metals	ppm for Fe-containing compounds	ppm for organic contaminants	ppm for some organic contaminants	Highly variable, can be very sensitive for specific compounds
Cost	Moderate-high	High	Moderate	Moderate-high	Extremely high
Portability	Benchtop and portable models	Benchtop only	Benchtop and portable models	Benchtop and portable models	Benchtop only
Advantages	Highly selective for specific elements, multi-element analysis is possible	Unique insights into the oxidation state and local environment of elements (esp. Fe), differentiates mineral forms	Broad spectral information for diverse compounds, non-destructive	Highly specific for molecular identification, non-destructive, in-situ measurements, complements FTIR	Highly specific for molecular identification, detailed structural information, non-destructive
Disadvantages	Destructive sample prep, limited to metallic elements	Limited elemental scope (primarily Fe), complex data analysis	Complex spectra require expertise, water interference, and quantitative analysis challenging for complex mixtures	Water interference, quantitative analysis challenging	Expensive and complex, limited sample size, and sensitivity for some compounds

environmental pollutants. Concluding with NMR, we appreciate its prowess in demystifying the complexities of organic structures in soils. Table 2 presents a comparison between these spectroscopy techniques. Collectively, these techniques exemplify the power and precision of modern analytical tools, each contributing uniquely to our comprehensive understanding of the environment.

4.1 Atomic absorption spectroscopy

Atomic absorption spectroscopy uses a light source, often a hollow cathode lamp or an electrodeless discharge lamp, to emit electromagnetic radiation [27]. This radiation is utilized to detect and measure the concentration of elements in liquid samples, and, for solid samples, they are typically dissolved or digested into a liquid form for analysis. The principle behind AAS is based on atoms in the sample absorbing ultraviolet or visible light, causing them to undergo a transition from a lower energy state to a higher one. The amount of light absorbed is directly proportional to the concentration of atoms in the sample, providing a means to quantify the element's presence. Flame atomic absorption spectrometry (FAAS) and graphite furnace atomic absorption spectrometry (GFAAS) are both subsets of AAS utilized for elemental detection [28]. As demonstrated in the schematic diagram in Fig. 3, FAAS introduces the sample into a flame for atomization and is generally suited for mg per kg level detection, requiring larger sample volumes and often facing more matrix interferences. In contrast, GFAAS, also known as electrothermal AAS, introduces the sample directly into a graphite furnace, undergoing a stepwise temperature increase for atomization. This method is highly sensitive, capable of detecting elements at μg per kg levels, and requires minimal sample volume. However, while GFAAS provides superior sensitivity, it usually processes samples at a slower rate than FAAS.

Atomic absorption spectrometry is widely used in the measurement of heavy metal-contaminated soils. For example, Sisay et al. applied FAAS to detect cadmium (Cd) and lead (Pb) in roadside soils [30]. Adedeji et al. applied

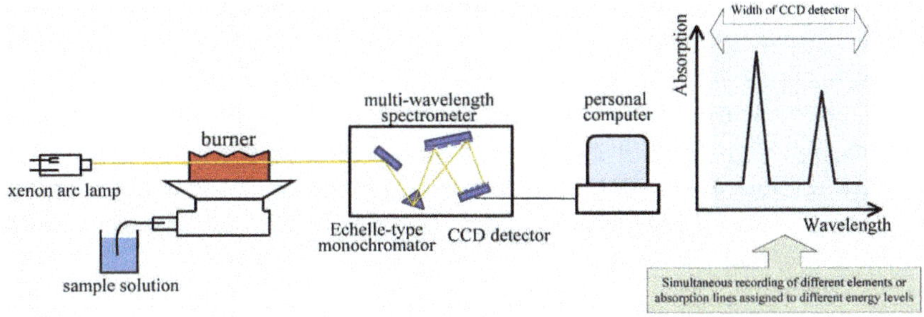

Figure 3 Schematic diagram of a FAAS with multi-wavelength spectrometer system [29].

FAAS for the detection of zinc (Zn), copper (Cu), Pd, and Cd in soil samples for land risk assessments [31]. Both FAAS and GFAAS were applied for the analysis of Cd, chromium (Cr), Cu, mercury (Hg), manganese (Mn), nickel (Ni), Pb, and Zn in the surface soil of public parks in Southern Ghana. However, traditional AAS techniques analyze one element at a time. If samples need to be tested for multiple elements, each element must be measured sequentially, which can be time-consuming. High-resolution continuum source AAS, employs a continuum radiation source, which emits a broad spectrum of wavelengths to offer the capability to simultaneously detect multiple metals [32]. This feature significantly speeds up multielement analysis and provides the capability to correct for spectral interferences, enhancing the precision and accuracy of measurements.

The preparation of soil samples for AAS is a detailed and sometimes cumbersome process. From the initial stages of drying to subsequent grinding and sieving, each step must be executed with precision to avoid contamination and ensure accuracy. Digestion of the samples further adds to the intricacy and time required for the procedure. Such meticulous preparation can be labour-intensive, demanding both attention and patience. In contrast, solid sampling AAS presents a notable advantage by facilitating the direct analysis of solid samples, eliminating the complexities associated with liquid sample preparation, and thus streamlining the overall process. Recently, Mochado et al. conducted a review of the applications of solid sampling [33]. The review highlighted the challenges associated with solid sampling, such as the need for sample homogenization. It also emphasized that the necessity for a higher energy source in solid sampling can accelerate wear and tear on the atomizer and other components of the AAS system.

In addition, in traditional AAS methods, organic solvents might be used for various purposes, including the stabilization of certain metal species. However, many organic solvents pose environmental and health risks. Deep eutectic solvents (DESs) have revolutionized liquid–liquid microextraction techniques due to their environmentally friendly and adaptable nature. Andruch et al. [34] recently reviewed the application of DESs in the application of ASS. In ultrasonic-assisted liquid-phase microextraction (UA-LPME), DESs benefit from ultrasonic waves, streamlining efficiency. Vortex-assisted LPME combines DESs with vertexing, leading to quick and efficient extraction. Meanwhile, dispersive liquid–liquid microextraction uses DESs as a sustainable alternative to traditional solvents, yielding superior extraction results. The versatility of DESs in these methods underscores their importance in contemporary analytical processes.

Atomic absorption spectrometry is known for its simplicity, cost-effectiveness, and low maintenance requirements, making it a popular choice for routine applications, particularly in analyzing specific heavy metals. It demonstrates excellent sensitivity and minimal interference in complex matrices, traits especially beneficial for the analysis of soil contaminants. However, there is a growing preference for inductively coupled plasma optical emission spectrometry (ICP-OES) in the scientific community for its ability to simultaneously analyze multiple elements, including non-metals, with lower detection limits and better resistance to matrix interferences. This makes it particularly suited for complex samples like soil contaminants, enhanced by its sophisticated data processing and adaptability for high-demand analyses.

4.2 Mössbauer spectroscopy

Mössbauer spectroscopy, based on the Mössbauer effect, enables a nucleus in a solid to emit and absorb gamma radiation without recoil, with the crystal lattice absorbing this momentum. The nuclear magnetic moment's interaction with the local magnetic field can cause magnetic splitting in the spectrum, revealing the sample's magnetic properties. Additionally, various oxidation states of an element can lead to shifts in the Mössbauer spectrum, termed isomer shifts.

Mössbauer spectroscopy is indeed a valuable tool for analyzing iron in soil samples. Given the natural abundance of iron in many soils and its importance in various geochemical processes, understanding its speciation and oxidation state is crucial [35]. Iron is involved in various geochemical processes, and its form can impact nutrient availability, soil colour, and more. The form of iron can significantly influence the mobility and bioavailability of contaminants in soils and waters. For instance, certain iron minerals can adsorb heavy metals or organic pollutants. An exemplary case is zero-valent iron (ZVI), which is iron in its elemental form. ZVI has shown great promise in the remediation of contaminated sites. When ZVI reacts with water, it produces iron hydroxides and molecular hydrogen. This molecular hydrogen can reduce many contaminants, including chlorinated hydrocarbons, transforming them into non-toxic or less toxic compounds [36]. Additionally, the iron hydroxides formed can adsorb heavy metals, thus immobilizing them and preventing their migration into the environment [36]. For example, Górka-Kostrubiec et al. applied Mössbauer spectroscopy for the identification of Fe phases bonding heavy metals in street dust [37].

4.3 Fourier-transform infrared

Fourier-transform infrared works by passing a beam of infrared light through a sample and measuring the amount of light absorbed by the sample at different wavelengths. Different compounds absorb light at specific wavelengths, which produces a unique 'fingerprint' or spectrum for each compound. Fourier-transform infrared can detect the presence of specific organic contaminants

Table 3 The functional group's characteristic wavelengths in the FTIR spectrum for different types of hydrocarbons and pesticides [38]

Analyte	Functional group	Wavelength (cm^{-1})
Alkanes	C–H stretch	2800–3000
	CH_2 bend	1465
	CH_3 bend	1375
Alkenes	C–H stretch	3100
	C=C stretch	1600–1680
Aromatics	C–H stretch	3050
	C=C stretch (rings)	1400–1600 (multiple peaks)
Halogenated pesticides	C–Cl stretch	600–800
	C–Br stretch	500–700
	C–F stretch	1000–1300
Organophosphates	P=O stretch	1250
	P–O–C stretch	1000–1200
	P–S stretch	700–900
Pesticides (with carbonyl group)	C=O stretch	1650–1750

in soil, such as hydrocarbons, pesticides, and other industrial chemicals. The resulting spectra can be compared to reference spectra to identify and quantify the contaminants. Table 3 summarizes the functional group's characteristic wavelengths in the FTIR spectrum for different types of hydrocarbons and pesticides.

There are different methods to introduce the sample into the FTIR instrument, like the potassium bromide (KBr) pellet method, attenuated total reflectance (ATR), and diffuse reflectance infrared Fourier transform (DRIFT). Using the traditional KBr pellet method, a dried and finely ground soil sample is mixed with KBr, compressed into a clear pellet, and then analyzed in the FTIR instrument. In contrast, ATR and DRIFT methods are more user-friendly. Soil samples, in these approaches, are either directly analyzed or amalgamated with reflective substances to generate the spectral data. Handheld or portable FTIR instruments have recently gained popularity in the assessment of soil contaminations due to their portability and the ability to provide on-site, real-time analysis [5, 38]. This allows for immediate decision-making and potential cost savings, as there's no need to send samples to a laboratory and wait for results. Recent applications, Wang et al applied a handheld FTIR for rapid field-based determination of petroleum hydrocarbons for land risk assessment [38]. Bell et al. applied a handheld FTIR for the determination of different polymers [39].

Recently, FTIR has been applied for the identification and quantification of microplastics. The applications have been summarized in the reviews by Veerasingam et al. [40] and Chen et al. [41]. Fourier-transform infrared photoacoustic spectroscopy (FTIR-PAS) is a technique that blends FTIR with the photoacoustic effect. A modulated infrared radiation source irradiates a sample. When the sample absorbs this radiation, it heats up and cools down periodically, producing acoustic waves. A microphone detects these waves, and the resulting signal undergoes Fourier transformation to produce a spectrum. FTIR photoacoustic spectroscopy was applied to detect polyethylene residues with different contaminants.

4.4 Raman

Raman and FTIR are vibrational spectroscopy techniques that measure the vibrational modes of molecules. Raman uses a laser to excite molecules, which then scatter the light at different frequencies, while FTIR measures how much light is absorbed by a molecule at different frequencies [42]. Fourier-transform infrared spectroscopy often requires special sample handling to avoid interference from water vapour and carbon dioxide. Raman spectroscopy is less sensitive to water and is therefore more suitable for studying aqueous samples. However, Raman spectroscopy can be more susceptible to interference from fluorescence in the sample, while FTIR spectroscopy is not. Some molecular vibrations are only active in one technique, so the best technique to use depends on the specific sample and application.

Surface-enhanced Raman spectroscopy has recently garnered significant attention in the detection of environmental contaminants, underlining its importance in monitoring and ensuring environmental safety. This technique leverages the enhancement of Raman signals on specially prepared metallic surfaces, typically gold or silver nanoparticles, to detect molecules at extremely low concentrations. Applications of SERS in environmental monitoring have grown rapidly due to its high sensitivity and specificity. Some of the key applications include the detection of pesticides and herbicides [43]; heavy metals [44]; and microplastics [45]. Numerous review articles delve into the applications of SERS in environmental detection, notably those by Terry et al. [46] and Ong et al. [47].

4.5 Nuclear magnetic resonance

The spectroscopy is an analytical technique that detects and analyzes the radiofrequency signals emitted by atomic nuclei, specifically carbon-13 (^{13}C) or hydrogen (^{1}H) nuclei when they are exposed to a magnetic field. Through the interpretation of ^{13}C and ^{1}H NMR data, the different hydrocarbon functional groups present in soil organic matter (SOM) can be identified and revealed [48]. ^{13}C NMR is also used to analyze the soil organic carbon (SOC) quality and its influence on the sorption of pesticides [49]. To investigate the influence of SOC quality on pesticide sorption, the researchers analyzed the humic acid fraction of SOC using ^{13}C NMR spectroscopy. This technique provides information about the chemical composition of organic compounds in the soil. Principal component regression was then employed to correlate the ^{13}C NMR spectra with the sorption coefficients of abamectin and atrazine, two commonly used pesticides. ^{1}H NMR spectroscopy was used as an analytical technique to assess dissolved organic matter released by hydrochar and biochar derived from pig manure [50]. The dissolved organic carbon was found to contain aliphatic compounds, including three humic acid-like substances and one protein-like substance.

5 Mass spectrometry

Mass spectrometry stands as a cornerstone in the analytical arena, indispensable for a broad spectrum of applications, particularly in soil science and environmental research. As a technique, it operates on the premise of segregating ions based on their mass-to-charge ratio, ultimately discerning elemental compositions and structural information. This review delves into the intricate details of several mass spectrometric modalities and their applications: gas chromatography–mass spectrometry (GC–MS), which, through its versatile detectors, can detect an array of organic compounds; Fourier transform ion cyclotron resonance mass spectrometry (FT-ICR-MS), renowned for its unparalleled resolution and its applications in characterizing SOM; liquid chromatography–mass spectrometry (LC–MS), a staple in assessing soil contaminants, especially those unamenable to GC; and inductively coupled plasma mass spectrometry (ICP-MS), a robust tool for trace and major elemental detection. Each of these modalities has been punctuated with advancements, unveiling greater depths in soil analysis, from elemental distribution to organometallic compounds and isotopic studies.

5.1 Gas chromatography-mass spectrometry

In GC, a vaporized sample is introduced to a column and carried through by an inert gas, e.g. helium. The sample's components separate based on their volatility and interactions with the column's stationary phase. Gas chromatography is extensively used to determine organic compounds including petroleum hydrocarbons [51], polycyclic aromatic hydrocarbons (PAHs) [52], PCBs [53], chlorinated solvents [54], phenolic compounds [55], and organochlorine pesticide residues [56]. Besides GC–MS, there are various GC detectors including flame ionization detector (FID), flame photometric detector (FPD), thermal conductivity detector (TCD), electron capture detector (ECD), photoionization detector (PID), and nitrogen-phosphorus detector (NPD). Table 4 provides the comparison of these popular GC techniques for soil contaminant analysis.

Table 4 Comparison of popular GC techniques for soil contaminant analysis

Method	Working principle	LOD	Cost	Portability	Advantages	Disadvantages
MS	Ionization, mass-to-charge ratios (*m/z*)	ppb-ppt	High	Limited	High sensitivity, broad range compounds	High cost/complexity, time-consuming
FID	Flame ionization of carbon/ hydrogen bonds	ppm	Low	Portable	Simple/robust, good for organics, inexpensive	Low sensitivity, limited selectivity, unknown ID
FPD	Flame photometry for S, P, halogens	ppm-ppb	Moderate	Portable	Selective for S, P, halogens, good for the environment	Limited to specific elements, interferences
TCD	Thermal conductivity difference	ppm	Low	Portable	Simple/robust, universal, inexpensive	Low sensitivity, limited differentiation
ECD	Electron capture for halogens/OCPs	ppb-ppt	Moderate-high	Limited	High sensitivity to halogens/ OCPs, selective	Radioactive source, not for all compounds
PID	Photoionization of organic compounds	ppm	Moderate	Portable	Broad range sensitivity, simple, inexpensive	Low sensitivity for specific compounds, interferences
NPD	N, P detector (chemiluminescence)	ppb-ppt	High	Limited	Highly sensitive to N, P compounds, selective	High cost, complex operation, limited to N, P compounds

Gas chromatography-FID measures ions formed during the combustion of the compound in a hydrogen/air flame, which is highly sensitive to hydrocarbons, including PAHs and BTEX (benzene, toluene, ethylbenzene, and xylenes) [57]. Gas chromatography-TCD measures the change in the thermal conductivity of the column effluent gas compared to a reference gas, which has been recently applied for petroleum hydrocarbons in soil samples [58]. The GC-ECD technique employs a radioactive beta emitter for ionizing the sample. When molecules capture these electrons, there's a reduction in the measured current. This method is particularly adept at detecting halogenated compounds, notably organochlorine pesticides [59] and PCBs [60]. Gas chromatography-PID applies ultraviolet light to ionize volatile organic compounds, such as BTEX, and has been developed as a compact device for near-real-time soil analysis and field deployment [61]. Gas chromatography-NPD is specifically a response to compounds containing nitrogen and/or phosphorus, such as atrazine (herbicides) and diazinon (pesticide) [62].

Recent advancements in GC–MS have incorporated pyrolysis, known as pyrolysis-GC–MS, a process where a sample undergoes thermal decomposition without oxygen. By quickly heating the sample to high temperatures, typically between 500°C and 900°C, complex organic materials are broken down into simpler, volatile compounds. Figure 4 illustrates the schematic diagram of a typical Py-GC–MS system.

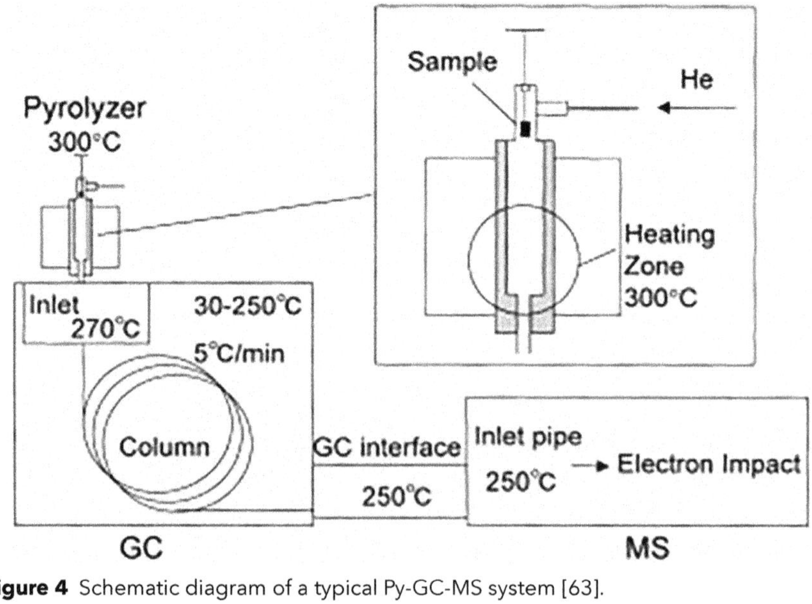

Figure 4 Schematic diagram of a typical Py-GC-MS system [63].

Pyrolysis-GC–MS has emerged as a valuable tool to determine the presence and type of microplastics in soils [64]. It thermally degrades microplastics into their monomeric units, which are then analyzed using GC–MS. Researchers can determine the type of plastic polymers, such as polyethylene, polypropylene, polystyrene present in the soil sample [65]. Beyond identifying the base polymer, Py-GC–MS can also detect various additives used in plastic production, such as plasticizers, flame retardants, antioxidants, or UV stabilizers [66]. This additional layer of information can help trace the origin or potential impacts of the microplastics. Pyrolysis-GC–MS has been applied by Kovacs et al. to investigate the impact of the accumulated pharmaceuticals on the quality of vegetables [67]. Parameters such as carbohydrate content, amino acid composition, volatile organic aroma compounds, and fatty acid methyl ester content were analyzed. In Chen et al., Py-GC–MS was applied to assess the lasting effects of severe wildfires on soil organic matter and the recovery of burned soils [68]. The study indicated that the burned soils had higher levels of benzene, indicating the presence of pyrogenic carbon from the fire, and soil aromaticity increased with fire severity.

5.2 Fourier transform ion cyclotron resonance mass spectrometry

Fourier transform ion cyclotron resonance mass spectrometry, based on the principles of ion cyclotron resonance and Fourier transform spectroscopy, offers high resolution and mass accuracy for a comprehensive characterization of SOM [69]. Zhang et al. provided a comprehensive overview of the application of FT-ICR-MS in analyzing SOM composition for studying soil degradation and restoration [69]. This review discusses extraction, enrichment, and purification methods used to prepare SOM samples for FT-ICR-MS analysis. The ionization techniques, mechanisms of FT-ICR-MS, and data analysis methods are also summarized.

Electrospray ionization (ESI) is a technique that transfers analyte molecules from the liquid phase to the gas phase without compromising their structural integrity, and it is commonly employed in conjunction with FT-ICR-MS. Fourier transform ion cyclotron resonance can identify compounds present in very low concentrations, providing detailed information on the molecular compositions. It can be applied to understand the transformation and decomposition of organic contaminants in soils. Recently, Geng et al. applied ESI-FT-ICR-MS to investigate the molecular transformation of SOM during the remediation of diesel-polluted soils [70]. Wang et al. applied the technique to study the molecular transformation of crude oil-contaminated soil after bio-electrochemical degradation [71].

The extraction methods can influence the compounds detected and their observed relative abundances. Goranov et al. compared sample preparation techniques for soil analysis using an ESI-FT-ICR-MS system and indicated that different sample preparation techniques can lead to variations in the detection of specific groups of molecules, such as differences in oxygenation, aromaticity, and molecular weight [72].

In normal FT-ICR MS instruments, the magnetic field strength typically ranges from 7 to 12 Tesla (T). However, 21 T FT-ICR MS refers to a system operating at a higher magnetic field strength of 21 T, which enables greater mass resolution and accuracy in the analysis. Roth et al. developed a positive-ion ESI 21 T FT-ICR MS system to overcome the challenges posed by soil pyrogenic organic matter from wildfires and achieve enhanced speciation of organic nitrogen [73].

5.3 Liquid chromatography-mass spectrometry

Liquid chromatography–mass spectrometry is a widely used analytical tool for the assessment of soil contamination, especially for compounds that are not amenable to gas chromatography due to their polarity, thermal instability, or large molecular sizes, such as herbicides and pesticides [74], per- and polyfluoroalkyl substances [75]. Once soil samples are prepared in solutions, LC techniques are employed to separate the sample mixtures into its individual components, based on their interactions with the stationary phase inside the column. Once eluted from the column, these separated components are introduced into various detectors, such as MS techniques, for determination.

To handle minute samples, nano-LC was developed to utilize narrow columns, typically ranging from 50 to 100 μm I.D., and operates at flow rates in the nanolitre per minute range [76]. There are numerous types of LC columns available, each designed to separate components based on different chemical or physical properties. The most common separation columns are such as reverse phase (RP), hydrophilic interaction (HI), and ion exchange (IE) columns. Reverse phase columns are non-polar, e.g. C18 and C30, and they separate molecules based on their hydrophobic interactions. This type of column has been applied to determine soil physicochemical properties, including acid dissociation constants, formation constants [77]; and soil biomarkers, such as neutral sugars and uranic acid [78]. Hydrophilic interaction columns are polar and the separation is based on the hydrophilic interactions. Recently, HILC was applied to detect organic acids [79], esters [80], and soil antimicrobial activity [81].

Ion exchange column is considered as an ion chromatography (IC) technique that separates molecules based on their charge. The stationary phase has either a positive or negative charge, attracting oppositely charged ions from the sample. Ion chromatography specifically targets the separation and measurement of ions, especially anions in solutions, including fluoride, chloride, nitrate, nitrite, sulphate, and phosphate ions, among others [82]. Combustion IC is a technique that combines the processes of combustion and ion chromatography to measure halides, sulphur, and other elements or compounds in solid samples, such as pesticides [83], carboxylic acids [84], and polyvinyl chloride [85]. During this combustion process, burned in a furnace with oxygen, the elements of interest in the sample are converted into their simplest ionic forms (e.g. sulphur becomes sulphate, chlorine becomes chloride) [84].

After the separation of LC techniques, various MS analyzers were applied for various targeted soil substances. Bugsel and Zwiener applied LC with Q TOF-MS for screening poly- and perfluoroalkyl substances in contaminated soil [86]. Liquid chromatography with MS/MS was applied for the determination of glyphosate and its metabolite AMPA [87] in soil. Liquid chromatography with FT-ICR-MS was applied for the determination of soil-dissolved organic matter and disinfection byproducts [88]. Fourier transform ion cyclotron resonance mass spectrometry is an MS technique in which ions are trapped in a magnetic field and oscillate at a characteristic cyclotron frequency. The resulting time-domain signal (free induction decay) is then transformed using a Fourier transform to produce a high-resolution mass spectrum in the frequency domain.

There are field-deployable LC–MS systems are recently developed. For example, Hemida et al. reported a small-footprint LC–MS system for on-site PFAS analysis [89]. Vargas Medina et al. reviewed the recent developments of chip-based LC–MS technology based on microelectromechanical systems and diverse microfabrication technologies [90].

5.4 Inductively coupled plasma mass spectrometry

Both ICP-MS and ICP-OES are analytical techniques used to detect and quantify trace and major elements, including metals and some non-metals, in soil samples [91]. Once the soil sample undergoes ionization in the plasma, the resulting ions are either distinguished by their mass-to-charge ratio using a mass spectrometer or their emitted light at specific wavelengths is analyzed through spectroscopy [92]. Inductively coupled plasma mass spectrometry offers a comprehensive capacity to measure a vast array of elements, including many that pose challenges for ICP-OES due to their detection limits or spectral interferences [93]. Notably, ICP-MS, especially with quadrupole (Q) mass analyzer, excels in detecting various heavy metals like Hg, Pb, thallium (Tl), uranium (U), and plutonium (Pu), even at incredibly low levels [94]. Additionally, it's adept at analyzing certain non-metals such as selenium (Se), As, and bromine (Br), with enhanced sensitivity [95].

Laser ablation (LA)-ICP-MS applies a focused laser beam to remove a small portion of the soil sample, which produces aerosol with small particles for ICP-MS analysis [96]. Laser ablation-ICP-MS offers the advantages of minimal sample preparation, detailed spatial information on elemental distribution, and the ability to perform trace and ultra-trace analysis of soil elements. For instance, this method has been employed to analyze trace elements in Fe mineralization processes, specifically in magnetite and pyrite [97, 98]. A recent review by Chew et al. delves into the geochronological applications of LA-ICP-MS imaging [99].

Meanwhile, GC and high-performance liquid chromatography (HPLC) are also coupled with ICP-MS for the analysis of organometallic compounds. Gas chromatography-ICP-MS has been applied for the determination of volatile organometallic compounds including mercury methylation [100] and volatile sulphur compounds [101] in soil samples. García-Bellido et al., reviews the recent advances in GC-ICP-MS [102]. High-performance liquid chromatography separates non-volatile organometallic compounds in soil solutions using a column coated with a stationary phase. For example, HPLC-ICP-MS was applied to the detection of hexavalent chromium (Cr) [103, 104], Hg speciation [105], and fluorinated compounds [106] in soil samples.

Several mass analyzers have been developed to improve the separation of isotopes and offer enhanced interference removal capabilities. These include ICP-MS/MS (or tandem ICP-MS), SF-ICP-MS (sector-field ICP-MS), MC-ICP-MS (multi-collector ICP-MS), and ICP-ToF-MS (time-of-flight ICP-MS).

The ICP-MS/MS instrument, featuring two serial quadrupoles, directs ions from the first quadrupole (acting as a selective mass filter) into a gas-filled collision/reaction cell, after which the resulting ions are filtered and detected by the second quadrupole [93]. For instance, recent applications of ICP-MS/MS techniques have been employed to ascertain the isotopes of sulphur [107], Pu [108], and U [109]. Sector-field ICP-MS integrates a magnetic sector field as a mass analyzer, so the ions are separated based on both their mass-to-charge ratio and their kinetic energy. Recently applications include the determination of the ultra-trace level of isotopes of neptunium (Np), Pu, Pb and Americium (Am) [110–112]. Conversely, MC-ICP-MS features multiple collectors, often in the form of Faraday cups or secondary electron multipliers, facilitating simultaneous isotope measurements. Sullivan et al. recently reviewed the application of MC-ICP-MS for Cu isotopic analysis [113]. Other isotopic studies for soil elements include Cd [114], K [115], silicon (Si) [116], and Zn [117]. In ICP-ToF-MS, a ToF mass analyzer determines the mass-to-charge ratio of ions by measuring the time they take to travel a fixed distance, capitalizing on the fact that lighter ions travel faster than heavier ones [118]. Recently, ICP-ToF-MS has been applied for the detection of multi-isotopes [118], nanoparticles [119], and rare earth elements [120] in soil samples.

6 Conclusion

Accurate monitoring and identification of soil contaminants are essential for safeguarding the environment and managing land resources effectively. Several analytical methods exist for determining and monitoring soil contaminants, each with its own set of advantages, limitations, and appropriateness for various scenarios.

For the imaging technologies, XRF imaging offers a non-destructive, rapid approach to map the spatial distribution of elements, primarily metals, within soil samples. Hyperspectral imaging paints a broader picture, providing valuable information about organic contaminants, albeit with limited specificity for individual compounds. Meanwhile, spectroscopy techniques delve deeper into the chemical composition of soil contaminants. Atomic absorption spectroscopy offers a simple and cost-effective way to quantify specific metals in soil but lacks the versatility to identify organic compounds. Fourier-transform infrared and Raman spectroscopy provide valuable insights into the functional groups and molecular vibrations present in organic matter and minerals, respectively. Mass spectrometry reigns supreme for unravelling the complex tapestry of soil contaminants. Gas chromatography–MS is a powerful tool for volatile and semi-volatile organic compounds, while ICP-MS unlocks the secrets of metals at ultra-trace levels. With unparalleled sensitivity and the ability to differentiate isotopes, these techniques offer a detailed molecular-level understanding of soil contamination. However, their high cost, complex operation, and often lengthy sample preparation processes impose limitations on routine monitoring.

In conclusion, selecting the optimal analytical method requires careful consideration of factors like the target contaminants, desired level of detail, budget, and available resources. For a comprehensive understanding of spatial distribution and element mapping, imaging techniques are invaluable. Spectroscopy techniques provide insights into the molecular composition of various contaminants, while mass spectrometry offers deep dives into specific compounds at trace levels. Ultimately, the most effective monitoring strategy often involves a judicious combination of these methods, allowing for a holistic understanding of the complex tapestry of soil contamination.

7 References

1. Li, C., et al., A review on heavy metals contamination in soil: effects, sources, and remediation techniques. *Soil and Sediment Contamination: An International Journal*, 2019. 28(4): pp. 380–394.
2. Guo, S., et al., Speciation, toxicity, microbial remediation and phytoremediation of soil chromium contamination. *Environmental Chemistry Letters*, 2020. 19(2): pp. 1413–1431.

3. Ullah, H., et al., A critical analysis of sources, pollution, and remediation of selenium, an emerging contaminant. *Environmental Geochemistry and Health*, 2023. 45: pp. 1359–1389.
4. Mukhopadhyay, R., et al., Nanomaterials for sustainable remediation of chemical contaminants in water and soil. *Critical Reviews in Environmental Science and Technology*, 2021. 52(15): pp. 2611–2660.
5. Wang, L., et al., The key factors for the fate and transport of petroleum hydrocarbons in soil with related in/ex situ measurement methods: an overview. *Frontiers in Environmental Science*, 2021. 9: p.756404.
6. Nhung, N.T.H., et al., A review of soil contaminated with dioxins and biodegradation technologies: current status and future prospects. *Toxics*, 2022. 10(6): p. 278.
7. Büks, F. and Kaupenjohann, M., Global concentrations of microplastics in soils: a review. *Soil*, 2020. 6(2): pp. 649–662.
8. Xiong, P., et al., A review of environmental occurrence, fate, and toxicity of novel brominated flame retardants. *Environmental Science & Technology*, 2019. 53(23): pp. 13551–13569.
9. Zhang, Y., et al., Impacts of farmland application of antibiotic-contaminated manures on the occurrence of antibiotic residues and antibiotic resistance genes in soil: a meta-analysis study. *Chemosphere*, 2022. 300: p. 134529.
10. Xu, Y., et al., Sorption of pharmaceuticals and personal care products on soil and soil components: influencing factors and mechanisms. *Science of The Total Environment*, 2021. 753: p. 141891.
11. Potts, P.J., *Portable X-ray flourescence spectrometry: capabilities for in situ analysis*. 2008, Cambridge: RSC Publishing.
12. Sieber, J.R., Portable X-ray fluorescence spectrometry-capabilities for in situ analysis. X-Ray Spectrometry, 2009. 38(2): p. 157.
13. Ravansari, R., et al., Portable X-ray fluorescence for environmental assessment of soils: not just a point and shoot method. *Environment International*, 2020. 134: p. 105250.
14. López-Núñez, R., Portable X-ray fluorescence analysis of organic amendments: a review. *Applied Sciences*, 2022. 12(14): p. 6944.
15. Acquafredda, P., XRF technique. *Physical Sciences Reviews*, 2019. 4(8). https://doi.org/10.1515/psr-2018-0171
16. Strotton, M., et al., Multielement Z-tag imaging by X-ray fluorescence microscopy for next-generation multiplex imaging. *Nature Methods*, 2023. 20(9): pp. 1310–1322.
17. van, K., P.M., et al., Methods for assessing laterally-resolved distribution, speciation and bioavailability of phosphorus in soils. *Reviews in Environmental Science and Bio/Technology*, 2022. 21(1): pp. 53–74.
18. van Veelen, A., et al., Root-induced soil deformation influences Fe, S and P: rhizosphere chemistry investigated using synchrotron XRF and XANES. *New Phytologist*, 2019. 225(4): pp. 1476–1490.
19. Meng, X., et al., Soil organic matter prediction model with satellite hyperspectral image based on optimized denoising method. *Remote Sensing*, 2021. 13(12): p. 2273.
20. Guo, L., et al., Mapping soil organic carbon stock by hyperspectral and time-series multispectral remote sensing images in low-relief agricultural areas. *Geoderma*, 2021. 398: p. 115118.

21. Lassalle, G., et al., Monitoring oil contamination in vegetated areas with optical remote sensing: a comprehensive review. *Journal of Hazardous Materials*, 2020. 393: p. 122427.
22. Stuart, M.B., et al., Hyperspectral imaging in environmental monitoring: a review of recent developments and technological advances in compact field deployable systems. *Sensors*, 2019. 19(14): p. 3071.
23. Cheng, L., et al., Stereoscopic hyperspectral remote sensing of the atmospheric environment: innovation and prospects. *Earth-Science Reviews*, 2022. 226: pp. 103958–103958.
24. Ma, J., et al., A global review of progress in remote sensing and monitoring of marine pollution. *Water*, 2023. 15(19): p. 3491.
25. Soni, A., et al., Hyperspectral imaging and machine learning in food microbiology: developments and challenges in detection of bacterial, fungal, and viral contaminants. *Comprehensive Reviews in Food Science and Food Safety*, 2022. 21(4): pp. 3717–3745.
26. Huang, H., et al., Hyperspectral imaging as a potential online detection method of microplastics. *Bulletin of Environmental Contamination and Toxicology*, 2020. 107(4): pp. 745–763.
27. Welz, B. and Sperling, M., *Atomic absorption spectrometry*. 2008, Weinheim: WILEY-VCH.
28. Butcher, D.J., Recent advances in graphite furnace atomic absorption spectrometry: a review of fundamentals and applications. *Applied Spectroscopy Reviews*, 2023: 59(2): pp. 247–275.
29. Toya, Y., et al., Correlation between the gas temperature and the atomization behavior of analyte elements in flame atomic absorption spectrometry estimated with a continuum-light-source spectrometer system. *Spectrochimica Acta Part B: Atomic Spectroscopy*, 2016. 125: pp. 146–151.
30. Sisay, B., et al., Analysis of cadmium and lead using atomic absorption spectrophotometer in roadside soils of Jimma town. *Journal of Analytical & Pharmaceutical Research*, 2019. 8(4): pp. 144–147.
31. Adedeji, O.H., et al., Assessing spatial distribution, potential ecological and human health risks of soil heavy metals contamination around a Trailer Park in Nigeria. *Scientific African*, 2020. 10: p. e00650.
32. Ferreira, S.L.C., et al., Atomic absorption spectrometry: a multi element technique. *TrAC Trends in Analytical Chemistry*, 2018. 100: pp. 1–6.
33. Machado, R.C., et al., Solid sampling: advantages and challenges for chemical element determination–a critical review. *Journal of Analytical Atomic Spectrometry*, 2020. 35(1): pp. 5–77.
34. Andruch, V., et al., Application of deep eutectic solvents in atomic absorption spectrometry. *Trends in Analytical Chemistry*, 2022. 147: pp. 116510–116510.
35. Abdrakhimov, V.Z. and Abdrakhimova, E.S., Study of the distribution of iron oxides in intershale clay and oil sludge porous filler with mossbauer spectroscopy. *Theoretical Foundations of Chemical Engineering*, 2019. 53(4): pp. 703–707.
36. Wu, Y., et al., Zero-valent iron-based technologies for removal of heavy metal(loid) s and organic pollutants from the aquatic environment: recent advances and perspectives. *Journal of Cleaner Production*, 2020. 277: p. 123478.
37. Górka-Kostrubiec, B., Świetlik, R., Szumiata, T., Dytłow, S. and Trojanowska, M., Integration of chemical fractionation, Mössbauer spectrometry, and magnetic

methods for identification of Fe phases bonding heavy metals in street dust. *Journal of Environmental Sciences*, 2022. 124: pp. 875–891.
38. Wang, L., et al., The application of rapid handheld FTIR petroleum hydrocarbon-contaminant measurement with transport models for site assessment: a case study. *Geoderma*, 2020. 361: p. 114017.
39. Bell, J., et al., Non-invasive identification of polymers in cultural heritage collections: evaluation, optimisation and application of portable FTIR (ATR and external reflectance) spectroscopy to three-dimensional polymer-based objects. *Heritage Science*, 2019. 7(1): p. 95.
40. Veerasingam, S., et al., Contributions of Fourier transform infrared spectroscopy in microplastic pollution research: a review. *Critical Reviews in Environmental Science and Technology*, 2020. 51(22): pp. 1–63.
41. Chen, Y., et al., Identification and quantification of microplastics using Fourier transform infrared spectroscopy: current status and future prospects. *Current Opinion in Environmental Science & Health*, 2020. 18: pp. 14–19.
42. Smith, E. and Dent, G., *Modern Raman spectroscopy: a practical approach*. 2019, Hoboken, NJ: John Wiley & Sons Ltd.
43. Sahin, F., et al., Machine learning-assisted pesticide detection on a flexible surface-enhanced raman scattering substrate prepared by silver nanoparticles. *ACS Applied Nano Materials*, 2022. 5(9): pp. 13112–13122.
44. Xu, G., et al., Examples in the detection of heavy metal ions based on surface-enhanced Raman scattering spectroscopy. *Nanophotonics*, 2021. 10(18): pp. 4419–4445.
45. Xu, G., et al., Surface-enhanced raman spectroscopy facilitates the detection of microplastics <1 μm in the environment. *Environmental Science & Technology*, 2020. 54(24): pp. 15594–15603.
46. Terry, L.R., et al., Applications of surface-enhanced Raman spectroscopy in environmental detection. *Analytical Science Advances*, 2022. 3(3–4): pp. 113–145.
47. Ong, T.T.X., et al., Surface enhanced Raman spectroscopy in environmental analysis, monitoring and assessment. *Science of The Total Environment*, 2020. 720: p. 137601.
48. Weng, Z.H., et al., Probing the nature of soil organic matter. *Critical Reviews in Environmental Science and Technology*, 2021. 52(22): pp. 4072–4093.
49. Novotny, E.H., et al., The quality of soil organic matter, accessed by 13C solid state nuclear magnetic resonance, is just as important as its content concerning pesticide sorption. *Environmental Pollution*, 2020. 266: pp. 115298–115298.
50. Song, C., et al., The comparison of dissolved organic matter in hydrochars and biochars from pig manure. *Science of The Total Environment*, 2020. 756: p. 144140.
51. Wang, L., et al., Rapid on-site detection of underground petroleum pipeline leaks and risk assessment using portable gas chromatography-mass spectrometry and solid phase microextraction. *Journal of Chromatography A*, 2023. 1696: p. 463980.
52. Majd, M. and Nojavan, S., Determination of polycyclic aromatic hydrocarbons in soil, tree leaves, and water samples by magnetic dispersive solid-phase extraction based on β-cyclodextrin functionalized graphene oxide followed by GC-FID. *Microchemical Journal*, 2021. 171: p. 106852.
53. Chu, H., et al., Developing an improved strategy for the analysis of Polychlorinated Dibenzo-p-Dioxins/Furans and Dioxin-like Polychlorinated Biphenyls in contaminated soils using a combination of a one-step cleanup method and gas

chromatography with triple quadrupole mass spectrometry. *Toxics*, 2023. 11(9): p. 738.
54. Wang, L., et al., Application of portable gas chromatography–mass spectrometer for rapid field based determination of TCE in soil vapour and groundwater. *Environmental Technology and Innovation*, 2021. 21: p. 101274.
55. Zhang, Y., et al., Effect of geographical location and soil fertility on main phenolic compounds and fatty acids compositions of virgin olive oil from Leccino cultivar in China. *Food Research International*, 2022. 157: p. 111207.
56. Ashesh, A., et al., Organochlorine pesticides in multi-environmental matrices of India: a comprehensive review on characteristics, occurrence, and analytical methods. *Microchemical Journal*, 2022. 177: p. 107306.
57. Dalvand, K. and Ghiasvand, A., Simultaneous analysis of PAHs and BTEX in soil by a needle trap device coupled with GC-FID and using response surface methodology involving Box-Behnken design. *Analytica Chimica Acta*, 2019. 1083: pp. 119–129.
58. Viggi, C.C., et al., Enhancing the anaerobic biodegradation of petroleum hydrocarbons in soils with electrically conductive materials. *Bioengineering*, 2023. 10(4): p. 441.
59. Jumepaeng, T. and Pongpun, N., Development of sample preparation method for organochlorine pesticides analysis in soil samples. *AIMS Environmental Science*, 2020. 7(5): pp. 380–386.
60. Šimić, I., et al., An optimized sample preparation and analysis method for the determination of polycyclic aromatic hydrocarbons and polychlorinated biphenyls in the atmospheric bulk deposition samples. *Journal of Chromatography A*, 2020. 1633: p. 461599.
61. Frausto-Vicencio, I., et al., Characterizing the performance of a compact BTEX GC-PID for near-real time analysis and field deployment. *Sensors*, 2021. 21(6): p. 2095.
62. Yuan, L., et al., Dissipation, residue, dietary, and ecological risk assessment of atrazine in apples, grapes, tea, and their soil. *Environmental Science and Pollution Research*, 2021. 28(26): pp. 35064–35072.
63. Ogumi, Z., et al., Pyrolysis/gas chromatography/mass spectroscopy analysis of the surface film formed on graphite negative electrode. *Journal of Power Sources*, 2001. 97–98: pp. 156–158.
64. Picó, Y. and Barceló, D., Pyrolysis gas chromatography-mass spectrometry in environmental analysis: focus on organic matter and microplastics. *TrAC Trends in Analytical Chemistry*, 2020. 130: pp. 115964.
65. Seeley, M.E. and Lynch, J.M., Previous successes and untapped potential of pyrolysis–GC/MS for the analysis of plastic pollution. *Analytical and Bioanalytical Chemistry*, 2023. 415: pp. 2873–2890.
66. Akoueson, F., et al., Identification and quantification of plastic additives using pyrolysis-GC/MS: a review. *Science of The Total Environment*, 2021. 773: p. 145073.
67. Kovacs, E.D., et al., Determination of the uptake of Ibuprofen, Ketoprofen, and Diclofenac by tomatoes, radishes, and lettuce by gas chromatography–mass spectrometry (GC–MS). *Analytical Letters*, 2020. 54(1–2): pp. 314–330.
68. Chen, H., et al., Characteristics of soil organic matter 14 years after a wildfire: a pyrolysis-gas-chromatography mass spectrometry (Py-GC-MS) study. *Journal of Analytical and Applied Pyrolysis*, 2020. 152: pp. 104922.

69. Zhang, X., et al., Application of Fourier transform ion cyclotron resonance mass spectrometry in deciphering molecular composition of soil organic matter: a review. *Science of the Total Environment*, 2021. 756: p. 144140.
70. Geng, W., et al., Unveiling molecular transformations of soil organic matter after remediation by chemical oxidation based on ESI-FT-ICR-MS analysis. *ACS ES&T Engineering*, 2023. 3(6): pp. 831–840.
71. Wang, H., et al., Molecular transformation of crude oil contaminated soil after bioelectrochemical degradation revealed by FT-ICR mass spectrometry. *Environmental Science & Technology*, 2020. 54(4): pp. 2500–2509.
72. Goranov, A.I., et al., Comparison of sample preparation techniques for the (−)ESI-FT-ICR-MS analysis of humic and fulvic acids. *Environmental Science & Technology*, 2022. 56(17): pp. 12688–12701.
73. Roth, H., et al., Enhanced speciation of pyrogenic organic matter from wildfires enabled by 21 T FT-ICR mass spectrometry. *Analytical Chemistry*, 2022. 94(6): pp. 2973–2980.
74. Kailani, M.H., et al., Monitoring of pesticides residues in soil samples from the southern districts of Jordan in 2016/2017. *Toxin Reviews*, 2019. 40(2): pp. 198–214.
75. Nassazzi, W., et al., A novel method for extraction, clean-up and analysis of per- and polyfluoroalkyl substances (PFAS) in different plant matrices using LC-MS/MS. *Journal of Chromatography B*, 2022. 1212: p. 123514.
76. Shan, L. and Jones, B.R., Nano-LC: an updated review. *Biomedical Chromatography*, 2022. 36(5): p. e5317.
77. Poole, C.F. and Atapattu, S.N., Determination of physicochemical properties of small molecules by reversed-phase liquid chromatography. *Journal of Chromatography A*, 2020. 1626: p. 461427.
78. Salas, E., et al., A rapid and sensitive assay to quantify amino sugars, neutral sugars and uronic acid necromass biomarkers using pre-column derivatization, ultra-high-performance liquid chromatography and high-resolution mass spectrometry. *Soil Biology & Biochemistry*, 2023. 177: p. 108927.
79. Wang, T., et al., Established methods and comparison of 10 organic acids based on reversed phase chromatography and hydrophilic interaction chromatography. *Cyta-Journal of Food*, 2022. 20(1): pp. 206–217.
80. Castellaneta, A., et al., HILIC-ESI-MS analysis of phosphatidic acid methyl esters artificially generated during lipid extraction from microgreen crops. *Journal of Mass Spectrometry*, 2021. 56(10): p. e4784.
81. Pahalagedara, A.S.N.W., et al., Antimicrobial activity of soil clostridium enriched conditioned media against bacillus mycoides, bacillus cereus, and pseudomonas aeruginosa. *Frontiers in Microbiology*, 2020. 11. https://doi.org/10.3389/fmicb.2020.608998
82. Michalski, R., Applications of ion chromatography for the determination of inorganic cations. *Critical Reviews in Analytical Chemistry*, 2009. 39(4): pp. 230–250.
83. Muhammad, N., et al., Ion chromatography coupled with fluorescence/UV detector: a comprehensive review of its applications in pesticides and pharmaceutical drug analysis. *Arabian Journal of Chemistry*, 2021. 14(3): p. 102972.
84. Michalski, R., et al., Ion chromatography and related techniques in carboxylic acids analysis. *Critical Reviews in Analytical Chemistry*, 2021. 51(6): pp. 549–564.

85. Huang, Z., et al., A micro sample pretreatment technique combined with ion chromatography and its application in the determination of polyvinyl chloride. *Journal of Chromatography A*, 2023. 1690: p. 463778.
86. Bugsel, B. and Zwiener, C., LC-MS screening of poly- and perfluoroalkyl substances in contaminated soil by Kendrick mass analysis. *Analytical and Bioanalytical Chemistry*, 2020. 412(20): pp. 4797–4805.
87. Zhang, P., et al., Direct determination of glyphosate and its metabolite AMPA in soil using mixed-mode solid-phase purification and LC-MS/MS determination on a hypercarb column. *Journal of AOAC International*, 2019. 102(3): pp. 952–965.
88. Han, L., et al., Direct non-target analysis of dissolved organic matter and disinfection by-products in drinking water with nano-LC-FT-ICR-MS. *Environmental Science*, 2023. 9(6): pp. 1729–1737.
89. Hemida, M., et al., Small-footprint, field-deployable LC/MS system for on-site analysis of per- and polyfluoroalkyl substances in soil. *Analytical Chemistry*, 2021. 93(35): pp. 12032–12040.
90. Vargas Medina, D.A., et al., Miniaturization of liquid chromatography coupled to mass spectrometry. 3. Achievements on chip-based LC–MS devices. *TrAC Trends in Analytical Chemistry*, 2020. 131: p. 116003.
91. Sangwan, P., et al., Soil as a tool of revelation in forensic science: a review. *Analytical Methods*, 2020. 12(43): pp. 5150–5159.
92. Moor, C., et al., Determination of heavy metals in soils, sediments and geological materials by ICP-AES and ICP-MS. *Microchimica Acta*, 2001. 136(3–4): pp. 123–128.
93. Thomas, R., *Practical guide to ICP-MS, a tutorial for beginners*. 2003, Boca Raton: CRC Press.
94. Lindahl, P., et al., Total dissolution of environmental samples for the determination of uranium and thorium by ICP-MS and alpha-particle spectrometry. *Applied Radiation and Isotopes*, 2022. 181: p. 110103.
95. Li, X., et al., Analytical methodologies for agrometallomics: a critical review. *Journal of Agricultural and Food Chemistry*, 2021. 69(22): pp. 6100–6118.
96. Günther, D. and Hattendorf, B., Solid sample analysis using laser ablation inductively coupled plasma mass spectrometry. *TrAC Trends in Analytical Chemistry*, 2005. 24(3): pp. 255–265.
97. Chen, F., et al., LA-ICP-MS trace element analysis of magnetite and pyrite from the Hetaoping Fe-Zn-Pb skarn deposit in Baoshan block, SW China: implications for ore-forming processes. *Ore Geology Reviews*, 2020. 117: p. 103309.
98. Liu, Y., et al., Trace element evolution of magnetite in iron oxide-apatite deposits: case study of Daling deposit, Eastern China. *Ore Geology Reviews*, 2022. 144: p. 104842.
99. Chew, D., et al., LA-ICP-MS imaging in the geosciences and its applications to geochronology. *Chemical Geology*, 2021. 559: p. 119917.
100. Bussan, D.D., et al., Mercury methylation potentials in sediments of an ancient cypress wetland using species-specific isotope dilution GC-ICP-MS. *Molecules*, 2022. 27(15): p. 4911.
101. Clases, D., et al., Quantitative speciation of volatile sulphur compounds from human cadavers by GC-ICP-MS. *Talanta*, 2021. 221: p. 121424.
102. García-Bellido, J., et al., Recent advances in GC-ICP-MS: focus on the current and future impact of MS/MS technology. *TrAC Trends in Analytical Chemistry*, 2020. 130: p. 115963.

103. Hamilton, E.M., et al., Online microdialysis-high-performance liquid chromatography-inductively coupled plasma mass spectrometry (MD-HPLC-ICP-MS) as a novel tool for sampling hexavalent chromium in soil solution. *Environmental Science & Technology*, 2021. 55(4): pp. 2422–2429.
104. Leśniewska, B. and Godlewska-Żyłkiewicz, B., Speciation of chromium in alkaline soil extracts by an ion-pair reversed phase HPLC-ICP MS method. *Molecules*, 2019. 24(6): p. 1172.
105. Favilli, L., et al., Strategies for mercury speciation with single and multi-element approaches by HPLC-ICP-MS. *Frontiers in Chemistry*, 2022. 10: p. 1082956.
106. Moirana, R.L., et al., Trends towards effective analysis of fluorinated compounds using inductively coupled plasma mass spectrometry (ICP-MS). *Journal of Analytical Methods in Chemistry*, 2021. 2021: 1–11.
107. Leyden, E., et al., A simple and rapid ICP-MS/MS determination of sulfur isotope ratios (34S/32S) in complex natural waters: a new tool for tracing seawater intrusion in coastal systems. *Talanta*, 2021. 235: p. 122708.
108. Hou, X., et al., Determination of femtogram-level plutonium isotopes in environmental and forensic samples with high-level uranium using chemical separation and ICP-MS/MS measurement. *Analytical Chemistry*, 2019. 91(18): pp. 11553–11561.
109. Wang, Y., et al., Determination of ultra-low 236U in environment samples using ICP-MS/MS measurement and chemical separation. *Talanta*, 2021. 224: p. 121882.
110. Huang, Z., et al., Simultaneous determination of ultra-trace level 237Np and Pu isotopes in soil and sediment samples by SF-ICP-MS with a single column chromatographic separation. *Microchemical Journal*, 2019. 148: pp. 597–604.
111. Zhang, W., et al., Determination of ultra-low level 241Am in soil and sediment using chemical separation and triple quadrupole inductively coupled plasma mass spectrometry measurement with He-NH3 as collision-reaction gas. *Spectrochimica Acta Part B: Atomic Spectroscopy*, 2021. 178: p. 106113.
112. Desem, C.U., et al., The utility of rapid throughput single-collector sector-field ICP-MS for soil Pb isotope studies. *Applied Geochemistry*, 2022. 143: p. 105361.
113. Sullivan, K., et al., Emerging applications of high-precision Cu isotopic analysis by MC-ICP-MS. *Science of The Total Environment*, 2022. 838: p. 156084.
114. Liu, M., et al., High-precision Cd Isotope measurements of soil and rock reference materials by MC-ICP-MS with double spike correction. *Geostandards and Geoanalytical Research*, 2019. 44(1): pp. 169–182.
115. Li, X., et al., An optimal separation method for high-precision K isotope analysis by using MC-ICP-MS with a dummy bucket. *Journal of Analytical Atomic Spectrometry*, 2020. 35(7): pp. 1330–1339.
116. Zeng, Z.-G., et al., Silicon isotope compositions of soil and sediment reference materials determined by MC-ICP-MS. *Geostandards and Geoanalytical Research*, 2021. 46(1): pp. 117–127.
117. Gu, J., et al., Isotopic fractionation of Zn during soil sorption as influenced by soil properties and the Zn–O bond distance: insights from EXAFS and MC-ICP-MS. *Applied Geochemistry*, 2022. 145: p. 105427.
118. Tian, X., et al., Simultaneous multi-element and multi-isotope detection in single-particle ICP-MS analysis: principles and applications. *TrAC Trends in Analytical Chemistry*, 2022. 157: p. 116746.

119. Meng, Z., et al., Single particle inductively coupled plasma time-of-flight mass spectrometry–a powerful tool for the analysis of nanoparticles in the environment. *Processess*, 2023. 11(4): p. 1237.
120. Balaram, V., Rare earth elements: a review of applications, occurrence, exploration, analysis, recycling, and environmental impact. *Geoscience Frontiers*, 2019. 10(4): pp. 1285–1303.

Chapter 8

Advances in risk assessment of contaminated agricultural soils

M.A. Ayanka Wijayawardena and Ravi Naidu, University of Newcastle, Australia and CRC for Contamination Assessment and Remediation of the Environment (crcCARE), Australia

1 Introduction

Advances in risk assessment of contaminated agricultural soils represent an important area of study in environmental science, particularly as it relates to food security and public health (Hu et al. 2019; Oliver and Gregory 2015). Agricultural soils, the foundation of global food production, are increasingly subjected to various forms of contamination, raising serious concerns about the safety and quality of food crops. The sources of these contaminants are manifold, ranging from industrial activities, mining operations, and urban runoff to agricultural practices themselves (Tóth et al. 2016). For example, the excessive use of fertilisers and pesticides causes many adverse effects on ecological and

http://dx.doi.org/10.19103/AS.2025.0151.12

human health (Kulathunga et al. 2019). Globally, it is estimated that millions of hectares of agricultural land are at risk of contamination, impacting food production and safety. The scale of this challenge highlights the importance of advanced risk assessment techniques, which are crucial for identifying contaminated sites, assessing the severity of contamination, and guiding remediation efforts. The United Nations Food and Agriculture Organization (FAO) reports that approximately 33% of the world's soils are moderately to highly degraded due to erosion, nutrient depletion, acidification, salinization, compaction, and chemical pollution (FAO 2021). The escalating crisis of soil pollution in the agricultural sector poses an unprecedented challenge to global food security and the sustainability of ecosystems. As the world's population is projected to reach nearly 10 billion by 2050, the demand for food, fiber, livestock feed, and biofuel is expected to increase dramatically, necessitating an almost 50% boost in agricultural production from 2012 levels. This surge in demand is particularly acute in regions such as sub-Saharan Africa and South Asia, where agricultural outputs must more than double to satisfy the estimated caloric needs. Meanwhile, the rest of the world is tasked with increasing its production by at least 30% (FAO 2021).

The infiltration of contaminants into the food chain through polluted soils, and some even leaching to groundwater environments that more than one third of the world's population relies on for their water needs, presents a grave risk to human health, agricultural productivity, and economic stability. In China, for instance, soil contamination is responsible for the annual loss of 10 million tonnes of crops, due to decreased yields or the inability to sell contaminated produce (Fu 2012; Wang et al. 2012; Zeng et al. 2007). This situation is not isolated. Across the globe, polluted soils threaten the livelihoods of millions who depend on agriculture for their survival. Furthermore, soil pollution has far-reaching effects on aquatic ecosystems. Contaminants leached into groundwater and surface waters disrupt aquatic life, leading to eutrophication, reduced water quality, and the loss of biodiversity. The transportation of pollutants through runoff and flooding extends the impact of soil pollution to coastal zones and oceans, exacerbating global environmental crises. Among these pollutants, heavy metals like cadmium (Cd) have gained significant attention due to their persistent and bioaccumulative nature, posing a substantial risk to human health. Cd, often categorised as a diffuse pollutant, can enter the soil through atmospheric deposition, sewage sludge, and phosphate fertilisers (Naidu et al. 1997). Once in the soil, Cd can be readily taken up by plants, accumulating in edible parts and thus entering the food chain. This accumulation raises concerns not just for crop quality but also for the long-term health of consumers, given Cd's potential to cause kidney damage, osteoporosis, and other serious health issues (Naidu et al. 1994; Naidu and Harter 1998; Siddique et al. 2021; Wijayawardena et al. 2016).

Diffuse pollution is often associated with broadacre farming (Lockie 1997). The nature of contaminants in agricultural soils can vary greatly, including organic compounds like pesticides and polycyclic aromatic hydrocarbons (PAHs), as well as inorganic elements like Cd, arsenic, lead (Pb), and mercury (Kulathunga et al. 2022b; Pelfrêne et al. 2011; Wijayawardena et al. 2020a; Zhao et al. 2007). Each of these contaminants behaves differently in the soil environment, influenced by factors such as soil pH, cation exchange capacity (CEC), organic matter content, and climatic conditions, which in turn affect their bioavailability and toxicity to plants and humans (Naidu et al. 2008; Vig et al. 2003). This complexity underscores the need for advanced risk assessment methodologies that can accurately predict and quantify the potential impacts of these contaminants on crop quality and human health (Klimkowicz-Pawlas and Debaene 2020; Taghavi et al. 2023).

Agricultural soil is a foundational element of global food production and ecosystem health. It serves several critical functions such as nutrient provision, water management, carbon sequestration, and supporting biodiversity. Soil is a reservoir of nutrients essential for plant growth. It supplies nitrogen, phosphorus, potassium, and other minerals to crops, making it vital for food security (Basak et al. 2021; Naidu and Rengasamy 1993). Proper nutrient management in agriculture ensures efficient use of resources and reduces the risk of nutrient runoff into rivers and oceans, which can cause harmful algal blooms and dead zones (Wurtsbaugh et al. 2019). Soil plays a key role in water absorption and retention, affecting both crop hydration and the availability of freshwater resources. It acts as a natural filter, improving water quality.

The pursuit of higher agricultural yields, however, is severely compromised by the intensifying degradation of land and water resources. Over recent years, the pressures from unsustainable land-use changes and water management practices have led to a decrease in the ecological functions and productive capacity of many agricultural systems (DeFries et al. 2004). Visible issues like erosion and deforestation are just the tip of the iceberg; below the surface, soil and water pollution are leading to serious environmental issues that need urgent attention fromregulatory bodies around the globe. This pollution results in salinisation, eutrophication, and a decline in soil health, crucial for biodiversity, carbon sequestration, and water filtration. Soil health, in particular, is vital for carbon capture and storage (Paz et al. 2020). Soil aids in carbon sequestration as it is a significant carbon sink, helping to mitigate climate change by storing carbon dioxide from the atmosphere (Rastogi et al. 2002). Soil erosion can lead to loss of fertile topsoil, reducing agricultural productivity and causing sedimentation in water bodies. Healthy soils with well-established vegetation act as natural erosion control mechanisms. The potential annual global carbon sequestration in soils could significantly mitigate climate change by offsetting a substantial portion of global fossil fuel emissions. This highlights the critical role of sustainable soil management practices in addressing both food security and climate change. Unsustainable agricultural practices not only disturb soil structure and deplete soil organic matter (SOM) but also contribute to the loss of soil biodiversity, further exacerbating the challenges of soil degradation.

Healthy soil supports a diverse range of organisms, from microorganisms to larger fauna, which are essential for maintaining ecological balance and enhancing soil fertility (Thomsen et al. 2012). Fertile soil is crucial for sustainable agriculture, enabling farmers to produce food efficiently and sustainably (Fig. 1). It helps in building resilience against climate extremes like droughts and floods.

Preserving and enhancing the quality of agricultural soil is therefore essential not only for current food production but also for maintaining ecological balance and ensuring long-term environmental sustainability. Addressing the issue of contaminated agricultural soil requires comprehensive strategies that include preventing contamination at the source, remediating polluted soils, and implementing sustainable agricultural practices to maintain soil health and

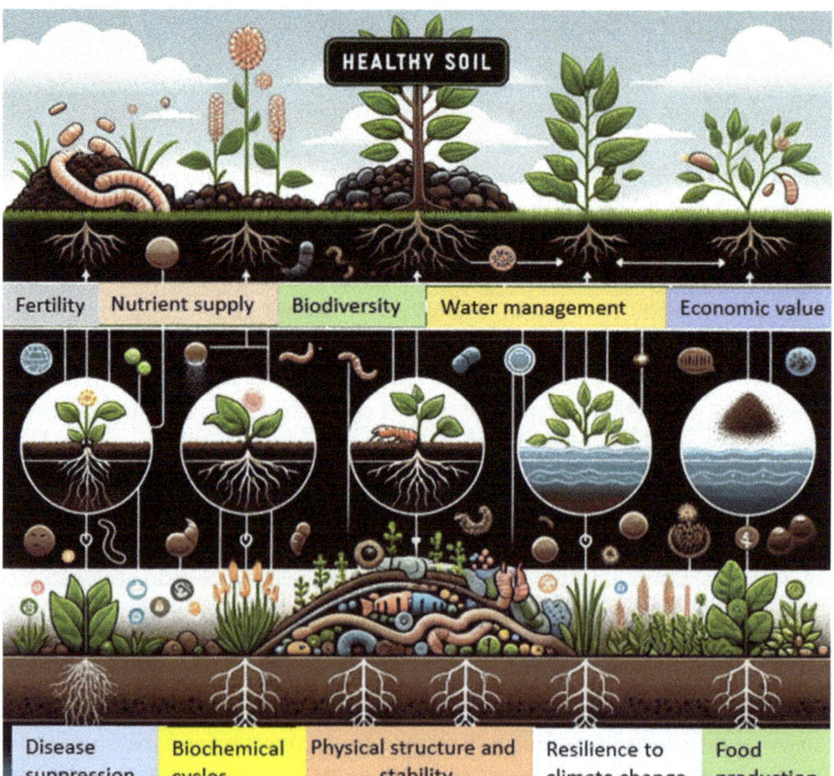

Figure 1 Importance of healthy agricultural soil.

prevent future contamination (Naidu et al. 2021). Agriculture is a significant contributor to the global economy. Productive soil supports livelihoods for millions of farmers and workers worldwide, contributing to economic stability and food security. Furthermore, many cultures and indigenous communities have deep connections to the land and its soil. Agricultural soil plays a central role in cultural practices, traditions, and spiritual beliefs.

Recent advancements in risk assessment techniques have focused on improving our understanding of the behaviour of contaminants in soil, their uptake by crops, and their subsequent effects on human health (Kulathunga et al. 2021b, 2022a; Liu et al. 2022; Naidu et al. 2003). These include sophisticated analytical methods for detecting and quantifying contaminants at trace levels, as well as the development of robust models that can predict contaminant transfer from soil to crops under various environmental conditions. Additionally, there is a growing emphasis on the holistic assessment of risks, taking into account not just the chemical properties of contaminants but also the socioeconomic aspects of agricultural practices and food consumption patterns (Arquette 2004; Dong et al. 2015; USEPA 2012).

Furthermore, the integration of Geographic Information Systems (GISs) and remote sensing technologies has revolutionised the way we assess and manage the risks associated with soil contamination (Kulathunga et al. 2022b; Poggio et al. 2009). These tools enable the mapping and monitoring of contaminated sites, providing valuable data for risk assessment and aiding in the formulation of targeted remediation strategies (Abowaly et al. 2021). Another promising area is the use of phytoremediation, which employs plants to extract, stabilise, or degrade contaminants, thus offering a sustainable approach to mitigating soil contamination (Cho et al. 2009; Esquenazi et al. 2009).

As previously mentioned, contaminated agricultural lands can lead to several detrimental impacts such as adverse human health effects, reduced soil fertility, environmental damage, and economic consequences. This dire situation highlights the urgent need for a paradigm shift towards more sustainable and regenerative agricultural practices. Integrated soil fertility management, conservation agriculture, and organic farming are examples of approaches that can restore soil health, enhance biodiversity, and improve water quality while increasing agricultural productivity (FAO 2021). Additionally, policies and investments that promote sustainable land use and water management, reduce chemical inputs, and support farmers in adopting eco-friendly practices are crucial for mitigating the impacts of soil pollution. The magnitude of the challenge posed by soil pollution requires a coordinated global response, incorporating scientific research, policy innovation, and community engagement to safeguard our soil resources. By prioritising soil health and sustainable agricultural practices, we can secure a healthier planet for future generations (FAO 2021). With growing concerns over soil pollution and its impacts on food safety and environmental health, innovative approaches are emerging to better understand and manage these risks. The advanced risk assessment of contaminated agricultural soils is a dynamic field that not only addresses the immediate concerns of food safety and crop quality but also contributes to the broader objective of sustainable agriculture and environmental protection. By understanding the sources, nature, and impacts of soil contaminants, and by developing effective strategies for risk management, we can ensure the health and well-being of current and future generations. In this chapter, we are focusing on recent advancements in the field of risk assessment for contaminated agricultural land.

The chapter highlights the following key points:

- Evolving Techniques: Advances in risk assessment have led to the development of sophisticated techniques, including high-resolution spatial analysis, remote sensing, and machine learning, which enhance our ability to identify, quantify, and mitigate contamination risks in agricultural settings.
- Multi-Contaminant Assessment: Modern risk assessment now includes a broader spectrum of contaminants, including pesticides, heavy metals, pathogens, and emerging pollutants, recognising the complexity of agricultural pollution.
- Holistic Perspective: Recent research emphasises a holistic approach, integrating ecological, human health, and socioeconomic factors to provide a comprehensive understanding of contamination risks and their potential consequences.
- Sustainable Solutions: The evolving field aims to not only assess risks but also inform sustainable land management practices, ensuring safe food production, protecting ecosystems, and safeguarding human well-being.

2 Hazard identification in contaminated agricultural soils

Hazard identification in agricultural soils is a critical environmental and public health issue that involves the detection and assessment of potential threats posed by soil contamination. Agricultural soils can become contaminated by a variety of pollutants, including heavy metals like lead, Cd, and mercury, pesticides, industrial chemicals, and even biohazards like bacteria and viruses (Kulathunga et al. 2022a; Wijayawardenaet al. 2017). These contaminants can originate from numerous sources such as industrial effluents, improper disposal of waste, excessive use of agricultural chemicals, and atmospheric deposition (Naidu et al. 2003).

The identification of these hazards is crucial as they can have profound impacts on crop quality and yield, food safety, human health, and the broader ecosystem. For instance, heavy metals can accumulate in crops, entering the food chain and posing health risks such as neurological damage and cancer (Wijayawardena et al. 2016). Similarly, pesticide residues in soils can lead to bioaccumulation in wildlife, disrupting ecosystems and biodiversity. The process of hazard identification in agricultural soils begins with soil sampling and analysis, which are essential for determining the presence and concentration of contaminants (Huq et al. 2003; Wijayawardena et al. 2020b). Advanced analytical techniques like gas chromatography, mass spectrometry, and atomic absorption spectroscopy are commonly employed for this purpose. However, the challenge lies not only in detecting contaminants but also in understanding their bioavailability, mobility, and potential to cause harm (Duan et al. 2014; Ng et al. 2013). This is where risk assessment models come into play, offering a means to evaluate the potential exposure of humans and wildlife to these contaminants and the likelihood of adverse health effects. These models consider various factors, including the type of contaminants, their concentration levels, exposure pathways, and the vulnerability of the exposed populations (Dong et al. 2016).

Beyond identification, managing the risks associated with soil contamination is essential. This involves implementing remediation strategies like phytoremediation, where plants are used to absorb contaminants, or soil washing, which physically removes pollutants (Naidu 2013; Wei et al. 2008). Equally important are preventive measures to avoid future contamination, such as promoting sustainable farming practices, regulating the use of agricultural chemicals, and enforcing strict industrial waste management protocols.

Moreover, policies and legislation play a critical role in managing soil contamination risks. Governments and international bodies need to establish standards and guidelines for permissible levels of soil contaminants, monitor compliance, and invest in research and development to innovate more effective methods for hazard identification and soil remediation (Naidu et al. 2021). Public awareness and education are also essential in this regard, as they can lead to more informed choices and practices among farmers and consumers.

As agriculture remains a backbone of many economies and a key factor in global food security, addressing soil contamination hazards is not just an environmental concern but a necessity for sustainable development and public health protection. Through collaborative efforts among scientists, policymakers, farmers, and the general public, effective management of soil contamination can be achieved, ensuring safe and productive agricultural practices for current and future generations (Biswas et al. 2018).

2.1 Identification of hazards to ecological receptors

The identification of hazards to ecological receptors, a crucial aspect of environmental risk assessment, involves understanding and predicting how pollutants or changes in the environment affect various organisms and ecosystems (Liu et al. 2022). Ecological receptors, which include a wide range of species from plants and insects to birds, mammals, and aquatic organisms, play diverse roles in their ecosystems and can be vulnerable to different types of hazards. These hazards may originate from chemical pollutants, such as pesticides, heavy metals, and industrial chemicals, or from physical changes like habitat destruction, climate change, and introduction of invasive species (Naidu et al. 2021; Wijayawardena 2014).

The process begins with identifying potential sources of hazards and understanding their nature and characteristics. Chemical hazards, for instance, require an understanding of the toxicity, persistence, and bioaccumulation potential of the contaminants. Meanwhile, physical and biological hazards might involve assessing changes in land use, climate variables, or the introduction of non-native species that can disrupt existing ecosystems. Once potential hazards are identified, the next step involves exposure assessment, which determines how ecological receptors are exposed to these hazards. This includes evaluating the pathways through which chemicals or other hazards reach the organisms, the duration and frequency of exposure, and the concentration or intensity of the hazard. For instance, in aquatic systems, fish and other organisms might be exposed to waterborne pollutants, whereas terrestrial animals might be exposed through soil, air, or food chain interactions (Fakhradini et al. 2021; Li et al. 2022). The sensitivity of different ecological receptors to various hazards is a key factor in hazard identification. Different species have varying tolerances to pollutants and environmental changes, influenced by factors like genetics, physiology, life history traits, and ecological roles. For instance, Hose (2005) found that there are significant differences in the sensitivities of surface and groundwater taxa to the pesticides Atrazine and Chlorpyrifos, which were detected when they used the species sensitivity distribution approach (SSD), indicating surface water quality guidelines are not always suitable for protecting or best managing the groundwater ecosystems.

2.2 Identification of hazards to humans – key studies and critical effect, weight of evidence

The identification of hazards to humans aims to recognise and evaluate risks associated with exposure to various hazards. This comprehensive process is crucial for developing strategies to protect human health and is grounded in a weight-of-evidence approach that involves the analysis of data from various sources including epidemiological studies, animal research, and mechanistic studies. A notable example is the identification of asbestos as a hazardous material, historically used in construction and manufacturing until key studies in the mid-twentieth century linked it to serious health effects like asbestosis, lung cancer, and mesothelioma (Borow et al. 1973; Selikoff et al. 1965). These findings, substantiated by a weight of evidence from numerous epidemiological studies and occupational health reports, led to its regulation and phase-out of use in many countries.

Another significant area of hazard identification concerns exposure to air pollutants, such as particulate matter (PM), nitrogen dioxide (NO_2), and sulphur dioxide (SO_2). Seminal studies like the Harvard Six Cities Study and subsequent research have consistently shown the critical effects of these pollutants on respiratory and cardiovascular health (Laden et al. 2006; Lepeule et al. 2012). This body of evidence, supported by findings from both long-term and short-term exposure studies, has been instrumental in establishing air quality standards worldwide.

Chemical hazards also represent a significant concern for human health. For instance, the identification of lead as a neurotoxin, particularly harmful to children's brain development, is a key example. Studies showing the effects of low-level lead exposure on cognitive development and behavioural problems in children have led to global efforts to reduce lead exposure, such as removing lead from gasoline and paints (Wijayawardena et al. 2017; Wijayawardena 2014). Similarly, research on the endocrine-disrupting effects of bisphenol A (BPA), used in plastics, has prompted regulatory actions and industry shifts towards BPA-free products (Escrivá et al. 2021; Kassotis et al. 2020; Kawa et al. 2021). In the pharmaceutical industry, thalidomide serves as a historical example of hazard identification. Initially marketed as a safe sedative and antiemetic, thalidomide was later found to cause severe birth defects (Stephens et al. 2000). The thalidomide tragedy led to the development of more stringent drug testing and approval processes, emphasising the importance of comprehensive hazard identification in drug safety (Melchert and List 2007).

The weight of evidence approach is critical in these identifications. It involves not just considering the results of individual studies but also evaluating the consistency of findings across different types of studies, the quality of the research methods used, and the plausibility of the biological mechanisms proposed (Biswas et al. 2018; Naidu et al. 2021). This approach ensures a more robust and reliable assessment of hazards, accounting for the complexities and uncertainties inherent in health risk assessments (Dong et al. 2015). Hazard identification also involves recognising the impact of lifestyle factors and social determinants of health (Kulathunga et al. 2019b, 2022a). For instance, studies on the health effects of smoking have not only focused on direct impacts like lung cancer and heart disease but also on how factors like advertising, socioeconomic status, and public policy influence smoking behaviour (Hiscock et al. 2012).

3 Conceptual site model

A conceptual site model (CSM) is an essential tool in risk assessment, particularly for contaminated agricultural soils. It serves as a visual and descriptive representation of potential environmental and human health risks resulting from contamination (Ludwig et al. 2008). The model helps in understanding the complexities of the site, including the sources of contamination, pathways through which contaminants are transported, and the receptors that might be affected (Hess et al. 2000).

3.1 Conceptual site model for ecological risk assessment for contaminated agricultural soils

At the core of a CSM for contaminated agricultural soils is the identification of specific contaminants of concern. These contaminants often originate from industrial activities, improper disposal of wastes, or agricultural practices such as the use of pesticides, fertilisers, use of biosolids, and secondary treated water (Kulathunga et al. 2022a). Common contaminants include heavy metals like Pb and Cd, organic chemicals like polychlorinated biphenyls (PCBs) and dioxins, and nitrates and phosphates from fertilisers (Huq and Naidu 2004). Understanding the nature and concentration of these contaminants is crucial for assessing the potential ecological risks.

The next step involves determining the pathways through which these contaminants are transported. In agricultural soils, major pathways include leaching into groundwater, surface runoff into nearby water bodies, and absorption by plants. Each of these pathways presents unique risks to different ecological receptors (Liu et al. 2022). For example, contaminants leaching into groundwater can affect groundwater organisms and other aquatic ecosystems linked to the groundwaters and potentially enter human drinking water sources, while those absorbed by plants can enter the food chain, affecting both wildlife and humans (Huq et al. 2006; Kulathunga et al. 2019a).

Receptor identification is another critical aspect of the CSM. Ecological receptors in agricultural areas typically include soil organisms, plants, wildlife. Each receptor has a different level of sensitivity and exposure to soil contaminants. For instance, earthworms living in the soil may be directly exposed to contaminants, while birds and mammals may be exposed indirectly through the consumption of contaminated plants or prey (Wijayawardena et al. 2017; Wijayawardena 2014).

Risk characterization in a CSM involves integrating information about the contaminants, pathways, and receptors to evaluate the potential adverse effects. This evaluation often requires the use of toxicity data, exposure models, and an understanding of ecological relationships in the affected area (Zhuang et al. 2009). The outcome is a risk estimate that helps in making informed decisions about the need for and type of remediation actions (Haijuan et al. 2008; Liu et al. 2022). Management and remediation strategies form the final component of a CSM. These strategies are tailored based on the risk assessment outcomes and may include soil remediation techniques such as phytoremediation, bioremediation, or soil excavation and disposal (Göhre and Paszkowski 2006; Shi et al. 2009). Additionally, management strategies may involve the use of buffer zones to prevent the spread of contaminants, changes in agricultural practices to reduce future contamination, and monitoring programmes to track the effectiveness of remediation efforts (Kidd et al. 2015; Wagena and Easton 2018).

An illustration of a CSM is shown in Fig. 2. It provides a structured approach to understanding the complex interactions between contaminants, environmental pathways, and ecological receptors.

3.2 Conceptual site model for human health risk assessment for contaminated agricultural soils

Using a CSM for human health risk assessment involves a systematic evaluation of potential health risks to individuals exposed to contaminated environments. The CSM provides a comprehensive and visual representation of the contamination scenario, integrating various data and information sources to portray the environmental conditions accurately. It is a dynamic tool, often updated as new information becomes available, ensuring an ongoing, accurate assessment of the risks.

The primary purpose of a CSM in human health risk assessment is to identify and understand the pathways through which humans can be exposed to contaminants. It considers various exposure pathways, including dermal contact, inhalation of dust or vapours, and ingestion of contaminated soil or crops. By visualising these pathways, the CSM helps in identifying the most significant routes of exposure, thereby focusing the risk assessment process (Butt et al. 2020). This targeted approach not only makes the assessment more efficient but also more relevant to the specific conditions of the contaminated site. In agricultural settings, soil contamination can arise from various sources, such as the use of pesticides, discharge of industrial wastes, or deposition of airborne contaminants. A CSM in this context would depict the distribution and concentration of contaminants in the soil, the potential for contaminants to migrate to other media (like groundwater or surface water), and the interaction

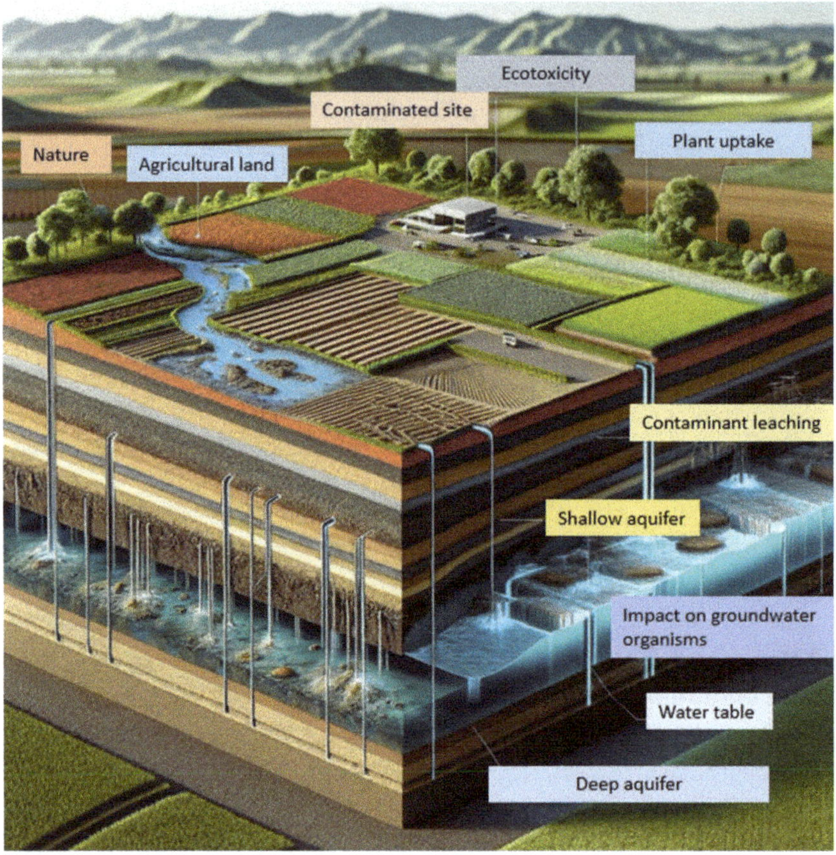

Figure 2 Illustration of a conceptual diagram depicting various elements of a contaminated agricultural site, such as polluted water bodies, contaminated soil areas, groundwater flow, a nearby river, and affected crops.

between these media and agricultural products (Fig. 2). This model is crucial in assessing the risk to human health, especially for individuals who directly interact with these environments, such as farmers, and those who consume the agricultural products.

The creation of a CSM involves an iterative process, starting with a preliminary model based on initial site information and evolving with additional data gathered from field investigations and laboratory analyses. This process ensures that the model accurately reflects the site's conditions and the contaminants' behaviour in different environmental media. The CSM acts as a communication tool, bringing together stakeholders such as environmental scientists, public health officials, and community members, to ensure a comprehensive understanding of the site and its potential risks

(Butt et al. 2016). One of the significant advantages of using a CSM in human health risk assessment is its ability to support decision-making. By providing a clear and visual representation of the contamination and its potential impacts, the CSM helps in identifying areas where risk reduction measures are most needed. This can guide remediation efforts, focusing resources on the most critical areas to reduce exposure and, consequently, health risks. For instance, if the CSM indicates a high risk of exposure through ingestion of contaminated crops, efforts can be directed towards soil remediation or changing agricultural practices (Robert et al. 2017).

Moreover, the CSM aids in public communication and engagement. By presenting complex environmental data in a more understandable format, it helps in explaining the potential health risks to non-specialist audiences, including local communities and policymakers. This transparency is crucial for building trust and ensuring community support for any necessary remediation or risk management actions (Bayliss et al. 2003).

4 Ecological exposure analysis in agricultural soils

Ecological exposure analysis in agricultural soils focuses on assessing the extent to which various environmental contaminants impact soil health and ecosystems (Hose 2005; Kulathunga et al. 2019b). It is a multifaceted field that combines elements of environmental science, chemistry, ecology, and risk assessment (Hope 2001).

Central to this discipline is the understanding that agricultural soils are not isolated entities; rather, they are integral components of larger ecological systems (Huq and Naidu 2004). This analysis typically involves evaluating the presence and concentration of various pollutants, such as pesticides, heavy metals, and organic contaminants, which can enter the soil through various means including agricultural practices, industrial activities, and natural processes (Huq et al. 2003; Kulathunga et al. 2022a).

The process starts with sampling and testing of soil, where scientists collect soil samples from different locations and depths to analyse the types and concentrations of contaminants present. This data is crucial for understanding the spatial distribution of pollutants and identifying potential hotspots of contamination (Cao et al. 2009; Qian et al. 1996). Advanced analytical techniques, like gas chromatography and mass spectrometry, are employed to detect and quantify the presence of these contaminants with high precision and accuracy (Rahman et al. 2014). One of the key challenges in ecological exposure analysis is understanding the behaviour of contaminants in the soil matrix. Factors such as soil composition, pH, temperature, and moisture content can significantly influence the mobility and bioavailability of pollutants (Naidu et al. 2003; Wijayawardena et al. 2015a). For example, heavy metals may bind tightly to clay particles or organic matter, reducing their mobility but potentially causing long-term contamination issues (Wijayawardena et al. 2015a, 2015b). Similarly, the degradation of organic contaminants like pesticides can be influenced by microbial activity in the soil, which is in turn affected by environmental conditions (Jacobsen and Hjelmsø 2014).

Another crucial aspect is assessing the impact of these contaminants on soil health and ecosystem functions. Soil microorganisms play a vital role in nutrient cycling, organic matter decomposition, and maintaining soil structure (Wang et al. 2007). Contaminants can disrupt these microbial communities, leading to reduced soil fertility and altered ecosystem services. Moreover, the uptake of pollutants by plants can lead to contaminated food crops, also impacting wildlife through the food chain (Megharaj and Naidu 2017; Naidu et al. 2021). Risk assessment models are often used in ecological exposure analysis to estimate the potential adverse effects on the environment (Dong et al. 2016). These models consider various exposure pathways, such as direct ingestion of soil, consumption of contaminated crops, and secondary exposure through the food chain (Kulathunga et al. 2022a). The risk assessment helps in formulating soil management and remediation strategies, guiding policymakers and stakeholders in making informed decisions to protect soil health and ensure sustainable agricultural practices (Naidu 2013).

Furthermore, ecological exposure analysis in agricultural soils is not static; it requires continuous monitoring and adaptation to changing environmental conditions and agricultural practices. Climate change, for instance, can alter patterns of precipitation and temperature, affecting the fate and transport of soil contaminants (Biswas et al. 2018; Huq et al. 2003). The integration of GISs and remote sensing technologies offers advanced tools for monitoring and mapping soil contamination over time and space, enabling a more dynamic and responsive approach to soil management (Fang et al. 2018).

4.1 Exposure assessment

Exposure assessment plays a critical role in understanding and managing risks associated with exposure to hazardous substances. It involves the identification, quantification, and evaluation of the intensity, frequency, and duration of human exposure to an agent, whether chemical, biological, or physical, in a specific environment. This process is important in risk assessment, informing regulatory decisions, public health policies, and the implementation of control strategies to protect human health.

The primary goal of exposure assessment is to determine the potential or actual exposure of individuals or populations to hazardous substances. This is achieved through a combination of data collection, modelling, and analysis (Dong et al. 2015, 2016). Data collection may involve measuring the concentration of contaminants in various media such as air, water, soil, and food. Direct measurement of exposure, such as personal air sampling, biomonitoring, or the use of wearable sensors, provides valuable data on the actual exposure levels experienced by individuals. Indirect methods, such as environmental sampling combined with data on human behaviour and activity patterns, are also commonly used to estimate exposure (Kulathunga et al. 2022a; Yan et al. 2018). Modelling plays a significant role in exposure assessment, especially when direct measurements are not feasible or when predictions of future exposure are needed. Models can range from simple calculations based on assumptions about exposure scenarios to complex computational simulations that take into account various factors such as the source of contamination, environmental transport and transformation processes, and human behaviour. These models help in understanding how contaminants move through the environment and how they come into contact with humans (Umeh et al. 2021).

One of the key challenges in exposure assessment is the variability and uncertainty inherent in measuring or predicting exposure levels (Dong et al. 2015). Variability arises from differences in individual behaviour, environmental conditions, and the properties of the contaminants themselves (Yan et al. 2018). Uncertainty can stem from limitations in data, measurement techniques, and model assumptions. Addressing these challenges requires a robust methodological approach and often involves the use of probabilistic methods to quantify uncertainty and variability (Dong et al. 2015). Exposure assessment is not only about measuring or estimating the levels of exposure but also involves interpreting these levels in the context of potential health effects. This requires an understanding of toxicology and dose-response relationships, which help in determining the levels of exposure that may lead to adverse health outcomes (Kulathunga et al. 2019b, 2022a; Naidu et al. 2003).

Moreover, exposure assessment is increasingly important in the context of environmental justice and public health. Certain populations or communities may be disproportionately exposed to environmental hazards due to factors such as socioeconomic status, location, and occupation. Understanding and addressing these disparities is essential for protecting vulnerable populations and ensuring equitable health outcomes. As a whole, exposure assessment is a multidisciplinary field that lies at the intersection of environmental science, public health, toxicology, and risk management (Megharaj and Naidu 2011; Naidu et al. 1997).

4.2 Toxicological mechanisms of ecological risk assessment of contaminated agricultural soils

Having an up to date knowledge on toxicological mechanisms is crucial for understanding and mitigating the potential adverse effects of soil pollutants on ecosystems and human health (Escher and Hermens 2002). This comprehensive assessment involves evaluating the likelihood of adverse ecological effects resulting from exposure to one or more environmental stressors, particularly contaminants present in agricultural soils (Liu et al. 2022). The core of this evaluation lies in understanding the toxicological mechanisms, which explain how these contaminants interact at the molecular, cellular, and organism levels, leading to potential harm (Scholz et al. 2008). Contaminated agricultural soils often contain a range of pollutants, including heavy metals, pesticides, and organic chemicals, each with distinct toxicological properties (Kulathunga et al. 2022a). The mechanisms by which these contaminants cause harm are diverse and complex, often involving interactions with biological molecules, alterations in cellular processes, and disruptions of physiological functions. For instance, heavy metals like Cd and Pb can bind to proteins and enzymes, disrupting their normal function and leading to cellular damage. Pesticides, on the other hand, may interfere with nerve function in insects and other wildlife, and persistent organic pollutants can accumulate in organisms, causing long-term toxic effects (Keifer and Firestone 2007).

One of the key aspects of understanding these mechanisms is the concept of bioavailability, which refers to the fraction of a contaminant that is available for uptake by organisms (Naidu et al. 2003; Ng et al. 2013). Bioavailability is influenced by various soil properties, such as pH, organic matter content, and moisture levels (Wijayawardena et al. 2015a; Yan et al. 2018). It determines the extent of exposure of soil-dwelling organisms to contaminants and, consequently, the potential for toxic effects. For example, certain soil conditions can increase the mobility of heavy metals, making them more readily available for uptake by plants, soil dwelling invertebrates and soil microorganisms, thereby entering the food chain and posing risks to higher organisms, including humans (Button et al. 2009; Wijayawardena et al. 2012). Toxicological mechanisms also include the understanding of dose-response relationships, which describe how the adverse effects of a contaminant change with the level of exposure (Duan et al. 2014; Liu et al. 2019). This is essential for setting regulatory standards and guidelines for soil contamination. Ecotoxicological studies, which involve testing the effects of soil contaminants on representative organisms, provide crucial data for these assessments (Liu et al. 2022; Wijayawardena et al. 2012). These studies help in identifying the concentrations of contaminants that cause harm to these organisms, information that is then extrapolated to assess risks to entire ecosystems.

Additionally, the synergistic and antagonistic interactions between different contaminants are significant in the ecological risk assessment of agricultural soils (Wijayawardena et al. 2017). Often, soils are contaminated with multiple pollutants, which may interact in ways that enhance or mitigate their individual toxic effects (Wijayawardena et al. 2016; Wijayawardena 2014).

Understanding these interactions is critical for accurate risk assessment, as the combined effects of multiple contaminants can be different from the sum of their individual effects. Moreover, ecological risk assessment must consider the potential for bioaccumulation and biomagnification of contaminants in food webs (Megharaj and Naidu 2011; Naidu et al. 2003). Certain contaminants, particularly those that are persistent and fat-soluble, can accumulate in organisms and become more concentrated as they move up the food chain (Liu et al. 2019; Umeh et al. 2019). This can lead to high levels of exposure and toxicity in top predators, including humans.

Understanding these toxicological mechanisms is crucial for assessing and managing the risks posed by soil contaminants to ecosystems and human health.

4.2.1 Plant toxicology and mechanisms

An up to date knowledge on plant toxicological mechanisms helps in understanding how various pollutants impact plant health and, consequently, entire ecosystems (Liu et al. 2019; Naidu et al. 2003). This comprehensive approach involves examining the interaction between plants and environmental stressors, primarily focusing on the physiological, biochemical, and molecular responses of plants to toxic substances. Understanding these mechanisms is essential for assessing the potential ecological risks posed by environmental contaminants, which include a wide range of substances such as heavy metals, pesticides, and organic pollutants (Kim et al. 2010).

At the core of plant toxicological mechanisms is the way in which plants respond to and manage the uptake and accumulation of toxic substances. Plants have evolved a variety of mechanisms to cope with environmental stressors, including the detoxification and sequestration of harmful compounds. For instance, when exposed to heavy metals like Cd or Pb, plants may activate specific transporters that compartmentalise these metals into vacuoles, reducing their toxicity (Song et al. 2003). Such transgenic plants may be useful for phytoremediation of Pb and Cd. Similarly, the production of chelating molecules such as phytochelatins that can bind to heavy metal(loid)s, can thereby decrease their availability and reactivity within the plant cells (Cobbett 2000). Another key aspect of plant toxicology is the generation of oxidative stress in response to contaminant exposure. Many pollutants can induce the production of reactive oxygen species (ROS) in plants, leading to oxidative damage to proteins, lipids, and nucleic acids (Czarnocka and Karpiński 2018). Plants, in turn, have developed antioxidant defence systems, including enzymes like superoxide dismutase, catalase, and various peroxidases, along with non-enzymatic antioxidants like ascorbate and glutathione, to mitigate this oxidative stress. The balance between the production of ROS and the efficiency of the antioxidant defence system is crucial in determining the plant's tolerance to contaminants (Demidchik 2017).

The impact of contaminants on plant growth and development is another important consideration in ecological risk assessment (Naidu et al. 2003). Toxic substances can affect various physiological processes in plants, including photosynthesis, nutrient uptake, and water balance. For example, certain pesticides may interfere with photosynthetic pathways, leading to reduced plant growth and productivity. These effects not only impact the plants directly exposed to contaminants but can also have cascading effects on the entire ecosystem, particularly if the affected plants play a key role in the ecosystem structure or function.

Furthermore, the interactions between plants and soil microorganisms are an integral part of plant toxicological mechanisms. Soil microorganisms, including bacteria and fungi, can influence the availability and toxicity of contaminants to plants. For instance, certain soil microbes can transform or degrade pollutants, thereby reducing their bioavailability to plants (Gadd 2010). In turn, the presence of contaminants can alter the microbial community composition and function, which can have further implications for plant health and soil ecology.

4.2.2 Invertebrate toxicology and mechanisms

Invertebrate toxicological mechanisms are important in ecological risk assessment, particularly in evaluating the impact of environmental contaminants on these diverse and ecologically significant organisms. Invertebrates, including insects, crustaceans, molluscs, and others, are often at the forefront of exposure to pollutants due to their varied habitats and roles in ecosystems (Liu et al. 2022; Wijayawardena 2014). Understanding the toxicological responses of these organisms to contaminants is crucial for assessing environmental risks and implementing effective conservation strategies. Toxicological mechanisms in invertebrates include a range of biochemical and physiological responses to environmental stressors, particularly contaminants like pesticides, heavy metal(loid)s, and industrial chemicals. These mechanisms are diverse, reflecting the vast diversity of invertebrate biology. One common response is the activation of detoxification pathways. Many invertebrates possess enzyme systems, such as cytochrome P450s, which metabolise toxic compounds, rendering them less harmful. This is particularly well-studied in insects, where resistance to pesticides often involves enhanced metabolic detoxification (Snyder 2000).

Another crucial aspect is the impact of contaminants on the nervous system of invertebrates. Many pollutants, especially neurotoxic pesticides, target the nervous system, disrupting normal neurotransmission. For example, organophosphates and carbamates inhibit acetylcholinesterase, an enzyme crucial for nerve function, leading to the accumulation of neurotransmitters and consequent nerve overstimulation (Fulton and Key 2001). Organophosphorus insecticides (OPs) produce toxicity by inhibiting the cholinesterase enzymes in the nervous system. Monitoring of acetylcholinesterase (AChE) inhibition has been widely used in terrestrial and freshwater aquatic systems as an indicator of OP exposure and effects. Similarly, pyrethroids and neonicotinoids interfere with ion channels in nerve cell membranes, disrupting neural signals. Due to the same reason, ion channels are molecular targets of neuroactive insecticides (Raymond-Delpech et al. 2005). Invertebrates also exhibit physical and behavioural adaptations to mitigate the impact of toxic substances. Some species can avoid contaminated areas or reduce feeding in polluted environments, minimising exposure (Wijayawardena et al. 2017). Morphological changes like thicker cuticles can also reduce the penetration of toxins. Additionally, some aquatic invertebrates can sequester or excrete heavy metals, reducing internal concentrations of these harmful substances (Rainbow 2018).

The reproductive and developmental processes in invertebrates are also affected by toxic exposure. Contaminants can cause direct toxicity to eggs and larvae, reduce fertility, or cause developmental abnormalities. These effects can have significant implications for population dynamics and ecosystem function, as invertebrates often form the basis of food webs and contribute to critical processes like pollination and decomposition (Wijayawardena et al. 2017).

Assessing the impact of contaminants on the immune function of invertebrates is another key component of ecological risk assessment. Many pollutants can suppress the immune system, making invertebrates more susceptible to diseases (Burgos-Aceves et al. 2021). This not only affects individual organisms but can also have cascading effects on the broader ecosystem, particularly if key species are impacted (Liu et al. 2022). Moreover, the ecological risk assessment must consider the potential for bioaccumulation and biomagnification of contaminants within invertebrate communities. Some pollutants, particularly those that are persistent and lipophilic, can accumulate in invertebrate tissues and become more concentrated as they move through food chains. This poses risks not only to the invertebrates themselves but also to higher organisms that feed on them, including birds, fish, and mammals (Liu et al. 2022).

5 Toxicological mechanisms of human risk assessment of contaminated agricultural soils

The toxicological mechanisms underpinning human risk assessment are critical for understanding how exposure to various soil contaminants can lead to both cancer and non-cancer risks (Dong et al. 2015; Yan et al. 2018). This comprehensive assessment involves examining the complex interactions between humans and a range of pollutants typically found in agricultural settings, such as pesticides, heavy metal(loid)s, and organic chemicals (Kulathunga et al. 2022a; Wijayawardena 2014). Understanding these mechanisms is vital for assessing potential health risks and implementing safety measures to protect both agricultural workers and the general population.

Evaluation of cancer risks associated with long-term exposure to specific contaminants is a very important part of this assessment. Many substances found in contaminated agricultural lands, such as certain pesticides and industrial chemicals, are known carcinogens (Punia et al. 2023). The toxicological mechanisms through which these carcinogens operate can vary. Some may directly damage DNA, leading to mutations that trigger oncogenesis (Kim et al. 2015). Others may act indirectly, for instance, by promoting chronic inflammation or hormonal imbalances, which in turn can lead to cancer. The assessment of cancer risk typically involves understanding these mechanisms at the molecular level, estimating exposure concentrations, and applying this knowledge to epidemiological models that predict cancer risk over different exposure periods.

In addition to cancer risks, a comprehensive risk assessment must also consider a range of non-cancer risks (Dong et al. 2015). These risks include a variety of adverse health effects that contaminants might cause aside from cancer. For example, exposure to heavy metal(loid)s like lead or arsenic in agricultural soils can lead to neurological damage, reproductive issues, or kidney damage (Huq et al. 2006; Jayasumana et al. 2015; Naidu et al. 2006). Pesticides, depending on their class and mode of action, can cause acute effects such as respiratory distress, skin irritation, and gastrointestinal problems, as well as chronic effects including endocrine disruption and impacts on cognitive development. Bioavailability plays a significant role in both cancer and non-cancer risk assessments (Naidu et al. 2003; Wijayawardena et al. 2016, 2020a). Not all contaminants in soil are readily absorbed by the human body; the degree to which they are available for absorption depends on factors like the chemical form of the contaminant, soil characteristics, and human behaviour (e.g. hand-to-mouth activities in children) (Naidu 2013; Wijayawardena et al. 2015b). This assessment helps in determining the actual dose of a contaminant that the body assimilates, which is crucial for accurate risk estimation.

The concept of threshold levels is also vital in non-cancer risk assessments. Unlike cancer risks, where any level of a carcinogen might potentially increase cancer risk, many non-cancer risks have threshold levels below which no adverse health effect is expected. This concept is crucial in setting regulatory standards and guidelines for soil contaminants. Furthermore, human risk assessment of contaminated agricultural lands often involves the consideration of vulnerable populations, such as children, pregnant women, and the elderly, who may be more susceptible to the effects of certain contaminants. This necessitates a more stringent evaluation of risks and often calls for lower acceptable exposure levels to protect these sensitive groups. For example, over the years, the Centres for Disease Control and Prevention (CDC) introduced a blood lead "reference value" to identify children with higher levels of lead in their blood compared to most children. In 1973 this blood Pb reference value was 40 micrograms per decilitre (µg/dL). With more evidence on toxicity and adverse effects of Pb on humans coming to light, in 1975 this value was reduced to 30 µg/dL. Then with the wealth of studies and epidemiological data on the adverse effects of Pb at different blood Pb concentrations in humans over the years, this reference value for Pb was reduced further along the time in 1980s, and was even further reduced down to 10 µg/dL in 1991. In 2012, CDC replaced the term "level of concern" with reference value based on 97.5th percentile of blood Pb levels in children. This resulted in new threshold of 5 µg/dL. CDC is now recommending that if a person has a blood lead level higher than 3.5 µg/dL, the source of exposure should be investigated further and reduced, particularly if the person is a child or a pregnant mother (Allwood et al. 2022). Figure 3 shows the CDC blood Pb reference values since 1970s. This timeline illustrates a significant decrease in the reference values over the decades, highlighting the evolving understanding and increased caution regarding Pb exposure.

The toxicological mechanisms involved in human risk assessment of contaminated agricultural lands are complex and multifaceted, requiring a deep understanding of both cancer and non-cancer risks associated with various soil contaminants (Dong et al. 2015). This assessment involves evaluating the molecular mechanisms of carcinogenesis, understanding the bioavailability

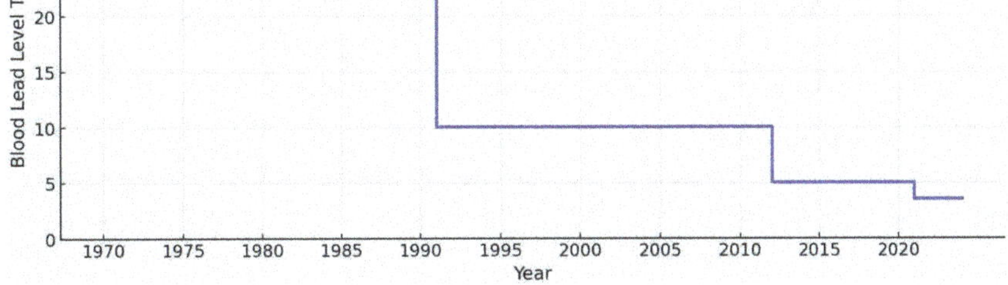

Figure 3 Centres for Disease Control and Prevention blood Pb level thresholds: historical perspective.

and absorption of contaminants, considering the threshold levels for non-cancer risks, and factoring in the susceptibility of vulnerable populations. Accurate risk assessment is essential for guiding regulatory actions, informing public health strategies, and ensuring the safe use of agricultural lands, thereby protecting the health of agricultural workers and the wider community (Li and Naidu 2005).

6 Mixed contaminant ecological effects on contaminated agricultural soils

The ecological effects of mixed contaminants in contaminated agricultural soils represent a significant environmental concern, characterised by the intricate and often unpredictable interactions among various pollutants (Kulathunga et al. 2021a; Wijayawardena 2014). In agricultural settings, soils frequently accumulate a combination of contaminants like pesticides, heavy metal(loid)s, and organic pollutants, each with distinct properties and toxicities. The presence of mixed contaminants can lead to synergistic, antagonistic, or additive effects, making the ecological consequences difficult to predict and manage (Naidu et al. 2003; Wijayawardena 2014).

Synergistic effects occur when the combined impact of multiple contaminants is greater than the sum of their individual effects (Wijayawardena et al. 2018). For example, certain heavy metal(loid)s might enhance the toxicity of organic pollutants by affecting their breakdown or bioavailability in the soil (Olaniran et al. 2013). Metal(loid)s can be distributed in the soil matrix by associating with the soil components in many ways including adsorption, precipitation, complexation, ion exchange or simply by being present in the structure of minerals. Irha et al. (2003) assessed the effect of heavy metal(loid)s (Cr, Pb, Cu, Cd) and PAH (Pyrene) on soil, and their results showed that the sensitivity of soil microbiology to toxic effects of contaminants correlated negatively with organic matter. Therefore, in such instances where organic matter is at reduced levels in the soil matrix, the heightened toxicity can severely disrupt soil microbial communities, which are essential for nutrient cycling, organic matter decomposition, and overall soil health. A decline in microbial activity can lead to diminished soil fertility and alter the ecological balance of the soil habitat (Ma et al. 2022; Megharaj and Naidu 2017).

Plants growing in soils with mixed contaminants may exhibit stunted growth, reduced yield, and impaired physiological functions, such as impaired photosynthesis and nutrient absorption (Siddique et al. 2021). These effects not only compromise plant health but also raise concerns about the bioaccumulation of contaminants in the food chain, posing risks to wildlife and human health. Furthermore, the combined toxicity of multiple pollutants can adversely affect soil invertebrates, such as earthworms and insects, which play vital roles in soil aeration, organic matter breakdown, and pest control (Wijayawardena et al. 2012, 2017). Negative impacts on these organisms can further disrupt ecological interactions and soil structure, leading to broader environmental consequences (Megharaj and Naidu 2011).

The ecological effects of mixed contaminants in agricultural soils are complex, with significant implications for soil health, plant growth, food safety, and overall ecosystem stability. Understanding and managing these combined effects is crucial for sustainable agricultural practices and environmental conservation, necessitating comprehensive research and targeted remediation strategies.

7 Mixed contaminant human health risk from contaminated agricultural soils

Worldwide, environmental legislations are often based on single contaminants, and therefore, when assessing risks associated with mixtures, it is likely that the ultimate decision relating to risk may not reflect the real risk. This necessitates the need for a paradigm shift in regulatory frameworks to incorporate the cumulative and interactive effects of contaminant mixtures. Robust, scientifically grounded methodologies must be developed to evaluate the synergistic and antagonistic impacts on ecosystems and human health, ensuring that legislation protects against not just individual chemical threats, but the complex reality of our environmental exposures (Wijayawardena et al. 2016).

The human health risks associated with mixed contaminant exposure from contaminated agricultural soils are a growing concern in environmental health. Agricultural lands, often subjected to a variety of chemicals including pesticides, fertilisers, and waste products, can accumulate a complex mixture of contaminants (Kulathunga et al. 2019b; Kuppusamy et al. 2016). These mixed contaminants pose a unique challenge for human health risk assessment due to their synergistic and cumulative effects (Wijayawardena et al. 2016). When different contaminants coexist in soil, their combined impact on human health can be significantly different from the effects of individual chemicals. For instance, heavy metal(loid)s like Pb and Cd can interact with organic pollutants such as pesticides, potentially exacerbating their toxicity. This synergism can lead to unforeseen health risks, complicating the assessment and management of contaminated sites. Moreover, the chronic exposure common in agricultural settings can lead to the accumulation of these substances in the human body, increasing the risk of long-term health effects (Kulathunga et al. 2022a; Wijayawardena et al. 2020a).

The health risks from mixed contaminants include a broad spectrum of potential effects. Acute exposures can lead to immediate symptoms like respiratory distress, skin irritation, or gastrointestinal issues. However, it is the chronic, low-level exposure that is particularly concerning. This can contribute to long-term health issues, including neurodevelopmental disorders in children, reproductive and endocrine system disruptions, and an increased risk of cancers (Dórea 2021). The diverse nature of these contaminants means they can affect multiple organ systems and biological processes. Furthermore, the risks are not evenly distributed across populations. Vulnerable groups such as children, pregnant women, the elderly, and agricultural workers may be at higher risk due to their increased exposure or sensitivity to these contaminants (Lowry 2020). Children, for instance, are more susceptible due to their developing bodies and behaviours like hand-to-mouth activity, which can increase soil ingestion (Faustman et al. 2000).

8 Advances in risk assessment techniques

Advances in risk assessment techniques have revolutionised the way we understand and manage risks in various fields. The evolution of these techniques over recent years has been driven by technological advancements, increased data availability, and a growing recognition of the complex nature of most risk scenarios. These developments have enabled more precise, comprehensive, and predictive risk assessments, enhancing our ability to make informed decisions and implement effective risk management strategies (Aven 2016; Goldstein 2005). One of the most significant advancements in risk assessment is the integration of big data analytics and machine learning (Li et al. 2022). The risk assessors now have access to vast amounts of information that can be analysed to identify patterns, trends, and potential risks. Machine learning algorithms, in particular, have become invaluable tools for processing and interpreting this data. They can learn from past events and outcomes to predict future risks, making risk assessments more proactive rather than reactive (Dong et al. 2015; Zhong et al. 2021).

Another notable advancement is the application of Geographic Information Technologies (GISs) in environmental and public health risk assessments (Kulathunga et al. 2022b). Geographic Information System technology allows for the spatial analysis of risk factors, such as the geographic distribution of pollutants, disease outbreaks, or natural hazards. This spatial perspective is crucial for understanding the environmental and societal impacts of these risks and for planning effective interventions (Shahid et al. 2018). The development of more sophisticated modelling techniques has also significantly enhanced risk assessment. Computational models can now simulate complex systems and scenarios, allowing for the analysis of multiple variables and their interdependencies (Wang et al. 2021). This is particularly important in fields like climate change and ecological risk assessment, where the interactions between various factors, such as emissions, weather patterns, and ecological responses, are complex and dynamic (Lal et al. 2023). These models enable a better understanding of potential future scenarios, aiding in the development of strategies to mitigate risks.

Advancements in sensor technology and the Internet of Things (IoT) have also transformed risk assessment (Wang et al. 2023). Real-time monitoring of environmental conditions, infrastructure, and other critical systems is now possible, providing immediate data that can be used to assess and manage risks. For instance, sensors set up on contaminated environments can provide real-time data on the level of contamination, and concentration fluctuations with time, at different locations, hence helping monitor potential contamination leaching or run-off (Fang et al. 2018; Siam et al. 2019).

The field of risk communication has also seen significant advancements. Communicating risk effectively to stakeholders, from policymakers to the general public, is crucial for risk management. Advances in digital communication, including social media and interactive web platforms, have enabled more engaging and accessible risk communication (Lundgren and McMakin 2018). This includes interactive risk maps and visualisations that help in translating complex risk data into understandable and actionable information. Finally, the integration of ethical considerations and stakeholder participation in risk assessment is an important advancement. There is a growing recognition that risk assessment is not just a technical process but also involves value judgements and ethical considerations. Engaging stakeholders, including affected communities, in the risk assessment process ensures that diverse perspectives and values are considered, leading to more equitable and acceptable risk management decisions.

9 Holistic risk assessment

Holistic environmental risk assessment is a comprehensive approach to evaluating the potential risks that various environmental factors pose to humans, wildlife, and ecosystems. Unlike traditional risk assessment methods that often focus on single factors or specific endpoints, holistic assessment considers the complex interactions between different environmental elements and their cumulative impact (Biswas et al. 2021; Dong et al. 2015). Therefore, this framework considers the effects of contaminants on the physical health of humans, and holistically takes into consideration the impacts on the natural world and on cultural, social, subsistence, economic, and spiritual practices as well (Arquette 2004). This approach is rooted in the understanding that environmental systems are interconnected and dynamic, and thus, a change in one element can have cascading effects on others.

This approach is in line with the World Health Organization's One Health initiative, which emphasises the interconnectedness of humans, animals, plants, and their environment (Queenan et al. 2017). In a holistic environmental risk assessment, multiple factors such as air and water quality, soil contamination, biodiversity, climate change, and human activities are evaluated in conjunction. This involves the integration of diverse data sources, including physical, chemical, biological, and socioeconomic information. The assessment process often employs advanced modelling techniques to predict how different factors interact over time and space, and how these interactions might affect ecological and human health (Arquette 2004). One of the key aspects of holistic assessment is its emphasis on the precautionary principle, which advocates for proactive measures to protect the environment in the face of scientific uncertainty. This approach acknowledges that our understanding of complex environmental systems is often incomplete, and therefore, decisions should err on the side of caution to prevent potential harm (Butt et al. 2016). Holistic environmental risk assessment also recognises the importance of stakeholder involvement and interdisciplinary collaboration. It involves experts from various fields such as ecology, toxicology, geology, and social sciences, ensuring a multidimensional analysis of risks. Furthermore, it encourages the participation of communities, policymakers, and other stakeholders in the decision-making process, promoting transparency and ensuring that diverse perspectives are considered. By adopting a holistic approach, environmental risk assessment can provide more accurate and comprehensive insights into the potential impacts of human activities and natural processes on the environment (Butt et al. 2016). This can lead to more effective and sustainable environmental management practices, ultimately contributing to the protection and preservation of our planet for future generations.

10 Conclusion

Agricultural soil stands at the crossroads of food security, biodiversity, climate regulation, water purity, and economic growth. Its preservation is critical to fostering a future that is both sustainable and resilient. The key to safeguarding the vitality of agricultural soil – and by extension, our collective future – lies in the strategic use of CSMs, better understanding of exposure risks, and better handling and effective remediation of the contaminants. The CSMs leverage site-specific data and scientific insights to navigate and mitigate the threats of soil contamination effectively. Through enhancing our understanding of contamination pathways and human exposure, CSMs play an important role in ensuring the well-being of ecosystems and communities, while optimising the economic returns from agricultural lands.

The process of identifying human health hazards and exposure risks employs a comprehensive weight-of-evidence approach, amalgamating data from various studies to pinpoint potential risks. The ongoing identification and analysis of hazards such as asbestos, air pollutants, and heavy metal(loid)s highlights the need for continuous adaptation in health risk assessment practices. This dynamism is fuelled by advancements in technology and new scientific discoveries, highlighting the critical need for persistent vigilance and research to protect health. As we navigate the complexities of emerging contaminants and technological breakthroughs, exposure assessment remains a important component in shaping public health policies, regulatory frameworks, and fostering an environment where safety and health prevail. Within this sphere, ecological exposure analysis is vital for the continuity of sustainable agricultural methods, ecosystem preservation, and safeguarding human health. This field's importance is ever-growing, demanding relentless pursuit of knowledge, innovation, and collaborative efforts across various sectors.

Moreover, plant toxicology plays a critical role in assessing and mitigating ecological risks, enhancing our comprehension of plant–environment interactions and ecosystem resilience to stressors. This area underpins the necessity for integrated approaches in ecological risk assessment, incorporating the study of food webs, contaminant dynamics, and ecological impacts. The evolution of risk assessment methodologies, buoyed by technological advancements such as machine learning, GIS, and IoT, offers unprecedented opportunities for addressing the various challenges posed by soil contamination. These tools are instrumental in refining decision-making processes and formulating strategies to counteract the complex and synergistic nature of risks in agriculture, emphasising the need for ongoing monitoring and adaptive management practices.

In summary, the continuous enhancement of risk assessment capabilities and a comprehensive view, and holistic risk assessment are reshaping our capacity to address the complex challenges of contaminated agricultural land and are revolutionising our approach to managing the intricate issues associated with contaminated agricultural lands.

11 Acknowledgements

The authors are grateful for the funding and facility support from the crc for Contamination Assessment and Remediation of the Environment (crcCARE) for the funding and the Global Centre for Environment Remediation, the University of Newcastle, Australia, for facilities which enabled us to successfully complete this chapter.

12 Where to look for further information

https://www.sciencedirect.com/science/article/abs/pii/S0065211316300402

https://link.springer.com/article/10.1007/s10653-019-00264-z

https://www.frontiersin.org/journals/water/articles/10.3389/frwa.2021.786487/full

13 References

Abowaly, M. E., Belal, A-A. A., Abd Elkhalek, E. E., Elsayed, S., Abou Samra, R. M., Alshammari, A. S., Moghanm, F. S., Shaltout, K. H., Alamri, S. A. and Eid, E. M. (2021) 'Assessment of soil pollution levels in North Nile Delta, by integrating contamination indices, GIS, and multivariate modeling', *Sustainability*, 13(14):8027.

Allwood, P. B., Falk, H. and Svendsen, E. R. (2022) 'A historical perspective on the CDC childhood lead poisoning prevention program', *American Journal of Public Health*, 112(S7):S635–S639.

Arquette, M. (2004) 'Holistic risk assessment: a new paradigm for environmental risk management', *Race, Poverty & the Environment*, 11(2):49–52.

Aven, T. (2016) 'Risk assessment and risk management: review of recent advances on their foundation', *European Journal of Operational Research*, 253(1):1–13.

Basak, B., Sarkar, B. and Naidu, R. (2021) 'Environmentally safe release of plant available potassium and micronutrients from organically amended rock mineral powder', *Environmental Geochemistry and Health*, 43:3273–3286.

Bayliss, P., Hogan, A., Walden, D., Boyden, J., Camilleri, C. and Begg, G. (2003) 'Uncertainty analysis of data, linking conceptual models to on-site management and communicating risk'. https://www.agriculture.gov.au/sites/default/files/documents/ir435.pdf

Biswas, B., Moore, F., Du, J., Wijayawardena, A. and Arabzadeh, M. (2021) 'Surface dispersal, emission source and human health risk assessment of heavy metal (loid)s in an active gas field, Southern Iran', *Research*, 1:2–46.

Biswas, B., Qi, F., Biswas, J. K., Wijayawardena, A., Khan, M. A. I. and Naidu, R. (2018) 'The fate of chemical pollutants with soil properties and processes in the climate change paradigm—a review', *Soil Systems*, 2(3):51.

Borow, M., Conston, A., Livornese, L. and Schalet, N. (1973) 'Mesothelioma following exposure to asbestos: A review of 72 cases', *Chest*, 64(5):641–646.

Burgos-Aceves, M. A., Abo-Al-Ela, H. G. and Faggio, C. (2021) 'Impact of phthalates and bisphenols plasticizers on haemocyte immune function of aquatic invertebrates: a review on physiological, biochemical, and genomic aspects', *Journal of Hazardous Materials*, 419:126426.

Butt, T. E., Akram, H., Mahammedi, C. and House, C. (2020) 'Conceptual site model: An intermediary between baseline study and risk assessment', *WIT Transactions on Engineering Sciences*, 129:101–108.

Butt, T. E., Javadi, A., Nunns, M. and Beal, C. (2016) 'Development of a conceptual framework of holistic risk assessment—Landfill as a particular type of contaminated land', *Science of the Total Environment*, 569:815–829.

Button, M., Jenkin, G. R., Harrington, C. F. and Watts, M. J. (2009) 'Arsenic biotransformation in earthworms from contaminated soils', *Journal of Environmental Monitoring*, 11(8):1484–1491.

Cao, H-c., Luan, Z-q., Wang, J-d. and Zhang, X-l. (2009) 'Potential ecological risk of cadmium, lead and arsenic in agricultural black soil in Jilin Province, China',

Stochastic Environmental Research and Risk Assessment, 23(1):57–64. https://doi.org/10.1007/s00477-007-0195-1.

Cho, Y., Bolick, J. A. and Butcher, D. J. (2009) 'Phytoremediation of lead with green onions (Allium fistulosum) and uptake of arsenic compounds by moonlight ferns (Pteris cretica cv Mayii)', *Microchemical Journal*, 91(1):6–8.

Cobbett, C. S. (2000) 'Phytochelatins and their roles in heavy metal detoxification', *Plant Physiology*, 123(3):825–832.

Czarnocka, W. and Karpiński, S. (2018) 'Friend or foe? Reactive oxygen species production, scavenging and signaling in plant response to environmental stresses', *Free Radical Biology and Medicine*, 122:4–20.

DeFries, R. S., Foley, J. A. and Asner, G. P. (2004) 'Land-use choices: balancing human needs and ecosystem function', *Frontiers in Ecology and the Environment*, 2(5):249–257.

Demidchik, V. (2017) 'Reactive oxygen species and their role in plant oxidative stress', in *Plant stress physiology*, edited by S. Shabala, 64–96, CABI, Wallingford, UK.

Dong, Z., Liu, Y., Duan, L., Bekele, D. and Naidu, R. (2015) 'Uncertainties in human health risk assessment of environmental contaminants: a review and perspective', *Environment International*, 85:120–132.

Dong, Z., Yan, K., Liu, Y., Naidu, R., Duan, L., Wijayawardena, A., Semple, K. T. and Rahman, M. M. (2016) 'A meta-analysis to correlate lead bioavailability and bioaccessibility and predict lead bioavailability', *Environment International*, 92:139–145.

Dórea, J. G. (2021) 'Exposure to environmental neurotoxic substances and neurodevelopment in children from Latin America and the Caribbean', *Environmental Research*, 192:110199.

Duan, L., Palanisami, T., Liu, Y., Dong, Z., Mallavarapu, M., Kuchel, T., Semple, K. T. and Naidu, R. (2014) 'Effects of ageing and soil properties on the oral bioavailability of benzo [a] pyrene using a swine model', *Environment International*, 70:192–202.

Escher, B. I. and Hermens, J. L. (2002) 'Modes of action in ecotoxicology: their role in body burdens, species sensitivity, QSARs, and mixture effects', *Environmental Science & Technology*, 36(20):4201–4217.

Escrivá, L., Zilliacus, J., Hessel, E. and Beronius, A. (2021) 'Assessment of the endocrine disrupting properties of bisphenol AF: a case study applying the European regulatory criteria and guidance', *Environmental Health*, 20(1):1–19.

Esquenazi, L., MontofréBacigalupo, Í., Cano, F., Mermut, A., Arocena, J. and Silla, O. (2009) *Phytoremediation of lead and zinc using Myoporum laetum and Brassica nigra in Antofagasta, Northern Chile*, Catena Verlag.

Fakhradini, S. S., Moore, F., Keshavarzi, B., Naidu, R., Wijayawardena, A., Soltani, N. and Rostami, S. (2021) 'Spatial distribution, partitioning, ecological risk and source apportionment of potential toxic elements in water and sediments of the Hoor Al-Azim wetland and their bioaccumulation in selected commercial fish species', *Marine Pollution Bulletin*, 172:112875.

Fang, C., Zhang, X., Dong, Z., Wang, L., Megharaj, M. and Naidu, R. (2018) 'Smartphone app-based/portable sensor for the detection of fluoro-surfactant PFOA', *Chemosphere*, 191:381–388.

FAO (The Food and Agriculture Organization of the United Nations). (2021) The state of the world's land and water resources for food and agriculture: systems at breaking point (SOLAW 2021). Accessed https://www.fao.org/documents/card/en/c/cb9910en.

Faustman, E .M., Silbernagel, S. M., Fenske, R. A., Burbacher, T. M. and Ponce, R. A. (2000) 'Mechanisms underlying Children's susceptibility to environmental toxicants', *Environmental Health Perspectives*, 108(S1):13–21.
Fu, G. (2012) 'Countermeasures for water and soil heavy metal pollution in China', *China Environmental Science*, 32(2):373–376.
Fulton, M. H. and Key, P. B. (2001) 'Acetylcholinesterase inhibition in estuarine fish and invertebrates as an indicator of organophosphorus insecticide exposure and effects', *Environmental Toxicology and Chemistry: An International Journal*, 20(1):37–45.
Gadd, G. M. (2010) 'Metals, minerals and microbes: geomicrobiology and bioremediation', *Microbiology*, 156(3):609.
Göhre, V. and Paszkowski, U. (2006) 'Contribution of the arbuscular mycorrhizal symbiosis to heavy metal phytoremediation', *Planta*, 223(6):1115–1122. https://doi.org/10.1007/s00425-006-0225-0.
Goldstein, B. D. (2005) 'Advances in risk assessment and communication', *Annual Review of Public Health*, 26:141–163.
Hess, G. R., Campbell, C. L., Fiscus, D. A., Hellkamp, A. S., McQuaid, B. F., Munster, M. J., Peck, S. L. and Shafer, S. R. (2000) 'A conceptual model and indicators for assessing the ecological condition of agricultural lands', *Journal of Environmental Quality*, 29(3):728–737.
Hiscock, R., Bauld, L., Amos, A., Fidler, J. A. and Munafò, M. (2012) 'Socioeconomic status and smoking: a review', *Annals of the New York Academy of Sciences*, 1248(1):107–123.
Hope, B. K. (2001) 'A case study comparing static and spatially explicit ecological exposure analysis methods', *Risk Analysis*, 21(6):1001–1010.
Hose, G. C. (2005) 'Assessing the need for groundwater quality guidelines for pesticides using the species sensitivity distribution approach', *Human and Ecological Risk Assessment*, 11(5):951–966.
Hu, Z., Li, J., Wang, H., Ye, Z., Wang, X., Li, Y., Liu, D. and Song, Z. (2019) 'Soil contamination with heavy metals and its impact on food security in China', *Journal of Geoscience and Environment Protection*, 7(5):168.
Huq, SI. and Naidu, R. (2004) 'Arsenic in ground water and contamination of the food chain: Bangladesh scenario', *Natural Arsenic in Ground Water: Occurrence, Remediation and Management* 95–101. https://doi.org/10.1201/9780203970829.sec2.
Huq, S. I., Rahman, A., Sultana, N. and Naidu, R. (2003) 'Extent and severity of arsenic contamination in soils of Bangladesh', *Fate of Arsenic in the Environment. Dhaka: Bangladesh University of Engineering and Technology* 69–84. https://scholar.google.com/scholar_lookup?title=Arsenic+contamination:+Bangladesh+perspective&author=M+Nuruzzaman&author=MF+Ahmed&publication_year=2003&.
Huq, S. M. I., Joardar, J., Parvin, S., Correll, R. and Naidu, R. (2006) 'Arsenic contamination in food-chain: Transfer of arsenic into food materials through groundwater irrigation', *Journal of Health, Population, and Nutrition*, 24(3):305.
Irha, N., Slet, J. and Petersell, V. (2003) 'Effect of heavy metals and PAH on soil assessed via dehydrogenase assay', *Environment International*, 28(8):779–782.
Jacobsen, C. S. and Hjelmsø, M. H. (2014) 'Agricultural soils, pesticides and microbial diversity', *Current Opinion in Biotechnology*, 27:15–20.
Jayasumana, C., Fonseka, S., Fernando, A., Jayalath, K., Amarasinghe, M., Siribaddana, S., Gunatilake, S. and Paranagama, P. (2015) 'Phosphate fertilizer is a main source

of arsenic in areas affected with chronic kidney disease of unknown etiology in Sri Lanka', *SpringerPlus*, 4(1):1–8.

Kassotis, C. D., Vandenberg, L. N., Demeneix, B. A., Porta, M., Slama, R. and Trasande, L. (2020) 'Endocrine-disrupting chemicals: Economic, regulatory, and policy implications', *The Lancet Diabetes & Endocrinology*, 8(8):719–730.

Kawa, I. A., Fatima, Q., Mir, S. A., Jeelani, H., Manzoor, S. and Rashid, F. (2021) 'Endocrine disrupting chemical Bisphenol A and its potential effects on female health', *Diabetes & Metabolic Syndrome: Clinical Research & Reviews*, 15(3):803–811.

Keifer, M. C. and Firestone, J. (2007) 'Neurotoxicity of pesticides', *Journal of Agromedicine*, 12(1). https://doi.org/10.1300/j096v12n01_03.

Kidd, P., Mench, M., Álvarez-López, V., Bert, V., Dimitriou, I., Friesl-Hanl, W., Herzig, R., Olga Janssen, J., Kolbas, A. and Müller, I. (2015) 'Agronomic practices for improving gentle remediation of trace element-contaminated soils', *International Journal of Phytoremediation*, 17(11):1005-–1037.

Kim, H. S., Kim, Y. J. and Seo, Y. R. (2015) 'An overview of carcinogenic heavy metal: molecular toxicity mechanism and prevention', *Journal of Cancer Prevention*, 20(4):232.

Kim, K-R., Owens, G., Naidu, R. and Kwon, S-I. (2010) 'Influence of plant roots on rhizosphere soil solution composition of long-term contaminated soils', *Geoderma*, 155(1–2):86–92.

Klimkowicz-Pawlas, A. and Debaene, G. (2020) 'Screening risk assessment of agricultural areas under a high level of anthropopressure based on chemical indexes and VIS-NIR Spectroscopy', *Molecules*, 25(14):3151.

Kulathunga, M., Wijayawardena, M. and Naidu, R. (2019a) 'Heavy metal exposure risk of commercial grown vegetables in Sri Lanka', Proceedings of the 8th International Contaminated Site Remediation Conference held in Adelaide convention Centre, Adelaide, South Australia, 8–12 September 2019, 52:308–309.

Kulathunga, M., Wijayawardena, M. and Naidu, R. (2021a) 'Dietary heavy metal (loid)s exposure and prevalence of chronic kidney disease of unknown aetiology (CKDu) in Sri Lanka', *Environmental Geochemistry and Health,* 1–12. https://doi.org/10.1007/s10653-021-01144-1.

Kulathunga, M., Wijayawardena, M. A. and Naidu, R. (2021b) 'Heavy metal (loid)s and health risk assessment of Dambulla vegetable market in Sri Lanka', *Environmental Monitoring and Assessment*, 193(4):1–10.

Kulathunga, M., Wijayawardena, M., Naidu, R., Wimalawansa, S. and Rahman, M. (2022a) 'Health risk assessment from heavy metals derived from drinking water and rice, and correlation With CKDu', *Frontiers in Water,* 3:786487. https://doi.org/10.3389/frwa.2021.786487.

Kulathunga, M. R. D. L., Wijayawardena, M. A. A., Naidu, R. and Wimalawansa, S. J. (2022b) 'Potential link between spatial variation and translocation characteristics of heavy metals in paddy topsoil and human health risks in a CKDu prevalent area of Sri Lanka', *Weeds-Journal of the Asian-Pacific Weed Science Society*, 4(1):62–75.

Kulathunga, M. R. D. L., Wijayawardena, M. A. A., Naidu, R. and Wijeratne, A. W. (2019b) 'Chronic kidney disease of unknown aetiology in Sri Lanka and the exposure to environmental chemicals: a review of literature', *Environmental Geochemistry and Health*, 41(5):2329-–2338.

Kuppusamy, S., Thavamani, P., Megharaj, M., Venkateswarlu, K. and Naidu, R. (2016) 'Agronomic and remedial benefits and risks of applying biochar to soil: current knowledge and future research directions', *Environment International*, 87:1–12.

Laden, F., Schwartz, J., Speizer, F. E. and Dockery, D. W. (2006) 'Reduction in fine particulate air pollution and mortality: extended follow-up of the Harvard Six Cities study', *American Journal of Respiratory and Critical Care Medicine*, 173(6):667–672.

Lal, A., Naidu, R. and Datta, B. (2023) 'Applications of machine learning models for solving complex groundwater modelling, monitoring and management problems', in *Groundwater in arid and semi-arid areas: monitoring, assessment, modelling, and management*, edited by Shakir Ali and Asaad Mater Armanuospp, 177–196, Springer.

Lepeule, J., Laden, F., Dockery, D. and Schwartz, J. (2012) 'Chronic exposure to fine particles and mortality: an extended follow-up of the Harvard Six Cities study from 1974 to 2009', *Environmental Health Perspectives*, 120(7):965–970.

Li, H., Zhou, Z., Long, T., Wei, Y., Xu, J., Liu, S. and Wang, X. (2022) 'Big-data analysis and machine learning based on oil pollution remediation cases from CERCLA database', *Energies*, 15(15):5698.

Li, J. and Naidu, R. (2005) 'Risk assessment of heavy metal contaminated soil in the vicinity of a lead/zinc mine', *Journal of Environmental Sciences*, 17(6):881–885.

Li, J., Wang, X., Yang, J., Liu, Y. and Naidu, R. (2022) 'Predicting the thresholds of metals with limited toxicity data with invertebrates in standard soils using quantitative ion character-activity relationships (QICAR)', *Journal of Hazardous Materials*, 423:126982.

Liu, S., Duan, L. and Naidu, R. (2019) 'Influence of soil properties and plant species on PFASs bioaccumulation', Proceedings of the 8th International Contaminated Site Remediation Conference held in Adelaide convention Centre, Adelaide, South Australia, 8–12 September 2019, 52:257.

Liu, Y., Bahar, M. M., Samarasinghe, S. C., Qi, F., Carles, S., Richmond, W. R., Dong, Z. and Naidu, R. (2022) 'Ecological risk assessment for perfluorohexanesulfonic acid (PFHxS) in soil using species sensitivity distribution (SSD) approach', *Journal of Hazardous Materials*, 439:129667.

Lockie, S. (1997) 'Chemical risk and the self-calculating farmer: diffuse chemical use in Australian broadacre farming systems', *Current Sociology*, 45(3):81–97.

Lowry, J. A. (2020) 'Assessment of exposures in vulnerable populations: exposure and response modelling for environmental contaminants through a lifetime', *Good Health and Well-being,* 38–50.

Ludwig, B., Kuka, K., Franko, U. and von Lützow, M. (2008) 'Comparison of two quantitative soil organic carbon models with a conceptual model using data from an agricultural long-term experiment', *Journal of Plant Nutrition and Soil Science*, 171(1):83–90.

Lundgren, R. E. and McMakin, A. H. (2018) *Risk communication: a handbook for communicating environmental, safety, and health risks,* John Wiley & Sons.

Ma, X., Tian, H., Liao, M., Huang, Y., Yang, R., Guo, Z., Megharaj, M. and He, W. (2022) 'Toxicity of antimony in 18 soils of China evaluated by soil dehydrogenase activity', *Soil Science Society of America Journal*, 86(3):703–713.

Megharaj, M. and Naidu, R. (2011) *Bioavailability and toxicity of contaminant mixtures to soil biota*, Elsevier.

Megharaj, M. and Naidu, R. (2017) 'Soil and brownfield bioremediation', *Microbial Biotechnology*, 10(5):1244–1249.

Melchert, M. and List, A. (2007) 'The thalidomide saga', *The International Journal of Biochemistry & Cell Biology*, 39(7–8):1489–1499.

Naidu, R. (2013) 'Recent advances in contaminated site remediation', *Water, Air, & Soil Pollution*, 224:1–11.

Naidu, R., Biswas, B., Willett, I. R., Cribb, J., Singh, B. K., Nathanail, C. P., Coulon, F., Semple, K. T., Jones, K. C. and Barclay, A. (2021) 'Chemical pollution: a growing peril and potential catastrophic risk to humanity', *Environment International*, 156:106616.

Naidu, R., Bolan, N. S., Kookana, R. S. and Tiller, K. (1994) 'Ionic-strength and pH effects on the sorption of cadmium and the surface charge of soils', *European Journal of Soil Science*, 45(4):419–429.

Naidu, R. and Harter, R. (1998) 'Effect of different organic ligands on cadmium sorption by and extractability from soils', *Soil Science Society of America Journal*, 62(3):644–650.

Naidu, R., Kookana, R. S., Sumner, M. E., Harter, R. D. and Tiller, K. (1997) 'Cadmium sorption and transport in variable charge soils: a review', *Journal of Environmental Quality*, 26(3):602–617.

Naidu, R. and Rengasamy, P. (1993) 'Ion interactions and constraints to plant nutrition in Australian sodic soils', *Soil Research*, 31(6):801–819. https://dx.doi.org/10.1071/SR9930801.

Naidu, R., Rogers, S., Gupta, V., Kookana, R. S., Bolan, N. S. and Adriano, D. (2003) *Bioavailability of metals in the soil plant environment and its potential role in risk assessment*, Science Publishers, Inc.

Naidu, R., Semple, K. T., Megharaj, M., Juhasz, A. L., Bolan, N. S., Gupta, S. K., Clothier, B. E. and Schulin, R. (2008) 'Bioavailability - New concepts and definitions', in *Developments in soil science*, edited by A. E. Hartemink, A. B. McBratney, and R. Naidupage, 7, Elsevier.

Naidu, R., Smith, E., Owens, G. and Bhattacharya, P. (2006) *Managing arsenic in the environment: From soil to human health*, CSIRO Publishing.

Ng, J. C., Juhasz, A., Smith, E. and Naidu, R. (2013) 'Assessing the bioavailability and bioaccessibility of metals and metalloids', *Environmental Science and Pollution Research*, 22:8802–8825.

Olaniran, A. O., Balgobind, A. and Pillay, B. (2013) 'Bioavailability of heavy metals in soil: Impact on microbial biodegradation of organic compounds and possible improvement strategies', *International Journal of Molecular Sciences*, 14(5):10197–10228.

Oliver, M. A. and Gregory, P. (2015) 'Soil, food security and human health: a review', *European Journal of Soil Science*, 66(2):257–276.

Paz, D. B, Henderson, K. and Loreau, M. (2020) 'Agricultural land use and the sustainability of social-ecological systems', *Ecological Modelling*, 437:109312.

Pelfrêne, A., Waterlot, C., Mazzuca, M., Nisse, C., Bidar, G. and Douay, F. (2011) 'Assessing Cd, Pb, Zn human bioaccessibility in smelter-contaminated agricultural topsoils (northern France)', *Environmental Geochemistry and Health*, 33(5):477–493.

Poggio, L., Vrscaj, B., Schulin, R., Hepperle, E. and Ajmone Marsan, F. (2009) 'Metals pollution and human bioaccessibility of topsoils in Grugliasco (Italy)', *Environmental Pollution*, 157(2):680–689. https://doi.org/10.1016/j.envpol.2008.08.009.

Punia, A., Dehal, L. and Chauhan, N. S. (2023) 'Evidence of the toxic potentials of agrochemicals on human health and biodiversity', in *One health implications of agrochemicals and their sustainable alternatives*, edited by Matthew C. Ogwu and Sylvester C. Izah, 105–133, Springer.

Qian, J., Shan, X-q., Wang, Z-j. and Tu, Q. (1996) 'Distribution and plant availability of heavy metals in different particle-size fractions of soil', *Science of the total environment*, 187(2):131–141. https://doi.org/10.1016/0048-9697(96)05134-0.

Queenan, K., Garnier, J., Nielsen, L. R., Buttigieg, S., Meneghi, D. d., Holmberg, M., Zinsstag, J., Rüegg, S., Häsler, B. and Kock, R. (2017) 'Roadmap to a One Health agenda 2030', *CABI Reviews*, (2017):1–17.

Rahman, M. A., Rahman, M. M., Reichman, S. M., Lim, R. P. and Naidu, R. (2014) 'Heavy metals in Australian grown and imported rice and vegetables on sale in Australia: Health hazard', *Ecotoxicology and Environmental Safety*, 100:53–60.

Rainbow, P. (2018) 'Heavy metal levels in marine invertebrates', *Heavy Metals in the Marine Environment*, 67–79.

Rastogi, M., Singh, S. and Pathak, H. (2002) 'Emission of carbon dioxide from soil', *Current Science*, 82(5):510–517.

Raymond-Delpech, V., Matsuda, K., Sattelle, B. M., Rauh, J. J. and Sattelle, D.B. (2005) 'Ion channels: molecular targets of neuroactive insecticides', *Invertebrate Neuroscience*, 5:119–133.

Robert, M., Thomas, A., Sekhar, M., Badiger, S., Ruiz, L., Raynal, H. and Bergez, J-E. (2017) 'Adaptive and dynamic decision-making processes: A conceptual model of production systems on Indian farms', *Agricultural Systems*, 157:279–291.

Scholz, S., Fischer, S., Gündel, U., Küster, E., Luckenbach, T. and Voelker, D. (2008) 'The zebrafish embryo model in environmental risk assessment–applications beyond acute toxicity testing', *Environmental Science and Pollution Research*, 15:394–404.

Selikoff, I. J., Churg, J. and Hammond, E. C. (1965) 'Relation between exposure to asbestos and mesothelioma', *New England Journal of Medicine*, 272(11):560–565.

Shahid, M., Niazi, N. K., Dumat, C., Naidu, R., Khalid, S., Rahman, M. M. and Bibi, I. (2018) 'A meta-analysis of the distribution, sources and health risks of arsenic-contaminated groundwater in Pakistan', *Environmental pollution*, 242:307–319.

Shi, W-y., Shao, H-b., Li, H., Shao, M-a. and Du, S. (2009) 'Co-remediation of the lead-polluted garden soil by exogenous natural zeolite and humic acids', *Journal of Hazardous Materials*, 167(1–3):136–140. https://doi.org/10.1016/j.jhazmat.2008.12.092.

Siam, M., Munna, J. I., Hasan, M. and Rahman, T. (2019) *Remote sensing kit for contamination event detection in water*, IEEE.

Siddique, A. B., Rahman, M. M., Islam, M. R., Mondal, D. and Naidu, R. (2021) 'Response of iron and cadmium on yield and yield components of rice and translocation in grain: Health risk estimation', *Frontiers in Environmental Science*, 9:716770.

Snyder, M. J. (2000) 'Cytochrome P450 enzymes in aquatic invertebrates: recent advances and future directions', *Aquatic Toxicology*, 48(4):529–547.

Song, W-Y., Ju Sohn, E., Martinoia, E., Jik Lee, Y., Yang, Y-Y., Jasinski, M., Forestier, C., Hwang, I. and Lee, Y. (2003) 'Engineering tolerance and accumulation of lead and cadmium in transgenic plants', *Nature Biotechnology*, 21(8):914–919.

Stephens, T. D., Bunde, C. J. and Fillmore, B. J. (2000) 'Mechanism of action in thalidomide teratogenesis', *Biochemical Pharmacology*, 59(12):1489–1499.

Taghavi, M., Darvishiyan, M., Momeni, M., Eslami, H., Fallahzadeh, R. A. and Zarei, A. (2023) 'Ecological risk assessment of trace elements (TEs) pollution and human health risk exposure in agricultural soils used for saffron cultivation', *Scientific reports*, 13(1):4556.

Thomsen, M., Faber, J. H. and Sorensen, P. B. (2012) 'Soil ecosystem health and services–Evaluation of ecological indicators susceptible to chemical stressors', *Ecological Indicators*, 16:67–75.

Tóth, G., Hermann, T., Da Silva, M. R. and Montanarella, L. (2016) 'Heavy metals in agricultural soils of the European Union with implications for food safety', *Environment International*, 88:299–309.

Umeh, A. C., Duan, L., Naidu, R., Esposito, M. and Semple, K. T. (2019) 'In vitro gastrointestinal mobilization and oral bioaccessibility of PAHs in contrasting soils and associated cancer risks: focus on PAH nonextractable residues', *Environment International*, 133:105186.

Umeh, A. C., Naidu, R., Shilpi, S., Boateng, E. B., Rahman, A., Cousins, I. T., Chadalavada, S., Lamb, D. and Bowman, M. (2021) 'Sorption of PFOS in 114 well-characterized tropical and temperate soils: application of multivariate and artificial neural network analyses', *Environmental Science & Technology*, 55(3):1779–1789.

USEPA (2012) *Technical overview of ecological risk assessment - analysis phase: ecological effects characterization* EPA U.

Vig, K., Megharaj, M., Sethunathan, N. and Naidu, R. (2003) 'Bioavailability and toxicity of cadmium to microorganisms and their activities in soil: a review', *Advances in Environmental Research*, 8(1):121–135.

Wagena, M. B. and Easton, Z. M. (2018) 'Agricultural conservation practices can help mitigate the impact of climate change', *Science of the Total Environment*, 635:132–143.

Wang, J., Lin, C., Chen, Y. and Liu, A. (2012) 'Cultivated land pollution at township level in China: situation, factors and measures', *China Land Science*, 26(2):25–30.

Wang, L., Cheng, Y., Gopalan, S., Luo, F., Amreen, K., Singh, R. K., Goel, S., Lin, Z. and Naidu, R. (2023) 'Review and Perspective: Gas Separation and Discrimination Technologies for Current Gas Sensors in Environmental Applications', *ACS Sensors*, 8(4):1373–1390.

Wang, L., Cheng, Y., Naidu, R. and Bowman, M. (2021) 'The key factors for the fate and transport of petroleum hydrocarbons in soil with related in/ex situ measurement methods: an overview', *Frontiers in Environmental Science*, 9:620.

Wang, Y., Shi, J., Wang, H., Lin, Q., Chen, X. and Chen, Y. (2007) 'The influence of soil heavy metals pollution on soil microbial biomass, enzyme activity, and community composition near a copper smelter', *Ecotoxicology and Environmental Safety*, 67(1):75–81. https://doi.org/10.1016/j.ecoenv.2006.03.007.

Wei, S., Teixeira da Silva, J. A. and Zhou, Q. (2008) 'Agro-improving method of phytoextracting heavy metal contaminated soil', *Journal of Hazardous Materials*, 150(3):662–668. https://doi.org/10.1016/j.jhazmat.2007.05.014.

Wijayawardena, A., Megharaj, M., Naidu, R. and Stojanovski, E. (2018) 'Chronic and reproductive toxicity of cadmium, zinc, and lead in binary and tertiary mixtures to the earthworm (Eisenia fetida)', *Journal of Soils and Sediments*, 18(4):1602–1609.

Wijayawardena, M., Liu, Y., Yan, K., Duan, L., Umeh, A., Naidu, R. and Semple, K. (2020a) 'Assessment of the oral bioavailability of organic contaminants in humans', *Bioavailability of Organic Chemicals in Soil and Sediment,* 191–218.

Wijayawardena, M., Liu, Y., Yan, K., Duan, L., Umeh, A., Naidu, R. and Semple, K. (2020b) 'Assessment of the oral bioavailability of organic contaminants in humans', in *Bioavailability of organic chemicals in soil and sediment*, edited by Jose J. Ortega-Calvo and John R. Parsons, 191–218, Springer.

Wijayawardena, M., Mallavarapu, M. and Naidu, R. (2012) *Bioavailability and toxicity of arsenic to earthworms in three soils*, CRC Press/Balkema.

Wijayawardena, M., Megharaj, M. and Naidu, R. (2012) 'Bioavailability and toxicity of arsenic to earthworms in three soils', in *Understanding the geological and medical interface of arsenic-as 2012*, edited by Jack C. Ng, Barry N. Noller, Ravi Naidu, Jochen Bundschuh, Prosun Bhattacharya, CRC Press.

Wijayawardena, M., Megharaj, M. and Naidu, R. (2016) 'Exposure, toxicity, health impacts, and bioavailability of heavy metal mixtures', *Advances in Agronomy*, 138:175–234.

Wijayawardena, M., Megharaj, M. and Naidu, R. (2017) 'Bioaccumulation and toxicity of lead, influenced by edaphic factors: Using earthworms to study the effect of Pb on ecological health', *Journal of Soils and Sediments*, 17(4):1064–1072.

Wijayawardena, M., Megharaj, M., Naidu, R. and Stojanovski, E. (2017) 'Chronic and reproductive toxicity of cadmium, zinc, and lead in binary and tertiary mixtures to the earthworm (Eisenia fetida)', *Journal of Soils and Sediments*, (18):1602–1609.

Wijayawardena, M., Naidu, R., Megharaj, M., Lamb, D., Thavamani, P. and Kuchel, T. (2015a) 'Influence of ageing on lead bioavailability in soils: a swine study', *Environmental Science and Pollution Research*, 22(12):8979–8988.

Wijayawardena, M., Naidu, R., Megharaj, M., Lamb, D., Thavamani, P. and Kuchel, T. (2015b) 'Using soil properties to predict in vivo bioavailability of lead in soils', *Chemosphere*, 138:422–428.

Wijayawardena, M. A., Megharaj, M. and Naidu, R. (2016) 'Exposure, toxicity, health impacts, and bioavailability of heavy metal mixtures', *Advances in Agronomy*, (138):175–234.

Wijayawardena, M. A., Naidu, R., Megharaj, M., Lamb, D., Thavamani, P. and Kuchel, T. (2017) 'Evaluation of relative bioaccessibility leaching procedure for an assessment of lead bioavailability in mixed metal contaminated soils', *Environmental Technology & Innovation*, 7:229–238.

Wijayawardena, M. A. A. (2014) *Bioavailability and toxicity of heavy metals and metal mixtures to ecological and human health*, PhD thesis. https://find.library.unisa.edu.au/discovery/fulldisplay?vid=61USOUTHAUS_INST:UNISA&docid=alma9916110104301831&context=L&lang=en.

Wurtsbaugh, W. A., Paerl, H. W. and Dodds, W. K. (2019) 'Nutrients, eutrophication and harmful algal blooms along the freshwater to marine continuum', *Wiley Interdisciplinary Reviews: Water*, 6(5):e1373.

Yan, K., Naidu, R., Liu, Y., Wijayawardena, A., Duan, L. and Dong, Z. (2018) 'A pooled data analysis to determine the relationship between selected metals and arsenic bioavailability in soil', *International Journal of Environmental Research and Public Health*, 15(5):888.

Zeng, X., Li, L. and Mei, X. (2007) 'Heavy metal content in soils of vegetable-growing lands in China and source analysis', *Scientia Agricultura Sinica*, 40(11):2507–2517.

Zhao, Y-F., Shi, X-Z., Huang, B., Yu, D-S., Wang, H-J., Sun, W-X., ÖBoern, I. and BlombÄCk, K. (2007) 'Spatial distribution of heavy metals in agricultural soils of an industry-based peri-urban area in Wuxi, China', *Pedosphere*, 17(1):44–51. https://doi.org/10.1016/s1002-0160(07)60006-x.

Zhong, S., Zhang, K., Bagheri, M., Burken, J. G., Gu, A., Li, B., Ma, X., Marrone, B. L., Ren, Z. J. and Schrier, J. (2021) 'Machine learning: new ideas and tools in environmental science and engineering', *Environmental Science & Technology*, 55(19):12741–12754.

Zhuang, P., McBride, M. B., Xia, H., Li, N. and Li, Z. (2009) 'Health risk from heavy metals via consumption of food crops in the vicinity of Dabaoshan mine, South China', *Science of the total environment*, 407(5):1551–1561. https://doi.org/10.1016/j.scitotenv.2008.10.061.

Chapter 9

Bioremediation: advances in using soil organic amendments to manage contaminants in agricultural soils

Gerhard Soja, University of Natural Resources and Life Sciences Vienna (BOKU), Austria

1 Introduction

Soil is the basis for global food production. More than 8 billion people directly or indirectly depend on plants growing in soil as fruit, field or garden crops to provide food for people or fodder for animals. The function of soil as the location for agricultural and forestry production is just one of the several essential soil functions: soil is also required for the extraction of raw materials; as the subsoil basis for buildings and infrastructure; as a master control of global water and elemental cycles; as an ecological regulator for material flows; and as a habitat for soil life. These soil ecosystem services are the main pillars of circular resource management and bio-based sustainable production systems (Anikwe and Ife, 2023).

http://dx.doi.org/10.19103/AS.2025.0151.14

To correctly accomplish all of these functions, good soil health is required. The Food and Agriculture Organization of the United Nations (FAO) has defined healthy soil as 'the ability to sustain productivity, diversity and environmental services of terrestrial ecosystems' (FAO, 2020). In agricultural systems, soil health can be maintained, improved or recovered through the implementation of sustainable soil management practices. The prevention, containment and remediation of soil contamination are important contributions to sustainable soil management and the protection of soil health. Applying the FAO's criteria to the health of European soils, the sobering report of the EUSO Soil Health Dashboard finds that 61% of EU soils are in an unhealthy state (EU Soil Observatory, 2024). This is as a result of a combination of different soil health indicators. This number may even be underestimated, as some soil degradation reasons has a lack of known data, including contaminants.

The UN Sustainable Development Goals (SDGs) have no specific goal for soil, but healthy soil is a cross-sectional issue for several SDGs that address land and soil directly or indirectly (EEA, 2019). The clearest reference to soil is contained in SDG 15, 'Life on Land'. Terrestrial ecosystems require healthy, uncontaminated soil that can be managed and used sustainably. Land degradation, biodiversity loss and desertification can only be successfully avoided if soil contamination is directly prevented. Regarding SDG 2, 'Zero Hunger', and SDG 3, 'Good Health and Well-Being', it is clear that healthy diets depend on uncontaminated soil substrates to prevent the uptake and bioaccumulation of pollutants in plants, which can pose risks to human health and negatively affect the health of farm animals or grazing animals. Contaminants may also migrate from soil to groundwater. SDG 6, 'Clean Water and Sanitation,' therefore, appeals to maintaining the filter function of the soil. Each filter performs its function best when the pollutant load is low in order to avoid reaching the breakthrough point. There is a large carbon storage pool size of soil (approximately two to three times larger than atmospheric storage; Xu et al., 2020). Therefore, the importance of a healthy soil in maintaining a large pool of carbon by sequestering CO_2 is included in SDG 13, 'Climate Action'. Finally, preventing soil contamination is part of SDG 12, 'Responsible Consumption and Production', and SDG 11, 'Sustainable Cities and Communities', as agricultural and horticultural production entities must be aware of their responsibility to prevent soil contamination from excess fertilizer, pesticides or wastes.

2 Contaminants in agricultural soils

Agricultural soils are managed systems, and any form of management can have both positive and negative impacts. In addition to the targeted deposition of fertilizers and plant protection chemicals that ensure satisfactory harvests, a diffuse deposition of background pollutants from the atmosphere cannot be avoided. These diffuse sources can originate from emissions of industrial activities such as power plants, mining, metal and ore processing or from the combustion of fossil fuels caused by transport. Landfills, waste incineration, fallow land and desertification of previously fertile areas may additionally contribute to the long-range transport of contaminants that are deposited on agricultural and forestry soils. Depending on the geological origin of soils, locally elevated concentrations of trace metals can result in undesirable uptake of these elements by crop plants, either directly from the soil or from irrigation water contaminated by excessive trace element concentrations, such as arsenic (Loukola-Ruskeeniemi et al., 2022).

The use of irrigation water is a management tool to prevent crop losses during drought periods. However, water quality requirements demand frequent monitoring to avoid contamination. In some arid regions, reusing treated wastewater is the only way to provide sufficient water for agricultural production without overtaxing groundwater resources. However, filtration and purification efforts present significant challenges in preventing accidental soil contamination (Oron and Demalach, 1987; Cornet and Weber, 2004). In addition to metals and metalloids, residues of pharmaceuticals, hormones and cosmetic products may also contaminate soil irrigated with inadequately treated wastewater.

In regions with rain-fed agriculture, fertilizer and pesticide applications are the main causes of farming-related soil contamination, either owing to excessive application rates or residues from previous applications that left accumulations of toxic metals like Cd or persistent pesticides like DDT (Bernal, 2017; Alvarenga et al., 2020). Essential nutrients, such as ammonium, nitrate or phosphate, may become contaminants. This can occur if the application rate exceeds plant uptake, causing the excess to migrate into the groundwater or be washed off the field in runoff after heavy rainfall, sometimes along with soil particles as soil erosion. These chemical species (nitrate, phosphate) occur as anion and are only weakly bound to soil organic matter which are especially prone to reach the groundwater or surface water. When a nutrient or trace element becomes a soil contaminant, concentrations are well above natural background levels; if the concentrations are high enough to endanger living organisms, soil microbiology or human health, they are considered as pollutants.

Trace elements are not degradable and may accumulate in the topsoil layers and rhizospheres of plants, which can lead to increased plant uptake, contamination of crop material and occasionally to the exceedance of legal limits or national standards. Elements such as Hg, Cd, Tl, U and Pb play no physiological role and are considered as undesirable in all cases. Other elements such as B, Cl, Co, Cu, Mn, Ni and Zn are essential micronutrients for plants and only excessive concentrations of them are harmful. For elements such as Zn, Cu, Ni and As, toxic concentrations for plants are lower than the levels required to protect animals or humans. Metalloids such as As, Sb and Se have characteristics that are both related to metals and to nonmetals. The term PTE (potentially toxic trace elements) can be used for all of these elements and considers that the toxicity depends on the specific conditions in terms of concentration, environmental medium, bioavailability and receptor (Alvarenga et al., 2020).

Organic contaminants in soil can be broken down biologically or through physico-chemical processes in the soil. This is an important difference from trace elements, which can only change their chemical speciation but always remain a constant in the mass balance. However, the degradation of organic contaminants is a time-varying process that may take days to decades, depending on the specific chemical, soil conditions, and the activity of the degrading microorganisms. While some of the deposited organic contaminants come from diffuse background deposition, e.g. polycyclic aromatic hydrocarbons (PAHs), dioxins or polychlorinated biphenyls, as by-products of industrial and other combustion process (Schroder et al., 1997; Wu et al., 2009; Augusto et al., 2010), another part of the contaminants consists of pesticides used as a management practice outside of organic farming (Chen et al., 2017).

The application of fertilizers may deposit not only excess nutrients in agricultural soils but also xenobiotics, which are also contaminants of the fertilizers. In mineral fertilizers, the concentrations of Cd and U are a concern (Schnug and Lottermoser, 2013; Carne et al., 2021) and have led to EU-wide limits for Cd concentrations in phosphate fertilizers. Since these fertilizers are typically made from phosphate rock, which may contain accompanying elements of geological origin, replacing mined phosphate fertilizer with recycled phosphate from secondary raw materials is an increasing desire (Hosseinian et al., 2023). The use of sewage sludge (biosolids) rich in nitrogen and phosphorus would be a clear solution, yet there are some disadvantages, including the risk of pollution that have led to restrictions or bans in many countries. The negative effects of sewage sludge are caused by potentially dangerous concentrations of trace elements, drug residues and by hygiene concerns. Other organic fertilizers such as manure, slurry, compost or digestate may also contribute to soil contamination, if their quality guidelines are not adhered to. In these materials, residues of veterinary drugs or pathogenic microorganisms may affect their applicability to agricultural soils (Kwon, 2011; Cekmecelioglu et al., 2005). PFASs (poly- and perfluoroalkyl substances) have caused great concern owing to their effects on human health, and sewage sludge is one of many different point and non-point sources that release PFAS into the environment (Zhou et al., 2024). Surprisingly, pyrolyzed sewage sludge proved to be one of the most efficient sorbents for PFAS in soil remediation

(Krahn et al., 2023). The increasing use of incompletely composted plastic bags for kitchen waste and the widespread use of plastic films in agriculture have contributed to a microplastic problem in soil or compost, limiting the direct use of compost in agricultural soil (Ruzicková et al., 2022). The problem of microplastics in soil may exacerbate the problems with other organic contaminants in soil, as they interact with microplastics and alter their sorption and migration behavior (He et al., 2022). Even weed seeds that have survived the composting process owing to too low a process temperature may become a source of compost contamination (Grundy et al., 1998).

This wide range of contamination risks to agricultural soils highlights the importance of both preventing future contamination problems and remediating existing ones. This section will focus on the options available to address problematic levels of contaminants in soil due to past atmospheric deposition or agricultural management actions.

Potential impacts of pollutants include:

1 ecosystem health;
2 animal and human health; and
3 local and international trade as most countries now have strict guidelines on food quality.

3 Remediation techniques for agricultural soils

The possibilities for soil remediation can be classified according to the place of use; the contaminant class that is the subject of remediation; or the technology chosen.

Characteristic of *off-site* (or *ex-situ*) methods is the removal of contaminated soil from the site of contamination to another site for remediation. Excavating, transporting and then treating the soil is typically a costly and ecologically destructive process. However, more advanced technological options can be used, and the remediation goal can be achieved quicker, by taking advantage of more sophisticated, centralized soil treatment equipment. But these treatments, either physical, chemical or a combination of both, often result in dead soil that needs to be revitalized and prepared for a reuse as a growing medium for plants. If the soil is excavated for a specific treatment at the same location without further transport, the treatment is conducted *on-site*. When the remediation goal is achieved, the same soil will be backfilled at the site of origin. However, this type of treatment is also disruptive and can significantly damage the microbiology of the soil and hence its soil health. Only *in-situ* methods do not require soil excavation and keep the soil structure more or less intact. However, they face the difficulty of bringing the remediating agents into contact with the contaminant, which requires sophisticated drilling, injection and/or distribution systems. Such methods are gentler on soil life than *on-site* or *ex-situ* methods. The fertility of the soil is maintained at least to a certain extent and can even be improved depending on the (bio-)remediation method (Bernal et al., 2020).

Remediation techniques can be based on physical, chemical and biological methods. In most cases, a combination of different methods is used to achieve optimal remediation success. Physical methods can be based on thermal treatment; soil washing and extraction methods, such as chelation; supercritical fluid extraction or steam-stripping; and vitrification or on encapsulation. Most often, these methods are biologically destructive and can leave a dead mineral soil, in the dearth of revitalization. When using chemical methods, it depends on the application rates and concentration of the chemicals used, which are deployed *in-situ, on-site* or *ex-situ*. In the case of soil contamination with trace elements, sometimes even simple and cheap methods, such as liming, can reduce the bioavailability of the elements of concern by increasing the pH (Gronflaten et al., 2005). In the case of organic contaminants, the treatment chemicals are strong oxidants or reducing agents, depending on the contaminant in question. While many organic chemicals can be mineralized oxidatively by adding permanganate, persulfate or hydrogen peroxide solutions, the remediation success of halogenated hydrocarbons is higher if they are treated with reducing agents that release hydrogen in exchange for halogen elements. In general, the oxidation of a pollutant is tantamount to a loss of electrons from the pollutant molecule. This requires the oxidant as an electron acceptor, while the reduction of a pollutant means an electron gain for the pollutant molecule, and requires an electron donor. When dealing with contaminants through oxidation, the effective electron transfer mechanisms (exemplified for persulfate application: Equation 1) can be supported by the formation of free radicals, which enable a higher oxidation potential. Free radicals arise from the activation of persulfate or hydrogen peroxide either at elevated temperatures or with divalent iron (Equation 2).

Equation 1: $S_2O_8^{2-} + 2\ H^+ + 2\ e^- \rightarrow 2\ HSO_4^-$

Equation 2: $S_2O_8^{2-} + 2\ Fe^{2+} \rightarrow 2\ SO_4^{-\bullet} + 2\ Fe^{3+}$

While most contaminant oxidants can be applied in the saturated zone as well as in topsoil or soil piles, ozone, as the only gaseous oxidant for soil remediation, is used in the vadose zone and also supports a radical mechanism (Krembs et al., 2010; Chang et al., 2022). Other chemical remediation methods are based on photochemical degradation using light-induced free radical formation and electrochemical methods (Niroumand et al., 2012; Dondi et al., 2016).

If organic contaminants must be remediated in soils, chemical or biological degradation methods are primarily used. If inorganic contaminants are the main problem, physical extraction, biological stabilization or phytoextraction methods may be applied. In all cases, the support of external organic amendments can increase the degradation or immobilization effect. The characteristic properties of the different amendments must be considered and the best options selected for each case.

When soil remediation or pollutant immobilization is based on the action of microorganisms and/or plants, the term bioremediation was coined. Phytoremediation is the synergistic use of plants to either remove contaminants from soil (phytoextraction) or to support microorganisms with root exudates to break down contaminants (phytodegradation). Stabilization (or immobilization) methods aim to reduce the ecotoxicity of the contaminants and promote microbiological degradation. Bioaugmentation is the introduction of separately cultured microorganism solutions with increased ability to degrade recalcitrant contaminants. Natural attenuation refers to the unsupported natural biological degradation process in the soil that is merely monitored. Bio-stimulation and bioventing support microbiology by adding nutrients and/or vitamins or by aerating deeper soil layers to alleviate oxygen deficiency for aerobic degradation mechanisms (Weil and Brady, 2017; Bernal et al., 2020). Even in bioremediation, a combination of methods is often the most promising method. Sánchez-Castro et al. (2023) emphasize the importance of combined strategies, including bioremediation and physical containment, to achieve better remediation performance. Other physical or chemical methods can help increase the bioavailability of contaminants by inducing partial degradation of large, difficult-to-attack molecules or by reducing the ecotoxicity of the contaminants. If external organic matter is added to support bioremediation, the microbiological degradation success can be significantly increased. The effective mechanisms of organic amendments to support pollutant reduction are described in the following sections.

4 Bioremediation: compost amendments

Compost is the product of aerobic biological degradation from organic matter, usually containing a variety of microorganisms, nutrients and various polymeric organic compounds, such as lignin and cellulose from plant material (Manyapu et al., 2022). Whereas, pre-decomposed organic materials like sewage sludge (biosolids), municipal waste, food waste, slurry or manure are also useful input materials for composting. The high value of compost produced from winery by-products for the remediation of soils with petroleum hydrocarbon contamination is highlighted by Dados et al. (2015). Efficient compost applications for the biodegradation of organic pollutants are enabled by the high microbial diversity and community composition of the degraders (Ren et al., 2018; Seo and Cho, 2021). However, compost can also support the remediation of inorganic contaminants: Beiyuan et al. (2018) describes a strong adsorption capacity of food waste compost for trace elements and for metal-chelating complexes, which come from phytoremediation enhanced by chelating agents. The high mobility of these complexes not only facilitates uptake by phytoremediation plants, but they could also migrate toward groundwater, requiring protection from accidental leaching. Plant materials are typically shredded before composting, but soil macrofauna such as earthworms, centipedes, slugs and springtails can also aid in the shredding process. During the microbiological degradation process, the polymers are broken down enzymatically, producing soluble organic carbon compounds, nutrients in ionic forms and sugars that can be utilized by microorganisms and plants. These nutrients also feed microorganisms that actively degrade organic soil contaminants. In addition, the environmental benefits of compost addition on soil quality will also support the contaminant decomposers (Table 1).

In controlled composting, strict control of the composting temperature, humidity and oxygen supply in the compost pile or windrow is essential for a catalyzed maturation of the compost and the killing of pathogens and weed

Table 1 Benefits and drawbacks of on-farm composting of agricultural residues

Economic benefits	Environmental benefits	Drawbacks
Financial income option of selling compost	Support of soil fertility and plant vigor; recycling of nutrients	Land requirements; investments in time and money for installations and equipment
Production of a usable product from waste	Improvement of soil water holding capacity	Odor, greenhouse gas emissions (methane) and rodents
Increase in crop yields	Plant disease suppression; savings in use of pesticides and mineral fertilizers	Management in unfavorable weather
Facilitation of organic agriculture	Soil erosion reduction	Diversion of organic carbon from the farm soil if the compost is sold off-farm
Reduction of herbicide costs	Rendering human, animal, and plant pathogens harmless	Potential loss of nitrogen as nitrous oxide and in seepage water
Lower waste disposal costs	Degradation of hormones, antibiotics, and pesticides	Variable nutrient concentrations; slow release and limited bioavailability of plant nutrients in compost
Higher revenues of farm product selling if the farm is run as an organic farm	Lower environmental impacts from compost (lower greenhouse gas emissions) versus uncontrolled rotting of raw feedstocks	Need for environmental permits and adherence to regulation; risk of being viewed as a commercial enterprise (rather than a farm)

Source: Adapted from: Rynk et al., 2022 (modified).

seeds. A successful composting process requires detailed planning and commitment, like growing plants. Although good compost products can be successfully achieved by any organic waste owner, the need to view composting as a separate enterprise with its specific requirements of time, investment, labor and infrastructure should not be neglected (Table 1; Rynk et al., 2022).

5 Bioremediation: biochar amendments for soil and environmental health

Biochar is the solid, porous product obtained after the thermochemical conversion of biomass at 350–1000°C in an oxygen-limited environment (= pyrolysis). Biochar can be used directly as a product; as a component of substrates; or composites for environmental management measures. Applications of biochar are characterized by their long-term carbon sink characteristics, unless the product is used thermally or as reducing agent in steel industry. However, replacing carbon from fossil fuels with carbon from biochar still offers an advantage in terms of the carbon footprint in the industrial process. When incorporated into the soil or building materials, the carbon is stored as a long-term carbon sink, which plays a significant role in helping to mitigate climate change. This long-term resistance against biological or physico-chemical degradation is caused by the highly condensed aromatic molecular structure of biochar, which is formed during pyrolysis of lignocellulosic feedstock.

To avoid contaminants in biochar, the guidelines of the European Biochar Certificate (EBC; EBC 2012–2023) define thresholds for organic contaminants and potentially toxic trace elements in the pyrolysis product. The versatility of biochar's uses has led to a sharp increase in production and application volumes in recent years (2023: approx. 52 000 t of production with a production capacity of approx. 95 000 t in Europe; EBI, 2023). Around 70% of these are produced as EBC-certified goods. At the same time, the topic of biochar is also being treated intensively in research (as of March 2024: >19 000 scientific papers with the term 'biochar' in the title can be retrieved from Web of Science). The focus of these studies is not only on the advantages of the use, production and properties of biochars but also on potential problems and risks of their applications. Such studies particularly emphasize the importance of avoiding misapplications and careless handling of this material, which can invalidate or negate the numerous advantages of biochar as an environmental technology. The overview in Table 2 summarizes the benefits and the feared (or actual) hazards of biochar applications and how these can be mitigated.

Micropores and mesopores in biochars, especially in wood-based biochars, as well as the associated high-specific surface areas in these materials, are partly responsible for the high sorption capacity for a variety of nutrients

Table 2 Benefits, perceived risks and risk mitigation of applying biochar in agricultural soils

Environmental/ economic benefits	Perceived risks	Risk mitigation
Support of circular economy to recover nutrients from agricultural and forestry residues	The use of biochar might promote competition for agricultural land, which is needed for food and feed production.	The guideline for the 'European Biochar Certificate' (EBC) contains lists of permitted input materials for the production of biochars. Most of these input materials are residues from agricultural, forestry and horticultural production, for which there is no other efficient use than incineration (in the best case: with thermal use) or rotting in the landscape.
Long-term carbon sequestration in soil with mean residence times of >100 years.	If all organic residues from agricultural/ forestry production are used for pyrolysis, the soil might lack of organic matter for humus formation.	The EBC guidelines prescribe the sustainable production of the primary product and the preservation of humus in the soil. If not considered, the certification requirements are not met.
Biochar as a negative emission technology for climate change mitigation.	The use of woody input materials for pyrolysis could promote deforestation and unsustainable forest management.	Woody input materials should primarily originate from (uncontaminated) waste or waste wood, or woody sieve residues from composting or screenings. In these cases, the wood is at the end of a cascadic use chain and has not been produced specifically for the purpose of pyrolysis. If wood is used that originates directly from forests, attention must be paid to the EBC guidelines: wood-based biochar should only be produced from forest wood if sustainable management of the forest in question can be proven by PEFC or FSC certificates or by comparable regional standards or laws.
Biochar in soil reduces pollutant uptake by crop plants.	Soil application of biochar with elevated pollutant concentrations might lead to permanent contamination of agricultural soils with heavy metals or polycyclic aromatic hydrocarbons (PAHs). Biochar could introduce dioxins into the soil.	The EBCE defines threshold values for pollutant concentrations of biochars depending on the intended use. This ensures that biochar does not add higher amounts of pollutants to the soil than when using compost or applying mineral fertilizers. Modern pyrolysis reactors have no difficulty in producing biochar that fulfills with all legal quality demands. Pyrolysis should not use chlorine-contaminated input material. The EBC directive contains threshold values for dioxins (and PCBs) in biochar, which must be considered.

Volatile pyrolysis products can be used for thermal energy or electricity generation.	Volatile compounds from the pyrolysis might be hazardous to health. In addition, the process is energetically inefficient due to heat losses.	An integral part of every pyrolysis reactor is the complete collection and post-combustion of volatile pyrolysis products. The resulting thermal energy can be used to supply energy to the reactor, the excess heat to dry input materials or to feed into district heating systems. A gas engine can produce electrical energy from the volatiles. A properly operated pyrolysis reactor therefore emits only harmless and odorless combustion gases and can be operated with high-energy efficiency.
Biochar in soil reduces contaminant migration toward groundwater and reduces runoff losses of contaminants and nutrients.	The addition of biochar to the soil might promote the breakdown of organic matter ('positive priming').	Biochar promotes the preservation and development of organic matter in the long term ('negative priming'). By promoting soil microbiological activity, the degradation of organic matter in the soil can increase in the short term after application (days to weeks). In the long term (from months to years), however, the inhibition of the degradation of organic matter in the soil predominates, as biochar has a stabilizing effect through aggregate conveyance (Majumder et al., 2019; Zhang et al., 2019).
Biochar can be used as nutrient carrier.	Biochar in soil might promote nitrogen fixation and competes with the nitrogen requirements of plant roots.	The high C/N ratio of freshly produced biochar requires supplementation with external nitrogen sources (compost, co-composted biochar, manure, slurry, (biogas) digestates, or mineral fertilizers). It is good agricultural practice to take nitrogen supplementation into account when applying biochar (Wang et al., 2019; Haider et al., 2020).
Biochar reduces greenhouse gas emissions of agricultural soil.	Biochar might promote greenhouse gas emissions from soil.	In the vast majority of studies, biochar application caused a significant reduction in greenhouse gas emissions (especially N_2O) (Liu et al., 2019; Zhang et al., 2020).
In poor, sandy or acidic soils biochar improves agricultural yields.	The cost of biochar might not justify the low or missing yield increases.	Field experiments in tropical areas typically show higher yield increases than in temperate climates, as soils in tropical areas more often suffer from unfavorable production conditions due to nutrient poverty and low pH (Jeffery et al., 2017). However, it has been reported in numerous experiments that biochar is not harmful in temperate climates, and that carbon sequestration, reduced nitrogen losses and improved water storage capacity of the soil can also be used to gain advantages in high-quality soils, which, however, do not become financially effective every year (Ye et al., 2020).

(Continued)

Table 2 (*Continued*)

Environmental/ economic benefits	Perceived risks	Risk mitigation
Darker soils warm more rapid in spring and accelerate seed germination.	The darker coloring of the soil after biochar application reduces albedo.	On the one hand, soil should be covered by plants for most of the year anyway, so that their albedo does not matter; but even on fallow land, biochar applications are hardly recognizable macroscopically already 1 year after incorporation (Zhang et al., 2018).
Pelletizing of biochar facilitates the use of existing farm machinery for applications. Production of slurries facilitates deep soil injections.	Dust pollution during the application of biochar could be a health hazard.	By quenching the biochar with water immediately after leaving the reactor, the water content is adjusted to about 30%, so that dust development is prevented. This measure is also in line with the guidelines of the EBC. Pellet production from biochar is state of the technology.
Biochar promotes soil aggregate formation and reduces soil erosion.	The surface application of biochar on soils could lead to undesirable environmental or health effects owing to wind or water erosion of small-scaled biochar particles.	Wind erosion can be avoided by moistening the biochar before application or by pelletizing. Run-off should not occur even during heavy precipitation if the biochar is incorporated right after application (with cultivator, tiller, rotary harrow, etc.) or if it is combined and distributed with slashed liquid manure.
Biochar supports soil microorganisms by providing nutrients and habitats.	Biochar could be harmful to soil life.	In response to biochar, the composition of microorganism consortia may shift, but the overall microbiological biomass tends to increase (Li et al., 2020). Earthworms seem to react differently depending on biochar and soil conditions, but in most studies, they were not significantly affected (Whalen et al., 2021).

and pollutants. The interactions and properties of biochar responsible for this depend on the substance adsorbed or bound. For metals such as cadmium, lead and zinc, the high cation exchange capacity of biochars is vitally important to bind these heavy metals. An increase in the pH value in the soil is also relevant for the speciation of a particular element. This can not only lead to immobilization of cationic heavy metals, such as cadmium, but can also result in mobilization of anionic metalloids, such as arsenic. Organic pollutants, such as PAHs, or pesticides, are predominantly bound to the aromatic carbon structure of biochar. The number of available binding sites depends heavily on the accessible specific surface area of the biochars. The behavior of uncharged organic pollutants is easier to estimate than it is for ionizable or permanently charged substances, where electrostatic interactions and pH also play an important role (Sigmund et al., 2018; Chaudhuri et al., 2024).

In addition to porosity, which enables a large internal surface area, negative surface charge and resistance to degradation are important properties that contribute to the suitability of biochar in soil remediation. While the surface of biochars produced at low pyrolysis temperatures (300–450°C) are rich in oxygen-containing functional groups (e.g. hydroxyl, carbonyl, carboxyl or phenolic groups) which facilitate the complexation and binding of cationic substances, at higher heating temperature (450–700°C) functional groups are lost and direct π-π electron or hydrophobic interactions between biochar surface and organic contaminants predominate (Sizmur et al., 2016; Xiao and Chen, 2016). Natural aging processes, post-pyrolysis treatments, atmospheric oxygen or microbial attacks can gradually generate additional functional groups in biochars that are prepared at high temperatures (Hilber et al., 2017).

6 Biochar: mechanisms of action in immobilizing trace elements in soil

The negative charges of functional groups on the surface of biochar, produced at low-to-medium temperatures, support the immobilization of cationic trace elements and metalloids through electrostatic attraction and chelation. Depending on the element in question, ion exchange and co-precipitation also contribute to the adsorption of metals. Electrostatic adsorption is related to surface charge, ion exchange is controlled by ash content and co-precipitation is related to alkaline groups and pH (Wang et al., 2018, 2022). Higher ash content is usually associated with a higher pH, which supports precipitation and immobilization of trace elements. High cation exchange capacity, large specific surface area and long residence times in soil are useful properties of biochars, which sorb elements from the surrounding soil pore water (Hilber et al., 2017).

Particular attention must be paid to elements like Cu, which form strong complexes with dissolved organic carbon (DOC). When biochar or biochar combinations with compost contribute to the DOC in the soil pore water, the Cu concentration in the pore water may increase temporarily but will decrease later (Wang et al., 2018; Soja et al., 2018). Cation immobilization is especially important for trace elements like Cd and Tl, which are more mobile in soil than other elements, to prevent their accumulation in crops or their migration into groundwater (Lu et al., 2022; Luo et al., 2020; Pipíska et al., 2017). Soil contamination with Hg can be particularly tricky because Hg is often transformed to the more toxic and bioavailable form methyl-Hg. Although biochar reduces the mobility of inorganic Hg and lowers the bioavailability and plant uptake of methyl-Hg, the roles of Fe and sulfate in soil, where they interact with the formation of methyl-Hg, and biochar need to be considered and still requires further study (Yang et al., 2021a).

The availability of field experiments to study immobilization effects of biochar over several years has confirmed that the sorption effects are effective for 2–3 years, and according to certain results, some are effective for 5 years (Wang et al., 2021). However, when severe changes in abiotic factors like flooding, acid rain or physico-chemical soil conditions occur, the immobilization effect of biochar may diminish earlier, requiring re-application and monitoring of remediation success (Taraqqi-A-Kamal et al., 2021).

The reduction of bioavailable metal concentrations in soil benefits soil microorganisms and plant roots as well as preventing toxic effects. In addition to the immobilization effect, biochar can also alleviate plant stress by trace metals, by activating the production of antioxidant enzymes, stress proteins and proline in the plants (Haider et al., 2022a). Biochar facilitates electron transfer between microbes, contaminants and soil organic matter, and supports the transformations of trace elements and the degradation of organic contaminants. The electron shuttle enhances both the interspecific electron transfer between microbial cells and the extracellular electron transfer between microbial cells and minerals; trace elements and organic matter, microorganisms and root exudates (Fan et al., 2023). This can facilitate the transformation of toxic element species to less toxic forms, e.g. from Cr(VI) to Cr(III) and from As(III) to As(V) (Zhu et al., 2017). Pure, unmodified biochar can even mobilize anionic trace elements instead of immobilizing them. In such cases, the production of 'designer biochars' is possible through modifications or activations, e.g. by incorporating Fe- or Mg-bridges on the surfaces, the sorption capacity of anions such as phosphate or of elements such as As or Sb is improved (Fristák et al., 2018; Fristák et al., 2021; Silvani et al., 2019). The risk of releasing toxic substances from biochar into the soil when adding biochar as remediation agent (He et al., 2019; Cheng et al., 2020) can be minimized if biochar production and the concentration of potential contaminants comply with quality guidelines and contaminant thresholds of the EBC (EBC, 2012–2023).

The specific site situations in contaminated soil often make it difficult to decide on the most suitable type of biochar for remediation. The simultaneous occurrence of several contaminating trace elements, the local physico-chemical and biological situation in the soil and the specific remediation goal must be considered. Decisions must be made regarding biomass feedstocks, modifications, mineral additives and pyrolysis conditions. The main modifications of feedstock and biochar can include magnetization, activation, the co-pyrolysis of multiple feedstocks; the production of composites and nanomaterial combinations; and the combinations with other natural or processed polymers or minerals (Liu et al., 2022a). The activation and modification treatments of biochar can multiply the adsorption capacity of pristine biochar (Boraah et al., 2022). When examining the removal efficiency of different biochars for Cr, Pb and As (Berslin et al., 2022), observations showed that the highest adsorption for each of the three elements was achieved by a different biochar. Mining soils are examples for completely different remediation requirements than agricultural soils: frequently the soil biology is severely damaged and the processes involved in ore processing or the recovery of rare earth elements affect soil texture, pH value, organic components and the bioavailability of nutrients and trace elements, requiring adapted remediation approaches (Gao et al., 2022; Chandra et al., 2023; Chen et al., 2023; Miu et al., 2022). After successful immobilization of potentially toxic elements (PTEs), the question remains as to their further whereabouts. Wu et al. (2024) advocates for the application of remediation techniques, including the recovery and regeneration from the remediated soil and a subsequent removal of the PTEs.

7 Biochar: mechanisms of action in immobilizing organic pollutants in soil

Organic pollutants in soil come in a broad range of varieties: they can be small molecules like phenols, or large molecules like petroleum hydrocarbons; aliphatic or aromatic; neutral or charged; more or less hydrophobic; and their toxicity and bioavailability can be high or low. Consequently, the sorption behavior and capacity of a given biochar for different organic pollutants differs by several orders of magnitude (Xiao and Chen, 2016). Accordingly, different sorption mechanisms play a role for the binding of organic pollutants to biochar. Pore filling, hydrogen bonding and single- or multi-layer adsorption of nonpolar pollutants dominates the sorption to biochars, which are produced at higher pyrolysis temperatures (500–750°C). Ionizable pollutants can also bind to oxygen-containing groups of biochars with lower pyrolysis temperatures via H-bonds. Hydration layers of metals and mineral contents of biochar can hinder pollutant sorption and competition from other organic pollutants.

For many hydrophobic organic contaminants, π-π electron-donor acceptor interactions are the dominant sorption mechanism. This is one of the most important noncovalent interaction mechanisms for π-conjugated materials (such as biochar) with molecules with π-structures in a benzene or aromatic ring (Xiao and Chen, 2016; Ahmed et al., 2018; Cheng et al., 2022). The versatility of pollutant sorption mechanisms in biochar may be key to the wide range of pollutants reported to benefit from soil remediation with biochar (Haider et al., 2022b).

The strong sorption of contaminants like PAH onto biochar has raised concerns as to whether this would also decrease bioavailability and thus the success of biodegradation in remediation projects. In fact, Bianco et al. (2021) have observed a decrease in PAH degradation during bioaugmentation and phytoremediation of sediments with a 5% addition of biochar. However, this decrease was more than compensated if nitrate was used as an electron acceptor, or if Fe^{2+} was co-applied for persulfate activation as an additional support for oxidative degradation. Also Liu et al. (2022b) emphasizes the acceleration of organic pollutant degradation when biochar is combined with persulfate to improve advances in oxidation processes.

8 Engineering biochars for specific remediation purposes

Advantages of biochar composites with hydrogels or digestates to improve bioaugmentation success were described by Anae et al. (2021) when soils were contaminated with mixtures of trace elements and PAHs. The immobilization of trace elements was important to increase the remediation effectiveness for the organic contaminant fraction. Such a complex mixture of pollutants requires mixtures of biochars with different properties, engineered biochars or combinations with other remediation additives for *in-situ* remediation (Yang et al., 2021b). The combinations must consider the requirements of the biodegrading microorganism consortia so that the addition of nutrients and/or vitamins can act as bio-stimulant and raise the degradation efficiency of biochar, which acts as an electron mediator to promote electron transfer (Dike et al., 2021; Zheng et al., 2022).

Among the various ways to engineer biochars for specific remediation purposes, the use of algal-based biochar or biochar from other aquatic biomasses has shown promising results (Singh et al., 2021; Verma et al., 2024). Modified biochar has been shown to be more effective than pristine biochar when soils contaminated with petroleum hydrocarbons needed to be remediated (Wei et al., 2024). Doping of biochar with hetero-atoms and various physical and chemical activation methods have been recommended by Katiyar et al. (2022) to increase the adsorption of antibiotics to biochar surfaces. Iron-modification of biochar with nano zero-valent iron has been shown to improve the remediation efficiency in soils contaminated by antibiotics, pesticides, dyes or PAHs (Chen et al., 2024). For the removal (or immobilization) of antibiotics and antibiotic-resistance genes, biochars produced from anaerobic digestates or as gasification char have also been described as promising sorbents (Fu et al., 2024; Rizwan et al., 2023).

Pan et al. also describe various activation and modification methods (2022) to improve pesticide sorption capacity. However, attention must be paid to why the pesticides are in the soil, e.g. whether they are soil herbicides, higher pesticide dosages may be required to achieve the desired effectiveness (Graber et al., 2012). However, if the pesticides are unwanted residues from previous applications, biochar could be an important means of minimizing the impact on the next generation of crop plants or microorganisms in the same field. This is not only owing to the adsorption capacity of biochar itself but also to biochar's performance as a vehicle to immobilize microbes used for pesticide contamination remediation (Yang et al., 2024).

One class of contaminants that has received particular attention in recent years are the PFASs owing to their extremely long residence times in soil. But successful applications of biochar have also been observed for these contaminants. Sun et al. (2023) reported the restoration of the bacterial community structure and function disrupted by PFAS contamination. Promising PFAS adsorption results were reported by (Rodrigo et al., 2022; Silvani et al., 2019) when gasification chars or Fe-modified biochars were used for a range of different PFAS. Both the soil type and the physico-chemical properties of each PFAS-congener influences the respective sorption and desorption behavior of biochar in contaminated soils (Askeland et al., 2020). Improved sorption performance of steam- or CO_2-activated biochars for various PFAS has also been observed (Sormo et al., 2021). For application in saturated soil systems, such as those in the groundwater zone, the production of biochar-alginate beads, using rice straw-derived biochar, showed promising results (Militao et al., 2023). Promising bioremediation treatments for PFAS have recently been reported, and bacteria with degradation ability for PFAS like *Acidimicrobium* sp. and *Pseudomonas parafulva* have been found (Shittu et al., 2023; Silva et al., 2022). Successful PFAS bioremediations involving biochar and competent microorganisms can be expected for the future.

9 Combining amendments for soil remediation: clay minerals, sewage sludge and chitosan

The numerous benefits of the two key soil (bio)remediation amendments, compost and biochar, suggest that a combination of these two main ingredients should result in even greater benefits than if they were applied separately. In fact,

Wu et al. (2017) describe numerous changes in physico-chemical properties, microbiological characteristics, gas emissions, organic matter composition, nutrient composition and cation exchange capacity when biochar is added to the composting process. To prevent suboptimal results when combining these two ingredients, Kammann et al. (2016) recommend following basic rules for producing a valuable biochar compost:

- The feedstock mixture should have a C/N-ratio of about 35;
- Addition of about 5% clay and about 1% rock powder;
- Avoidance of anaerobic conditions;
- Windrow piles should not exceed a diameter of 3 m and should be turned at least every 3 days during the heat rotting phase for oxygenation; and
- The materials should have low bulk density to facilitate aeration and liquid absorption, and the water content should be around 40% (Kammann et al., 2016).

If composts are enriched with clay minerals, sewage sludge composts can also support the immobilization of trace elements in contaminated soils and reduce the plant concentrations (van Herwijnen et al., 2007). A reduction in bioavailable trace metal concentrations is also reported by Sabir et al. (2015) when investigating the remediation effectiveness of manure and compost, but the results differed depending on the trace element examined.

A further promising amendment for soil remediation is chitosan. It is produced from chitin, one of the most common polysaccharides found in nature, which can be obtained from the exoskeletons of crustaceans, mollusks or insects, as well as from the cell walls of certain algae and fungi. Chitosan is the N-deacetylated form of chitin and can be produced by the chemical or enzymatic processing of chitin (El-Araby et al., 2024). Chitosan and its derivatives are used in a variety of ways in the food industry; pharmaceuticals and cosmetics; and agriculture, medicine, and environmental technology. In the latter area of application, the good binding efficiency of cationic trace elements to chitosan has led to the use of chitosan for wastewater cleaning or soil remediation (Kamari et al., 2011). By further modifying chitosan with Mo, Mg, Fe or phosphates, the immobilization potential and cation exchange capacity could be further improved (Tripathi et al., 2017; Li et al., 2024). Complexes of chitosan with FeS increased the reduction efficiency for Cr(VI), thereby reducing the toxicity to soil microorganisms (Shang et al., 2023).

Given the success of using biochar, compost and chitosan individually for soil remediation, it is not surprising that products have been developed that contain all three components together to exploit synergistic effects. Liu et al. (2023) report improved trace metal immobilization, nutrient availability and plant growth after the application of such a composite product. Organic contaminants such as PAH can also be combated oxidatively with chitosan in the form of a composite with urea and persulfate (Tang et al., 2023). These recent developments show that tailored remediation product development by modifying and combining basic materials like biochar, compost and chitosan also supports microbiological contaminant degraders and can deliver the best results when remediating contaminated soil and making it productive again.

10 Conclusion and future trends

The number of contaminated or degraded sites is increasing worldwide, at the same time the rapidly growing global population, which is dependent on a vast and adequate food supply. Therefore, every effort should be made to make contaminated agricultural land fertile again and restore their productivity. The physical, chemical and biological toolbox of innovative remediation tools has already reached an impressive size; yet all too often, 'dig and dump' methods are preferred over new physico-chemical and biological remediation methods. Future research should aim in two directions:

- Transferring knowledge about innovative remediation methods in authorities and engineering companies.
- Faster development of tailor-made solutions for site and pollutant mixture-specific remediation tasks.

The first research direction is a task for science communicators, science policymakers and education institutions. On one hand, practitioners need to understand and be aware of the contamination risks associated with certain crop and soil management practices. On the other hand, authorities should be open to new developments in remediation options, provided they are based on science and experience, even if there are few or no case studies using the same technology. If a new (bio)remediation technology can demonstrate its contaminant removal efficiency in a first step, under laboratory conditions, and in a second step, under confined protected field conditions – the hurdles for the third step of full-scale remediation should not be set too high. This is because the need to develop a site-specific solution to a complex mixture of contaminants under difficult subsurface conditions often leads to a novel technological remediation process, that is not necessarily comparable to previous treatments and results achieved elsewhere. The development of guidelines and multi-criteria decision trees should enable a rapid selection and screening process to adapt the selected (bio)remediation technique to the local situation and achieve the best possible degradation and/or detoxification success. Such a framework will enable authorities to approve not only safe but also most advanced remediation techniques, leading to a greater availability of decontaminated land for agricultural production.

11 Where to look for further information

- Zhang, C.L. (2019) *Soil and Groundwater Remediation: Fundamentals, Practices, and Sustainability*. Wiley.
- De Albergaria, J.T.V.S. and Nouws, H.P.A. (eds.) (2021) *Soil Remediation: Applications and New Technologies*. CRC Press.
- Parray, J.A., et al. (eds.) (2021) *Soil Bioremediation: An Approach Towards Sustainable Technology*. Wiley Blackwell.
- Condori Apaza, R.M., et al. (2022) *Soil Pollution: Soil, Contamination and Soil Bioremediation*. Our Knowledge Publishing.
- Sanchez-Hernandez, C. (ed.) (2021) *Bioremediation of Agricultural Soils*. CRC Press.

12 References

Ahmed, M.B., et al. (2018) Sorption of hydrophobic organic contaminants on functionalized biochar: Protagonist role of π-π electron-donor-acceptor interactions and hydrogen bonds. *Journal of Hazardous Materials* 360, 270–278.

Alvarenga, P., et al. (2020) Protecting agricultural soils from contamination. Minipaper 1: Agricultural sources of soil contamination. EIP-AGRI, https://ec.europa.eu/eip/agriculture/sites/default/files/fg37_mp1_sources_of_contamination_in_agricultural_soils_final.pdf (accessed 2024-02-28).

Anae, J., et al. (2021) Recent advances in biochar engineering for soil contaminated with complex chemical mixtures: Remediation strategies and future perspectives. *Science of the Total Environment* 767, 144351.

Anikwe, M.A.N. and Ife, K. (2023) The role of soil ecosystem services in the circular bioeconomy. *Frontiers in Soil Science* 3, 1209100

Askeland, M., et al. (2020) Biochar sorption of PFOS, PFOA, PFHxS and PFHxA in two soils with contrasting texture. *Chemosphere* 249, 126072.

Augusto, S., et al. (2010) Lichens as an integrating tool for monitoring PAH atmospheric deposition: a comparison with soil, air and pine needles. *Environmental Pollution* 158(2), 483–489.

Beiyuan, J., et al. (2018) Interactions of food waste compost with metals and metal-chelant complexes during soil remediation. *Journal of Cleaner Production* 192: 199–206.

Bernal, M.P. (2017) Grand challenges in waste management in agroecosystems. *Frontiers in Sustainable Food Systems* 1. https://doi.org/10.3389/fsufs.2017.00001

Bernal, M.P., et al. (2020) Protecting agricultural soils from contamination. Minipaper 3: Biological remediation of contaminated agricultural soils. https://ec.europa.eu/eip/agriculture/sites/default/files/fg37_mp3_remediation_final.pdf (accessed 2024-02-28).

Berslin, D., et al. (2022) Remediation of emerging metal pollutants using environment friendly biochar- Review on applications and mechanism. *Chemosphere* 290, 133384.

Bianco, F., et al. (2021) The addition of biochar as a sustainable strategy for the remediation of PAH-contaminated sediments. *Chemosphere* 263, 128274.

Boraah, N., et al. (2022) Attributes of wood biochar as an efficient adsorbent for remediating heavy metals and emerging contaminants from water: a critical review and bibliometric analysis. *Journal of Environmental Chemical Engineering* 10(3), 107825.

Carne, G., et al. (2021) Mass balance approach to assess the impact of cadmium decrease in mineral phosphate fertilizers on health risk: the case-study of French agricultural soils. *Science of the Total Environment* 760, 143374.

Cekmecelioglu, D., et al. (2005) Feedstock optimization of in-vessel food waste composting systems for inactivation of pathogenic microorganisms. *Journal of Food Protection* 68(3), 589–596.

Chandra, S., et al. (2023) The role of modified biochar for the remediation of coal mining-impacted contaminated soil: a review. *Sustainability* 15(5), 3973.

Chang, Y.C., et al. (2022) The effect of different in situ chemical oxidation (ISCO) technologies on the survival of indigenous microbes and the remediation of petroleum hydrocarbon-contaminated soil. *Process Safety and Environmental Protection* 163, 105–115.

Chaudhuri, S., et al. (2024) The efficacy of Pb, As(v) and Sb(III) removal by biochar is determined by solution chemistry. *Environmental Science-Water Research & Technology*. 10, 912–921, https://doi.org/10.1039/d3ew00726j

Chen, A.Q., et al. (2024) Applications and synergistic degradation mechanisms of nZVI-modified biochar for the remediation of organic polluted soil and water: a review. *Science of the Total Environment* 911, 168548.

Chen, H.M., et al. (2023) Application of biochar for ion-adsorption of rare earth contaminated soil remediation: a review. *Sustainability* 15(10), 7934.

Chen, L.G., et al. (2017) Sources, atmospheric transport and deposition mechanism of organochlorine pesticides in soils of the Tibetan Plateau. *Science of the Total Environment* 577, 405–412.

Cheng, S., et al. (2020) Application research of biochar for the remediation of soil heavy metals contamination: a review. *Molecules* 25(14), 3167.

Cheng, Y.Z., et al. (2022) Preparation of novel N-doped biochar and its high adsorption capacity for atrazine based on π-π electron donor-acceptor interaction. *Journal of Hazardous Materials* 432, 128757.

Cornet, P. and Weber, B. (2004) Water reuse for irrigation from waste water treatment plants with seasonal varied operation modes. *Water Science and Technology* 50(2), 47–53.

Dados, A., et al. (2015) Rapid remediation of soil heavily contaminated with hydrocarbons: a comparison of different approaches. *Annals of Microbiology* 65(1), 241–251.

Dike, C.C., et al. (2021) Can biochar be an effective and reliable biostimulating agent for the remediation of hydrocarbon-contaminated soils? *Environment International* 154, 106553.

Dondi, D., et al. (2016) Heterogeneous photochemistry: solar energy conversion and environmental remediation. *International Journal of Photoenergy* 2016, 2083696.

EBC (2012–2023) *European Biochar Certificate - Guidelines for a Sustainable Production of Biochar*. Carbon Standards International (CSI), Frick, Switzerland. (http://european-biochar.org). Version 10.3 from 5th April 2022.

EBI European Bbiochar Industries Consortium (2023) Market Report 2022/2023. http://www.biochar-industry.com/market-overview/ (accessed 2024-03-01)

EEA European Environment Agency: Soil and United Nations Sustainable Development Goals. https://www.eea.europa.eu/signals-archived/signals-2019-content-list/infographics/soil-and-united-nations-sustainable/view (accessed 2024-02-27).

El-Araby, A., et al. (2024) Chitosan, chitosan derivatives, and chitosan-based nanocomposites: eco-friendly materials for advanced applications (a review). *Frontiers in Chemistry* 11, 1327426.

EU Soil Observatory: EUSO Soil Health Dashboard. https://esdac.jrc.ec.europa.eu/esdacviewer/euso-dashboard/ (accessed 2024-02-27).

Fan, C.F., et al. (2023) A critical review of the interactions between rhizosphere and biochar during the remediation of metal(loid) contaminated soils. *Biochar* 5(1), 87.

FAO Food and Agriculture Organization of the United Nations: IPS Soil Letters 1/2020. https://www.fao.org/3/cb1110en/cb1110en.pdf (accessed 2024-02-27).

Fristák, V., et al. (2018) Sorption separation of Eu and As from single-component systems by Fe-modified biochar: kinetic and equilibrium study (vol 14, pg 521, 2017). *Journal of the Iranian Chemical Society* 15(2), 511–511.

Fristák, V., et al. (2021) Preparation and characterization of novel magnesium composite/walnut shells-derived biochar for As and P sorption from aqueous solutions. *Agriculture-Basel* 11, 714.

Fu, Z., et al. (2024) A comprehensive review on the preparation of biochar from digestate sources and its application in environmental pollution remediation. *Science of the Total Environment* 912, 168822.

Gao, Y.N., et al. (2022) Biochar as a potential strategy for remediation of contaminated mining soils: mechanisms, applications, and future perspectives. *Journal of Environmental Management* 313, 114973.

Graber, E.R., et al. (2012) High surface area biochar negatively impacts herbicide efficacy. *Plant and Soil* 353(1–2), 95–106.

Gronflaten, L.K., et al. (2005) Influence of liming and vitality fertilization on trace element concentrations in Scots pine forest soil and plants. *Forest Ecology and Management* 213, 261–272.

Grundy, A.C., et al. (1998) The effect of temperature on the viability of weed seeds in compost. *Compost Science & Utilization* 6(3), 26–33.

Haider, F.U., et al. (2022a) Biochar application for the remediation of trace metals in contaminated soils: Implications for stress tolerance and crop production. *Ecotoxicology and Environmental Safety* 230, 113165.

Haider, F.U., et al. (2022b) Biochar application for remediation of organic toxic pollutants in contaminated soils; An update. *Ecotoxicology and Environmental Safety* 248, 114322.

Haider, G., et al. (2020) Mineral nitrogen captured in field-aged biochar is plant-available. *Scientific Reports* 10, 13816.

He, L.Z., et al. (2019) Remediation of heavy metal contaminated soils by biochar: mechanisms, potential risks and applications in China. *Environmental Pollution* 252, 846–855.

He, S.Y., et al. (2022) Interactions of microplastics and soil pollutants in soil-plant systems. *Environmental Pollution* 315, 120357.

Hilber, I., et al. (2017) The different faces of biochar: contamination risk versus remediation tool. *Journal of Environmental Engineering and Landscape Management* 25(2), 86–104.

Hosseinian, A., et al. (2023) Phosphorus recovery methods from secondary resources, assessment of overall benefits and barriers with focus on the Nordic countries. *Journal of Material Cycles and Waste Management* 25(5), 3104–3116.

Jeffery, S., et al. (2017) Biochar boosts tropical but not temperate crop yields. *Environmental Research Letters* 12, 053001.

Kamari, A., et al. (2011) Binding of heavy metal contaminants onto chitosans: an evaluation for remediation of metal contaminated soil and water. *Journal of Environmental Management* 92(10), 2675–2682.

Kammann, C., et al. (2016) Combining Biochar and Organic Amendments. In: Shackley, S., et al. (eds.), *Biochar in European Soils and Agriculture*. Earthscan.

Katiyar, R., et al. (2022) Efficient remediation of antibiotic pollutants from the environment by innovative biochar: current updates and prospects. *Bioengineered* 13(6), 14730–14748.

Krahn, K.M., et al. (2023) Sewage sludge biochars as effective PFAS-sorbents. *Journal of Harzardous Materials* 445, 130449.

Krembs, F.J., et al. (2010) ISCO for groundwater remediation: analysis of field applications and performance. *Ground Water Monitoring and Remediation* 30(4), 42–53.

Kwon, J.W. (2011) Mobility of veterinary drugs in soil with application of manure compost. *Bulletin of Environmental Contamination and Toxicology* 87(1), 40–44.

Li, Y.T., et al. (2024) Cd-contaminated soil remediation using MgFe2O4/chitosan carbon aerogel beads. *Journal of Environmental Chemical Engineering* 12(1), 111918.

Li, X.N., et al. (2020) Biochar increases soil microbial biomass but has variable effects on microbial diversity: A meta-analysis. Sci Total Environ. 749, 141593.

Liu, M., et al. (2023) Modified biochar/humic substance/fertiliser compound soil conditioner for highly efficient improvement of soil fertility and heavy metals remediation in acidic soils. *Journal of Environmental Management* 325, 116614.

Liu, M.S., et al. (2022a) A critical review of biochar-based materials for the remediation of heavy metal contaminated environment: Applications and practical evaluations. *Science of the Total Environment* 806, 150531.

Liu, T.H., et al. (2022b) Applications and influencing factors of the biochar-persulfate based advanced oxidation processes for the remediation of groundwater and soil contaminated with organic compounds. *Science of the Total Environment* 836, 155421.

Liu, X., et al.. (2019) Impact of biochar application on yield-scaled greenhouse gas intensity: a meta-analysis. *Sci. Total Environ*. 656, 969–976.

Loukola-Ruskeeniemi, K., et al. (2022) Risk management for arsenic in agricultural soil-water systems: lessons learned from case studies in Europe. *Journal of Hazardous Materials* 424, 127677.

Lu, Y.C., et al. (2022) Efficient remediation of cadmium contamination in soil by functionalized biochar: recent advances, challenges, and future prospects. *Processes* 10(8), 1627.

Luo, Z.A.X., et al. (2020) Thallium contamination in agricultural soils and associated potential remediation via biochar utilization. *Biochar* 2(1), 33–46.

Majumder, S., et al. (2019) The impact of biochar on soil carbon sequestration: meta-analytical approach to evaluating environmental and economic advantages. *Journal of Environmental Management* 250, 109466.

Manyapu, V., et al. (2022) Role of psychrotrophic bacteria and cold-active enzymes in composting methods adopted in cold regions. *Advances in Applied Microbiology* 121, 1–26. https://doi.org/10.1016/bs.aambs.2022.08.001

Militao, I.M., et al. (2023) PFAS removal from water by adsorption with alginate-encapsulated plant albumin and rice straw-derived biochar. *Journal of Water Process Engineering* 53, 103616.

Miu, B.A., et al. (2022) Bringing life back into former mining sites: a mini-review on soil remediation using organic amendments. *Sustainability* 14(19), 12469.

Niroumand, H., et al. (2012) The performance of electrochemical remediation technologies in soil mechanics. *International Journal of Electrochemical Science* 7(6), 5708–5715.

Oron, G. and Demalach, J. (1987) Reuse of domestic waste-water for irrigation in arid zones: a case-study. *Water Resources Bulletin* 23(5), 777–783.

Pan, L.X., et al. (2022) Modified biochar as a more promising amendment agent for remediation of pesticide-contaminated soils: modification methods, mechanisms, applications, and future perspectives. *Applied Sciences* 12(22), 11544.

Pipíska, M., et al. (2017) Sorption separation of cobalt and cadmium by straw-derived biochar: a radiometric study. *Journal of Radioanalytical and Nuclear Chemistry* 311(1), 85–97.

Ren, X.Y., et al. (2018) The potential impact on the biodegradation of organic pollutants from composting technology for soil remediation. *Waste Management* 72, 138–149.

Rizwan, M., et al. (2023) Sustainable manufacture and application of biochar to improve soil properties and remediate soil contaminated with organic impurities: a systematic review. *Frontiers in Environmental Science* 11, 1277240.

Rodrigo, P.M., et al. (2022) Batch and fixed bed sorption of low to moderate concentrations of aqueous per- and poly-fluoroalkyl substances (PFAS) on Douglas fir biochar and its Fe_3O_4 hybrids. *Chemosphere* 308, 136155.

Ruzicková, J., et al. (2022) Fate of microplastics during composting and their leachability. *Sustainable Chemistry and Pharmacy* 30, 100867.

Rynk, R., et al. (2022) *Why Compost? The Composting Handbook*. Academic Press. https://doi.org/10.1016/B978-0-323-85602-7.00001-7

Sabir, M., et al. (2015) Contrasting Effects of Farmyard Manure (FYM) and compost for remediation of metal contaminated soil. *International Journal of Phytoremediation* 17(7), 613–621.

Sánchez-Castro, I., et al. (2023) Past, present and future trends in the remediation of heavy-metal contaminated soil - Remediation techniques applied in real soil-contamination events. *Heliyon* 9(6), e16692.

Schnug, E. and Lottermoser, B.G. (2013) Fertilizer-derived uranium and its threat to human health. *Environmental Science & Technology* 47(6), 2433–2434.

Schroder, J., et al. (1997) Measurement of atmospheric deposition of polychlorinated dibenzo-p-dioxins (PCDDs) and dibenzofurans (PCDFs) to a soil. *Atmospheric Environment* 31(18), 2983–2989.

Seo, Y. and Cho, K.S. (2021) Effects of plant and soil amendment on remediation performance and methane mitigation in petroleum-contaminated soil. *Journal of Microbiology and Biotechnology* 31(1), 104–114.

Shang, C., et al. (2023) Remediation of Cr(VI) contaminated soil by chitosan stabilized FeS composite and the changes in microorganism community. *Chemosphere* 327, 138517.

Shittu, A.R., et al. (2023) The effects of per- and polyfluoroalkyl substances on environmental and human microorganisms and their potential for bioremediation. *Arhiv Za Higijenu Rada I Toksikologiju-Archives of Industrial Hygiene and Toxicology* 74(3), 167–178.

Sigmund, G., et al. (2018) Influence of compost and biochar on microbial communities and the sorption/degradation of PAHs and NSO-substituted PAHs in contaminated soils. *Journal of Hazardous Materials* 345, 107–113.

Silva, A.R., et al. (2022) Bioremediation of Perfluoroalkyl Substances (PFAS) by anaerobic digestion: effect of PFAS on different trophic groups and methane production accelerated by carbon materials. *Molecules* 27(6), 1895.

Silvani, L., et al. (2019) Can biochar and designer biochar be used to remediate per- and polyfluorinated alkyl substances (PFAS) and lead and antimony contaminated soils? *Science of the Total Environment* 694, 133693.

Singh, A., et al. (2021) Engineered algal biochar for contaminant remediation and electrochemical applications. *Science of the Total Environment* 774, 145676.

Sizmur, T., et al. (2016) Application of Biochar for Soil Remediation. In: Guo, M., et al. (eds.), *Agricultural & Environmental Applications of Biochar. Advances & Barriers*. SSSA Special Publication 63, 295–324.

Soja, G., et al. (2018) Compost and biochar interactions with copper immobilisation in copper-enriched vineyard soils. *Applied Geochemistry* 88, 40–48.

Sormo, E., et al. (2021) Stabilization of PFAS-contaminated soil with activated biochar. *Science of the Total Environment* 763, 144034.

Sun, T., et al. (2023) Biochar remediation of PFOA contaminated soil decreased the microbial network complexity. *Journal of Environmental Chemical Engineering* 11(1), 109239.

Tang, X.J., et al. (2023) Efficient remediation of PAHs contaminated site soil using the novel slow-release oxidant material. *Chemical Engineering Journal* 472, 144713.

Taraqqi-A-Kamal, A., et al. (2021) Biochar remediation of soil: linking biochar production with function in heavy metal contaminated soils. *Plant Soil and Environment* 67(4), 183–201.

Tripathi, N., et al. (2017) Evaluation of modified chitosan for remediation of zinc contaminated soils. *Journal of Geochemical Exploration* 182, 180–184.

van Herwijnen, R., et al. (2007) Remediation of metal contaminated soil with mineral-amended composts. *Environmental Pollution* 150(3), 347–354.

Verma, M., et al (2024) Remediation of emerging pollutants using biochar derived from aquatic biomass for sustainable waste and pollution management: a review. *Journal of Chemical Technology and Biotechnology* 99, 330–342.

Wang, J., et al. (2021) Analysis of the long-term effectiveness of biochar immobilization remediation on heavy metal contaminated soil and the potential environmental factors weakening the remediation effect: a review. *Ecotoxicology and Environmental Safety* 207, 111261.

Wang, M.M., et al. (2018) Review on utilization of biochar for metal-contaminated soil and sediment remediation. *Journal of Environmental Sciences* 63, 156–173.

Wang, Y., et al. (2022) Remediation of heavy-metal-contaminated soils by biochar: a review. *Environmental Geotechnics* 9(3), 135–148.

Wang, Y.C., et al. (2019) A quantitative understanding of the role of co-composted biochar in plant growth using meta-analysis. *Sci. Total Environ.* 685, 741–752.

Wei, Z., et al. (2024) Biochar-based materials as remediation strategy in petroleum hydrocarbon-contaminated soil and water: Performances, mechanisms, and environmental impact. *Journal of Environmental Sciences* 138, 350–372.

Weil, R.R. and Brady, N.C. (2017) *The Nature and Properties of Soils*, 15th Global edition. Pearson.

Whalen, J.K., et al. (2021) Earthworm populations are stable in temperate agricultural soils receiving wood-based biochar. *Pedosphere* 31, 398–404.

Wu, H.P., et al. (2017) The interactions of composting and biochar and their implications for soil amendment and pollution remediation: a review. *Critical Reviews in Biotechnology* 37(6), 754–764.

Wu, Y., et al. (2024) Biochar application for the remediation of soil contaminated with potentially toxic elements: Current situation and challenges. *Journal of Environmental Management* 351, 119775.

Wu, Y.L., et al. (2009) Atmospheric deposition of polychlorinated dibenzo-p-dioxins and dibenzofurans on the soils in the vicinity of municipal solid waste incinerators. *Journal of Environmental Science and Health Part a-Toxic/Hazardous Substances & Environmental Engineering* 44(13), 1327–1334.

Xiao, X. and Chen, B.L. (2016) Interaction Mechanisms between Biochar and Organic Pollutants. In: Guo M, He Z and Uchimiya SM (eds.), Agricultural & Environmental Applications of Biochar. Advances & Barriers. SSSA Special Publication 63, 225–257.

Xu, S., et al. (2020) Changing soil carbon: influencing factors, sequestration strategy and research direction. *Carbon Balance Management*, 15, 2. https://doi.org/10.1186/s13021-020-0137-5

Yang, Q., et al (2021a) Remediation of mercury-contaminated soils and sediments using biochar: a critical review. *Biochar* 3(1), 23–35.

Yang, Y.Y., et al. (2021b) Application of biochar for the remediation of polluted sediments. *Journal of Hazardous Materials* 404, 124052.

Yang, Y., et al. (2024) Pesticide contamination remediation by biochar-immobilized microorganisms: a review. *International Journal of Environmental Science and Technology* 21, 2225–2238.

Ye, L.L., et al. (2020) Biochar effects on crop yields with and without fertilizer: a meta-analysis of field studies using separate controls. *Soil Use and Management* 36, 2–18.

Zhang, L.Y., et al. (2019) Biochar amendment effects on the activities of soil carbon, nitrogen, and phosphorus hydrolytic enzymes: a meta-analysis. *Environmental Science and Pollution Research*, 26, 22990–23001.

Zhang, Q., et al. (2020) Quantifying the effects of biochar application on greenhouse gas emissions from agricultural soils: a global meta-analysis. *Sustainability* 12, 3436.

Zhang, Y.Y., et al. (2018) Response of surface albedo and soil carbon dioxide fluxes to biochar amendment in farmland. *Journal of Soils Sediments* 18, 1590–1601.

Zheng, X.M., et al. (2022) The effects of biochar and its applications in the microbial remediation of contaminated soil: A review. *Journal of Hazardous Materials* 438, 129557.

Zhou, T., et al. (2024) Occurrence, fate, and remediation for per-and polyfluoroalkyl substances (PFAS) in sewage sludge: A comprehensive review. *Journal of Hazardous Materials* 466, 133637.

Zhu, X.M., et al. (2017) Effects and mechanisms of biochar-microbe interactions in soil improvement and pollution remediation: a review. *Environmental Pollution* 227, 98–115.

Chapter 10

Clay mineral-based additives for the in situ bioremediation of hydrocarbon-contaminated agricultural soils

Solmaz Bidast and Santosh Kumar Paul, University of Newcastle, Australia; Ravi Naidu, University of Newcastle, Australia and CRC for Contamination Assessment and Remediation of the Environment (crcCARE); and Bhabananda Biswas[1], University of Newcastle, Australia/ CRC for Contamination Assessment and Remediation of the Environment (crcCARE), Australia/University of Greifswald, Germany

1 Introduction

Agricultural soils can be contaminated with various compounds as a result of their both direct (point source) and indirect inputs (non-point source). Contaminated soils adversely affect soil productivity, natural ecosystems, food quality, and human health (Behera et al., 2021). The contamination in the agricultural landscapes has spread significantly in recent decades, driven by industrial development, military interests, mining activities, intensive agriculture, and natural disasters (Ambaye et al., 2022; Oerke, 2006). This contamination continues to spread as these drivers remain prevalent and, in some cases, are intensifying. A range of manufactured chemicals, such as toxic metal(loid)s, dyes, industrial waste, pesticides, fertilisers, and hydrocarbons, cause these contaminations (Khodabakhshloo et al., 2021; Naidu et al., 2021). Of all contaminants of concern, petroleum hydrocarbons (PHCs) rank highly due to their broad exposure pathways in soil and water, and severe toxicological impact on ecological receptors (Figure 1).

1 Corresponding author: Bhaba.Biswas@newcastle.edu.au (BB).

http://dx.doi.org/10.19103/AS.2025.0151.15

Figure 1 Sources of petroleum hydrocarbons (PHCs) in agricultural soils.

PHCs are a group of organic compounds derived from crude oil. They include alkanes, cycloalkanes, aromatic hydrocarbons, and other complex mixtures of partially refined and degraded hydrocarbons. They are the primary components of fuels such as gasoline, diesel, and jet fuel and are also found in lubricating oils, solvents, and many industrial products (Petrov, 2012). PHCs are used as raw materials for primary energy sources and various consumer products (Ambaye et al., 2022). These compounds are also used as additives in pesticide formulation to improve storage durability and pest control efficacy (Usman et al., 2022; Uddin et al., 2021; Paul and Naidu, 2022). Excessive and indiscriminate use of these pesticides and many other industrial applications of PHCs are major sources of soil and water contamination with various PHCs and associated toxic co-contaminants (Figure 1). Given the contamination of agricultural soils and associated risks to the food chain from PHCs, the remediation of PHCs is necessary. Various techniques, including physical, chemical, and biological approaches, are widely used to remediate PHCs-contaminated agricultural soils, each with its own prospects and limitations (Kuppusamy et al., 2017). The details of these techniques can be read elsewhere (Verma and Haritash, 2019; Kuppusamy et al., 2017). In this context, bioremediation offers cost-effective and environmentally friendly remediation of PHCs-contaminated areas (Patel et al., 2020). The use of microbial populations in bioremediation typically includes bacteria, yeast, algae, and fungi, whereas plants, including both crop and non-crop varieties, are also used. Despite the favourable aspect of microbe-assisted bioremediation, the challenge is that they may require additives to function more efficiently to access, extract, and degrade PHCs. The types and mode of application of these additives are a relatively new research area. Mineral-based additives are among them. In this chapter, we focus on clay mineral-based additives and their potential for bioremediation in agricultural soil.

2 Additives for enhanced bioremediation of PHCs

The process of using extra material or chemical substances to enhance bioremediation is commonly referred to as biostimulation. As part of this process, the use of additives is often necessary. In the case of agricultural soil, *in situ* bioremediation is a preferred method, considering that the soil fertility and food safety are not compromised as a result of the application of biostimulating additives (Liu et al., 2024). Therefore, direct use of soil-compatible minerals or mineral-assisted compounds can be effective and preferred additives for the application in contaminated agricultural soils.

2.1 Raw clay mineral as additives

Clay minerals are naturally occurring phyllosilicates, often deposited in the rock and subsurface soil environment. In soil, clay fractions are associated with the smallest size particles (<2 μm) in the formation of soil texture (Biswas et al., 2024; Warr, 2022). Clay minerals play a key role in the soil systems, including for abiotic and biotic functions (Churchman, 2018; Fomina and Skorochod, 2020). Therefore, these natural minerals are deemed to be useful additives, fulfilling the criteria of soil-compatible additives (Churchman and Velde, 2019). They can also be used for specific purposes, such as providing microbial support materials to improve bioremediation (Fomina and Skorochod, 2020; Biswas et al., 2015).

The utilisation of clay minerals to support hydrocarbon-degrading bacteria has primarily been investigated in small-scale experiments using pure cultures in aqueous media (Khodijah Chaerun et al., 2005). This has resulted in limited research on the potential benefits of using clay minerals as a soil amendment to enhance the bioremediation of contaminated agricultural soil. It is worth noting that the addition of clay minerals in contaminated agricultural soils alone may not be considered a one-stop solution for enhancing bioremediation. It requires careful consideration of understanding the complex interactions between hydrocarbons, clay minerals, and degrading microorganisms in soils (Biswas et al., 2015). The bioavailability of hydrocarbons plays a key role in determining whether they can be readily degraded by microbes or become sequestered in soil matrices. The latter may result in limited biodegradation of contaminants in the soils (McAllister and Semple, 2010; Semple et al., 2003).

The interplay between clay minerals and organic matter (OM) in soil has significant effect on hydrocarbons adsorption onto them, and that enables the breakdown of these compounds by bacteria and other biota (Khan et al., 2018). In this case, the rate of biodegradation can be influenced by clay additives. Several physical processes, including attachment, diffusion, and sorption, can enhance the breakdown of PHC compounds in soil (Figure 2). The presence of OM-clay association, particularly in agricultural soils, contributes to these processes. For example, the clay surface-bound PHCs can be accessible to the active degrader of PHCs. Conversely, the dispersal of PHCs into the micropores of the clay minerals and OM can be a slow process of degradation, largely from the inaccessibility of these organic compounds to the microbes. However, a suitable arrangement of clay particles in soil can promote microbes to form biofilm. Once a biofilm is established with the support of clay particles, microbes can initiate biodegradation pathways supported by their biosurfactants. This pathway is beneficial, as biosurfactants enhance the availability of PHCs in the system (Figure 2) (Biswas et al., 2015).

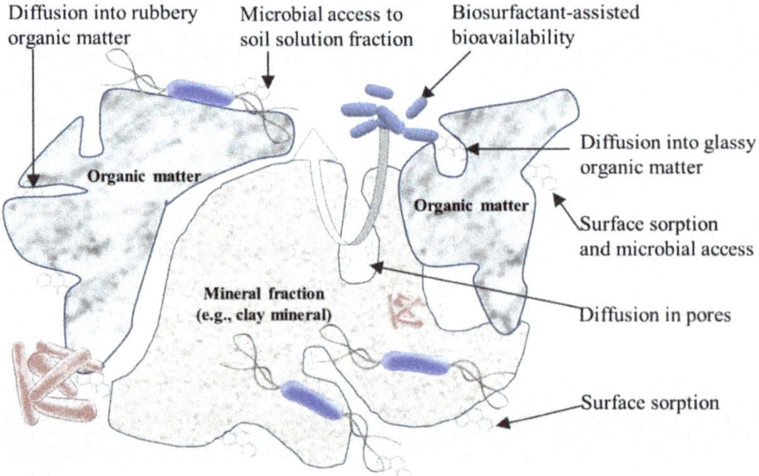

Figure 2 Summary of the physical behaviour of hydrocarbons in soils and the microbial degradation potentially in the presence of raw clay mineral's active surfaces. Schematic drawing is modified after Khan et al. (2018).

The effectiveness of clay minerals as biostimulants cannot be generalised for all types of clay minerals. Gan et al. (2022) used kaolinite and montmorillonite to test the biodegradation of benzo[a]pyrene (BaP) in colloidal solution at bench scale. They reported that the montmorillonite supported microbial growth and promoted BaP biodegradation, while kaolinite did the opposite. In the presence of kaolinite, the bioaccessibility of BaP decreased, resulting in poor degradation outcome. Conversely, montmorillonite's amphoteric functional groups created favourable conditions for microbial growth, allowing cells to attach to the mineral surfaces and form biofilms, thereby enhancing BaP degradation. Regardless of these multifaceted roles of raw clay minerals, they may require additional modifications using "fit-for-purpose" chemicals, such as chemical surfactants, which also serve as additives for bioremediation.

2.2 Surfactant-modified clay minerals

Due to their hydrophobic nature and low solubility in water, PHCs adhere to soil particles, especially those in the OM domain. This hinders microorganisms' access to them, resulting in poor biodegradation outcomes (Ben Ayed et al., 2014; Ling et al., 2023).

Surfactants enhance the solubility of PHCs, which facilitates their separation from soil particles and makes them more accessible to microorganisms for degradation (Staninska-Pięta et al., 2019; Mohanty and Mukherji, 2013). In this case, surfactant acts as an additive to the enhanced bioremediation. Nevertheless, application of surfactant directly to agricultural soils could potentially pose ecotoxicity risks, leading to unpredictable long-term bioremediation outcomes (Biswas et al., 2019). By using clay minerals, scientists have been aiming to deliver surfactant to the contaminated site, resulting in increased bioavailability of PHCs, minimised toxicity of surfactant, and improved microbial growth and functions.

This surfactant-treated clay is often known as "organoclay" to describe that the clay mineral has been converted into an organically modified material (Leyva-Ramos et al., 2021; de Paiva et al., 2008). Organoclays are deemed to be good adsorbents for a range of pollutants, including hydrocarbons in water and soil (Leyva-Ramos et al., 2021; de Paiva et al., 2008). Practical applications of organoclays have been reported for soil amendment to contain contaminants from leaching (Wiles et al., 2005; Brixie and Boyd, 1994). A recent review article discussed various surfactant-modified clays that have been studied for environmental remediation, particularly in removing heavy metals from soil and water (Qi et al., 2023). Many of these organoclays function primarily as adsorbents, and their use as bioremediation support materials remains limited. Reports describing the organoclay as a bioremediation additive in agricultural soil is limited or may not be publicly available. The use of organoclays helped minimise the surface and groundwater toxicity of herbicides in agricultural soil by containing them within the soil column (Gámiz et al., 2010). Yet, numerous studies on organoclay amendments have overlooked the ecotoxicity of the leached organoclay or associated surfactant, rendering their application in agricultural soils uncertain and inconclusive. However, while the biocompatibility of organoclay is a concern, Biswas et al. (2019) proposed that the loading of surfactant into the interlayer is a critical consideration (Figure 3). They suggested that using a surfactant equivalent to the clay mineral's cationic exchange capacity (CEC) or less than that was desirable to avoid leaching

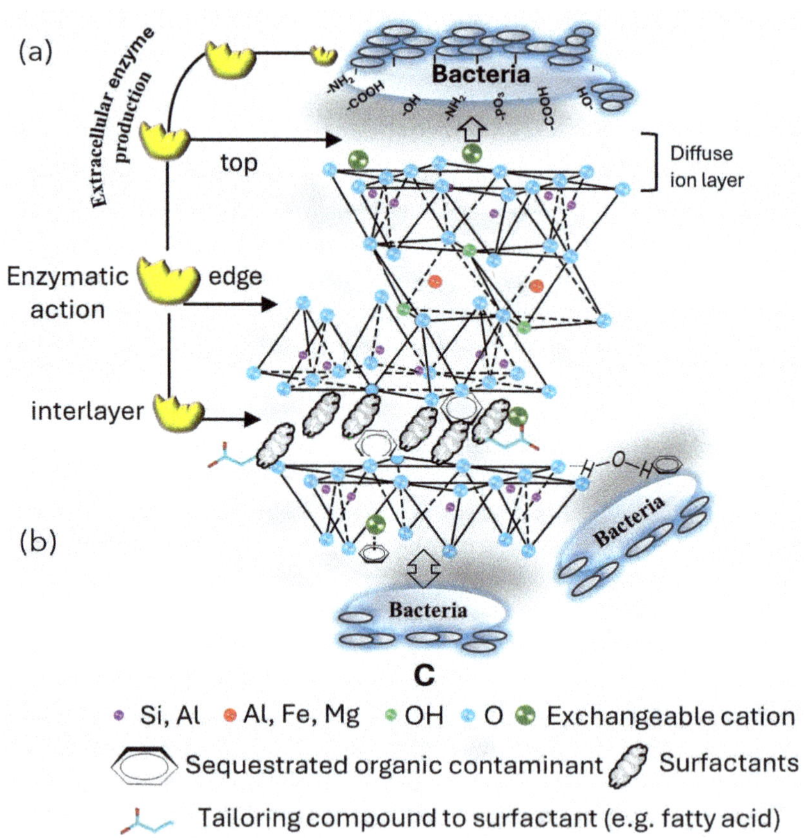

Figure 3 Mechanism of biocompatible (surface-tailored) organoclay and its application in contaminated soil. When bacteria form a hutch-like structure compatible with the modified clay, the extracellular enzymes produced by the bacterial biofilm can penetrate the interlayer. This process helps accessing the sequestrated organic contaminants and enhances degradation. A, B, and C represent the electrostatic attachment of bacteria via cation bridging, hydrogen bonding, and van der Waals interactions, respectively. This scheme is modified after Biswas et al. (2019). The sizes of the figure components (clay mineral structure, bacteria, and chemical compounds) in the sketch are not to scale.

of surfactant from the organoclay into the soil matrix. These authors further reported that the organoclay might need further modifications with secondary additives, such as palmitic acid to make the organoclay useful for remediating hydrocarbon-heavy metal co-contaminants. Fatty acids (e.g. palmitic acid), in this case, can immobilise heavy metals, thereby reducing their bioavailability in soil. As a result, microorganisms are exposed to a less toxic environment, which supports the degradation of hydrocarbons. Additionally, they can facilitate biofilm formation by utilising the clay particles in the soil system (Figure 3).

2.3 Clay-assisted biosurfactant and enzyme delivery

As an alternative to chemical surfactants described in section 2.2, researchers have explored the use of bio-based surfactants for similar purposes (Mulligan, 2021; Inobeme et al., 2021; Singh et al., 2021). However, the delivery of biosurfactants is relatively costly, which often leads to low acceptance of *in situ* applications in remediating contaminated agricultural soils. To address this issue, one effective strategy can be the use of clay mineral-based delivery systems for the biosurfactant molecules (Biswas et al., 2019). By using this approach, biosurfactants can be protected and maintained for long-term effectiveness in the soil. When considering the practicality of using biosurfactants as additives, it is more advantageous to employ bacteria that produce biosurfactants instead of directly using biosurfactant as a chemical substance. In addition, immobilising these bacteria onto clay minerals and delivering this composite to the remediating sites can be considered a sustainable approach. This sustainability is with the expectation that the active degraders, that are already present in the contaminated soil, can utilise clay-supported biosurfactant for a longer period (Purnomo et al., 2023; Biswas et al., 2017).

There is direct inoculation of bacteria or clay-supported bacteria as part of enhanced bioremediation of PHC-contaminated soil (Biswas et al., 2015). However, the discussion surrounding this has limited scope in this chapter. Rather, additives that can enhance the bioremediation capacity of the existing microbial community have been considered. Enzymes are an example of such additives. Enzymes are protein complexes that act as catalysts in chemical reactions by reducing activation energy and enhancing the rate of reaction. In bioremediation processes, enzymes serve as bio-oxidation catalysts that facilitate biotransformation reactions (Fan and Krishnamurthy, 1995; Schwarzenbach et al., 1993). However, biodegradation by existing microbial populations can be hindered by various rate-limiting factors such as the population density of specific microorganisms, temperature, pH, moisture, oxygen supply, and nutrient availability (Bollag, 1992; Fan and Krishnamurthy, 1995; Peixoto et al., 2011). Additionally, high concentrations of PHCs and their metabolites may inhibit microbial activity.

To address these challenges, the application of enzymes can be beneficial. Enzyme delivery in a more sustainable manner, preferably as a material-supported form such as clay-immobilised enzymes, can enhance bioremediation processes. Immobilised enzymes are more cost-effective compared to their free forms, as they exhibit greater stability in soil systems, thereby facilitating long-term application (Ruggiero et al., 1989; Sarkar et al., 1989; Peixoto et al., 2011). This stability helps overcome the adverse conditions that limit microbial biodegradation. As a result, immobilised enzymes represent a viable option for effective and sustainable bioremediation of PHC-contaminated agricultural soils.

2.4 Clay-based fertiliser and clay-biochar composite

Raw clay, including bentonite, can be included in the compost, and its application in soil can boost biodegradation of petroleum hydrocarbons (Qv et al., 2024). These authors reported that the bentonite and humic-acid modified bentonite added to the composting systems significantly reduced petroleum pollutant concentration by 79.09% and 79.67%, respectively. This degradation effect was almost double the control (i.e. only compost). Bentonite stimulated microbial enzyme production, modulated microbial community structure, increased nitrogen content, and enhanced biodegradation of petroleum hydrocarbons in the composting process (Qv et al., 2024). Indeed, Datta et al. (2020) also reported that bentonite, when combined with manure, enhanced soil microbial properties like microbial biomass carbon, dehydrogenase activity, and microbial community abundance. These changes resulted in enhanced soil health and quality. Bentonite's high cation exchange capacity helps in nutrient retention and availability in the soil, supporting microbial growth and function (Datta et al., 2020). This microbial growth and function is necessary for achieving enhanced *in situ* bioremediation of PHCs. Apart from these organic composites, other forms of clay-based organic amendments, such as those including biochar, are deemed to be highly relevant for direct use in agricultural soils.

Understanding the compatibility of biochar and clay in agricultural soils, their composite has recently emerged as a functional material that serves as an adsorbent, fertiliser, and biodegradation stimulant. Biochar has the potential to interact significantly with clay and microorganisms (Lin et al., 2023). Biochar produced through slow pyrolysis, which results in a larger specific surface area and more microporous structures, can provide shelters and facilitate the transport of nutrients for the microbial community. This, in turn, promotes the colonisation and immobilisation of microorganisms on the surface and within the pores of biochar (Anyika et al., 2015; Lin et al., 2023; Yang et al., 2022). In addition to providing soil nutrients, biochar has also showed potential in bioremediation processes. It supports microbial populations, releases contaminant-degrading enzymes, and protects beneficial microorganisms from the immediate toxicity of surrounding contaminants. Moreover, it promotes biofilm formation, which is essential for the enhanced bioremediation of hydrocarbons (Mukherjee et al., 2022; Sun et al., 2021; Zhou et al., 2021). Consequently, biochar can stimulate the activity of hydrocarbon-metabolising bacteria, leading to significant changes in the bacterial community structure and promoting the biodegradation of hydrocarbons in soil (Li et al., 2024). In these cases, when clay is incorporated into biochar, the immobilisation of functional microorganisms is promoted, which can enhance the degradation capacity of the composite as a whole (Sun et al., 2023). And that can be achieved through enhancing the surface properties of the composites. For example, clay-biochar composites are produced through various activation and functionalisation methods. These approaches enhance the physical, chemical, and biological properties of biochar, resulting in cost-effective products that combine the benefits of biochar and modified clay minerals (Panahi et al., 2020). These composites exhibit unique characteristics due to their high carbon content, multi-pore structure, and compatibility, incorporating the remarkable features of both biochar and clay minerals. Lin et al. (2022) found that the specific surface area of attapulgite-biochar composites can be nearly 1.68 times that of pristine biochar. The orderly or staggered coverage of attapulgite in a fibrous, needle-like structure on the biochar surface significantly increased its specific surface area. These composites have shown higher adsorption capacities for hydrocarbons compared to pristine biochar or clay minerals alone. For instance, the attapulgite-biochar composite exhibited an approximately 12% higher removal rate of petroleum hydrocarbons compared to biochar only, indicating improved adsorption performance due to the presence of dispersed attapulgite on the biochar surfaces (Lin et al., 2022). However, achieving unequivocal properties of this complex composite is not conclusive. One would argue that this is due to the varieties of biochar produced and the types of clay minerals added to it. For example, as opposed to the finding of Lin et al. (2022), Arif et al. (2022) reported a reduction in specific surface area of the clay-biochar composite. This reduction is because of the blockage of biochar pores by montmorillonite-type clay minerals. Therefore, there is a necessity for developing and optimising these composites for site-specific applications. In the case of bioremediation additives derived this biochar-clay composite, the desired surface properties that support the active microbial function are critical. These properties need to be studied extensively to ensure effective performance.

3 Cost-benefit analysis of clay-based additives

Clay minerals are cost-effective, widely available, and naturally occurring substances found in many regions globally. They have been successfully utilised for the remediation of various contaminants (Biswas et al., 2024). The cost-benefit analysis of certain products varies depending on the target applications. This includes considerations, such as application methods (bulk vs. small scale) and frequency of use. There is no assessment available for clay-based additives for PHC remediation in agricultural soils. However, one would assume the following scenario: a ton of kaolin clay may cost U$160, while that for bentonite U$100. If raw clay is suitable as a biodegradation stimulatant, and the application rate is 1% on top 20 cm of surface soil of 1 acre, the cost of raw clay would be ~US$1,000. The cost may go up when the clay needs to be activated or modified due to physicochemical processes. However, these additional expenses would be counterbalanced by the benefits, such as: (i) avoiding the costs associated with *ex situ* cleanup of PHCs, and (ii) by incorporating the clay-based additives with common agrichemicals like fertilisers. To estimate the cost accurately, we need to study the benefits and losses caused by the clay-based additives. Estimation of the cost without this study would be premature. For example, one way to reduce greenhouse gas (GHG) emissions from biochar is by combining it with clay materials (Hanafi et al., 2024), which can also serve as an additive for bioremediation (Mukherjee et al., 2022; Sun et al., 2021; Zhou et al., 2021). The cost-benefit assessment in this situation should consider the input cost for PHC cleanup and the sustainability benefits from desirable reduced GHG emissions.

4 Conclusion

The contamination of agricultural soils with PHCs presents serious environmental and health challenges. While bioremediation is a cost-effective and environmentally friendly solution, its effectiveness is limited by various factors, such as microbial population density, temperature, pH, moisture, oxygen supply, and nutrient availability.

This chapter explored the use of clay minerals and their composites as additives that can be biostimulants to enhance bioremediation. Raw clay minerals can support hydrocarbon-degrading bacteria, but their effectiveness depends on the specific interactions within the soil. Surfactant-modified clays, or organoclays, increase PHC bioavailability, but their synthesis must ensure that the immobilised surfactant is degraded or protected to avoid secondary pollution risks. There is scope for further research in this aspect. Biosurfactants and enzymes, delivered through clay-supported systems, provide promising biocatalysts for PHC degradation. These immobilised enzymes are more stable and cost-effective compared to their free counterparts. Clay-based organic fertilisers and clay-biochar composites, particularly relevant for agricultural applications, also show enhanced adsorption and biodegradation capacities.

However, in this case also, optimisation is needed to address overall constraints and benefits, including the fate of biochar and fertilisers in the soil system.

5 Acknowledgement

The authors thank crcCARE Australia and its partner organisations for their funding support. BB acknowledges the Alexander von Humboldt Foundation for financing his modified clay project at the University of Greifswald, Germany (https://www.humboldt-foundation.de/en/).

6 References

Ambaye, T.G., et al. 2022. Remediation of soil polluted with petroleum hydrocarbons and its reuse for agriculture: Recent progress, challenges, and perspectives. *Chemosphere* 293, 133572.

Anyika, C., et al. 2015. The impact of biochars on sorption and biodegradation of polycyclic aromatic hydrocarbons in soils–a review. *Environmental Science and Pollution Research* 22, 3314–3341.

Arif, M., et al. 2022. Carbon dioxide activated biochar-clay mineral composite efficiently removes ciprofloxacin from contaminated water-Reveals an incubation study. *Journal of Cleaner Production* 332, 130079.

Behera, I.D., et al. 2021. Enhanced biodegradation of total petroleum hydrocarbons by implementing a novel two-step bioaugmentation strategy using indigenous bacterial consortium. *Journal of Environmental Management* 292, 112746.

Ben Ayed, H., et al. 2014. Characterization and stability of biosurfactant produced by Bacillus mojavensis A21 and its application in enhancing solubility of hydrocarbon. *Journal of Chemical Technology & Biotechnology* 89, 1007–1014.

Biswas, B., et al. 2015. Bioremediation of PAHs and VOCs: Advances in clay mineral–microbial interaction. *Environment International* 85, 168–181.

Biswas, B., et al. 2017. Structural changes in smectite due to interaction with a biosurfactant-producing bacterium Pseudoxanthomonas kaohsiungensis. *Applied Clay Science* 136, 51–57.

Biswas, B., et al. 2019. Biocompatible functionalisation of nanoclays for improved environmental remediation. *Chemical Society Reviews* 48, 3740–3770.

Biswas, B., et al. 2024. The multidisciplinary science of applied clay research: a 2021–2023 bibliographic analysis. *Applied Clay Science* 258, 107471.

Bollag, J.M. 1992. Decontaminating soil with enzymes. *Environmental Science & Technology* 26, 1876–1881.

Brixie, J.M. and Boyd, S.A. 1994. Treatment of contaminated soils with organoclays to reduce leachable pentachlorophenol. *Journal of Environmental Quality* 23, 1283–1290.

Churchman, G.J. 2018. Game changer in soil science. Functional role of clay minerals in soil. *Journal of Plant Nutrition and Soil Science* 181, 99–103.

Churchman, G.J. and Velde, B. 2019. *Soil Clays: Linking Geology, Biology, Agriculture, and the Environment*. CRC Press.

Datta, R., et al. 2020. Bentonite-Based Organic Amendment Enriches Microbial Activity in Agricultural Soils, Land. 10.3390/land9080258.

de Paiva, L.B., et al. 2008. Organoclays: properties, preparation and applications. *Applied Clay Science* 42, 8–24.

Fan, C.-Y. and Krishnamurthy, S. 1995. Enzymes for enhancing bioremediation of petroleum-contaminated soils: a brief review. *Journal of the Air & Waste Management Association* 45, 453–460.

Fomina, M. and Skorochod, I. 2020. Microbial Interaction with Clay Minerals and Its Environmental and Biotechnological Implications, Minerals. 10.3390/min10100861.

Gámiz, B., et al. 2010. Organoclays as soil amendments to increase the efficacy and reduce the environmental impact of the herbicide fluometuron in agricultural soils. *Journal of Agricultural and Food Chemistry* 58, 7893–7901.

Gan, X., et al. 2022. Influence of kaolinite and montmorillonite on benzo[a]pyrene biodegradation by Paracoccus aminovorans HPD-2 and the underlying interface interaction mechanisms. *Pedosphere* 32, 246–255.

Hanafi, M., et al. 2024. Feasibility of biochar for low-emission soft clay stabilization using CO2 curing. *Transportation Geotechnics* 49, 101370.

Inobeme, A., et al. 2021. Chapter 6 - Ecorestoration of Soil Treated with Biosurfactant During Greenhouse and Field Trials, in: Inamuddin, A. (Eds.), *Green Sustainable Process for Chemical and Environmental Engineering and Science*. Elsevier, pp. 89–105.

Khan, M.A.I., et al. 2018. Toxicity assessment of fresh and weathered petroleum hydrocarbons in contaminated soil- a review. *Chemosphere* 212, 755–767.

Khodabakhshloo, N., et al. 2021. Organically functionalized bentonite for the removal of perfluorooctane sulfonate, phenanthrene and copper mixtures from wastewater. *Applied Clay Science* 200, 105883.

Khodijah Chaerun, S., et al. 2005. Interaction between clay minerals and hydrocarbon-utilizing indigenous microorganisms in high concentrations of heavy oil: implications for bioremediation. *Clay Minerals* 40, 105–114.

Kuppusamy, S., et al. 2017. Remediation approaches for polycyclic aromatic hydrocarbons (PAHs) contaminated soils: technological constraints, emerging trends and future directions. *Chemosphere* 168, 944–968.

Leyva-Ramos, R., et al. 2021. Organoclays. Fundamentals and Applications for Removing Toxic Pollutants from Water Solution, in: Moreno-Piraján, J.C., et al. (Eds.), *Porous Materials: Theory and Its Application for Environmental Remediation*. Springer International Publishing, pp. 341–363.

Li, Q., et al. 2024. Examining diverse remediation mechanisms of biochar in soil contaminated with polycyclic aromatic hydrocarbon (PAH) of various ring structures: a global meta-analysis. *Science of The Total Environment* 171178. http://dx.doi.org/10.2139/ssrn.4654345

Lin, M., et al. 2022. Interfacial chemical behaviors and petroleum hydrocarbon removal performances of the biochar-mineral composites prepared by one-step pyrolysis. *Colloids and Surfaces A: Physicochemical and Engineering Aspects* 655, 130217.

Lin, M., et al. 2023. Biochar-clay, biochar-microorganism and biochar-enzyme composites for environmental remediation: a review. *Environmental Chemistry Letters* 21, 1837–1862.

Ling, H., et al. 2023. Surfactant-enhanced bioremediation of petroleum-contaminated soil and microbial community response: a field study. *Chemosphere* 322, 138225.

Liu, Y., et al. 2024. Green Adsorbents for Environmental Remediation: Synthesis Methods, Ecotoxicity, and Reusability Prospects, Processes. 10.3390/pr12061195.

McAllister, L.E. and Semple, K.T. 2010. Role of Clay and Organic Matter in the Biodegradation of Organics in Soil, in: Barton, L.L., et al. (Eds.), *Geomicrobiology: Molecular and Environmental Perspective*. Springer Netherlands, pp. 367–384.

Mohanty, S. and Mukherji, S. 2013. Surfactant aided biodegradation of NAPLs by Burkholderia multivorans: comparison between Triton X-100 and rhamnolipid JBR-515. *Colloids and Surfaces B: Biointerfaces* 102, 644–652.

Mukherjee, S., et al. 2022. Biochar-microorganism interactions for organic pollutant remediation: Challenges and perspectives. *Environmental Pollution* 308, 119609.

Mulligan, C.N. 2021. Sustainable remediation of contaminated soil using biosurfactants. Frontiers in Bioengineering and Biotechnology 2021, 9.

Naidu, R., et al. 2021. Chemical pollution: a growing peril and potential catastrophic risk to humanity. *Environment International* 156, 106616.

Oerke, E.-C. 2006. Crop losses to pests. *The Journal of Agricultural Science* 144, 31–43.

Panahi, H.K.S., et al. 2020. A comprehensive review of engineered biochar: production, characteristics, and environmental applications. *Journal of Cleaner Production* 270, 122462.

Patel, A.B., et al. 2020. Polycyclic aromatic hydrocarbons: sources, toxicity, and remediation approaches. *Frontiers in Microbiology* 11, 562813.

Paul, S.K. and Naidu, R. 2022. Layered aluminosilicate nanoskeletons: the structure and properties of nanoherbicide formulations. *Advances in Agronomy* 175, 301–345.

Peixoto, R., et al. 2011. Petroleum-degrading enzymes: bioremediation and new prospects. *Enzyme Research* 2011, 475193.

Petrov, A.A. 2012. *Petroleum Hydrocarbons*. Springer Science & Business Media.

Purnomo, A.S., et al. 2023. The application of biosurfactant-producing bacteria immobilized in PVA/SA/bentonite bio-composite for hydrocarbon-contaminated soil bioremediation. *RSC Advances* 13, 21163–21170.

Qi, J., et al. 2023. Applicability of Clay/Organic Clay to Environmental Pollutants: Green Way–An Overview, Applied Sciences. 10.3390/app13169395.

Qv, M., et al. 2024. Bentonite addition enhances the biodegradation of petroleum pollutants and bacterial community succession during the aerobic co-composting of waste heavy oil with agricultural wastes. *Journal of Hazardous Materials* 462, 132655.

Ruggiero, P., et al. 1989. Detoxification of 2, 4-dichlorophenol by a laccase immobilized on soil or clay. *Soil Science* 147, 361–370.

Sarkar, J.M., et al. 1989. Immobilization of enzymes on clays and soils. *Soil Biology and Biochemistry* 21, 223–230.

Schwarzenbach, R.P., et al. 1993. Vapour pressure. *Environmental Organic Chemistry* 56–75. https://doi.org/10.1002/0471649643

Semple, K.T., et al. 2003. Bioavailability of hydrophobic organic contaminants in soils: fundamental concepts and techniques for analysis. *European Journal of Soil Science* 54, 809–818.

Singh, P., et al. 2021. *Biosurfactant Enhanced Sustainable Remediation of Petroleum Contaminated Soil, Biosurfactants for a Sustainable Future*, pp. 119–138. Willey. https://doi.org/10.1002/9781119671022.ch6

Staninska-Pięta, J., et al. 2019. The impact of natural and synthetic surfactants on bacterial community during hydrocarbon biodegradation. *International Biodeterioration & Biodegradation* 142, 191–199.

Sun, P., et al. 2023. Biochar/Clay Composite Particle Immobilized Compound Bacteria: Preparation, Collaborative Degradation Performance and Environmental Tolerance, Water. 10.3390/w15162959.

Sun, Y., et al. 2021. Roles of biochar-derived dissolved organic matter in soil amendment and environmental remediation: a critical review. *Chemical Engineering Journal* 424, 130387.

Uddin, S., et al. 2021. Petroleum hydrocarbon pollution in sediments from the Gulf and Omani waters: status and review. *Marine Pollution Bulletin* 173, 112913.

Usman, M., et al. 2022. Fenton oxidation for soil remediation: a critical review of observations in historically contaminated soils. *Journal of Hazardous Materials* 424, 127670.

Verma, M. and Haritash, A. 2019. Degradation of amoxicillin by Fenton and Fenton-integrated hybrid oxidation processes. *Journal of Environmental Chemical Engineering* 7, 102886.

Warr, L.N. 2022. Earth's clay mineral inventory and its climate interaction: a quantitative assessment. *Earth-Science Reviews* 234, 104198.

Wiles, M.C., et al. 2005. Matrix-immobilized organoclay for the sorption of polycyclic aromatic hydrocarbons and pentachlorophenol from groundwater. *Chemosphere* 59, 1455–1464.

Yang, C., et al. 2022. Soil pore structure changes induced by biochar affect microbial diversity and community structure in an Ultisol. *Soil and Tillage Research* 224, 105505.

Zhou, Y., et al. 2021. Patterns of heavy metal resistant bacterial community succession influenced by biochar amendment during poultry manure composting. *Journal of Hazardous Materials* 420, 126562.

Chapter 11

Phytoremediation: advances in soil contaminant immobilization/phytostabilization combining soil amendments and vegetative cover

Paula Alvarenga[1], Luísa Louro Martins and Miguel Mourato, University of Lisbon[2], Portugal

1 Introduction

Of all the contaminants that can affect soils, the most common to all anthropogenic activities are heavy metals and metalloids, also called potentially toxic elements, or simply trace elements, because they are present at low concentrations (mg kg^{-1}, or less) in most soils, plants, and living organisms (He et al., 2005; Kabata-Pendias, 2011; Song et al., 2022; Briffa et al., 2020). Despite some controversy about each of these designations (Duffus, 2002; Gustin et al., 2021; Zhang et al., 2022), in the scope of this chapter, we will use the designation heavy metals, as more recently defended by Gustin et al. (2021), specifying the elements of interest whenever necessary. Examples of these are arsenic (As), cadmium (Cd), chromium (Cr), copper (Cu), mercury (Hg), lead (Pb), selenium (Se), and zinc (Zn) (Wuana and Okieimen, 2011), a group that also includes As and Se, which are metalloids. In addition to the fact that heavy metals are natural elements in the Earth's crust, and some are essential for plants

1 Corresponding author: palvarenga@isa.ulisboa.pt
2 LEAF, Associate Laboratory TERRA, School of Agriculture, University of Lisbon, Lisboa, Portugal.

http://dx.doi.org/10.19103/AS.2025.0151.16

and animals, others are non-essential, toxic at low concentrations, and cannot be broken down, i.e. they are nonbiodegradable (He et al., 2005; Kabata-Pendias, 2011; Bolan et al., 2014; Briffa et al., 2020). This characteristic means that, in the remediation of soils contaminated with heavy metals, we often must choose between techniques that allow their immobilization/stabilization in soil, reducing the risk associated with their presence, and techniques that allow the effective removal of these elements from the contaminated environment, extracting them from the soil, and reducing their concentration to acceptable levels, normally below legislated limit values (Kidd et al., 2009; Khalid et al., 2017; Bolan et al., 2014; Zhou et al., 2017; Burges et al., 2018). The latter is usually preferable when considering agricultural soils because of the continued and unacceptable risk of transference of heavy metals to the human food chain (He et al., 2005). This perspective gave rise to a specific phytoremediation technology, non-disruptive soil multifunctionality, that is called phytoextration, where the plants are used to extract heavy metals from the contaminated soil, accumulating them in aboveground plant parts, that can be collected and processed adequately, as contaminated biomass (Kuppens et al., 2015; Robinson et al., 2015), eventually recovering the heavy metals and valorizing the biomass in an adequate manner (Kidd et al., 2009; Wang et al., 2020). Of course, care must be taken to minimize the ingestion of the aboveground plant parts by animals.

Unfortunately, when the areas to be remediated were affected by mining activities and processing metal ores, the reality is quite different (Naidu, 2013; Brown and Chaney, 2016), and phytoextration is not workable, for multiple reasons, namely the low bioavailability of heavy metals to be extracted from the soil and the long-time needed to reach the desired concentrations. One major limitation of phytoextration is the fact that the plants which are adequate for this strategy should be accumulators or hyperaccumulators, and these are frequently wild plants, difficult to produce using conventional agricultural practices, producing low plant biomass, and with the ability to accumulating only one heavy metal, frequently not the most problematic (Mahar et al., 2016; Egendorf et al., 2020). Robinson et al. (2015) calculated that bioaccumulation coefficients higher than 10 would be necessary to reduce the total heavy metal concentration in soil by 50% within 25 years, under ideal phytoextraction conditions. That is why some authors defend that phytoremediation strategies need, typically, more time than others to achieve the remediation objectives (i.e. <10 years) when using only plants to lower the contaminants' concentrations in soils (Míguez et al., 2020).

There are numerous remediation technologies that work fast and can be applied to those heavily contaminated soils, both *in-situ* and *ex-situ*, but the remediation costs are considerably high, usually prohibitive to be applied for large areas, and very disruptive to the soil ecosystem functions (Naidu,

2013; Brown and Chaney, 2016; Song et al., 2022). That is why, alongside phytoextraction, a different phytoremediation strategy has evolved, mainly applied to heavily contaminated soils, namely former industrial areas, and abandoned mine sites (Fig. 1a), which is phytostabilization (Mendez and Maier, 2008; Bolan et al., 2014). In this strategy, the remediation objectives are to establish a vegetation cover with metal-tolerant plants, sometimes called metallophytes (Baker et al., 2010; Bacchetta et al., 2015), and promote the *in situ* immobilization of heavy metals with an adequate mixture of soil amendments, with a high binding capacity to heavy metals (Brown and Chaney, 2016), working together toward a different goal: the *in-situ* immobilization/stabilization of contaminants in soil, especially heavy metals, lowering their phytoavailability, mobility, and overall bioavailability, and avoiding their dispersal and entrance into different biotic and abiotic compartments, leading steadily to the restoration of the ecosystem functions (Kidd et al., 2009; Epelde et al., 2009; Burges et al., 2016, 2018).

When phytostabilization is being considered, one crucial aspect is that contaminants should be locked in the soil at the root–soil interface (or rhizosphere), or in the root itself (translocation does not occur) (Mahar et al.,

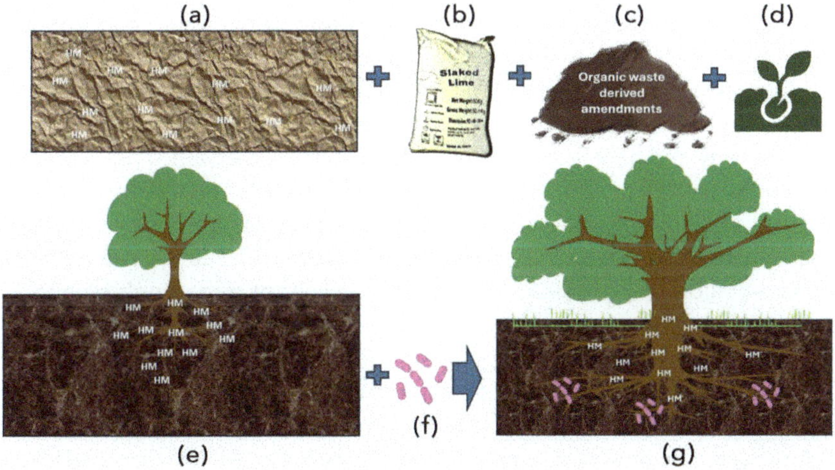

Figure 1 (A) Heavily contaminated soils, from former industrial areas, with no plant cover, are of special interest to be used for immobilization/phytostabilization strategies. It is often necessary the combination of (B) inorganic and (C) organic amendments, to improve soil properties, and assist in the chemical stabilization of heavy metals in the phytostabilization process. (D) Excluder plants should be chosen, (E) that will lock and immobilize heavy metals in the root or in the rhizosphere (choosing perennials or crops whose biomass can be properly valorized is important). (F) Increase the microbial community of the rhizosphere to help in the phytoremediation process. (G) Rhizosphere microorganisms can help the plant to adapt to contaminated soil, enhancing root growth, and increasing mineral uptake.

2016), which can be achieved if (1) appropriate agricultural amendments are used, both inorganic (Fig. 1b) and organic (Fig. 1c), allowing the effective immobilization of contaminants in soil, reducing their phytoavailability; (2) excluder plants are selected (Fig. 1d), avoiding the accumulation of contaminants in their edible parts; and (3) non-edible crops are selected (Ghosh and Maiti, 2021), choosing perennials or crops whose biomass can be properly valorized (Fig. 1e) without the risk of the transference of contaminants from one abiotic compartment to the other.

Considering the different phytoremediation strategies, the choice for phytostabilization as a solution for the remediation of heavy metals contaminated soils often arises as a conundrum solution: how can it be considered a soil remediation technology if it does not remove the contaminants from the contaminated soil? In light of the existing legislation and guideline values, mostly based on total trace element concentrations, the soil will remain with contaminants' total concentrations above intervention values, even with the strategy fully implemented (Bolan et al., 2014). Therefore, an immobilization/phytostabilization project for a specific situation of soil contamination must evidence that the risk of the presence of heavy metals in the soil is acceptable, even without significantly reducing total heavy metals concentration.

Since metal bioavailability is the fraction of the total concentration that can interact with the biological target (Harmsen, 2007), it is necessary to assess heavy metals bioavailability and soil health. The drawbacks of this strategy are the lack of consensus around the methods to measure metals bioavailability and how to assess soil health (Alvarenga et al., 2018). Moreover, the immobilized heavy metals may become plant available with time, because of natural weathering processes or by the breakdown of high-molecular-weight organic-metal complexes, increasing the risk (Bolan et al., 2014). Therefore, the implementation of a phytostabilization strategy needs a thorough monitoring plan, and correction measures may have to be considered in that plan (Bolan et al., 2014).

Therefore, important factors to the success of a phytostabilization strategy are (1) the selection of appropriate soil amendments, (2) appropriate crops, (3) strategies for their biomass valorization, and the (4) right parameters to monitor the risk and efficacy of the remediation strategy. All in all, these can be considered gentle remediation options (GROs) (Kidd et al., 2009, 2015; Cundy et al., 2013; Quintela-Sabarís et al., 2017; Touceda-González et al., 2017a; Touceda-González et al., 2017b) and allow the remediation of soils without the need to choose more harsh techniques, disruptive of soils properties, less invasive and suitable for *in situ* application in large contaminated sites, and eventually allowing the integration of non-agricultural contaminated soils in the productive circuit, enlarging the area dedicated to agriculture (Lacalle et al., 2020; Moreira et al., 2021).

The main objective of this chapter is to describe relevant studies and recent advances in these matters, pointing solutions to the phytostabilization of soils contaminated with heavy metals and, if possible, also with organic pollutants, and agricultural soil, besides the obvious industrial and mining-affected soils.

2 Types of contaminated soil for phytostabilization

2.1 Mining and industrial areas

In phytostabilization, the combination of organic or inorganic additives (chemical stabilization) with metal-tolerant plants is sometimes termed aided- or assisted phytostabilization because it consists of a combined process (Alvarenga et al., 2009a, 2009b; Zornoza et al., 2015; Garaiyurrebaso et al., 2016). In phytostabilization, a simultaneous reduction of the mobility/bioavailability of the contaminants in soil is achieved, thus reducing leaching and heavy metals transference through the trophic chain, while also improving soil microbial properties and enabling plant establishment (revegetation) in extensive areas, like former industrial sites and abandoned mining areas (Valente et al., 2012; Clemente et al., 2015; Garaiyurrebaso et al., 2016; Duarte et al., 2023).

That is why, by far, the higher number of phytostabilization studies considered soils from abandoned mines and other industrial sites (Kumpiene et al., 2012; Burges et al., 2018; Nagy et al., 2023; Nedjimi, 2021), with a heavy load of contamination, too high for complete removal of contaminants from the soil to be assumed within a reasonable time scale. Phytostabilization provides conditions for future natural attenuation processes by reducing the bioavailability of toxic heavy metals, even without their removal from soil (Valente et al., 2012), i.e. a risk-based land management approach (Thavamani et al., 2015). That is why, different groups of investigators have dedicated their studies to the phytostabilization of mine sites, especially, in the remediation of abandoned mines in the Iberian Pyrite Belt (Alvarenga et al., 2008b, 2009a, 2009b; Clemente et al., 2010; Pardo et al., 2011; Karami et al., 2011; Pardo et al., 2014a, 2014b; García-Sánchez et al., 2015; Lebrun et al., 2018), and in other similar environments like in soils affected by steelmill wastes (Oustriere et al., 2016) or by a former Pb–Zn smelter (Pavel et al., 2014).

Pyritic environments encompass other problems associated with heavy metals contamination, namely soil acidity, low organic matter (OM) content, and fertility status, besides severe physical constraints related to a deteriorated structure (e.g. low water-holding capacity, low infiltration rate, and difficulty for root penetration) and, as a consequence to all these summed problems, the complete depletion of vegetation cover and the continuous generation of acid mine drainage (Kefeni et al., 2017; Alvarenga et al., 2021), which can enlarge the contaminated area and affect the hydrologic system in the surroundings. Some of these phytostabilization strategies considered the revegetation of mine tailings and other reactive mine wastes deposited in these abandoned mines, or their use in the development of technosols, which were the most difficult projects to be accomplished (Pardo et al., 2017; Karaca et al., 2018; Nagy et al., 2023; Wang et al., 2017).

2.2 Agricultural soils

The problem of heavy metal pollution in soils is not specific to mine and industrial soils, agricultural soils worldwide struggle with this problem, which is very dramatic when it affects food safety. One example is, for instance, the case reported by Arco-Lázaro et al. (2016) in Segovia, Central Spain, where the use of groundwater for irrigation has elevated concentrations of As with geological origin (above 10 sg L^{-1}). Arsenic is naturally present in the environment in two main soluble forms: arsenate (As (V), the predominant form in oxidizing aerobic conditions) and arsenite (As (III), predominant form under reducing, anaerobic conditions) (Arco-Lázaro et al., 2016; Kabata-Pendias, 2011). This is even a bigger problem in some parts of the world, like in India and Bangladesh, where the groundwater available for irrigation is also contaminated with As and not only because of geological origins (Ghosh and Bhattacharyya, 2004; Hossain et al., 2021; Sengupta et al., 2021). Rice is the staple food for a huge part of the human population, mainly in these countries, and it is traditionally produced in flooded soils (anaerobic conditions), contaminating rice grains with As and, sometimes, with Cd and Pb (Ghosh and Bhattacharyya, 2004; Sengupta et al., 2021). Moreover, not only the accumulation of As in rice grain is problematic but also its speciation, with the inorganic As species (As(III) + As(V)) being considered more toxic than the organic-As species (Šlejkovec et al., 2021). The predominance of these species is also affected by the irrigation regime, and Moreno-Jiménez et al. (2014) verified a reduction in inorganic As in rice grain under a non-flooded regime (sprinkler irrigation). The adaptation to non-flooded irrigation regimes could thus be a way to solve situations of water scarcity problems that we are facing now and may be aggravated in the near future (Farhat et al., 2021). Because of that rice production has been gradually adapting to deficit irrigation practices (aerobic conditions, e.g. sprinkler irrigation, alternate wetting, and drying), alternatively to flooding irrigation (Sengupta et al., 2021; Wu et al., 2021; López-Piñeiro et al., 2007, 2011; Peña et al., 2019). However, the anaerobic or aerobic systems of water management in rice cultivation may not only affect rice production but also affect metals and metalloids concentration in the rice grains, in opposite directions, with a direct impact on food safety (Moreno-Jiménez et al., 2014; Maneepitak et al.,

2019; Sengupta et al., 2021). For instance, a change to the mentioned deficit irrigation systems allowed a decrease in the concentration of As in the rice grain, with a gain in food safety, but with an undesired increase in the bioavailability of metals, like Cd, as a response to the aerobic systems (Moreno-Jiménez et al., 2014; Wu et al., 2021) (Spanu et al., 2020; Moreno-Jiménez et al., 2014; Sengupta et al., 2021; Wu et al., 2021; Ishfaq et al., 2020), a situation that needs to be controlled.

As will be discussed later in this chapter, the most important agricultural practice for the immobilization of metals in soil is the application of amendments. The application of organic amendments with exogenous sources of OM could be, per se, a conservation agriculture practice used by farmers to counterbalance the OM depletion in certain environments, e.g. in Mediterranean agricultural systems (Alvarenga et al., 2020). Waste-derived materials, from locally available sources, are often used to increase the sustainability of this practice, like compost from olive mill waste (López-Piñeiro et al., 2007; López-Piñeiro et al., 2011; Peña et al., 2019). Composting this waste produces a more stabilized and mature material and it is able to establish organo-metal chelates, reducing the risk of metal uptake by rice grain when using a sprinkler irrigation system. Therefore, the application of 80 Mg ha^{-1} olive mill waste compost, with simultaneous use of sprinkler irrigation, proved to be effective in decreasing As accumulation in the rice grain, because of the aerobic conditions, and, at the same time, decreasing Cd accumulation, because of the compost application (Alvarenga et al., 2022). The same group also used biochar produced from holm-oak pruning residues (pyrolysis at 550°C, 48 h) in a single application (28 Mg ha^{-1}) as an amendment to immobilize heavy metals in different rice production systems in a Mediterranean climate (Martín-Franco et al., 2023), and they were able to decrease As, Cd, and Pb accumulation in rice grains under different production systems allowing simultaneous control of metals and of As accumulation in rice and increase its commercial value.

Cadmium is, in fact, a serious problem in farmed soils, in other cases, because of the application of phosphate fertilizers, threatening the safe production of food crops (Nziguheba and Smolders, 2008; Gómez et al., 2023). Passivator amendments may help to cope with this problem some that may be used in common, like rice husk biochar and limestone, in different proportions, as evaluated by Niu et al. (2022), to avoid Cd accumulation in wheat grain, helping to cope with food safety problems in China, which faces serious problems of agricultural soils contamination with Cd. Limestone is a cheap and effective inorganic soil amendment, able to counterbalance soil acidity, pH increase being the the main effect responsible for the Cd immobilization in soil, thus reducing Cd accumulation in crops (Niu et al., 2022).

In line with this idea, Qi et al. (2023) concluded that chemical immobilization is one of the most effective technologies for remediating soils contaminated with heavy metals and that the selected material and the ratio of its use is of utmost importance for the remediation efficiency. They have dedicated a review to the use of biochar, phosphate, lime, metal oxides, and clay minerals to immobilize Cd and Pb, two of the most problematic heavy metals in agricultural soils (Qi et al., 2023). They have conducted a meta-analysis of 489 independent observations. One of their most important conclusions was that Cd immobilization in agricultural systems can be achieved by the application of lime, which is most effective when the initial bioavailable Cd content was between 0.1 mg kg^{-1} and 1 mg kg^{-1}, whereas the immobilization effects of lime and phosphate were similar when the initial bioavailable Cd content raises to 10 mg kg^{-1}. Phosphate was the most effective material for Pb immobilization (Qi et al., 2023). The effects on both elements are different: while phosphate mostly leads to the precipitation of Pb, lime leads to a reduction in Cd and Pb bioavailability, because it increases soil pH.

2.3 Urban and peri-urban areas elective for urban agriculture

Urban population is predictably increasing, demanding an urgent increase in food and biomass production in urban or peri-urban areas, with shorter supply chains, lowering energy consumption in material transports and CO_2 emissions, and assuring food security aspects in the city during crisis events (e.g. pandemics, fossil fuels shortage, and strikes) (Coles and Costa, 2018). This can be done through a set of soilless and building-integrated agriculture options, e.g. vertical farming, intensive green roofs, indoor agriculture, and hydroponics (Skar et al., 2020).

However, it is important to keep in mind that the decline of large industrial areas left vacant land in the peri-urban belts of many cities around the world, which is an economic, social, and environmental concern, especially when they have low-level of contaminated soils that would be important to integrate in a productive circuit of urban farms and productive gardens, i.e. soil-based urban agriculture (Barbillon et al., 2019; Skar et al., 2020). This strategy could also be considered more adequate for nutrient recycling from some organic waste streams, equally representative to produce for both food and biomass production, and undoubtedly with added quantitative potential (Clemente et al., 2015; Basta et al., 2016; Sarker et al., 2019).

Therefore, to increase the area dedicated to urban agriculture, it would be of utmost importance to start using vacant soils in the urban and peri-urban areas of the cities. These soils may have elevated toxicity levels due to past industrial activities or to the proximity to high-traffic roads (Barbillon et al., 2019; Leitão et al., 2018), but several nature-based solutions (NBSs), combining phytoremediation and bioremediation, could be an interesting option. These NBS, or soil biological remediation techniques, require soil enrichment with

OM, nutrients, and an increase in biological activity, which can be a win–win solution for biowastes generated in the city. Amending these soils with carbon waste-derived materials (e.g. compost, digestate, sewage sludge, and biochar) would fit into the circularity concept, recycling nutrients and, ultimately, allowing an increase in the soil organic C pool, a contribution to adapt and mitigate the climate change in cities (Lal et al., 2018; Sarker et al., 2019).

The phytomanagement technologies proposed to treat heavy metal contaminated soils, which will be specified in Section 2.4 could, in the short to medium term, contribute to the gain of new peri-urban areas dedicated to agriculture. These remediated soils have already demonstrated their merit in producing non-food crops for biomass-processing technologies (e.g. biofuel and bioenergy, ecomaterials, and chemistry resources), avoiding the depletion of primary raw materials (Moreira et al., 2021) and avoiding the problems of ethics connected with the production of biofuels in agricultural lands (Thompson, 2012). It would be important to evaluate the possibility of their safe use to produce food and feed with an adequate combination of immobilization/phytostabilization of the contaminants.

That was the idea explored by Míguez et al. (2020), the phytomanagement of urban vacant lands, reducing the competition for soil and water resources in the production of non-food crops and creating, at the same time, new urban green spaces. In their case, they produced a biofuel crop (*Brassica napus*). Since these soils have physical, chemical, and biological constraints to plant growth, they have incorporated compost produced from the organic fraction of wastes produced in the city (50 t ha^{-1}), allowing the desired resources circularity and the improving soils' properties: reducing compaction, increasing water-holding capacity, increasing OM content, providing some nutrients, and increasing soil microbial activity essential to restore soil health (Lacalle et al., 2018; Gómez-Sagasti et al., 2018).

2.4 The evolution to a phytomanagement concept

The genesis of the two phytoremediation strategies that were presented in the introduction section, phytoextraction and phytostabilization, was quite different and were very clearly presented by different authors (Kidd et al., 2009; Burges et al., 2018), and Kidd et al. (2015) presented an important study explaining why these 'gentle remediation' options need a boost to finish being called 'promising phytotechnologies' by so many authors over so many years (Suresh and Ravishankar, 2004; Moreira et al., 2021): agronomic management practices are needed, and some recent projects (namely PhytoSUDOE and PHY_2SUDOE), both Interreg Sudoe Projects, with demonstration sites, may help.

It is impossible to approach recent advances in phytotechnologies without acknowledging the evolution in the terminology, and in the philosophy associated with soil contaminants' immobilization/phytostabilization and its turn into a phytomanagement strategy, where profitable crop production and soil amendments are combined to obtain environmental, economic, and long-term social gains (Evangelou et al., 2013; Burges et al., 2018). Phytomanagement evidences the importance of the plant biomass destination in the whole process not discarding plants' ability to extract some heavy metals from the soil, lowering its total concentration over time. In fact, the whole idea is that phytomanagement can be applied as part of an integrated site risk management, producing valuable crops (Cundy et al., 2016; Kidd et al., 2015).

Phytomanagement overpasses somehow the conventional designations which we have defined previously, combining both phytotechnologies, with a sustainable and profitable site use (Burges et al., 2018) and it uses both plant abilities, their capacity is not only to block some elements in the root zone but also to gradually uptake some contaminants, allowing a progressive decontamination of the soil. In the phytomanagement, besides plants, the microbial community of the rhizosphere can be increased (Fig. 1f), and those rhizosphere microorganisms can help the plant to adapt to the contaminated soil, because rhizosphere microorganisms and enhance root-growth regulators can promote plant growth and adaptation to the contaminated soil, increasing mineral uptake and controlling phytopathogens (Kidd et al., 2021). For instance, arbuscular mycorrhizal fungi (AMF) can help the plant to create an extensive extraradical hyphal network, which explores the soil, absorbing nutrients, and improving nutrition in low-quality soils, helping their establishment (Fig. 1g). In some cases, these rhizosphere microorganisms degrade organic contaminants in soil, especially when there is also a mixed contamination of heavy metals with organic contaminants, and more recently, by the addition of specific soil amendments (zero-valent iron nanoparticles), it was possible to enhance organic contaminants' decomposition, e.g. diesel hydrocarbons and lindane (Lacalle et al., 2018; Hidalgo et al., 2022).

3 Soil amendments for immobilization of contaminants

Only a small fraction of the total heavy metals that exist in soil are bioavailable, depending on soil properties like pH and OM content, among others. The mobility and availability of heavy metals are controlled by many chemical and biochemical processes, the most important are precipitation–dissolution, adsorption–desorption, complexation–dissociation, and oxidation–reduction (He et al., 2005; Palansooriya et al., 2020). Controlling soil properties is of pivotal importance in all phytoremediation processes and that is especially true for phytostabilization, because the conditions for heavy metals immobilization must be assured, not all the conditions are equally important for each element, and in some environments, e.g. pyritic mine tailings, heavy metals immobilization tend to change over time, because the soils, or even some of the materials which are deposited at the site (e.g. tailings), are very reactive and were deposited for long periods in the area (Alvarenga et al., 2004, 2012, 2013). One thing is for sure, all these processes are affected by soil pH and by biological processes (He et al., 2005), which makes the control of soil pH and soil microbial activity so important.

The application of organic or inorganic additives to soil works through different mechanisms in the immobilization of heavy metals, some more important than others, depending on the element which is being considered and on the amendments that are being used, and different authors have been able to thoroughly explain the processes which are involved (Stouraiti et al., 2002; Palansooriya et al., 2020; Xu et al., 2021; Bolan et al., 2014):

- pH control: using alkaline additives, avoiding acidity, by the addition of lime, hydrated lime, limestone, cement, biochar, and fly ash;
- chemical precipitation as low solubility-salts, like hydroxides, sulfides, silicates, and phosphates;
- organic complexation of metal cations, which can occur when using materials with many oxygen-containing functional groups, which can act as ligands, like –COOH and –OH;
- sorption, a surface interaction mechanism which occur by the application of natural additives (e.g. clays, peat moss, and saw dust), or synthetic materials (e.g. fly ash, zeolite, activated carbon, and metal oxides), and that can be non-specific, when the charge of the ions balances the charge of the soil particles by electrostatic attraction, whereas the specific adsorption involves chemical bonding, i.e. electron sharing; and
- ion exchange, a mechanism involved at some degree in almost all stabilization processes.

Different soil amendments have been evaluated to achieve heavy metals immobilization, which is of pivotal importance in phytostabilization processes, and their main division is if the amendments are organic or inorganic as a material or if a combination of materials are being used or are necessary. Usually, their combination is mostly precise because some of the materials are limited in their benefits and their combination potentiates their capacity to immobilize heavy metals providing, at the same time, OM and nutrients and improving physical and/or biological properties of the soils. Table 1 summarizes the most relevant advantages and limitations of selected amendments that will be discussed further.

Another important issue is if those amendments are waste derived, which can be considered a win–win strategy, since it's achieved at the same time a remediation of contaminated soil and a valorization of a waste that present management problems and are, in some cases, produced in excessive

Table 1 Summary of the most relevant advantages and limitations of selected amendments which were mentioned in the text. Their limitations are the justification for their combined use

	Advantages	Limitations
Compost	• Increases soil OM • Provides plants' essential macro- and micro-nutrients • Increases soil cation exchange capacity (CEC) • Contribute to heavy metal immobilization by providing organic ligands • Increases soil microbial activity • Improves soil structure • Increases water-holding capacity • Reduces soil compaction • Facilitates root penetration • Enhances the development of a vegetation cover • If produced from a local organic waste, contributes to the circular economy	• May have limitations in increasing soil pH, and in maintaining it in the desired range to have heavy metal immobilization • If the soluble organic carbon (SOC) compounds are very representative in the compost, it may increase some heavy metals mobility • Compost quality needs to be analysed and comply with the legislation in force in each country
Lime, hydrated lime, limestone	• Increases soil pH, reducing heavy metal bioavailability, by chemical precipitation • Low cost • Market available, with a 'common agricultural practice' developed • Already used by farmers	• Amendments which only work as chemical passivators. Very limited in recovering other soil chemical, physical, and biological properties
Biochar	• Obtained from the pyrolysis of almost any type of biomass (sometimes wastes), contributing to the circular economy • Increase SOC sequestration • Increasing water-holding capacity • Increases CEC • It has an alkaline pH • Carbonaceous material, porous structure, effective to immobilize of heavy metals • Possible to be used in soils with heavy metals and hydrocarbon spills (hydrophobicity characteristics)	• Alone, they might not be sufficiently effective to remediate acid mine soils
Biomass ashes	• Very alkaline (pH ranging from 9 to 13) • Rich in some plant nutrients, e.g., K, Mg, Ca • Produced in very high quantities, with a tendency to increase, because biomass is a renewable energy source	• Poor in OM • Poor in some plant nutrients, e.g., N and P • Fine powder material, difficult to apply • High salinity
Red mud	• High pH • Good immobilization potential for heavy metals (richness in iron–aluminium oxide active functional groups)	• Only work as chemical passivators of heavy metals. Very limited in recovering other soil properties

amounts in specific areas (Clemente et al., 2015). This recycling is, therefore, an alternative to its landfilling, a situation which is highly recommended by all End-Waste Management Policies Worldwide.

In phytostabilization, organic amendments are important, not only because they play a role in heavy metals immobilization, contributing to an increase in the cation exchange capacity (CEC), providing organic ligands, eventually contributing to control in soil pH, but, especially, because organic amendments improve all chemical, physical, and biochemical conditions necessary for plant growth (Kumpiene et al., 2009, 2019; Pardo et al., 2014b; Galende et al., 2014; Peña et al., 2015; Zornoza et al., 2015; Zornoza et al., 2016; Basta et al., 2016; Zhou et al., 2017; Al-Lami et al., 2019). The application of organic amendments plays an important role to improve soil structure. Soils elective for phytostabilization, especially if they are from abandoned mine sites, suffer from physical constraints that inhibit revegetation, especially because they are very compact (Valente et al., 2012): very coarse fragments or, on the other hand, very fine tails and the lack of clay minerals, which, combined with a low OM content gave rise to soils with bad structure, highly compacted (Tordoff et al., 2000). The incorporation of OM, especially compost (Table 1), leads to an improvement of soil structure, which is of pivotal importance to the improvement of all physical soil properties, increasing the water-holding capacity and reducing soil compaction, which is a very important restriction to the penetration of plant roots and the establishment of a vegetation cover (Somerville et al., 2018). The increase in soil OM, by the addition of organic amendments, is not only important from a chemical point of view, providing essential nutrients to the plants, important to establish a vegetative cover, but it is also important to increase soil microbial activity (Alvarenga et al., 2009b; Santos et al., 2016; Lwin et al., 2018; Alam et al., 2020). In fact, the addition of essential nutrients to the crop which is intended to be installed could be provided by conventional fertilizers, but the benefits of adding OM are far more important that just providing nutrients.

On the other hand, inorganic amendments (e.g. lime, hydrated lime, and limestone) are mainly applied to the soil with the main purpose of controlling soil pH (Table 1). As the problem in question is about soil contamination with heavy metals, their salts solubility in soil are mainly ruled by pH, with a decrease in the mobility of most toxic metals with an increase in soil pH (Brady and Weil, 2017). Therefore, in some situations, an alleviation of soil acidity by the application of proper soil amendments, like limestone (Hussain Lahori et al., 2017), will contribute to the immobilization of heavy metals, which are affecting the soil. That is why some assisted-phytostabilization projects, applied to abandoned pyrite mines, started with the idea of combining both types of amendments: the addition of alkaline amendments that would increase the pH of soil and, therefore, reduce the availability of metals, and the application of organic materials that would improve the conditions to plant establishment (Madejon et al., 2006; Alvarenga et al., 2008a, 2008b, 2008c; Lebrun et al., 2018; Kumpiene et al., 2019; Palansooriya et al., 2020). The term aided-phytostabilization or assisted-phytostabilization were reported since then, and different groups of investigators started independent studies with the same objectives: using waste-derived amendments, organic and inorganic, combined with plants in the immobilization of heavy metals in abandoned mines (Clemente et al., 2005; Madejon et al., 2006; Alvarenga et al., 2008a, 2009a) and former industrial sites (Pavel et al., 2014; Hussain Lahori et al., 2017).

Biochar is an amendment that has been widely used in immobilization/phytostabilization projects of heavy metal-contaminated soils, by different groups of investigators (Table 1) (Kiran and Prasad, 2019; Ghosh and Maiti, 2021; Duarte et al., 2023). It is a solid material obtained from the thermochemical decomposition of almost any type of biomass (sometimes wastes) in an oxygen-limited environment (pyrolysis), leading to a carbonaceous material, with a highly porous structure (Ghosh and Maiti, 2021). The research in biochar in the last years has focused on its role as a soil-ameliorating agent to improve soil fertility, water-holding capacity, pH, nutrients, carbon sequestration, microbial activities, and pollution remediation (Bhattacharyya et al., 2024; Ghosh and Maiti, 2021). Biochar has shown to have remarkable success as a soil amendment (Bhattacharyya et al., 2024), but alone might not be so effective to remediate contaminated mine soil, thus a synergy between biochar and phytoremediation is reported to be more effective in the immobilization of a set of other heavy metals. Biochar soil application leads to an improvement of the soil microbial activity, thus influencing the interaction between biochar, plants, and microbes, which can alter the metal's behaviour in the soil (Kiran and Prasad, 2019).

One interesting characteristic of biochar as an amendment in the immobilization of heavy metals is the possibility of its use in soils with high hydrophobicity, because of a mixed contamination also with organics, e.g. petroleum-based acid tar, like the example provided by Edenborn et al. (2015), where a surface application of hardwood biochar to a soil contaminated with Pb, As, and antimony (Sb) was able to promote a long-term restoration of the site, resulting in successful growth of wild-rye-grass.

In most studies aiming for the evaluation of organic amendments to ameliorate mine soils, waste-derived materials are usually considered. It is possible to group the waste-derived amendments in different groups. The most representative being the compost produced from the organic fraction of urban wastes (Clemente et al., 2005, 2012; Pérez-de-Mora et al., 2006; Alvarenga et al., 2009a, 2009b, 2009c, 2013, 2014; de Varennes et al., 2010a; Pardo et al., 2014a) and sewage sludge obtained in the wastewater treatment plants (Mora et al., 2005; Pérez-de-Mora et al., 2006; Alvarenga et al., 2008a, 2009b, 2009c;

Madejón et al., 2009; Pérez-de-Mora et al., 2011; Burgos et al., 2010). Of course, when the compost that is produced and applied to the soil is produced from green waste or agricultural wastes, its quality is usually higher, because the risk of previous contamination of the contamination of the compost itself is lower (Alvarenga et al., 2008b , 2009a, 2009c, 2013; Madejón et al., 2009; Clemente et al., 2010; Karami et al., 2011).

Other important groups of waste-derived organic amendments are those available locally, with problems over overproduction and seasonality, specific to some geographic areas (e.g. from agro-industries and animal production) (Clemente et al., 2015). As common examples in the Iberian Peninsula, it is possible to find studies with the application of compost from olive mill waste of olive oil extraction (locally called *'alperujo'*), which has been used with success in phytostabilization studies by different groups of investigators (Pardo et al., 2011; Clemente et al., 2015; Burgos et al., 2010), sometimes composted with other materials (Clemente et al., 2006). Another example is the rice husk ash, a cost-effective agricultural by-product, produced from the burning of rice husk, which includes 50% cellulose, 25–30% lignin, 15–20% hydrated silica, and the rest of the mineral components are alkalis and trace elements, representing a low-cost adsorbent able to immobilize toxic metals from contaminated soils (Kiran and Prasad, 2019). The authors attribute its bio-adsorption affinity to the complexation of the toxic metal ions to the functional groups, like carboxyl, imidazole, hydroxyl, sulphydryl, phosphate, amino, sulphate, phenol, thioether, and carbonyl, in the material (Kiran and Prasad, 2019).

An important alkaline residual waste, from the sugar manufacturing process, is sugar beet sludges, which can provide OM while increasing the pH at the same time. They were used by several investigators, alone or in combination with other materials, to amend the soils affected by mines in the Iberian Pyrite Belt (Pérez-de-Mora et al., 2006; Alvarenga et al., 2008a; Madejón et al., 2009; Pérez-de-Mora et al., 2011; Clemente et al., 2015; Hinojosa et al., 2008; Burgos et al., 2010).

Countries with intensive animal production, especially pigs, cows, and poultry, have problems to deal with large amounts of animal slurries and manures, like pig slurry (Pardo et al., 2011, 2014a; Clemente et al., 2012; Zornoza et al., 2015), cow manure or slurry (Galende et al., 2014; Clemente et al., 2015), poultry manure (Galende et al., 2014), and composted horse manure (Nakamaru and Martín Peinado, 2017). These are materials which have already been used as organic amendments in the phytostabilization of mine soils.

Biomass ashes (Table 1) may also be recycled to remediate acidic mine soils, as a liming agent, increasing the pH and buffering capacity of acid soils, avoiding the use of traditional agricultural lime, since they are very alkaline (pH ranging from 9 to 13) (Ram and Masto, 2014; Demeyer et al., 2001). Moreover, they are considered a sustainable way to recycle plant nutrients such as K, Mg, and Ca, saving non-renewable sources, such as some important minerals (Demeyer et al., 2001). Sometimes, biomass ashes can be combined with sewage sludges, e.g., with the biologic sludges from the wastewater treatment plant when they are produced in the paper mill industry, enriching them in OM, P, and N, and experiments for their co-granulation and stabilization have been evaluated by some authors, providing a very interesting amendment to be used in the remediation of mine soils (Galende et al., 2014; Alvarenga et al., 2019). The simultaneous use of both wastes in the process of granulation helps to overcome the deficiencies of OM, N, P, and K of the biomass ash, while turning it into a material easier to apply to the soil than the usual fine powder of some ashes (Alvarenga et al., 2019). However, the same authors identified some phytotoxicity problems, which they have attributed to the high salinity of the biomass ash, after soil application (Alvarenga et al., 2019).

Red mud (Table 1) (Lee et al., 2017) is an industrial by-product of the alumina industry, deriving from the digestion of crushed bauxite with caustic soda, that has shown some positive results as heavy metals sorbent (Garau et al., 2007). The red mud immobilization capabilities for heavy metals are based on its richness in iron–aluminium oxide active functional groups and high pH. For instance, Lee et al. (2017) reported 4 years of monitoring activity of a reclamation of mine-degraded agricultural soils, in Korea, amended with red mud, zeolite, and lime on heavy metal immobilization. The soils were mainly affected by As, Cd, Cu, Pb, and Zn, associated with metal mining. These 4 years of monitoring revealed that the levels of contamination at surface soil could be maintained at acceptable levels, although at some sites, the metal levels in crops exceeded legislative limits, evidencing that there is a potential for safe agricultural operations on large sites in a cost-effective manner, if the appropriate control of surface soil contamination and adequate agronomic management is undertaken. Meirong et al. (2012) also evaluated the effects of the application of red mud on the pH, electrical conductivity (EC), and the immobilization of Pb, Zn, and Cd in a paddy soil, contaminated by mining activities (incubation experiment for 30, 60, and 90 days with the red mud dosage of 4%(W/W)), and observed significant immobilization for all the contaminants.

More uncommon amendments have also been evaluated by other investigators, frequently with the main aim of recycling wastes (organic or inorganic) which are not so frequently used as agricultural soil amendments but could be used in the rehabilitation of mine soils, avoiding their landfilling, like drinking water treatment sludges (Alvarenga et al., 2018), polyacrylate polymers (de Varennes et al., 2010b), hydroxyapatite, chematite and caghemite (Arenas-Lago et al., 2019), rockwool industrial waste, agriculture wastes (plant remains + strawberry substrate) and wastes from liquor distillation of *Arbutus unedo* L. fruits (Santos et al., 2014, 2019).

In some studies, inorganic and organic additives were used in combination, complementing each other in their specific deficiencies (Somerville et al., 2018; Kumpiene et al., 2019; Parra et al., 2016; Paradelo et al., 2019), and very important reviews have been published gathering these informations (Palansooriya et al., 2020).

However, the rehabilitation of mining areas is a long process in which the dynamics of the heavy metals are uncertain, therefore, periodic monitoring and adaptive management is necessary (Epelde et al., 2009, 2014b; Basta et al., 2016; Karan et al., 2016; Garaiyurrebaso et al., 2016; Khalid et al., 2017; Bolan et al., 2011). Moreover, soil ameliorants used to increase heavy metals immobilization may need to be reapplied periodically to maintain their effectiveness (Pérez-de-Mora et al., 2011; Bolan et al., 2011). Recently, Xu et al. (2021) have published a systematic review on the different types of materials that can be used in the chemical stabilization of heavy metals in soils from studies published in the last decade (2010–2020), on the factors influencing their success, and on the assessment of the stabilization efficiency. Among other important conclusions, they have pointed to the main bottle-neck of the immobilization process in the long-term, which is much dependent on the temperature, on the acid deposition, and on the microbial activity, which needs to be controlled in the long term (Xu et al., 2021).

Other situations were identified where the application of amendments, biochar, and green-waste compost led to an increase in the soluble fraction, or mobilization of As and Cu in the soil pore water, whereas Zn and Cd were significantly immobilized, decreasing multi-contaminated soil phytotoxicity, like the situations identified by Beesley et al. (2010), when they added green waste-derived compost and verified that the elements behaved differently. This may happen when the organic amendments that are added, leading to an increase of the soluble organic compounds, with low molecular weight, establishing complexation with the metallic cations and bringing them to the solution (Beesley et al., 2010). Nevertheless, biochar application counteracted the compost effect and led to a decrease in polycyclic aromatic hydrocarbons (PAHs) toxicity which were also affecting the soil (an urban soil, also affected by organics).

4 Selecting appropriate crops for phystostabilization of contaminated soil

To successfully implement the use of phytostabilization processes in contaminated soils, adequate plants must be chosen that are able to grow in the adverse conditions that these soils present. Several different plant species have been shown to be able to tolerate contaminated soils, and thus being candidates for phytoremediation processes, but some of them are food crops and thus should not be used in these procedures as they present an increased risk of food chain contamination. Another factor to be taken into account is that sometimes plants are subjected to more than one type of environmental stress (soil contamination, salinity, drought, high temperatures, etc.) which makes it more difficult to evaluate and select proper plant candidates for these remediation processes (Mendez and Maier, 2008).

Good candidates for phytostabilization processes include plants that have an extensive root system are able to grow fast and able to tolerate the presence of the contaminants and that minimize the translocation of contaminants between the roots and the shoots (Alvarenga et al., 2008a). It should be noted that there is always some translocation of the contaminants to the aerial parts, dependent on the type of plant and contaminant and of the diverse environmental conditions and this should be always be taken into account in the planning and implementation of phytostabilization processes. Another relevant consideration is the use of native species in order to avoid the introduction of potentially invasive ones that could have negative consequences in the local ecosystems (Zhang et al., 2012). In contaminated sites (like old mining areas, for example), native plants have adapted to the stressful environment over time and can be good candidates for phytoremediation (Pošćić et al., 2024). Not only do different species have different accumulation properties that can make them more or less adequate for these remediation processes, but even different ecotypes might be more adequate due to the adaptative mechanisms of the plants in different environments.

Previous studies (Wu et al., 2018; Alvarenga et al., 2008a; Zhang et al., 2012) have considered four indexes to be taken into consideration in the selection of a plant for phytostabilization purposes with regard to heavy metal contamination (although this can be generalized for different contaminants):

- contaminant concentration in the roots (should be high, compared to other plant parts);
- translocation factor (TF), i.e. the ratio between contaminant concentration in the shoots in relation to the roots (should be less than 1);
- bioaccumulation coefficient (usually represented as BCF), i.e the ratio between contaminant concentration in the shoots in relation to the content in the soil (should be less than 1); and
- root biomass (should be as high as possible).

Other authors also report the importance of plants used for phytostabilization having a high bioconcentration factor (confusingly using the same abbreviation of BCF), the ratio between contaminant concentration in roots in relation to the soil rather than the absolute root concentration (Yoon et al., 2006).

Although there are many publications regarding the adequation of different plants for phytoremediation processes, they usually focus on phytoextraction (that is the subject of another chapter in this book). We will thus focus here on the use of plants in phytostabilization processes.

The large majority of the published studies deal with phytostabilization processes in regard to inorganic contaminants like heavy metals, frequently in contaminated soils near mines (Lebrun et al., 2021). Even so, there are some reports on the potential of using plants to stabilize organic contaminants, like per- and polyfluoroalkyl substances (PFASs) (Nason et al., 2024).

Some published studies involve pot or hydroponic experiments under controlled conditions (sometimes in unrealistically high concentrations of the contaminants) and usually a single contaminant is analysed and although this is understandable to isolate the toxic effects, most real-world contaminations are due to multiple contaminants. However, most phytostabilization studies deal with real situations usually using contaminated soils from mining areas, thus allowing much more definitive conclusions to be taken on the capability of certain plant species to be used in this type of phytomanagement process.

4.1 Herbaceous plants

Herbaceous ornamental plants present good characteristics that allow their use in phytoremediation processes like high adaptability, relatively large biomass and quick growth periods. Some species are also widely used, often in gardens, because of their nice and bright colours.

Wu et al. (2018) studied several ornamental plants for possible Cd bioremediation and identified *Sida rhombifolia* (also known as arrowleaf sida, of the family Malvaceae) as a candidate for phytostabilization processes. *Celosia argentea*, also known as plumed cockscomb or silver cock's comb, is a plant of the Amaranthaceae family that has been shown to be adequate for the phytostabilization of heavy metals contaminated soils in a pot experiment with different concentrations of Cu, Cr, Zn, and Pb (Hussain et al., 2024). Although the plants showed reduced biomass in contaminated soils, the authors claim that the plants are adequate for phytostabilization of heavy metals as they accumulate more in the roots than in the aerial parts. Curiously, in the previously cited study, Wu et al. (2018) also tested *C. argentea* and reported that it has medium tolerance for Cd but did not identify *C. argentea* use in phytostabilization.

In a study regarding the potential of local plants for phytoremediation processes in a mine area in Iran (Nouri et al., 2011), the authors identified several species as adequate for phytostabilization of different metals: *Scrophularia scoparia* for Pb, *Centaurea virgata*, *Echinophora platyloba*, and *Scariola orientalis* for Zn, and *Centaurea virgata* and *Cirsium congestum* for Mn. This study highlights the fact that the phytostabilization potential is highly dependent on plant species and type of contaminant.

Crinum bulbispermum plants (of the Amaryllidaceae family) have also been reported to accumulate different metals (Co, Cd, Cu, Mo, and Zn), with little or no translocation to the upper plant parts (Clarke et al., 2024). The authors thus conclude that the plant grown in contaminated sites can even be used for medicinal purposes, besides the use in phytostabilization. Other medicinal plants have also been described as having the potential for phytostabilization of contaminated soils (Pandey et al., 2023). The use of medicinal plants in phytoremediation processes must be carefully analysed to avoid the possibility of contamination of the plant parts that are used for medicinal purposes.

Plants of the *Lupinus* genus, like *Lupinus luteus* (yellow lupin) or *Lupinus albus* (white lupin), although being used as food or feed have been reported as having the potential to be used in phytostabilization due to the accumulation of metals mainly in the roots (Dary et al., 2010; Martínez-Alcalá et al., 2012; Garau et al., 2021). As these are annual crops, there is always the risk of releasing the contaminants absorbed by the roots back into the soil, after root decomposition, and this is a challenge that needs to be considered (Martínez-Alcalá et al., 2012).

The genus *Brassica* (of the Brassicaceae family) includes important agricultural and horticultural crops that are reported to be tolerant to different heavy metals and thus adequate for phytoremediation processes (Mourato et al., 2015; Bortoloti and Baron, 2022). While most are herbaceous annual plants, others are small shrubs. Several studies have been performed to evaluate the potential of different Brassica species (like *Brassica juncea* or *Brassica napus*) in the phytostabilization of contaminated soils, either by themselves or with the help of soil amendments (Visconti et al., 2020; Pérez-Esteban et al., 2014; Lacalle et al., 2023; Naveed et al., 2021).

4.2 Grasses

Several grass species have also been evaluated as having potential for phytoremediation due to their high tolerance to the presence of contaminants, fast growth and relatively high biomass (Hou et al., 2020; Xie and van Zyl, 2023). Rabêlo et al. (2021) have made an extensive review of the suitability of grasses in the phytoremediation of contaminated soils.

Plants from several genus of the Poaceae family have been reported to have adequate properties to use for phytostabilization purposes. Studies have been carried out on the ability of *Pennisetum purpureum* (also known as Napier grass or elephant grass) to stabilize dumpsites for fly ash, due to their tolerance to heavy metals (Gajaje et al., 2024). Another member of the Poaceae family that has been widely tested in phytostabilization processes due to its metal tolerance, is *Lolium perenne*, also called perennial ryegrass (Masotla et al., 2023). Although the plant has good characteristics for use in phytoremediation of contaminated soils, it can be improved by the addition of exogenous substances, like indoleacetic acid (Zhu et al., 2023), the addition of biosolids (Kacprzak et al., 2014; Alvarenga et al., 2009b) or other substances like biochar to the soil (Qiu et al., 2024). Species from the *Miscanthus* (or silvergrass) genus have also been studied for the same purposes (Nsanganwimana et al., 2021; Zgorelec et al., 2020). *Sorghum bicolor* and *Dactylis glomerata* are other members of the Poaceae family that have high tolerance to heavy metals and are thus candidates for use in phytostabilization (Anbuganesan et al., 2024; Visconti et al., 2020).

4.3 Shrubs

Shrubs have good characteristics for phytoremediation processes as they present a good canopy cover and also a deep root system that helps to prevent erosion (Mendez and Maier, 2008).

Tamburini et al. (2023) reported *Pistacia lentiscus* L. (mastic shrub) as an excluder metallophyte proposed for phytostabilization of highly contaminated mining sites in Sardinia, Italy. Previously, other endemic plants from the same location like *Helichrysum microphyllum* have been reported to have the potential to be used in phytostabilization strategies, using local species well adapted to the respective environment (Bacchetta et al., 2018).

Some halophyte plants, like those of the genus *Atriplex*, have also been considered to be adequate for phytostabilization due to their deep root system and ability to thrive not only in saline conditions but along with other contaminants (Eissa, 2015).

Plants of the genera Erica (Ericaceae family), like *Erica australis* and *Erica andevalensis*, have also been reported to be tolerant to different metals and able to grow in contaminated mining sites, (Pérez-López et al., 2014). These plants have thus been considered to have potential in phytostabilization techniques.

4.4 Trees

Deep-rooting woody species can be effective in reducing leaching and stabilizing contaminants, also reducing erosion, as long as they don't translocate them at a large scale to the upper plant parts (Mendez and Maier, 2008). In a study of a contaminated site in Spain (Domínguez et al., 2008), it was confirmed that white poplar (*Populus alba*) are able to extract metals from the soil to the shoots and are thus not suited to phytostabilization processes. Other tree species like Holm oak (*Quercus ilex*) and wild olive tree (*Olea europaea* var. sylvestris) were shown to be more adequate in this respect, by the same authors. In a study with several species, Evangelou et al. (2013) concluded that birch (*Betula pendula*) could be suitable for phytostabilization of soils with high Cd and Zn, while poplars (*Populus monviso*) and willows (*Salix viminalis*) could be used to stabilize soils with high Cu and Pb.

5 Monitoring immobilization/phytostabilization effectiveness

When using physicochemical methods of soil remediation, they are more effective in removing the contaminant from the soil, but often more expensive and sometimes environmentally disruptive, in the sense that they damage the soil's overall multifunctionality (Song et al., 2022). On the contrary, when using immobilization/phytostabilization, total and immediate removal of the contaminant does not occur, but a decrease in the heavy metals' bioavailability is achieved and the soil is again able to deliver ecosystem services (Epelde et al., 2009, 2014a, 2014b; Burges et al., 2016; Alvarenga et al., 2018). Therefore, this can be seen as a risk-based land management strategy, based on risk reduction, but the benefits and risks of its implementation should be assessed in an integrated way, using a set of chemical, biochemical and ecotoxicological/ecological parameters, able to demonstrate the recovery of soil functionality and the reduction of environmental risks. This is the only way to defend this soil remediation strategy, in which the decrease of the total concentration of contaminants in soil is not achieved, rather their immobilization and the recovery of soil health (Garbisu et al., 2011; Epelde et al., 2014a, 2014b; Alvarenga et al., 2018).

That is why it is now commonly accepted, with evidence from laboratory and field studies, that biological effects are not related to the total concentration of a contaminant in the soil, but rather to its bioavailable fraction. According to Alexander (2000), bioavailability represents the accessibility of a chemical for assimilation and its potential for toxicity. Therefore, organisms respond only to the fraction of the contaminants that are biologically available (bioavailable), which are dependent on a diverse set of factors, such as soil properties and various processes varying with time and on the biological receptors (Harmsen, 2007; Alvarenga et al., 2018; ISO, 2018).

The assumption of this fact from the International Organization for Standardization (ISO) was very important, and the delivery of the Guideline on 'Requirements and guidance for the selection and application of methods for the assessment of bioavailability of contaminants in soil and soil materials' (ISO, 2018), was crucial at the time. This ISO Guidance does not define a precise method to assess bioavailability but sets the framework for its necessity and for its use in risk evaluation. In fact, the bioavailable fraction of heavy metals in soils depends on soil properties and on different processes that vary with time, and on the behaviour of the target organisms (ISO, 2018; Harmsen, 2007).

According to the ISO Guideline, bioavailability may be assessed in two complementary ways by (1) chemical methods (e.g. using single and sequential extraction procedures), which determine a specific available fraction of a well-defined class of contaminants; and (2) by biological methods, with exposure of organisms to soils, evaluating effects on organisms and/or bioaccumulation (Peña et al., 2019; ISO, 2018; Alvarenga et al., 2012, 2008c; Peijnenburg et al., 2007). The ISO standard only assumes these possibilities, but without defining a precise protocol (ISO, 2018).

The fraction obtained using a chemical extraction procedure can be functionally defined (e.g. plant-available, mobile, mobilizable, or in exchangeable forms), or operationally defined (by the procedures or extractants used to isolate them) (Gupta et al., 1996; Peijnenburg et al., 2007). However, these methods are empirical, and highly dependent on the extraction protocol used, when in fact, the real bioavailability should be measured in 'real scenarios', or by putting organisms in contact with the contaminants and observing effects on organisms, which is in fact not possible in most of the cases in laboratorial conditions. Of course, the use of a single chemical extraction procedure in bioavailability studies is attractive from a laboratory-operational point of view: it is relatively simple to perform and to interpret, it is cost-effective, and, in general, very reproducible (Peijnenburg et al., 2007). However, it is difficult to establish a protocol that could be generalized by all the investigators, despite some efforts that have been made toward that objective (Gutiérrez et al., 2016). Single extractions using extractants such as deionized water, neutral salts, low-molecular-weight organic acids and chelating agents, like diethylene triamine pentaacetic acid (DTPA) or ethylene diamine tetraacetic acid (EDTA), have been used for metals (Zornoza et al., 2012; Galende et al., 2014; Pardo et al., 2014a), while in the case of metalloids, like, As, extractions have been made with $(NH_4)_2SO_4$ $NaHCO_3$, usually 0.5 M, which have shown to correlate well with As concentrations in plants growing in mining soils (Clemente et al., 2012; Arco-Lázaro et al., 2016; Moreno-Jiménez et al., 2013). The 0.01 M $CaCl_2$-extractable heavy metals concentration has been used in many countries and widely accepted as being able to extract the mobile, or readily phytoavailable fraction of heavy metals by many authors (Houba et al., 1996; Pueyo et al., 2004; Alvarenga et al., 2013). Lacalle et al. (2023) in a phytostabilization study, with both mining and agricultural soils, amended with compost, were able to evidence a relation between a reduction in the $CaCl_2$-extractable As, Cd, Cu, Pb, and Zn and a stimulation of microbial activity and a decrease in toxicity data. Studies like this are important to validate the use of simpler protocols, like chemical extraction, in the evaluation of risk assessment.

Brown and Chaney (2016) also defended that to evaluate the efficacy of metal mining-impacted sites, in most cases, measures of metal availability and ecosystem functions are related, and they give examples: surveys of small mammals on restored sites provide information on metal availability as well as on the suitability of the restoration of the habitat function.

Nowadays, the concept of availability is applied in the risk assessment of contaminants in some environmental risk assessment (ERA) projects and can be tailored to the specific protection goals. Depending on the intended use of a soil or soil material, soil characterization for different purposes (e.g. assessment of habitat and retention functions, risk assessment, and compliance with regulatory values) may include chemical testing and ecotoxicological testing with selected representative test organisms (Alvarenga et al., 2018). These tests will, in many cases, be soil- or site-specific at a given point in time and cannot be extrapolated to other soils or points in time where other factors may control bioavailability (Alvarenga et al., 2018; Harmsen, 2007; ISO, 2018).

However, regulatory contexts in different countries are still very conservative, and typically assume that the total concentration of a contaminant present in soil is available for uptake by organisms, including man, and, as such, risks are still overestimated in many environmental projects. Therefore, most of the regulatory documents have their limit values based on the determination of the total concentration of the contaminants (Alvarenga et al., 2018). It may be argued that the remediation criteria and risk evaluation should be supported by contaminants bioavailability assessment, but again, analytically speaking, standardization is easier to achieve when total concentrations are being considered and, therefore, limit values and guideline values are mostly based on total concentrations (Rodrigues et al., 2009a, 2009b; Mourinha et al., 2022).

In a very practical approach (Brown and Chaney, 2016), the effectiveness of a phytostabilization plan needs to be assessed at different levels: (1) evaluating soils' chemical properties related to its fertility and capacity to bear a plant cover (e.g. OM content, inorganic and organic N, extractable P and K, CEC, and concentrations of other macro and micronutrients); (2) soil's physical properties related to its structure, root penetration, water-holding capacity and infiltrability; (3) total and extractable heavy metal fractions to evaluate their (bio) availability and soil retention function; and (4) ecotoxicity assessment, using a battery of bioassays with organisms from different trophic levels to evaluate the soil habitat function (Brown and Chaney, 2016; Harmsen, 2007; ISO, 2018; Alvarenga et al., 2018; Xu et al., 2021).

Several authors defend that regulations on soil protection should have a holistic view of soil health in remediation projects, demanding that, besides chemical, and physical, indicators, contaminants' concentrations, and their bioavailability also biological and microbial indicators should be used as indicators of soil health (Garbisu et al., 2011; Epelde et al., 2014a, 2014b;

Lehmann et al., 2020). For instance, due to their sensitivity, rapid response and environmental relevance, microbial parameters, that estimate the abundance-biomass, activity and diversity of the soil microbiome should be used as bioindicators of soil health (Garbisu et al., 2011; Epelde et al., 2014a, 2014b). That is why, several authors defend their use in the assessment of remedial actions, and the capacity to protect soil health (Volchko et al., 2013, 2014; Gutiérrez et al., 2016).

More recently, advanced remote sensing techniques, using satellite images, are also able to help monitor the success of phytostabilization of mine sites, as Karan et al. (2016) have evidenced. In their study, they have evaluated different vegetation indices (e.g. Normalized Difference Vegetation Index (NDVI) and Normalized Difference Moisture Index (NDMI)), trying to relate these vegetation health indicators with the soil moisture content of the study area using remote sensing techniques. This study was applied to the remediation of an open-cast mine coal mine in India, images from the years 2000 to 2015 were used, and an increase of 21% was observed in the vegetation cover of the reclaimed sites, and NDVI results also indicated that vegetation health improved over the years (Karan et al., 2016).

6 Conclusion and future trends

- It is very important, with all the phytoremediation strategies, that ecosystem services are gradually restored in the contaminated sites, e.g. nutrient cycling, carbon storage, water flow regulation, erosion control, water purification, fertility maintenance, etc. which is a major outcome of these technologies relatively to all other non-biological remediation technologies (Burges et al., 2018).
- As Brown and Chaney (2016) have demonstrated, it is all a question of economy and sustainability. This technology is cheaper and the environmental impacts, globally, are lower; it can be shown by looking at how much we would spend with both options.
- We must move from lab scale and greenhouse to field scale and to demonstrate which agronomic practices are still missing to make the phytotechnologies feasible, not only a 'promising' technology (Kidd et al., 2015).

7 Acknowledgements

The authors acknowledges the support of FCT – Fundação para a Ciência e a Tecnologia, I.P., under the project UIDB/04129/2020 of LEAF-Linking Landscape, Environment, Agriculture and Food, Research Unit.

8 References

Alam M, et al. (2020) The effects of organic amendments on heavy metals bioavailability in mine impacted soil and associated human health risk. *Scientia Horticulturae* 262: 109067.

Al-Lami MK, et al. (2019) Amendment-assisted revegetation of mine tailings: improvement of tailings quality and biomass production. *International Journal of Phytoremediation* 21: 425–434.

Alexander M. (2000) Aging, bioavailability, and overestimation of risk from environmental pollutants. *Environmental Science & Technology* 34: 4259–4265.

Alvarenga P, et al. (2004) Elemental uptake and root-leaves transfer in Cistus Ladanifer L. growing in a contaminated pyrite mining area (Aljustrel-Portugal). *Water, Air, and Soil Pollution* 152: 81–96.

Alvarenga P, et al. (2008a) Evaluation of composts and liming materials in the phytostabilization of a mine soil using perennial ryegrass. *Science of the Total Environment* 406: 43–56.

Alvarenga P, et al. (2008b) Assessment of chemical, biochemical and ecotoxicological aspects in a mine soil amended with sludge of either urban or industrial origin. *Chemosphere* 72: 1774–1781.

Alvarenga P, et al. (2008c) Evaluation of tests to assess the quality of mine-contaminated soils. *Environmental Geochemistry and Health* 30: 95–99.

Alvarenga P, et al. (2009a) Reclamation of a mine contaminated soil using biologically reactive organic matrices. *Waste Management & Research* 27: 101–111.

Alvarenga P, et al. (2009b) Organic residues as immobilizing agents in aided phytostabilization: (I) effects on soil chemical characteristics. *Chemosphere* 74: 1292–1300.

Alvarenga P, et al. (2009c) Organic residues as immobilizing agents in aided phytostabilization: (II) Effects on soil biochemical and ecotoxicological characteristics. *Chemosphere* 74: 1301–1308.

Alvarenga P, et al. (2012) A contribution towards the risk assessment of soils from the São Domingos Mine (Portugal): chemical, microbial and ecotoxicological indicators. *Environmental Pollution* 161: 50–56.

Alvarenga P, et al. (2013) A study on As, Cu, Pb and Zn (bio)availability in an abandoned mine area (São Domingos, Portugal) using chemical and ecotoxicological tools. *Environmental Science and Pollution Research* 20: 6539–6550.

Alvarenga P, et al. (2014) The effect of compost treatments and a plant cover with agrostis tenuis on the immobilization/mobilization of trace elements in a mine-contaminated soil. *International Journal of Phytoremediation* 16: 138–154.

Alvarenga P, et al. (2018) Chemical and ecotoxicological effects of the use of drinking-water treatment residuals for the remediation of soils degraded by mining activities. *Ecotoxicology and Environmental Safety* 161: 281–289.

Alvarenga P, et al. (2019) Use of wastes from the pulp and paper industry for the remediation of soils degraded by mining activities: chemical, biochemical and ecotoxicological effects. *Science of the Total Environment* 686: 1152–1163.

Alvarenga P, et al. (2020) Chapter 5 - Managing organic amendments in agroecosystems to enhance soil carbon storage and mitigate climate change. In: Prasad MNV and Pietrzykowski M (eds) *Climate Change and Soil Interactions*. Elsevier, 89–141.

Alvarenga P, et al. (2021) Assessment of the environmental impact of acid mine drainage on surface water, stream sediments, and macrophytes using a battery of chemical and ecotoxicological indicators. *Water* 13: 1436.

Alvarenga P, et al. (2022) Combined use of olive mill waste compost and sprinkler irrigation to decrease the risk of As and Cd accumulation in rice grain. *Science of the Total Environment* 835: 155488.

Anbuganesan V, et al. (2024) Synergistic effect of biochar and plant growth promoting bacteria improve the growth and phytostabilization potential of Sorghum bicolor in Cd and Zn contaminated soils. *Rhizosphere* 29: 100844.

Arco-Lázaro E, et al. (2016) Arsenic(V) adsorption-desorption in agricultural and mine soils: effects of organic matter addition and phosphate competition. *Environmental Pollution* 216: 71–79.

Arenas-Lago D, et al. (2019) Is nanoremediation an effective tool to reduce the bioavailable As, Pb and Sb contents in mine soils from Iberian Pyrite Belt? *CATENA* 176: 362–371.

Bacchetta G, et al. (2015) Use of native plants for the remediation of abandoned mine sites in mediterranean semiarid environments. *Bulletin of Environmental Contamination and Toxicology* 94: 326–333.

Bacchetta G, et al. (2018) Metal tolerance capability of helichrysum microphyllum Cambess. subsp. tyrrhenicum Bacch., Brullo & Giusso: a candidate for phytostabilization in abandoned mine sites. *Bulletin of Environmental Contamination and Toxicology* 101: 758–765.

Baker AJM, et al. (2010) Metallophytes: the unique biological resource, its ecology and conservational status in Europe, central Africa and Latin America. In: Batty LC and Hallberg KB (eds) *Ecology of Industrial Pollution*. Cambridge University Press, 7–40.

Barbillon A, et al. (2019) Health risks assessment in three urban farms of Paris region for different scenarios of urban agricultural users: a case of soil trace metals contamination. *Agricultural Sciences* 10: 352–370.

Basta NT, et al. (2016) Restoring ecosystem function in degraded urban soil using biosolids, biosolids blend, and compost. *Journal of Environmental Quality* 45: 74–83.

Beesley L, et al. (2010) Effects of biochar and greenwaste compost amendments on mobility, bioavailability and toxicity of inorganic and organic contaminants in a multi-element polluted soil. *Environmental Pollution* 158: 2282–2287.

Bhattacharyya PN, et al. (2024) Biochar as soil amendment in climate-smart agriculture: opportunities, future prospects, and challenges. *Journal of Soil Science and Plant Nutrition* 24: 135–158.

Bolan NS, et al. (2011) Chapter four - phytostabilization: a Green approach to contaminant containment. In: Sparks DL (ed) *Advances in Agronomy*. Academic Press, 145–204.

Bolan NS, et al. (2014) Remediation of heavy metal(loid)s contaminated soils – To mobilize or to immobilize? *Journal of Hazardous Materials* 266: 141–166.

Bortoloti GA and Baron D. (2022) Phytoremediation of toxic heavy metals by Brassica plants: A biochemical and physiological approach. *Environmental Advances* 8: 100204.

Brady NC and Weil RR. (2017) *The Nature and Properties of Soils*. Pearson.

Briffa J, et al. (2020) Heavy metal pollution in the environment and their toxicological effects on humans. *Heliyon* 6: 24–26.

Brown SL and Chaney RL. (2016) Use of amendments to restore ecosystem function to metal mining-impacted sites: tools to evaluate efficacy. *Current Pollution Reports* 2: 91–102.

Burges A, et al. (2018) From phytoremediation of soil contaminants to phytomanagement of ecosystem services in metal contaminated sites. *International Journal of Phytoremediation* 20: 384–397.

Burges A, et al. (2016) Enhancement of ecosystem services during endophyte-assisted aided phytostabilization of metal contaminated mine soil. *Science of the Total Environment* 562: 480–492.

Burgos P, et al. (2010) By-products as amendment to improve biochemical properties of trace element contaminated soils: effects in time. *International Biodeterioration & Biodegradation* 64: 481–488.

Clarke VC, et al. (2024) Crinum bulbispermum, a medicinal geophyte with phytostabilization properties in metal-enriched mine tailings. *Plants* 13: 79.

Clemente R, et al. (2005) Uptake of heavy metals and As by Brassica juncea grown in a contaminated soil in Aznalcollar (Spain): the effect of soil amendments. *Environmental Pollution* 138: 46–58.

Clemente R, et al. (2006) A remediation strategy based on active phytoremediation followed by natural attenuation in a soil contaminated by pyrite waste. *Environmental Pollution* 143: 397–406.

Clemente R, et al. (2010) Trace element mobility in a contaminated soil two years after field-amendment with a greenwaste compost mulch. *Environmental Pollution* 158: 1644–1651.

Clemente R, et al. (2012) The use of a halophytic plant species and organic amendments for the remediation of a trace elements-contaminated soil under semi-arid conditions. *Journal of Hazardous Materials* 223–224: 63–71.

Clemente R, et al. (2015) Food byproducts as amendments in trace elements contaminated soils. *Food Research International* 73: 176–189.

Coles R and Costa S. (2018) Food growing in the city: exploring the productive urban landscape as a new paradigm for inclusive approaches to the design and planning of future urban open spaces. *Landscape and Urban Planning* 170: 1–5.

Cundy AB, et al. (2013) Developing principles of sustainability and stakeholder engagement for "gentle" remediation approaches: the European context. *Journal of Environmental Management* 129: 283–291.

Cundy AB, et al. (2016) Brownfields to green fields: realising wider benefits from practical contaminant phytomanagement strategies. *Journal of Environmental Management* 184: 67–77.

Dary M, et al. (2010) "In situ" phytostabilisation of heavy metal polluted soils using Lupinus luteus inoculated with metal resistant plant-growth promoting rhizobacteria. *Journal of Hazardous Materials* 177: 323–330.

de Varennes A, et al. (2010a) Enzymatic activity of a mine soil varies according to vegetation cover and level of compost applied. *International Journal of Phytoremediation* 12: 371–383.

de Varennes A, et al. (2010b) Amendment of an acid mine soil with compost and polyacrylate polymers enhances enzymatic activities but may change the distribution of plant species. *Water, Air, and Soil Pollution* 208: 91–100.

Demeyer A, et al. (2001) Characteristics of wood ash and influence on soil properties and nutrient uptake: an overview. *Bioresource Technology* 77: 287–295.

Domínguez MT, et al. (2008) Trace element accumulation in woody plants of the Guadiamar Valley, SW Spain: a large-scale phytomanagement case study. *Environmental Pollution* 152: 50–59.

Duarte B, et al. (2023) Cistus ladanifer metal uptake and physiological performance in biochar amended mine soils. *South African Journal of Botany* 153: 246–257.

Duffus JH. (2002) "Heavy metals" a meaningless term? (IUPAC Technical Report). *Pure and Applied Chemistry* 74: 793–807.

Edenborn SL, et al. (2015) Influence of biochar application methods on the phytostabilization of a hydrophobic soil contaminated with lead and acid tar. *Journal of Environmental Management* 150: 226–234.

Egendorf SP, et al. (2020) The limits of lead (Pb) phytoextraction and possibilities of phytostabilization in contaminated soil: a critical review. *International Journal of Phytoremediation* 22: 916–930.

Eissa MA. (2015) Impact of compost on metals phytostabilization potential of two halophytes species. *International Journal of Phytoremediation* 17: 662–668.

Epelde L, et al. (2009) Evaluation of the efficiency of a phytostabilization process with biological indicators of soil health. *Journal of Environmental Quality* 38: 2041–2049.

Epelde L, et al. (2014a) Adaptive long-term monitoring of soil health in metal phytostabilization: ecological attributes and ecosystem services based on soil microbial parametersxs. *International Journal of Phytoremediation* 16: 971–981.

Epelde L, et al. (2014b) Microbial properties and attributes of ecological relevance for soil quality monitoring during a chemical stabilization field study. *Applied Soil Ecology* 75: 1–12.

Evangelou MWH, et al. (2013) Metal uptake and allocation in trees grown on contaminated land: implications for biomass production. *International Journal of Phytoremediation* 15: 77–90.

Farhat YA, et al. (2021) Altered arsenic availability, uptake, and allocation in rice under elevated temperature. *Science of the Total Environment* 763: 143049.

Gajaje K, et al. (2024) Phytostabilization of fly ash from a coalmine in Botswana and biovalorisation of the recovered Napier grass (Pennisetum purpureum Schumach.). *International Journal of Phytoremediation*: 1–14.

Galende MA, et al. (2014) Field assessment of the effectiveness of organic amendments for aided phytostabilization of a Pb–Zn contaminated mine soil. *Journal of Geochemical Exploration* 145: 181–189.

Garaiyurrebaso O, et al. (2016) Long-term effects of aided phytostabilisation on microbial communities of metal-contaminated mine soil. *FEMS Microbiology Ecology* 93.

Garau G, et al. (2007) Influence of red mud, zeolite and lime on heavy metal immobilization, culturable heterotrophic microbial populations and enzyme activities in a contaminated soil. *Geoderma* 142: 47–57.

Garau M, et al. (2021) Combining grass and legume species with compost for assisted phytostabilization of contaminated soils. *Environmental Technology & Innovation* 22: 101387.

Garbisu C, et al. (2011) Assessment of soil quality using microbial properties and attributes of ecological relevance. *Applied Soil Ecology* 49: 1–4.

García-Sánchez M, et al. (2015) Effect of digestate and fly ash applications on soil functional properties and microbial communities. *European Journal of Soil Biology* 71: 1–12.

Ghosh AK and Bhattacharyya P. (2004) Arsenate sorption by reduced and reoxidised rice soils under the influence of organic matter amendments. *Environmental Geology* 45: 1010–1016.

Ghosh D and Maiti SK. (2021) Biochar assisted phytoremediation and biomass disposal in heavy metal contaminated mine soils: a review. *International Journal of Phytoremediation* 23: 559–576.

Gómez V, et al. (2023) Trace elements in fertilizers used in Peru. *OnLine Journal of Biological Sciences* 23: 219–225.

Gómez-Sagasti MT, et al. (2018) How valuable are organic amendments as tools for the phytomanagement of degraded soils? The knowns, known unknowns, and unknowns. *Frontiers in Sustainable Food Systems* 2. https://doi.org/10.3389/fsufs.2018.00068

Gupta SK, et al. (1996) The importance of mobile, mobilisable and pseudo total heavy metal fractions in soil for three-level risk assessment and risk management. *Science of the Total Environment* 178: 11–20.

Gustin MS, et al. (2021) The term "heavy metal(s)": history, current debate, and future use. *Science of the Total Environment* 789: 147951.

Gutiérrez M, et al. (2016) Abandoned PbZn mining wastes and their mobility as proxy to toxicity: A review. *Science of the Total Environment* 565: 392–400.

Harmsen J. (2007) Measuring bioavailability: from a scientific approach to standard methods. *Journal of Environmental Quality* 36: 1420–1428.

He ZLL, et al. (2005) Trace elements in agroecosystems and impacts on the environment. *Journal of Trace Elements in Medicine and Biology* 19: 125–140.

Hidalgo J, et al. (2022) Zero-valent iron nanoparticles and organic amendment assisted rhizoremediation of mixed contaminated soil using Brassica napus. *Environmental Technology & Innovation* 28: 102621.

Hinojosa MB, et al. (2008) Effects of pyrite sludge pollution on soil enzyme activities: ecological dose–response model. *Science of the Total Environment* 396: 89–99.

Hossain M, et al. (2021) Arsenic dynamics in paddy soil under traditional manuring practices in Bangladesh. *Environmental Pollution* 268: 115821.

Hou X, et al. (2020) Potential phytoremediation of soil cadmium and zinc by diverse ornamental and energy grasses. *BioResources* 15: 616–640.

Houba VJG, et al. (1996) State of the art and future developments in soil analysis for bioavailability assessment. *Science of the Total Environment* 178: 21–28.

Hussain U, et al. (2024) Phytoremediation of heavy metals spiked soil by Celosia argentea L.: effect on plant growth and metal stabilization. *Environ Sci Pollut Res Int* 31: 15339–15347.

Hussain Lahori A, et al. (2017) Potential use of lime combined with additives on (im) mobilization and phytoavailability of heavy metals from Pb/Zn smelter contaminated soils. *Ecotoxicology and Environmental Safety* 145: 313–323.

Ishfaq M, et al. (2020) Alternate wetting and drying: a water-saving and ecofriendly rice production system. *Agricultural Water Management* 241: 106363.

ISO IOfS. (2018) *Soil Quality - Requirements and Guidance for the Selection and Application of Methods for the Assessment of Bioavailability of Contaminants in Soil and Soil Materials*. Geneva, Switzerland.

Kabata-Pendias A. (2011) *Trace Elements in Soils and Plants*. CRC Press.

Kacprzak M, et al. (2014) Efficacy of biosolids in assisted phytostabilization of metalliferous acidic sandy soils with five grass species. *International Journal of Phytoremediation* 16: 593–608.

Karaca O, et al. (2018) Mine tailing disposal sites: contamination problems, remedial options and phytocaps for sustainable remediation. *Reviews in Environmental Science and Bio/Technology* 17: 205–228.

Karami N, et al. (2011) Efficiency of green waste compost and biochar soil amendments for reducing lead and copper mobility and uptake to ryegrass. *Journal of Hazardous Materials* 191: 41–48.

Karan SK, et al. (2016) Assessment of the capability of remote sensing and GIS techniques for monitoring reclamation success in coal mine degraded lands. *Journal of Environmental Management* 182: 272–283.

Kefeni KK, et al. (2017) Acid mine drainage: prevention, treatment options, and resource recovery: a review. *Journal of Cleaner Production* 151: 475–493.

Khalid S, et al. (2017) A comparison of technologies for remediation of heavy metal contaminated soils. *Journal of Geochemical Exploration* 182: 247–268.

Kidd P, et al. (2009) Trace element behaviour at the root–soil interface: Implications in phytoremediation. *Environmental and Experimental Botany* 67: 243–259.

Kidd P, et al. (2015) Agronomic practices for improving gentle remediation of trace element-contaminated soils. *International Journal of Phytoremediation* 17: 1005–1037.

Kidd PS, et al. (2021) Beneficial traits of root endophytes and rhizobacteria associated with plants growing in phytomanaged soils with mixed trace metal-polycyclic aromatic hydrocarbon contamination. *Chemosphere* 277: 130272.

Kiran BR and Prasad MNV. (2019) Biochar and rice husk ash assisted phytoremediation potentials of Ricinus communis L. for lead-spiked soils. *Ecotoxicology and Environmental Safety* 183: 109574.

Kumpiene J, et al. (2009) Microbial biomass, respiration and enzyme activities after in situ aided phytostabilization of a Pb- and Cu-contaminated soil. *Ecotoxicology and Environmental Safety* 72: 115–119.

Kumpiene J, et al. (2012) Arsenic fractionation in mine spoils 10 years after aided phytostabilization. *Environmental Pollution* 166: 82–88.

Kumpiene J, et al. (2019) In situ chemical stabilization of trace element-contaminated soil – Field demonstrations and barriers to transition from laboratory to the field – A review. *Applied Geochemistry* 100: 335–351.

Kuppens T, et al. (2015) Techno-economic assessment of fast pyrolysis for the valorization of short rotation coppice cultivated for phytoextraction. *Journal of Cleaner Production* 88: 336–344.

Lacalle RG, et al. (2018) Brassica napus has a key role in the recovery of the health of soils contaminated with metals and diesel by rhizoremediation. *Science of the Total Environment* 618: 347–356.

Lacalle RG, et al. (2020) Biological methods of polluted soil remediation for an effective economically-optimal recovery of soil health and ecosystem services. *Journal of Environmental Science and Public Health* 4: 112–133.

Lacalle RG, et al. (2023) Phytostabilization of soils contaminated with As, Cd, Cu, Pb and Zn: Physicochemical, toxicological and biological evaluations. *Soil & Environmental Health* 1: 100014.

Lal R, et al. (2018) The carbon sequestration potential of terrestrial ecosystems. *Journal of Soil and Water Conservation* 73: 145A–152A.

Lebrun M, et al. (2018) Eco-restoration of a mine technosol according to biochar particle size and dose application: study of soil physico-chemical properties and phytostabilization capacities of Salix viminalis. *Journal of Soils and Sediments* 18: 2188–2202.

Lebrun M, et al. (2021) Rehabilitation of mine soils by phytostabilization: does soil inoculation with microbial consortia stimulate Agrostis growth and metal(loid) immobilization? *Science of the Total Environment* 791: 148400.

Lee S-H, et al. (2017) Reclamation of mine-degraded agricultural soils from metal mining: lessons from 4 years of monitoring activity in Korea. *Environmental Earth Sciences* 76: 720.

Lehmann J, et al. (2020) The concept and future prospects of soil health. *Nature Reviews Earth & Environment* 1: 544–553.

Leitão TE, et al. (2018) Environmental quality in urban allotment gardens: atmospheric deposition, soil, water and vegetable assessment at LISBON city. *Water, Air, & Soil Pollution* 229: 31.

López-Piñeiro A, et al. (2007) Changes in organic matter and residual effect of amendment with two-phase olive-mill waste on degraded agricultural soils. *Science of the Total Environment* 378: 84–89.

López-Piñeiro A, et al. (2011) Long-term impacts of de-oiled two-phase olive mill waste on soil chemical properties, enzyme activities and productivity in an olive grove. *Soil and Tillage Research* 114: 175–182.

Lwin CS, et al. (2018) Application of soil amendments to contaminated soils for heavy metal immobilization and improved soil quality–a critical review. *Soil Science and Plant Nutrition* 64: 156–167.

Madejon E, et al. (2006) Soil amendments reduce trace element solubility in a contaminated soil and allow regrowth of natural vegetation. *Environmental Pollution* 139: 40–52.

Madejón E, et al. (2009) Trace elements, pH and organic matter evolution in contaminated soils under assisted natural remediation: a 4-year field study. *Journal of Hazardous Materials* 162: 931–938.

Mahar A, et al. (2016) Challenges and opportunities in the phytoremediation of heavy metals contaminated soils: a review. *Ecotoxicology and Environmental Safety* 126: 111–121.

Maneepitak S, et al. (2019) Effects of water and rice straw management practices on water savings and greenhouse gas emissions from a double-rice paddy field in the Central Plain of Thailand. *European Journal of Agronomy* 107: 18–29.

Martín-Franco C, et al. (2023) Effects of fresh and field-aged holm-oak biochar on As, Cd and Pb bioaccumulation in different rice growing environments. *Science of the Total Environment* 887: 164012.

Martínez-Alcalá I, et al. (2012) Efficiency of a phytoimmobilisation strategy for heavy metal contaminated soils using white lupin. *Journal of Geochemical Exploration* 123: 95–100.

Masotla MKL, et al. (2023) Extraction potential of Lolium perenne L. (Perennial Rye Grass) for metals in landfill soil: its tolerance and defense strategies. *Minerals* 13: 873.

Meirong F, et al. (2012) Effects of red mud on the remediation of Pb, Zn and Cd in heavy metal contaminated paddy soil. *Agricultural Science & Technology* 2: 388–392.

Mendez MO and Maier RM. (2008) Phytostabilization of mine tailings in arid and semiarid environments--an emerging remediation technology. *Environ Health Perspect* 116: 278–283.

Míguez F, et al. (2020) In situ phytomanagement with Brassica napus and bio-stabilised municipal solid wastes is a suitable strategy for redevelopment of vacant urban land. *Urban Forestry & Urban Greening* 47: 126550.

Mora APd, et al. (2005) Changes in enzyme activities and microbial biomass after "in situ" remediation of a heavy metal-contaminated soil. *Applied Soil Ecology* 28: 125–137.

Moreira H, et al. (2021) Phytomanagement of metal(loid)-contaminated soils: options, efficiency and value. *Frontiers in Environmental Science* 9.

Moreno-Jiménez E, et al. (2013) Arsenic and selenium mobilisation from organic matter treated mine spoil with and without inorganic fertilisation. *Environmental Pollution* 173: 238–244.

Moreno-Jiménez E, et al. (2014) Sprinkler irrigation of rice fields reduces grain arsenic but enhances cadmium. *Science of the Total Environment* 485–486: 468–473.

Mourato M, et al. (2015) Effect of heavy metals in plants of the Genus Brassica. *International Journal of Molecular Sciences* 16: 17975.

Mourinha C, et al. (2022) Potentially toxic elements' contamination of soils affected by mining activities in the Portuguese sector of the iberian pyrite belt and optional remediation actions: a review. *Environments* 9: 11.

Nagy A, et al. (2023) Composted sewage sludge utilization in phytostabilization of heavy metals contaminated soils. *International Journal of Phytoremediation* 25: 1510–1523.

Naidu R. (2013) Recent advances in contaminated site remediation. *Water, Air, & Soil Pollution* 224: 1705.

Nakamaru YM and Martín Peinado FJ. (2017) Effect of soil organic matter on antimony bioavailability after the remediation process. *Environmental Pollution* 228: 425–432.

Nason SL, et al. (2024) A comprehensive trial on PFAS remediation: hemp phytoextraction and PFAS degradation in harvested plants. *Environmental Science: Advances* 3: 304–313.

Naveed M, et al. (2021) Co-composted biochar enhances growth, physiological, and phytostabilization efficiency of brassica napus and reduces associated health risks under chromium stress. *Frontiers in Plant Science* 12. https://doi.org/10.3389/fpls.2021.775785

Nedjimi B. (2021) Phytoremediation: a sustainable environmental technology for heavy metals decontamination. *SN Applied Sciences* 3: 286.

Niu Z, et al. (2022) Effects of application of rice husk biochar and limestone on cadmium accumulation in wheat under glasshouse and field conditions. *Scientific Reports* 12: 21929.

Nouri J, et al. (2011) Phytoremediation potential of native plants grown in the vicinity of Ahangaran lead–zinc mine (Hamedan, Iran). *Environmental Earth Sciences* 62: 639–644.

Nsanganwimana F, et al. (2021) Potentials of Miscanthus x giganteus for phytostabilization of trace element-contaminated soils: ex situ experiment. *Ecotoxicology and Environmental Safety* 214: 112125.

Nziguheba G and Smolders E. (2008) Inputs of trace elements in agricultural soils via phosphate fertilizers in European countries. *Science of the Total Environment* 390: 53–57.

Oustriere N, et al. (2016) Aided phytostabilization of a trace element-contaminated technosol developed on steel mill wastes. *Journal of Hazardous Materials* 320: 458–468.

Palansooriya KN, et al. (2020) Soil amendments for immobilization of potentially toxic elements in contaminated soils: a critical review. *Environment International* 134: 105046.

Pandey V, et al. (2023) Phyto-exclusion of Pb and Cd by different genotypes of Andrographis paniculata (Burm. F.) nees: a novel approach for safe cultivation. *Industrial Crops and Products* 191: 115977.

Paradelo R, et al. (2019) Soil physical properties of a Luvisol developed on loess after 15 years of amendment with compost. *Soil and Tillage Research* 191: 207–215.

Pardo T, et al. (2011) Effects of compost, pig slurry and lime on trace element solubility and toxicity in two soils differently affected by mining activities. *Chemosphere* 84: 642–650.

Pardo T, et al. (2014a) Efficiency of soil organic and inorganic amendments on the remediation of a contaminated mine soil: II. Biological and ecotoxicological evaluation. *Chemosphere* 107: 101–108.

Pardo T, et al. (2014b) Evaluation of the phytostabilisation efficiency in a trace elements contaminated soil using soil health indicators. *Journal of Hazardous Materials* 268: 68–76.

Parra A, et al. (2016) Evaluation of the suitability of three Mediterranean shrub species for phytostabilization of pyritic mine soils. *CATENA* 136: 59–65.

Pardo T, et al. (2017) Phytostabilisation of severely contaminated mine tailings using halophytes and field addition of organic and inorganic amendments. *Chemosphere* 178: 556–564.

Pavel P-B, et al. (2014) Aided phytostabilization using Miscanthus sinensis×giganteus on heavy metal-contaminated soils. *Science of the Total Environment* 479–480: 125–131.

Peijnenburg WJGM, et al. (2007) Monitoring metals in terrestrial environments within a bioavailability framework and a focus on soil extraction. *Ecotoxicology and Environmental Safety* 67: 163–179.

Peña A, et al. (2015) Improving the mining soil quality for a vegetation cover after addition of sewage sludges: inorganic ions and low-molecular-weight organic acids in the soil solution. *Journal of Environmental Management* 150: 216–225.

Peña D, et al. (2019) Effects of olive mill wastes with different degrees of maturity on behaviour of S-metolachlor in three soils. *Geoderma* 348: 86–96.

Pérez-de-Mora A, et al. (2006) Microbial community structure and function in a soil contaminated by heavy metals: effects of plant growth and different amendments. *Soil Biology and Biochemistry* 38: 327–341.

Pérez-de-Mora A, et al. (2011) Phytostabilization of semiarid soils residually contaminated with trace elements using by-products: sustainability and risks. *Environmental Pollution* 159: 3018–3027.

Pérez-Esteban J, et al. (2014) Phytostabilization of metals in mine soils using Brassica juncea in combination with organic amendments. *Plant and Soil* 377: 97–109.

Pérez-López R, et al. (2014) Erica andevalensis and Erica australis growing in the same extreme environments: phytostabilization potential of mining areas. *Geoderma* 230–231: 194–203.

Pošćić F, et al. (2024) Dianthus sylvestris subsp. sylvestris as a promising candidate for phytostabilization of copper-contaminated post-mining sites in Alpine ecosystems. *Nordic Journal of Botany* 2024: e04199.

Pueyo M, et al. (2004) Assessment of CaCl2, NaNO3 and NH4NO3 extraction procedures for the study of Cd, Cu, Pb and Zn extractability in contaminated soils. *Analytica Chimica Acta* 504: 217–226.

Qi S, et al. (2023) Optimization strategies for Cd and Pb immobilization in soil using meta-analysis combined with numerical modeling. *Pedosphere* 33: 61–73.

Qiu J, et al. (2024) Influence of biochar addition and plant management (cutting and time) on ryegrass growth and migration of As and Pb during phytostabilization. *Science of the Total Environment* 913: 169771.

Quintela-Sabarís C, et al. (2017) Assessing phytotoxicity of trace element-contaminated soils phytomanaged with gentle remediation options at ten European field trials. *Science of the Total Environment* 599–600: 1388–1398.

Rabêlo FHS, et al. (2021) Are grasses really useful for the phytoremediation of potentially toxic trace elements? A review. *Frontiers in Plant Science* 12.

Ram LC and Masto RE. (2014) Fly ash for soil amelioration: a review on the influence of ash blending with inorganic and organic amendments. *Earth-Science Reviews* 128: 52–74.

Robinson BH, et al. (2015) Phytoextraction: where's the action? *Journal of Geochemical Exploration* 151: 34–40.

Rodrigues SM, et al. (2009a) A review of regulatory decisions for environmental protection: Part I–Challenges in the implementation of national soil policies. *Environment International* 35: 202–213.

Rodrigues SM, et al. (2009b) A review of regulatory decisions for environmental protection: Part II–The case-study of contaminated land management in Portugal. *Environment International* 35: 214–225.

Santos ES, et al. (2014) Effects of organic/inorganic amendments on trace elements dispersion by leachates from sulfide-containing tailings of the São Domingos mine, Portugal. Time evaluation. *Geoderma* 226–227: 188–203.

Santos ES, et al. (2016) Chemical quality of leachates and enzymatic activities in Technosols with gossan and sulfide wastes from the São Domingos mine. *Journal of Soils and Sediments* 16: 1366–1382.

Santos ES, et al. (2019) Rehabilitation of mining areas through integrated biotechnological approach: technosols derived from organic/inorganic wastes and autochthonous plant development. *Chemosphere* 224: 765–775.

Sarker AH, et al. (2019) A framework for integrating agriculture in urban sustainability in Australia. *Urban Science* 3: 50.

Sengupta S, et al. (2021) Deficit irrigation and organic amendments can reduce dietary arsenic risk from rice: introducing machine learning-based prediction models from field data. *Agriculture, Ecosystems & Environment* 319: 107516.

Skar SLG, et al. (2020) Urban agriculture as a keystone contribution towards securing sustainable and healthy development for cities in the future. *Blue-Green Systems* 2: 1–27.

Šlejkovec Z, et al. (2021) Arsenic speciation and elemental composition of rice samples from the Slovenian market. *Food Chemistry* 342: 128348.

Somerville PD, et al. (2018) Effects of deep tillage and municipal green waste compost amendments on soil properties and tree growth in compacted urban soils. *Journal of Environmental Management* 227: 365–374.

Song P, et al. (2022) Recent advances in soil remediation technology for heavy metal contaminated sites: a critical review. *Science of the Total Environment* 838: 156417.

Spanu A, et al. (2020) Effect of the irrigation method and genotype on the bioaccumulation of toxic and trace elements in rice. *Science of the Total Environment* 748: 142484.

Stouraiti C, et al. (2002) Reduction of Pb, Zn and Cd availability from tailings and contaminated soils by the application of Lignite Fly Ash. *Water, Air, and Soil Pollution* 137: 247–265.

Suresh B and Ravishankar GA. (2004) Phytoremediation–A novel and promising approach for environmental clean-up. *Critical Reviews in Biotechnology* 24: 97–124.

Tamburini E, et al. (2023) Metal phytostabilization by mastic shrub (Pistacia lentiscus L.) and its root-associated bacteria in different habitats of Sardinian abandoned mining areas (Italy). *Environmental Science and Pollution Research* 30: 122107–122120.

Thavamani P, et al. (2015) Risk based land management requires focus beyond the target contaminants–A case study involving weathered hydrocarbon contaminated soils. *Environmental Technology & Innovation* 4: 98–109.

Thompson PB. (2012) The agricultural ethics of biofuels: the food vs. fuel debate. *Agriculture* 2: 339–358.

Tordoff GM, et al. (2000) Current approaches to the revegetation and reclamation of metalliferous mine wastes. *Chemosphere* 41: 219–228.

Touceda-González M, et al. (2017a) Aided phytostabilisation reduces metal toxicity, improves soil fertility and enhances microbial activity in Cu-rich mine tailings. *Journal of Environmental Management* 186: 301–313.

Touceda-González M, et al. (2017b) Microbial community structure and activity in trace element-contaminated soils phytomanaged by Gentle Remediation Options (GRO). *Environmental Pollution* 231: 237–251.

Valente T, et al. (2012) Natural stabilization of mine waste-dumps – Evolution of the vegetation cover in distinctive geochemical and mineralogical environments. *Journal of Geochemical Exploration* 123: 152–161.

Visconti D, et al. (2020) Use of Brassica juncea and Dactylis glomerata for the phytostabilization of mine soils amended with compost or biochar. *Chemosphere* 260: 127661.

Volchko Y, et al. (2013) Incorporating the soil function concept into sustainability appraisal of remediation alternatives. *Journal of Environmental Management* 129: 367–376.

Volchko Y, et al. (2014) Using soil function evaluation in multi-criteria decision analysis for sustainability appraisal of remediation alternatives. *Science of the Total Environment* 485–486: 785–791.

Wang L, et al. (2020) Field trials of phytomining and phytoremediation: a critical review of influencing factors and effects of additives. *Critical Reviews in Environmental Science and Technology* 50: 2724–2774.

Wang N, et al. (2017) Biochar increases arsenic release from an anaerobic paddy soil due to enhanced microbial reduction of iron and arsenic. *Environmental Pollution* 220: 514–522.

Wu M, et al. (2018) Screening ornamental plants to identify potential Cd hyperaccumulators for bioremediation. *Ecotoxicology and Environmental Safety* 162: 35–41.

Wu Q, et al. (2021) Water management of alternate wetting and drying combined with phosphate application reduced lead and arsenic accumulation in rice. *Chemosphere* 283: 131043.

Wuana RA and Okieimen FE. (2011) Heavy metals in contaminated soils: a review of sources, chemistry, risks and best available strategies for remediation. *ISRN Ecology* 2011: 20.

Xie L and van Zyl D. (2023) Identification of grass species candidates for phytostabilization and enhanced metal(loid)s immobilisation using cost-effective amendments on sulfidic mine tailings. *International Journal of Mining, Reclamation and Environment* 37: 489–503.

Xu D-M, et al. (2021) Chemical stabilization remediation for heavy metals in contaminated soils on the latest decade: available stabilizing materials and associated evaluation methods–A critical review. *Journal of Cleaner Production* 321: 128730.

Yoon J, et al. (2006) Accumulation of Pb, Cu, and Zn in native plants growing on a contaminated Florida site. *Science of the Total Environment* 368: 456–464.
Zgorelec Z, et al. (2020) Cadmium and Mercury phytostabilization from soil using Miscanthus × giganteus. *Scientific Reports* 10: 6685.
Zhang S, et al. (2012) Cd accumulation and phytostabilization potential of dominant plants surrounding mining tailings. *Environmental Science and Pollution Research* 19: 3879–3888.
Zhang X, et al. (2022) "Potentially toxic element"–something that means everything means nothing. *Environmental Science & Technology* 56: 11922–11925.
Zhou R, et al. (2017) Remediation of Cu, Pb, Zn and Cd-contaminated agricultural soil using a combined red mud and compost amendment. *International Biodeterioration & Biodegradation* 118: 73–81.
Zhu C, et al. (2023) Foliar spraying of indoleacetic acid (IAA) enhances the phytostabilization of Pb in naturally tolerant ryegrass by limiting the root-to-shoot transfer of Pb and improving plant growth. *PeerJ* 11: e16560.
Zornoza R, et al. (2012) The effect of former mining activities on contamination dynamics in sediments, surface water and vegetation in El Avenque Stream, SE Spain. *Water, Air, & Soil Pollution* 223: 519–532.
Zornoza R, et al. (2015) Main factors controlling microbial community structure and function after reclamation of a tailing pond with aided phytostabilization. *Geoderma* 245–246: 1–10.
Zornoza R, et al. (2016) Microbial growth and community structure in acid mine soils after addition of different amendments for soil reclamation. *Geoderma* 272: 64–72.

Chapter 12

Phytoremediation: advances in soil contaminant phytoextraction by accumulator and hyperaccumulator plants

David W. M. Leung, University of Canterbury, New Zealand

1 Introduction

Agricultural activities have often resulted in the addition of heavy metal contaminants, particularly cadmium (Cd) and copper (Cu), to agricultural and horticultural soils. For example, the link between the use of cadmium-containing phosphorus fertilisers and a rise in Cd content in agricultural soils in New Zealand has prompted the development of the government- and industry-led voluntary management strategies for this pollution problem since 2011 (Gray and Cavanagh, 2023). Basically, the fertiliser industry and the government have endorsed the sustainable farming goal (over the next 100 years) with safeguards against the toxic effects of excessive soil Cd on human health. This is thought to be achieved with the soil Cd level not exceeding 1 mg/kg. A soil Cd map in NZ has been established based on the analysis of thousands of soil samples from all over NZ taken from depths of 7.5 cm to 15

http://dx.doi.org/10.19103/AS.2025.0151.17

cm. Whether there are restrictions on phosphate fertiliser application or not (type of phosphate fertiliser with lower Cd contamination, rates and need for some risk management practices) at a particular rural site is dependent on 5 categories or "tiers" of soil Cd contents (Abraham, 2020). For example, at the tier 0, soil Cd content is less than 0.6 mg Cd/Kg, and there is no restriction to phosphate fertiliser application. At the highest tier (tier 4), soil Cd content has exceeded 1.8 mg Cd/Kg, phosphate fertiliser application is not allowed pending on some further investigations.

Water is a limited resource on farmland in many parts of the world where crop production is often only possible using irrigation with waste water, such as sewage effluent, a source of heavy metal pollution (Lottermoser, 2012). Another common problem of growing crop plants is that they may be susceptible to diseases and pests, and it is necessary to use pesticides for effective plant disease control, for example, in vineyards. Some of the more effective pesticides are Cu-based chemicals or complexed with Cu, and the continuous use of these pesticides has contributed significantly to a rise in the Cu content, for example, in viticultural soils of many different countries (Widmer and Norgove, 2023). Cd and Cu pollution of agricultural soils is a typical example of diffuse pollution of the water environment. These chemical pollutants may enter surface waters, groundwater or as runoffs to rivers, and can be found in the sediments of rivers, etc. (Gallego et al., 2024). Remediation of farm soils with heavy metal pollution issue seems to be an attractive strategy to disrupt the undesirable leaching of heavy metals to waterways.

Agricultural soils are vital to food security globally. In response to the degradation of agricultural soil quality by rising heavy metal contents in soils, it is not surprising to have government policy and regulations that would slow down or minimise the further addition of heavy metals to agricultural soils. For example, changes in the farms and vineyards could aim to reduce excessive use of fertilisers and pesticides. Conceivably, the possible immediate or short-term impacts from these regulatory measures may include the following: the contaminant metal content in soil would take longer to reach a level that could harm the crops, or that is available for excessive uptake into plant food, posing public health concerns to the consumers. However, the long-term threat to agricultural soils for sustained food security remains the same because the heavy metal contaminants persist in agricultural soil. The eventual build-up of heavy metal contaminants in agricultural soil to dangerous levels is inevitable. In some soils, the levels of heavy metal contaminants might have already been too high. For example, in some viticultural soils, the Cu content might be higher than 100 mg kg^{-1} depending on soil depth (Widmer and Norgove, 2023). In the soil of a former vineyard in France, the total soil Cu concentration was higher than 1000 mg kg^{-1} (Michaud et al., 2007).

In many countries, including Australia and New Zealand, liming of agricultural soils has been used to help immobilise Cd and Cu from being taken up by plants or entering surface runoff (Mubeen et al., 2023; Xu et al., 2021). The addition of liming material such as calcium carbonate, calcium hydroxide and calcium sulphate brings the soil pH close to neutrality and is thought to result in the reduction of Cd uptake into rice grains, potato tubers and ryegrass, etc. However, there are also some reports that liming may not always be effective in this way, and there are many disadvantages, particularly arising from over-liming (He and Singh, 1994; Kirkham, 2006; Jawad et al., 2014; Morton and Ag, 2019). Overliming might render some essential trace metals, such as zinc, manganese and iron in sandy soils not available for plants to uptake. Besides, liming may be a recurring on-farm cost burden as liming material might have to be applied repeatedly along with the continuous reliance on phosphate fertilisers contaminated with Cd and Cu-containing sprays. Liming is also not a 100% climate change-friendly practice as it can contribute directly (from soil reaction to liming releasing carbon dioxide) and indirectly (methane and nitrous oxide from the growth of ruminant animals) to the emission of greenhouse gases (Morton and Ag, 2019).

A long-term solution would require the cleaning up or the removal of heavy metal contaminants of the polluted agricultural soil away from the contaminated farmland. There are conventional environmental engineering technologies to clean up soil such as excavation of contaminated soil with heavy machineries, etc. (Nnaji et al., 2023; Wang et al., 2024). It would be prohibitively expensive to then replace, after excavation of contaminated soil, with clean topsoil. The environmental engineering methods should be applicable to the remediation of abandoned mine sites, but they are not ideal to rid heavy metal contaminants in farmland soil, as there will be expensive disruption to crop growing while remediation work is underway. Besides, the soil microbial communities and structure supporting plant growth could be irreversibly changed with unknown consequences for crop production using conventional environmental engineering remediation methods.

Phytoremediation, particularly the subset of techniques called phytoextraction, is an attractive alternative for agricultural soil remediation. Phytoextraction is the use of plants that are capable of removing heavy metal contaminants from agricultural soil and keeping the contaminants inside their own tissues. If the metal accumulator plants, particularly those called hyperaccumulator plants, that can accumulate relatively large quantities of heavy metals in the above-ground parts, such as the leaves and stems, then these tissues could be harvested and moved off the site of contamination for disposal or further processing. Ideally, the plants selected for the phytoextraction purpose can be cultivated for intercropping and are compatible with the crops of interest in the same field with polluted soil (Widmer and Norgove, 2023). It may take several years to clean up a polluted site. Therefore, the phytoextraction technology could be more acceptable at the farm level, as while agricultural soil is being remedied, the food crops can still be grown for the upkeeping of farmers' livelihood.

There have been over 6000 publications (refereed papers, conference proceedings etc.) from phytoremediation research on soil and water contamination with heavy metals and metalloids since 1989 as revealed in a number of systematic bibliographic reviews (Li et al., 2019; Widmer and Norgove, 2023; Wang et al., 2024; Zhang et al., 2024). Although these systematic studies do not distinguish phytoextraction as a subset of phytoremediation and do not explicitly highlight the trends, frontiers and hotspots in phytoextraction research, it seems that most of the research publications are related to phytoextraction of heavy metals. It is also clear from these bibliometric surveys that phytoextraction research has attracted global efforts, as usable agricultural soils are vital but are limited resources worldwide. However, there seems to be a paucity of phytoextraction research aiming to remediate heavy-metal contaminated agricultural soils while permitting crop production without interruption. Most published phytoextraction research findings are related to studies to remediate heavy metals in soils and water from mining-related or industrial activities. In addition, there are investigations on the use of plants to help restore the abandoned mine sites or reclaim farmlands lost due to spill-over contaminants from nearby mining and industrial activities. These latter types of studies are important to restore and even increase land with usable agricultural soils and therefore better safeguard the sustainability of food security globally. In addition, these studies are generally helpful to obtain a

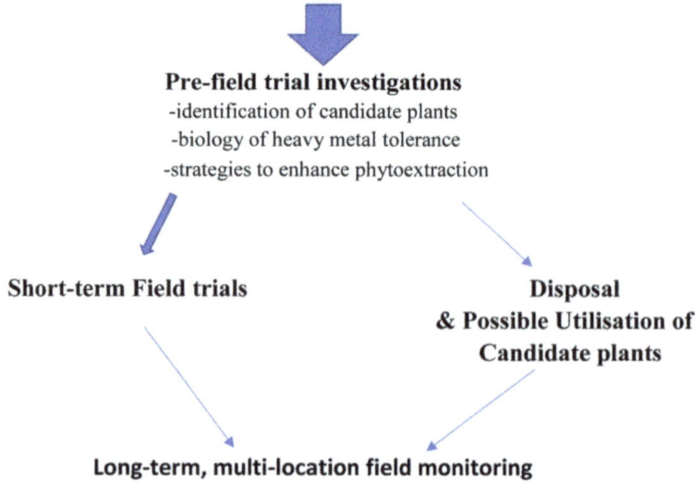

Figure 1 Workflow of phytoextraction investigations.

better understanding of the fundamental plant characteristics and strategies that are important for heavy metal phytoextraction. However, a comprehensive review of these studies is beyond the scope of the present chapter. Here, the primary objective is to discuss selected studies with impacts on phytoextraction research that have emerged from the major contemporary flow of research themes on phytoextraction of Cd and Cu contaminants in agricultural and viticultural soils (Figure 1).

2 Recent findings affecting phytoextraction experiment methodology

For the identification of candidate plants for phytoextraction, the basic research methodology is to compare the response of different plant species or different varieties to soil artificially spiked with different concentrations of heavy metals such as Cd or Cu, singly or in combination. The heavy metal contaminants used in phytoextraction experiments can be made up as metal salt solutions such as cadmium chloride ($CdCl_2$) or copper sulfate ($CuSO_{4.}5H_2O$) in the lab and then added to potting mix or soil. Alternatively, soils from known heavy metal-contaminated sites could be dug up and used in pot trials.

The pot trials experiments aiming to identify candidate plants are often allowed to run for a fixed time period of varying durations (Table 1). It is not clear nor explicitly explored what would be an appropriate time to compare the different plants with regard to their phytoremediation or phytoextraction potential. In a pot trial, the time course of uptake of Cu and other metals by four different plants (sunflower, red cabbage, Indian mustard and Cineraria) was compared (Gani et al., 2024). It was revealed that there were two phases of metal uptake: an initial slow phase followed by a rapid uptake in all fourplants

Table 1 Examples of heavy metal phytoextraction experiments that were run for varying durations before the plants were harvested for determination of phytoextraction efficiency.

Plant	Phytoextraction	Study period	References
Industrial hemp	Cd, Pb and Ni	2 years from seed	Testa et al. (2023)
Maize	Cu	42 days from seed	Zand et al. (2020)
Artemisia absinthium	Cu	4 months of fast growing seedlings	Ghazaryan et al. (2022)
Salix mucronata	Cd, Cu and Pb	27 months from cuttings	El-Mahrouk et al. (2019)
Brassica juncea (L.) Czern and *Bidens alba* (L.) DC. var. *radiata*	Cu	65 days from seeds	Ariyskanon and Winaipanich (2006)
Sunflower	Cu	57 days	Andreazza, et al. (2015)
Rapeseed	Cu	90 days	Luo et al. (2024)
Eucalyptus	Cu	120 days	Negrini et al. (2024)

studied. The duration of the initial slow phase was variable, dependent on the plants. For the initial slow uptake of Cu, the duration was 45 days for sunflower and red cabbage, but only 30 days for Indian mustard and Cineraria. This is interesting, but the reason and significance of this phytoextraction characteristic are not clear. Unfortunately, the study did not appear to analyse the heavy metal contents in the roots and shoots separately. Therefore, it is not clear if there might be a change in the rhizosphere environment or the physiology of the plants, enabling a rapid uptake following the initial slow period of uptake.

The relative sensitivity of plants to a heavy metal can be assessed in an ecotoxicological test in which a common practice is to grow one or two plants in a pot with different concentrations of a heavy metal. However, there is new evidence that has practical implications for phytoextraction investigations. It was found that the apparent tolerance (in terms of biomass) of *Arabidopsis thaliana* (Columbia wild type) to a high level of Cu (175 μg g^{-1}) in a pot of soil was dependent on the plant density (Dingus et al., 2023). When there was one plant in a pot, the biomass of the plant under 175 μg g^{-1} Cu stress was only about 25% of the control (plant grown in soil without added Cu). When there were three or more plants in a pot, the biomass of the plants under Cu stress was about 60% of the control. It seems that Arabidopsis is a plant trying to avoid Cu uptake by secretion of citrate which binds to Cu in soil to prevent its uptake into plants. Therefore, in this example of heavy metal resistance based on reducing metal uptake, plant density was an important factor. It is also possible that plant density could be an important factor in a comparative study of the phytoextraction efficiency of different metal accumulator or hyperaccumulator plants to a heavy metal. It seems worthwhile for plant density to be taken into consideration in phytoextraction experiments.

3 Biology of metal tolerance

For an accumulator or hyperaccumulator plant to be able to accumulate more heavy metals in its tissues, a biological pre-requisite is the ability to tolerate the presence of the heavy metals in the plant body without injuries or substantial reduction in growth. To obtain a better understanding of the biology of increased metal tolerance, a useful approach is to compare, preferably two genotypes or varieties with similar genetic background but differ in their metal contents and relatively sensitivity to the same heavy metals. In a recent study, two genotypes of *Brassica napus* L. (rapeseed), H2009 and ZYZ16, with contrasting tolerance to Cu were identified from evaluating the growth and Cu accumulation in 32 different genotypes when they were grown under a range of concentrations of Cu (Luo et al., 2024). The seed yield of H2009 grown in contaminated soil (with 95.4 mg kg^{-1} of bioavailable Cu) was 75% of the control

(H2009 grown in uncontaminated soil (with 1.98 mg kg^{-1} bioavailable Cu). In comparison, the seed yield of ZYZ16 grown in the contaminated soil was only 36% of the control (ZYZ16 grown in the uncontaminated soil). Moreover, H2009 accumulated 9-fold more Cu in the contaminated soil than the control, while ZYZ16 only accumulated 2-fold more Cu in the contaminated soil than the control. Based on these parameters, it is clear that H2009 was tolerant to Cu while ZYZ16 more sensitive to Cu contamination. A comparative study of these two rapeseed genotypes has yielded insights into the biochemical and molecular differences associated with differential tolerance to Cu.

The roots of the Cu-tolerant H2009 genotype were found to have a lower level of malondialdehyde (MDA), a marker of oxidative stress, than the Cu-sensitive ZYZ16 genotype (Luo et al., 2024). However, there was little difference in the MDA content in the leaves of the two genotypes (Luo et al., 2024). There was also little difference in the proline content as well as antioxidative enzyme activities, including superoxide dismutase, catalase, peroxidase and ascorbate peroxidase in the leaves of the two genotypes. Interestingly, the roots accumulated about three times more Cu than the leaves in both genotypes. Another major biochemical difference between the Cu-tolerant and Cu-sensitive genotype was a significant increase in the pectin and hemicellulose content in the cell wall of the roots of H2009 but not in ZYZ16 when grown in the Cu-contaminated soil. This appears to be an important mechanism to keep excessive Cu from triggering harmful level of oxidative stress in the cytsol of the Cu-tolerant genotype. Transcriptome analysis also showed enhanced expression of genes for cell wall binding, chelation and Cu transport in the Cu-tolerant genotype.

There is a comprehensive review of the genes associated with Cd uptake, transport and detoxification in rice (Ai et al., 2022). Furthermore, in evaluating 212 rice accessions, a quantitative trait locus (QTL) named as *CAL1* (cadmium accumulating in leaf) was identified to be associated with Cd accumulation preferentially in the leaf rather than the grains (Luo et al., 2018). The CAL1 protein seems to chelate Cd and facilitate Cd transport in the xylem. Cd accumulated in the rice leaf sheaths may be stored in vacuoles, a well-known detoxification pathway in plants, and the transport may be mediated by the ATP-binding cassette or ABC transporters (AI et al., 2022).

4 Strategies to promote uptake of copper and cadmium into accumulator or hyperaccumulator plants

It seems reasonable to start developing strategies to enhance phytoextraction of heavy metals from contaminated soil using ways that can increase the tolerance of accumulator and hyperaccumulator plants to the heavy metals.

Based on our basic understanding of metal tolerance that is related to oxidative stress protection and other mechanisms, exogenous application of natural and synthetic agents that could mitigate oxidative stress and/or other aspects of metal toxicity has attracted many investigations. It is clear that some agents might have great metal toxicity mitigation, but they might not necessarily result in more bioaccumulation of heavy metals in the treated plants. For example, melatonin (N-acetyl-5-methoxytryptamine) is chemically related to the plant hormone indole-3-acetic acid, has been shown to have protective effects against various abiotic stressors when applied exogenously to plants (Zhao et al., 2022). When 3-week-old hydroponic-grown basil seedlings (*Ocimum basilicum* L) were grown in sandy soil with excess Cu (supplied in the form of 50 μM and 150 μM copper oxide) for 14 days, all the growth parameters (shoot length, root length, shoot fresh and dry weights) were reduced compared to the control (hydroponic nutrient solution without any augmented Cu (Oloumi et al., 2024). After 1 week of growth in the Cu-contaminated sandy soil, 100 μM melatonin was added to the irrigation water, and the growth of the basil seedlings was improved in Cu-contaminated soil compared to those irrigated with water without melatonin. This improved Cu resistance was, however, correlated with reduced Cu uptake, not increased phytoextraction of Cu.

Useful strategies to boost heavy metal tolerance in plants might also be derived from a scrutiny of the metabolites that participate in heavy metal tolerance. However, it is also important to bear in mind that improving metal tolerance by application of a target metabolite to plants grown in heavy metal-contaminated soil might not be translated into increased phytoextraction of heavy metals into plant tissues. For example, it has been suggested that nicotinamide (NIC), a chemical moiety of the important enzyme cofactors, NADH and NADPH, has a low ecotoxicology risk when applied to mitigate metal stress in plants grown in heavy metal-contaminated soil (Li et al., 2024b). In their study, *Pistia stratiotes* L., a Cd accumulator, was grown in a Hoagland nutrient solution supplemented with 10 mg L^{-1} Cd^{2+} with or without 200 μM NIC for 5 days. Plant growth (biomass) under Cd culture was higher in the presence of NIC, suggesting that NIC addition was able to mitigate Cd toxicity. The Cd content in the root was reduced while that in the leaves increased. However, NIC addition did not result in increased total Cd uptake, but instead, the translocation factor was increased by almost 3-fold (from 0.5 to 1.3 when in the NIC application). Therefore, NIC was effective in promoting translocation of Cd from the roots to the leaves, but not in increasing overall phytoextraction of Cd from soil. It was also reported that NIC enhanced Mn, Cu and Cd tolerance but not the phytoextraction capability of *Salix* plants (Ohlsson et al., 2008).

4.1 Fertilisers and chelators

The phytoremediation potential of *Artemisia absinthium* was investigated in Cu-contaminated soil with bioavailable Cu content that was 150-fold higher than that in the unaffected soil (Ghazaryan et al., 2022). The growth of the above-ground parts and the roots was reduced when the plant was grown in the Cu-contaminated soil. The above-ground parts accumulated 60 mg Cu/kg dry weight while the roots accumulated 360 mg Cu/Kg dry weight. Addition of ammonium nitrate (0.1 g/kg soil) to the Cu-contaminated soil promoted plant growth slightly (by 6%) compared to plant growth in the Cu-contaminated soil without ammonium nitrate supplementation. The ammonium nitrate addition did not affect the uptake of Cu into the above-ground parts and the roots, suggesting that ammonium nitrate had no chelating activity to facilitate increased uptake of Cu. The possibility of enhancing Cu transport and accumulation in the above-ground parts was investigated by spiking the Cu-contaminated soil with ammonium nitrate in addition to EDTA, citric acid or malic acid singly or in combination. It was found that the use of these chelators singly or in combination could lead to a 2-fold or greater increase in the Cu content of the above-ground parts, while the growth of the above-ground parts was slightly lower than that of the plants grown in the contaminated soil without these added chelators. Therefore, it seems worthwhile to investigate if the addition of a plant growth promoter such as a fertiliser like ammonium nitrate, together with a chelator, would be a practical strategy to enhance the phytoextraction of Cu or other heavy metals into the above-ground parts.

4.2 Plant growth regulators

Hyperaccumulators are known to accumulate heavy metals in their above-ground tissues. Very often, many well-known hyperaccumulators have small biomass. Plant growth promotion has been investigated as a practical strategy to overcome this natural limitation of phytoextraction using hyperaccumulators. In a representative study, the effects of nine different plant growth regulators (PGRs) on *Sedum alfredii* Hance, a Zn/Cd co-hyperaccumulator found in an old mine site in Quzhou, Zhejiang province of China, were investigated (Chen et al., 2022). The plants derived from cuttings of the seedlings found in the old mine site were grown for 3 months in pots with soil from the mine site. Exogenous application of only the following PGRs was found to promote the growth (dry weight increase) of *S. alfredii* more than the control when grown in the mine site soil: strigolactone (SL) and gibberellic acid (GA). However, it was not clear about the effects of the individual PGRs on the dry weights of the shoot and the root.

Exogenous application of the following PGRs was found to stimulate the uptake of more Cd into the roots of *S. alfredii* than into the control when grown in the mine site soil: ethrel (ETH, ethylene), abscisic acid (ABA) and cytokinin (CKs). In contrast, exogenous application of the following PGRs resulted in reduced levels of Cd in the roots compared with the control: SL, indole-3-acetic acid (IAA, auxin), GA, brassinolide (Br), jasmonic acid (JA) and salicylic acid (SA).

Exogenous application of the following PGRs was found to stimulate the increased translocation of Cd into the shoots of *S. alfredii* compared to the control when grown in the mine site soil: ABA, ETH and SL, while ABA stimulated the highest Cd translocation rate to the shoots. Application of ABA appears to be a promising strategy for treatment to improve hyperaccumulation of Cd by *S. alfredii* as it stimulated both higher Cd uptake into the roots and the shoots. Since ABA did not promote whole plant dry biomass (growth), the hypothesis that plant growth promotion was necessary to enhance hyperaccumulation of Cd into the shoots of *S. alfredii* is in doubt. More work is needed to evaluate this hypothesis further, and the effect and the mechanism of ABA on hyperaccumulation in other plants.

4.3 Gamma-butyric acid addition as a soil amendment

Gamma-aminobutyric acid (GABA) is a natural amino acid not normally used in protein synthesis, but has been found to be important in the regulation of many changes throughout plant growth and development. Since it is naturally occurring in all organisms including microbes in the soil, it has been added as a environmentally friendly soil additive or as foliar applications to protect plants against various abiotic stressors such as heat stress (Razik et al., 2021). Exogenous application of GABA to 1-month-old *Coreopsis grandiflora* plants (a Cd accumulator) grown for 1 month in Cd-contaminated soil supplemented with 0 to 8 g GABA kg^{-1} soil was also found to enhance the phytoextraction of Cd into the shoots in a dose-dependent manner but without any significant effect on the biomass of the shoot and root (Huang et al., 2023). There was no significant difference in the Cd concentrations in the roots with increasing applied GABA concentrations, however.

Further investigations into the possible mechanisms of GABA on increased bioaccumulation of Cd in *C. grandiflora* showed that the added GABA was able to increase bioavailability of Cd in the soil. in addition, GABA changed the many physicochemical properties, including decreased soil pH of the soil which was thought to be mainly responsible for increased Cd bioavailability and variations in the abundance of rhizobacteria. Many rhizobacteria, particularly *Pseudomonas* and *Sphingomonas*, associated with Cd mobilisation were found to increase after GABA addition.

Therefore, adding GABA to soil appears to bring about many changes in the rhizosphere, favouring increased Cd phytoextraction. It would be worthwhile to validate this in other studies involving applying GABA to other plants grown in Cd, Cu or other heavy metal-contaminated soils of different origins or compositions.

4.4 Nanoparticles

Addition of nanomaterials such as titanium dioxide nanoparticles (TiO_2 NPs) at low levels (100 mg kg^{-1} soil) can decrease Cd stress in plants and has been shown to promote bioaccumulation of Cd in the root and shoot of soybean plants (Singh and Lee, 2016) and into the aerial parts of *Eichhornia crassipes* and *Salvinia molestatwo* (Harikumar and Megha, 2017) when the growth media were contaminated with Cd. Furthermore, the enhanced effect of a combination of adding TiO_2 NPs and plant growth-promoting rhizobacteria (PGRB) to soil on biomass growth of *Trifolium repens* after 80 days was reported (Zhan et al., 2020). The nanomaterial (500 mg kg^{-1}) in the absence of the addition of PGRB enhanced the accumulation of Cd in the root and in the shoot. This was further enhanced with the addition of the rhizobacteria. An issue that has attracted some concern is regarding the biosafety of releasing nanoparticles large scale in the open environment.

5 Field studies

It is obvious that for cleaning up an abandoned old mine site, it is a matter of planting a metal accumulator or hyperaccumulator at the site. It would not be practical to clean up farmland with a heavy metal contamination issue by stopping cropping activities in favour of planting accumulator or hyperaccumulator, which are often weedy plants or difficult to cultivate. Intercropping of a metal accumulator or hyperaccumulator with a crop plant in farmland with metal-contaminated agricultural soil might offer a practical solution to cleaning up the contaminated agricultural soil without disrupting the farmers' ability to grow a target crop. An intercropping study was carried out in pot trials of growing *Sedum plumbizincicola* (a Cd hyperaccumulator) together in the same pots with wheat plants on a Cd-contaminated soil for 87 days (Zou et al., 2021). The wheat plants accumulated less Cd and decreased biomass under intercropping with *Sedum* plants. Both the biomass and the ability to accumulate Cd seemed to decline as the planting density of *Sedum* plants increased when intercropping with wheat. This suggests that when intercropping with a hyperaccumulator, crop plants can be grown with reduced metal content, and that is a desirable outcome with less metal ending up in the food chain. It is also clear that there are many complex aspects of plant-plant interactions and plant-soil changes under intercropping. In addition, the different combinations of crop plants and accumulator plants should be investigated for optimal intercropping outcomes, encompassing little to no disruption to the production of safe food plants while metal contaminants have been removed and bioaccumulated in the above-ground parts of the accumulators or hyperaccumulators. Moreover, the challenge remains for these studies to be carried out under field conditions, preferably in multiple locations (Widmer and Norgove, 2023).

In a typical acidic rice paddy soil contaminated with Cd in the Guangdong Province, China, the following three Cd accumulators that are fast-growing, with high biomass and easy cultivation were grown in a microplot area (2 x 3 m, total area 6 m^2, with spacing of 80 cm): *Amaranthus hypochondriacus* (AH), *Celosia argentea* (CA) and *Pfaffia glomerata* (PG). AH under monocropping conditions did not do well, whereas CA and PG did not show any phytotoxicity symptoms (chlorosis, leaf curling, etc.) and seemed to grow well in the field (Huang et al., 2024).

The best arrangement was the intercropping between AH and CA as far as total Cd uptake per plot was concerned. The increased Cd removal from soil under intercropping resulted in more Cd accumulation in the above-ground parts (leaves and stems of AH but not CA). There was no difference in the Cd contents in the roots of AH and CA compared to when they were under monocropping conditions.

The intercropping between AH and PG led to a reduction in the Cd uptake by both plants compared to under monocropping conditions, suggesting that interactions of plants under intercropping conditions are complex and could influence the efficiency of phytoextraction under field conditions. It is also important to consider and investigate the differences in the rhizosphere between monocropping and intercropping arrangements (Huang et al., 2024).

Growing metal accumulator plants could provide metal and metalloid exposure and spread pathways to the receptors in the local ecological food chain (Boyd, 2004; Jeong et al., 2015; Reynolds and Pilon-Smits, 2018). For example, a likely exposure pathway would be that some small mammals, herbivores and birds might consume or ingest metal-loaded foliages of hyperaccumulator plants as food. Field studies of metal accumulators and hyperaccumulators in active farms are needed to understand the potential ecological consequences of these plants on the likely receptors present in the farms. The effectiveness of some management measures, such as the use of nets, fencing or some high-tech sound deterrence system or futuristic predator-like drones, should also be evaluated/researched.

6 Post-phytoextraction: Disposal and utilisation options

6.1 Bioenergy crops

Since the principal purpose of phytoextraction is to reduce/remove heavy metal contaminants from agricultural soil, a further workflow after phytoextraction is to harvest the candidate plants from the field. In this way, the contaminants have been removed from the crop production field. The further processing options after harvest require some careful thoughts and innovative thinking with the view of adding value to phytoextraction of heavy metals. One notion that has attracted a considerable interest is to use candidate plants, particularly bioenergy crops, that can accumulate heavy metals which can then be processed to generate some forms of bioenergy (biogas production etc.) after harvest (Pandey et al., 2016; Wang et al., 2021).

Based on mining the published scientific databases, a multi-criteria matrix analysis was developed to assist the selection of candidate bioenergy plants with the best phytoremediation performance (Amabogha et al., 2023). A short list of eight plants, including *Helianthus annuus* (sunflower), *Brassica juncea* (Indian mustard), *Glycine max* (soybean), *Populus spp.* (poplar), *Panicum virgatum* (switchgrass) and *Miscanthus sinensis* (silvergrass), was drawn up for a systematic analysis using the decision matrix. These are mainly considered to have a combination of desirable characteristics for phytoextraction and bioenergy production: a high proportion of their biomass as lignin and cellulose, their desirable calorific value (the amount of heat energy released when combusted), metal tolerance and accumulation. Using this decision matrix, sunflower and silvergrass were picked as the most promising candidates for phytoremediation and post-harvest bioenergy production.

6.2 Pretreatment of plant biomass after phytoextraction

An issue of concern is the presence of a large amount of Cd or Cu in the plant biomass after phytoextraction for safe disposal or handling in further processing of the harvested plant material, for example, for bioenergy (bioethanol) production. This has recently attracted some studies looking for a solution to this issue. For example, some cultivars of *Sorghum bicolor* L. Moench (sweet sorghum) were previously estimated to generate around 6000 L of bioethanol per hectare and were also found to be able to phytoextract Cd from contaminated soil and accumulate 148 g/ha Cd in their stalks (Xiao et al., 2021). Furthermore, a pretreatment with 0.5% sulfuric acid for 0.5 h at 121°C of the harvested sweet sorghum biomass after phytoextraction of Cd from polluted farmland in Hunan Province, China, was found to be a highly beneficial step before bioethanol production from Cd-loaded plant biomass (Xiao et al.,

2024). After the acid pretreatment step, 98.5% of the Cd was removed from the biomass, suggesting that no further special (no further additional cost-incurring) precaution was necessary in the downstream handling of the pretreated biomass. Moreover, since a considerable amount of lignin was also removed during the pretreatment step, the cellulose (with reduced lignin interference) in the residue biomass was more accessible to enzymes used for saccharification (generation of sugars) for bioethanol production by fermenting yeasts. It has been estimated that more sugars (4.7-fold) could be produced from pretreated biomass for bioethanol production than from biomass that was not pretreated.

It has been suggested that the plant biomass after phytoextraction could be processed into biochars, which are thought to be beneficial to crop production and to the health of the agricultural soils (Allohverdi et al., 2021). Biochar is produced using a thermal process called pyrolysis that is basically heating biomass at a high temperature of above 500°C in the absence of oxygen (therefore no burning) to decompose the biopolymers. The main byproducts from the pyrolysis of biomass are the bio-oil and gases that could be used to generate energy (Ibrahim, 2020). There is a huge reduction in the volume of the input biomass to the residue biochar after pyrolysis. Therefore, this could reduce the space of a physical facility needed for a large-scale disposal and further downstream processing of accumulator or hyperaccumulator plants after phytoextraction of heavy metals from contaminated fields. Since during pyrolysis, the heavy metals in the phytoextractor plants would be retained in the biomass, there is still a need to find a way to remove the heavy metals from the biomass. In a recent proof-of-concept study, it was found that the biomass of the hyperaccumulator plant, *Sedum plumbizincicola*, obtained from a heavy metal-contaminated field was pretreated with chlorinating agents (bleaching) to remove heavy metals before pyrolysis at 750°C (Li et al., 2024a). The Zn content in the residue biochar was found to be only 10% of the biochar without the bleaching pretreatment and was considered a safe level of Zn in an ecological risk assessment. Moreover, unexpectedly, the bioavailability of phosphorus in the biochar from the bleaching pretreatment was enhanced 22-fold compared to the biochar without the bleaching pretreatment. This seems to be a promising option for the disposal and processing of other biomass of phytoextractors, but it is necessary to validate if this bleaching pretreatment would also work for removal of Cd and Cu from phytoextractor biomass.

Accumulation of heavy metals from agricultural soils mainly into the roots and relatively little transport to the aboveground parts of the accumulator plants might not be useful for the removal of heavy metals away from the agricultural land area. However, provided these accumulators are also tolerant to the heavy metals accumulated in their root tissues, they could still be useful or safe for economic purposes such as the production of food and plant-based useful compounds, including pharmaceuticals. For this purpose, it is necessary to first confirm that there are heavy metals reaching the seeds or fruits or whatever edible parts of the accumulator plants.

Silybum marianum (L.) Gaertn (also known as milk thistle of the Asteraceae family) is a herb known to produce a group of bioactive compounds called silymarin in its seeds. *S. marianum* is known in folklore medicine for the treatment of liver diseases. It has been found to be capable of growing in Cu-contaminated soil while the quality of silymarin produced was not affected, and in fact, the quantity of silymarin was raised in response to the Cu content in the soil (Papadimou et al., 2024). This was correlated with the accumulation of Cu mainly in the roots (19.7 mg kg^{-1}) compared to only 0.7 mg kg^{-1} in the seeds. Although there was no accompanying growth data of the roots and the plants grown in Cu-contaminated soil, the roots appeared to be tolerant to Cu accumulation as the plants were able to produce seeds with no apparent effect on silymarin production. Interestingly, recent studies showed that exogenous application of silymarin to wheat and maize plants grown under Cd stress seemed to be able to protect the plants from Cd toxicity (Alharby et al., 2021; El-Sappah et al., 2023).

Cultivation of industrial hemp (*Cannabis sativus* L.) may be controversial, but the plant is well recognised to be a multifunctional plant yielding different economically valuable products, including the above-ground parts of the plant that are rich in plant fibres for the manufacturing of ropes and the flowers containing cannabinoids for medicinal uses. In addition, there was a study on whether the industrial hemp plant was able to grow in Cu-contaminated soils (with a high level of Cu or 200 mg kg^{-1}). After 18 weeks of growth, the Cu content in the root was about 15 mg kg^{-1} of dry biomass but the middle and the upper leaves contained less than 1 mg Cu kg^{-1} of dry biomass. Therefore, Cu is accumulated mainly in their roots (Vasilou et al., 2024). The pattern of Cu distribution in the industrial plant was consistent with the results showing that the fresh weight and height of the above-ground parts and the chlorophyll levels as well as the production of cannabidiol were not affected when the plant was grown in the Cu-contaminated soil. Therefore, there is a scientific basis that industrial hemp could be first purposedly cultivated in Cu-contaminated soil to phytoextract Cu into their roots. Then the whole plants could be a useful resource for further processing to recover Cu from the roots, whereas the above-ground plant fibres and the flowers with little Cu contaminant could be safe to make ropes and medicinal cannabinoids, respectively.

7 Conclusion and future trends

Phytoextraction of heavy metals from contaminated soil is a subset of phytoremediation techniques to use heavy metal accumulator and hyperaccumulator plants to clean up the heavy metal-contaminated soil. At present, most of the heavy metal phytoextraction investigations have been at the pre-field trials using pots with heavy metal-contaminated soil. A time course study of the Cu phytoextraction process is rare but has brought into the question of the implications of the two phases of heavy metal uptake into plants. Another study has shown that plant density should be taken into consideration in future phytoextraction experiments. In comparative studies of genotypes with contrasting tolerance to Cd or Cu, useful insights have been gained about metal tolerance in plants at the biochemical level, for example, oxidative stress regulation, cell wall and vacuole compartmentation of heavy metals, as well as at the molecular level involving specific genes. A prominent research trend has been to search for ways, including exogenous application of plant growth regulators such as abscisic acid (ABA) to boost Cd or Cu phytoextraction ability of plants. There are some limited field trials carried out, including investigations of the intercropping of a metal hyperaccumulator with wheat. More multi-year and multi-location intercropping field trials are needed if the ideal scenario of applying phytoextraction technology to clean up agricultural soils without disrupting usual crop production in the farm is to be realised. A recent hot research trend is to find ways for safer disposal and/or post-phytoextraction processing of the heavy metal accumulator and hyperaccumulator plants with the view of adding value to the plants besides cleaning up heavy metal pollution. What seems to be largely underexplored at present is the potential impact of powerful genetic engineering and plant genome editing technology on the development of heavy metal phytoextraction technology that is applicable for farmland phytoremediation.

8 Where to look for further information

For a general comparative overview of phytoextraction with other related technologies for remediation of heavy metal-contaminated soil:

- Nnaji, N.D., Onyeaka, H., Miri, T. and Ugwa, C. (2023), Bioaccumulation for heavy metal removal: a review, *SN Applied Sciences*, 5: 125.

For a comprehensive overview of the genes associated with phytoextraction and the development of food-safe rice cultivars:

- Ai, H., et al. (2022), Advances in molecular mechanisms underlying cadmium uptake and translocation in rice, *Frontiers in Plant Science*, 13: 1003953.

The following are comprehensive bibliographic articles using statistical software to mine the online databases such as Web of Science for thousands of research papers, conference proceedings, etc. They provide an analysis of multiple perspectives, including the trends, hot spots, challenges and opportunities in phytoremediation research. Useful keywords in searching the literature on phytoremediation and the top labs in the world working on phytoremediation can also be found in these articles:

- Li, C., et al. (2019), Phytoremediation of heavy metal pollution: a bibliometric and scientometric analysis from 1989 to 2018, *International Journal of Environmental Research and Public Health*, 16: 4755.
- Wang, C., et al. (2024) Farmland phytoremediation in bibliometric analysis, *Journal of Environmental Management*, 351: 119971.
- Zhang, K., et al. (2024), Trends in phytoremediation of heavy metals-contaminated soils: a Web of Science and CiteSpace bibliometric analysis, *Chemosphere*, 352: 141293.

The following article is a bibliometric study with the sole focus on phytoremediation of copper in viticultural soils:

- Widmer, J. and Norgove, L. (2023), Identifying candidates for the phytoremediation of copper in viticultural soils: a systematic review, *Environmental Research*, 216: 114518.

9 References

Abraham, E. (2020), Cadmium in New Zealand agricultural soils, *New Zealand Journal of Agricultural Research*, 63: 202–219.

Ai, H., et al. (2022), Advances in molecular mechanisms underlying cadmium uptake and translocation in rice, *Frontiers in Plant Science*, 13: 1003953.

Alharby, H.F., et al. (2021), Silymarin-enriched biostimulant foliar application minimizes the toxicity of cadmium in maize by suppressing oxidative stress and elevating antioxidant gene expression, *Biomolecules* 11: 465.

Allohverdi, T., et al. (2021), A review on current status of biochar Uses in agriculture, *Molecules*, 26: 5584.

Amabogha, O.N., et al. (2023), Combining phytoremediation with bioenergy production: developing a multi-criteria decision matrix for plant species selection, *Environmental Science and Pollution Research*, 30: 40698–40711.

Andreazza, R., et al. (2015), Phytoremediation of vineyard copper-contaminated soil and copper mining waste by a high potential bio-energy crop (*Helianthus annus* L.), *Journal of Plant Nutrition*, 38: 1580–1594.

Ariyskanon, N. and Winaipanich, B. (2006), Phytoremediation of copper contaminated soil by *Brassica juncea* (L.) Czern and *Bidens alba* (L.) DC. var. radiata, *Journal Science Research Chulalongkorn University*, 31: 49–56.

Boyd, R.S. (2004), Ecology of metal hyperaccumulation, *New Phytologist*, 162: 563–567.

Chen, Z., et al. (2022), Roles of exogenous plant growth regulators on phytoextraction of Cd/Pb/Zn by *Sedum alfredii* Hance in contaminated soils, *Environmental Pollution*, 293: 118510.

Dingus, A., et al. (2023), Arabidopsis response to copper is mediated by density and root exudates: Evidence that plant density and toxic soils can shape plant communities, *American Journal of Botany*, 111: e16285.

El-Mahrouk, E.S.M., et al. (2019), Phytoremediation of cadmium-, copper-, and lead-contaminated soil by *Salix mucronata* (Synonym Salix safsaf), *HortScience*, 54: 1249–1257.

El-Sappah, A.H., et al. (2023), Interplay of silymarin and clove fruit extract effectively enhances cadmium stress tolerance in wheat (*Triticum aestivum*), *Frontiers in Plant Science*, 14: 1144319

Gallego, J.L., et al. (2024), Occurrence and toxicological relevance of pesticides and trace metals in agricultural soils, sediments, and water of the Sogamoso River basin, Colombia, *Chemosphere*, 354: article 141713.

Gani, A., et al. (2024), An empirical investigation on the elimination of heavy metals using bioremediation method for selected plant species, *Physics and Chemistry of the Earth*, 134: 103568.

Ghazaryan, K.S., et al. (2022), Phytoremediation of copper-contaminated soil by *Artemisia absinthium*: comparative effect of chelating agents, *Environmental Geochemistry Health*, 44: 1203–1215.

Gray, C.W. and Cavanagh, J.A.E. (2023), The state of knowledge of cadmium in New Zealand agricultural systems: 2021, *New Zealand Journal of Agricultural Research*, 66: 285–335.

Harikumar, P.S. and Tharayil, M. (2017), Treatment of cadmium, lead and copper using phytoremediation enhanced by titanium dioxide nanoparticles, *Eco Chronicle*, 12: 25–31.

He, Q.B. and Singh, B.R. (1994), Crop uptake of cadmium from phosphorus fertilisers: yield and cadmium content, *Water, Air and Soil Pollution*, 74: 251–265.

Huang, Y., et al. (2023), Gamma-aminobutyric acid enhances cadmium phytoextraction by *Coreopsis grandiflora* by remodeling the rhizospheric environment, *Plants*, 12: 1484.

Huang, R., et al. (2024), Phytoremediation and environmental effects of three Amaranthaceae plants in contaminated soil under intercropping system, *Science of the Total Environment*, 914: 169900.

Ibrahim, H.A.H. (2020), Introductory chapter: pyrolysis, In: *Recent advances in pyrolysis*, ed. By Ibrahim, H.A.H. IntechOpen, Available from: http://dx.doi.org/10.5772/intechopen.77528.

Jawad, I.T., et al. (2014), Soil stabilization using lime: Advantages, disadvantages and proposing a potential alternative, *Research Journal of Applied Sciences, Engineering and Technology*, 8: 510–520.

Jeong, S., et al. (2015), Increased ecological risk due to the hyperaccumulation of As in *Pteris cretica* during the phytoremediation of an as-contaminated site, *Chemosphere*, 122: 1–7.

Kirkham, M.B. (2006), Cadmium in plants on polluted soils: Effects of soil factors, hyperaccumulation, and amendments, *Geoderma*, 137: 19–32.

Li, C., et al. (2019), Phytoremediation of heavy metal pollution: a bibliometric and scientometric analysis from 1989 to 2018, *International Journal of Environmental Research and Public Health*, 16: 4755.

Li, X., et al. (2024a), Integrated ultrastructural, physiological, transcriptomic, and metabolomic analysis uncovers the mechanisms by which nicotinamide alleviates cadmium toxicity in *Pistia stratiotes* L., *Journal of Hazardous Materials*, 467: 133702.

Li, Z., et al. (2024b), Attempts to obtain clean biochar from hyperaccumulator through pyrolysis: Removal of heavy metals and transformation of phosphorus, *Journal of Hazardous Materials*, 468: 133837.

Lottermoser, B.G. (2012), Effect of long-term irrigation with sewage effluent on the metal content of soils, *Environmental Geochemistry and Health*, 34: 67–76.

Luo, J.S., et al. (2018), A defensin-like protein drives cadmium efflux and allocation in rice, *Nature Communications*, 9: 645.

Luo, T., et al. (2024), Phytoremediation of copper-contaminated soils by rapeseed (*Brassica napus* L.) and underlying molecular mechanisms for copper absorption and sequestration, *Ecotoxicology and Environmental Safety*, 273: 116123.

Michaud, A.M., et al. (2007), Copper uptake and phytotoxicity as assessed in situ for durum wheat (*Triticum turgidum durum* L.) cultivated in Cu-contaminated, former vineyard soils, *Plant Soil*, 298: 99–111.

Morton, J. and Ag, M. (2019), *Lime use on New Zealand pastoral farms*. The Fertiliser Association of New Zealand, 32pp.

Mubeen, S., et al. (2023), Agricultural strategies to reduce cadmium accumulation in crops for food safety, *Agriculture*, 13: 471.

Negrini, A., et al. (2024), Copper bioconcentration and bioaccumulation in different Eucalyptus species and their impact on plant physiology, *Journal of Plant Nutrition*, 47: 413–422.

Nnaji, N.D., et al. (2023), Bioaccumulation for heavy metal removal: a review, *SN Applied Sciences*, 5: 125.

Ohlsson, A.B., et al. (2008), Increased metal tolerance in *Salix* by nicotinamide and nicotinic acid. *Plant Physiology and Biochemistry*, 46: 655–664.

Oloumi, H., et al. (2024), Zinc and copper toxicity in basil (*Ocimum basilicum* L.) seedlings: role of melatonin in mitigating stress, *Plant Stress*, 11: 100365.

Pandey, V.C., et al. (2016), Energy crops in sustainable phytoremediation, *Renewable and Sustainable Energy Reviews*, 54: 58–73.

Papadimou, S.G., et al. (2024), Dual role of the hyperaccumulator *Silybum marianum* (L.) Gaertn. in circular economy: Production of silymarin, a valuable secondary metabolite, and remediation of heavy metal contaminated soils, *Sustainable Chemistry and Pharmacy*, 38: 101454.

Razik, E.S.A., et al. (2021), γ-Aminobutyric acid (GABA) mitigates drought and heat stress in sunflower (*Helianthus annuus* L.) by regulating its physiological, biochemical and molecular pathways, *Physiologia Plantarum*, 172: 505–527.

Reynolds, R.J.B. and Pilon-Smits, E.A.H. (2018), Plant selenium hyperaccumulation-ecological effects and potential implications for selenium cycling and community structure, *Biochim Biophys Acta Gen Subj*, 1862: 2372–2382.

Singh, J. and Lee, B.K. (2016), Influence of nano-TiO_2 particles on the bioaccumulation of Cd in soybean plants (*Glycine max*): A possible mechanism for the removal of Cd from the contaminated soil, *Journal of Environmental Management*, 170: 88–96.

Testa, G., et al. (2023), Phytoremediation of cadmium-, lead-, and nickel-polluted soils by industrial hemp, *Agronomy*, 13: 995.

Vasilou, C., et al. (2024), Phytoremediation & valorization of Cu-contaminated soils through *Cannabis sativa* (L.) cultivation: a smart way to produce cannabidiol (CBD) in Mediterranean soils, *Waste and Biomass Valorization*, 15: 1711–1724.

Wang, S., et al. (2021), Evaluation of biogas production potential of trace element-contaminated plants via anaerobic digestion, *Ecotoxicology and Environmental Safety*, 208: 111598

Wang, C., et al. (2024), Farmland phytoremediation in bibliometric analysis, *Journal of Environmental Management*, 351: 119971.

Widmer, J. and Norgove, L. (2023), Identifying candidates for the phytoremediation of copper in viticultural soils: A systematic review, *Environmental Research*, 216: 114518.

Xiao, M.Z., et al. (2021), A sustainable agricultural strategy integrating Cd-contaminated soils remediation and bioethanol production using sorghum cultivars, *Industrial Crops and Products*, 162: 113299.

Xiao, M.Z., et al. (2024), Advancing safe bioenergy production from cadmium-contaminated sweet sorghum through mild synergistic pretreatment, *Journal of Cleaner Production*, 448: 141695.

Xu, L., et al. (2021), The combination of lime and plant species effects on trace metals (copper and cadmium) in soil exchangeable fractions and runoff in the red soil region of China, *Frontiers in Environmental Sciences*, 9: 638324.

Zand, A.D., et al. (2020), Application of titanium dioxide nanoparticles to promote phytoremediation of Cd-polluted soil: contribution of PGPR inoculation, *Bioremediation Journal*, 24: 1–19.

Zhang, K., et al. (2024), Trends in phytoremediation of heavy metals-contaminated soils: a Web of Science and CiteSpace bibliometric analysis, *Chemosphere*, 352: 141293.

Zhao, C., et al. (2022), Melatonin is a potential target for improving horticultural crop resistance to abiotic stress, *Scientia Horticulturae*, 291: 110560.

Zou, J., et al. (2021), Phytoremediation potential of wheat intercropped with different densities of *Sedum plumbizincicola* in soil contaminated with cadmium and zinc, *Chemosphere*, 276: 130223.

Chapter 13

Managing contaminants in farmed soils: case studies from China

Fang Wang, Chinese Academy of Sciences, China, University of Chinese Academy of Sciences, China, and Technical University of Munich, Department of Chemistry, Germany; Yu Liu, Yuhao Fu, Leilei Xiang, Xin Jiang, and Yongming Luo, Chinese Academy of Sciences, China and University of Chinese Academy of Sciences, China

1 Introduction

Soil is a crucial but non-renewable resource that underpins numerous life-supporting functions on Earth and serves as a fundamental platform for agriculture (Clunes et al., 2021). However, various human activities have resulted in the degradation or contamination of up to 40% of global soils, with these levels continuing to escalate (Sarkar et al., 2021). This situation presents a formidable challenge to sustainable agriculture, especially in regions experiencing rapid industrialization and urban development. In China, decades of impressive economic growth have coincided with heightened soil pollution, largely attributed to the extensive production and use of chemicals. China is a major producer and consumer of heavy metals such as cadmium (Cd), lead (Pb), and arsenic (As) (Li et al., 2014). These metals accumulate in soils primarily through industrial activities like mining, smelting, and manufacturing processes. As the world's largest consumer of pesticides, China extensively applies these chemicals to protect crops and boost agricultural productivity. For instance, in

http://dx.doi.org/10.19103/AS.2025.0151.19

2021, approximately 1.8 million tons of pesticides were used (Alavanja, 2009). Improper application practices and insufficient regulatory oversight contribute to pesticide residues accumulating in soils, impacting soil biodiversity and potentially contaminating groundwater.

China also leads the world in the production and use of chemical fertilizers, which are essential for high-yield crop production (Guo et al., 2022). However, excessive and improper use can lead to nutrient imbalances, soil acidification, and the accumulation of harmful substances such as heavy metals, nitrates, and phosphates in soils (Atafar et al., 2010). This contributes to environmental degradation and reduces soil fertility over time. Besides, industrial activities produce a variety of chemical pollutants, including persistent organic pollutants (POPs), polycyclic aromatic hydrocarbons (PAHs), and various toxic chemicals (Xiang et al., 2022a). These pollutants infiltrate soils through atmospheric deposition, wastewater discharge, and inadequate waste management practices, contaminating agricultural lands and posing risks to ecosystems and human health. Additionally, contaminants such as pharmaceuticals, personal care products, and microplastics are increasingly being detected in soils, further complicating the issue of soil contamination (Wang et al., 2024a). These substances originate from sources such as agricultural runoff, sewage sludge, and urban waste, and their long-term effects on soil health and crop safety are not yet fully understood. The cumulative impact of these factors has led to widespread soil contamination across China, affecting millions of hectares of arable land and presenting significant challenges for sustainable agriculture, food security, and environmental protection.

A survey conducted by the Ministry of Environmental Protection and the Ministry of Land and Resources revealed that approximately 16.1% of surveyed soil sites in China are polluted, with heavy metals being the most prevalent contaminants (Zhao et al., 2015). Notably, Cd levels surpass national standards at 7% of surveyed sites. Provinces such as Hunan, Guangxi, and Sichuan are particularly affected, with alarming concentrations of Cd in paddy soils exceeding the threshold value (Wang et al., 2023a). For instance, some areas in Hunan Province have Cd levels 300 times higher than the threshold value (He, 2014). The ramifications of soil contamination are profound, posing significant risks to both ecological balance and human health. Contaminated soils have resulted in elevated levels of heavy metals in staple foods like rice and vegetables, directly jeopardizing food safety. Localized pollution hotspots near industrial areas, such as Shuangqiao and Liuyang in Hunan Province, have notably impacted public health, necessitating extensive soil remediation efforts on a large scale (Zhao, 2020). Addressing this complex issue is challenging due to the widespread nature of contamination and the technical complexities involved in remediation. The

Chinese government has responded with initiatives like the "Soil Pollution Prevention and Control Action Plan" and various remediation projects (Lu et al., 2015; Zhou et al., 2023). However, these efforts are resource-intensive and require sustained commitment, advanced technologies, and long-term strategies to achieve effectiveness. In light of these challenges, there is a critical need to intensify efforts toward developing sustainable strategies for soil remediation and implementing practices that prevent soil contamination in China. This entails leveraging innovative technologies and fostering sustainable agricultural practices to safeguard both the environment and public health over the long term.

In this chapter, we explore several case studies from China on managing contaminants in farmed soils, focusing on the extent of soil contamination, the effectiveness of remediation strategies, and the ongoing efforts to develop sustainable practices. By analyzing these case studies, we can gain valuable insights into the challenges and solutions associated with managing contaminated farmed soils. This analysis not only sheds light on China's specific situation but also provides broader implications for similar efforts worldwide. Understanding these approaches can help inform global strategies for soil remediation, enhance agricultural sustainability, and mitigate the environmental and health impacts of soil contamination.

2 Contaminants in farmed soils in China

China, with its extensive agricultural land comprising 1.918 billion mu (127.87 million hectares) (Ministry of Natural Resources, 2021), faces increasing challenges related to soil contamination. Despite having only 7% of the world's farmlands, China supports 22% of the global population (Huang et al., 2019). This disparity places immense pressure on the country's agricultural resources and highlights the critical importance of maintaining soil health. Contaminants in farmed soils have far-reaching implications for food safety, environmental health, soil ecological balance, and the sustainability of agricultural practices (Sun et al., 2018; Lin et al., 2022b). This section will explore the sources of contamination in agricultural practices, introduce the common types of contaminants found in farmed soils, and discuss their distribution across China. By understanding the origins and impacts of these contaminants, we can better address the challenges they pose and work towards more sustainable and safe agricultural practices.

2.1 Sources of contamination in agricultural practices

In China, the contamination of agricultural soil primarily stems from a range of human activities, notably those linked to industrialization and farming practices. Industrial activities contribute to soil pollution through the discharge of pollutants, improper waste management, and the use of contaminated water for irrigation. Within agricultural practices, the main sources of soil contamination include the excessive use of chemical fertilizers and pesticides, the application of sewage sludge and livestock manure, and the presence of atmospheric deposition from nearby industrial emissions.

2.1.1 Agricultural chemicals

The use of chemical fertilizers and pesticides has been crucial for enhancing agricultural productivity in China (Wu et al., 2018). However, their widespread application has led to significant soil pollution, adversely affecting soil health and agricultural sustainability. The extensive and ongoing use of these agrochemicals is a major contributor to soil contamination, which poses serious risks to both the environment and long-term agricultural viability (Liu et al., 2021). According to data from the Food and Agriculture Organization, China is the largest consumer of chemical fertilizers globally, applying approximately 50.79 million tons annually as of 2022 (Textor, 2023). This intense use of fertilizers has led to severe soil issues, including nutrient imbalances and the accumulation of harmful elements such as Cd, copper (Cu), and As, particularly from phosphate fertilizers (Lin et al., 2022b). Additionally, an 18-year field experiment in the red soil of southern China showed that soil pollution from nitrogen fertilizers led to increased acidification, with soil pH decreasing over time from 5.7 to 4.2–4.5 (Cai et al., 2015).

Similarly, pesticide use in China is extensive, with an estimated application of around 1.7 million tons per year (China, 2023). This high volume of pesticide use has resulted in substantial soil contamination. Research indicates that persistent organic pollutants from pesticides can accumulate in the soil, with residues of commonly used pesticides like glyphosate and chlorpyrifos detected in up to 80% of soil samples from agricultural areas (Mu et al., 2023). The long-term presence of these chemicals disrupts soil microbial communities and reduces soil fertility, thereby impacting crop yields and ecosystem health. Organochlorine pesticides (OCPs) were extensively used in agricultural pest control but are persistent in the environment, allowing them to accumulate in soil. Moreover, OCPs are bioaccumulative and toxic to non-target species, posing significant risks to human health and ecosystems (Chen et al., 2023). Historically, large quantities of OCPs were produced in China, and despite their use being banned for decades, these pollutants remain prevalent in agricultural soils (Zhang et al., 2018). These practices, while aiming to enhance agricultural productivity, inadvertently contribute to the degradation of soil quality, necessitating the adoption of more sustainable farming practices to mitigate these adverse effects.

2.1.2 Industrial discharges

In China, rapid industrialization has profoundly impacted agricultural soils, primarily through the release of pollutants. Factories often discharge untreated or inadequately treated wastewater into nearby water bodies, which can spread to adjacent farmlands through flooding, irrigation, or groundwater contamination, allowing pollutants to infiltrate the soil and affect crops (Qing et al., 2015). This process significantly elevates the concentrations of heavy metals such as Pb, chromium (Cr), and mercury (Hg) in agricultural soils. A case study in Huizhou, Guangdong Province, highlights this issue, where industrial activities caused severe soil contamination, particularly with Pb and Hg, adversely affecting crop quality and posing health risks to local communities (Cai et al., 2012).

In regions surrounding mines, waste rocks and sludge from mining operations are common sources of heavy metal contamination. For instance, in Chenzhou City, Hunan Province, the surrounding agricultural soils showed high levels of Cd, Pb, and Zn due to the deposition of mining waste, leading to substantial soil degradation and crop contamination (Ma et al., 2015). Additionally, coal and petroleum production have been identified as significant sources of PAHs in agricultural soils. In Shanxi Province, a study revealed that agricultural lands near coal production sites had PAH concentrations several times higher than the national safety limits, illustrating the severe impact of industrial emissions on soil quality (Zhao et al., 2014). These pollutants not only degrade soil health but also accumulate in crops, entering the food chain and posing long-term health risks to humans and wildlife. These case studies underscore the urgent need for stringent regulatory measures and sustainable industrial practices to mitigate soil contamination and protect agricultural productivity and environmental health.

2.1.3 Irrigation with wastewater

Wastewater is increasingly used to irrigate agricultural land and for urban greening in China, driven by water scarcity resulting from climate change, regional droughts, population growth, and pollution (Pan et al., 2014). While this practice offers several economic and environmental advantages, it also presents significant challenges. Currently, 3.3 million hectares of arable land in China are contaminated with heavy metals and organic pollutants due to wastewater irrigation (Pan & Chu, 2018). For example, a case study in Tianjin revealed that long-term irrigation with wastewater led to the accumulation of heavy metals such as Cd, As, and Pb in agricultural soils (Meng et al., 2016). This resulted in elevated concentrations of these metals in vegetables grown in these soils, posing potential health risks to consumers. A similar situation was observed in Beijing, where the concentrations of Cu, Ni, Pb, Zn, and Cd in plants grown in wastewater-irrigated soils exceeded the permissible limits set by the State Environmental Protection Administration of China and the World Health Organization (Khan et al., 2008).

Beyond heavy metals, wastewater can introduce organic pollutants, including pharmaceuticals, personal care products, and industrial chemicals (Li et al., 2018). For example, in the Wangyang River basin, a typical wastewater irrigation area in Hebei Province, significant levels of phthalates (PAEs) were found in soils, with Di(2-ethylhexyl) phthalate (DEHP) and di-n-butyl phthalate (DnBP) being the dominant PAE species (Zhang et al., 2015). Another study in a representative wastewater irrigation farmland in suburban Beijing found high concentrations of PAHs and low levels of OCPs in the soils.

The implications of wastewater irrigation are profound. It compromises soil health and threatens food safety and public health through the transfer of contaminants from soil to crops. The increased presence of antibiotics in the soil can lead to the development and spread of antibiotic-resistant bacteria, posing a significant challenge to public health (Pan & Chu, 2018; Sun et al., 2018). Addressing the challenges posed by wastewater irrigation requires comprehensive strategies, including stringent regulations, advanced treatment technologies, and sustainable water management practices to mitigate the adverse effects on agricultural soil and ensure long-term soil health and food security.

2.1.4 Atmospheric deposition

Atmospheric deposition has increasingly emerged as a significant contributor to agricultural soil pollution in China, with profound implications for soil health and crop productivity (Liu et al., 2020c). This process involves airborne pollutants, such as heavy metals, nitrogen compounds, and sulfur oxides, settling onto soil surfaces through mechanisms like precipitation or dry deposition (Shao et al., 2021). Recent studies have highlighted the alarming extent of soil contamination related to atmospheric deposition, especially in regions with intensive industrial and agricultural activities (Liu et al., 2020c). These heavy metals, deposited from industrial emissions and other sources, accumulate in soils, impacting soil quality and crop safety. Organic pollutants such as polycyclic aromatic hydrocarbons (PAHs) and per- and polyfluoroalkyl substances (PFASs) are also of concern, as they can contaminate agricultural soils through atmospheric deposition (Sun et al., 2018; Wang et al., 2024b). For instance, in Lanzhou, an industrial city, the annual mean concentration of $\sum$16PAHs in the atmosphere was 478 ± 640 ng/m^3, reaching a maximum of 2180 ng/m^3 (Ma et al., 2018b). Similarly, atmospheric concentrations of PFASs (C4–C12) averaged 4900 pg/m^3, with PFOA at 1900 pg/m^3 and PFBA at 1800 pg/m^3 near Fuxin in the Liaoning Province (Wang et al., 2020a). These pollutants present a growing risk due to their persistence in the environment and potential health effects.

A study conducted on wet nitrogen and sulfur deposition at three sites in the Yangtze River Delta has shown that deposition from agricultural and industrial sources significantly raises soil nitrate and sulfate levels (Chen et al., 2019). This increased deposition contributes to soil acidification, nutrient imbalances, and overall degradation of soil quality. The study revealed that elevated nitrogen and sulfur inputs led to changes in soil chemical properties and reduced soil fertility, impacting crop yields and ecosystem health. Another notable case is the research conducted by Huang (Huang et al., 2009), which analyzed heavy metal contamination in the Yangtze River Delta region, revealing that atmospheric deposition of metals such as Cd, Cu, Pb, and Zn, originating from industrial processes, has led to elevated concentrations in soils. The contamination of soils with these metals affects not only crop safety but also poses risks to human health throughout the food chain. The study highlighted the need for stringent control measures to reduce industrial emissions and mitigate their impact on soil environments.

These studies underscore the critical need for comprehensive policies to mitigate atmospheric emissions and monitor soil quality. Addressing the complex interactions between atmospheric deposition and soil pollution in China requires a multi-faceted approach, including stricter emission regulations, improved monitoring systems, and enhanced soil management practices to protect agricultural productivity and public health.

2.2 Common types of contaminants found in farmed soils and their distribution

In China, farmed soils are subject to contamination by a diverse array of pollutants, with the types and concentrations of these contaminants varying significantly based on regional industrial activities and agricultural practices. Common contaminants include heavy metals such as Cd, As, Pb, and Hg, which are often introduced through industrial discharges, mining activities, and the application of contaminated fertilizers and pesticides. Additionally, POPs such as polychlorinated biphenyls (PCBs) and OCPs are prevalent in some areas due to historical use and improper disposal. Nutrient imbalances from excessive use of nitrogen and phosphorus-based fertilizers also contribute to soil degradation. The distribution of these contaminants is influenced by factors such as proximity to industrial zones, historical land use, and local agricultural practices. The spatial variability of soil contamination underscores the need for targeted remediation strategies and stringent regulatory measures to mitigate the impacts on soil health and agricultural productivity.

2.2.1 Heavy metals and arsenic

Heavy metals are significant contaminants in Chinese agricultural soils, posing severe risks due to their toxicity, persistence, and potential for bioaccumulation. Among these metals, Cd is notable for its high mobility from soil into plant tissues, adversely impacting plant growth and food security. Plants with high Cd accumulation capabilities can introduce this metal into the human food chain, leading to considerable health risks from prolonged dietary exposure (Wang et al., 2019). Approximately 37% of soil samples surpass the risk screening value of the Chinese soil environmental quality standard (GB15618-2018) (0.3 mg/kg, pH ≤ 7.5). Regional variations in Cd concentrations are notable, with higher levels found in Central, Southwest, and East China (Yang et al., 2018). Hg contamination, largely stemming from industrial discharges and coal combustion, poses a severe threat to soil quality and human health. Elevated Hg levels are particularly problematic in West, South, and Northeast China (Li et al., 2020). The average Hg concentrations in all regions were below the risk screening value (GB 15618-2018) (paddy soil, 0.5 mg/kg; other soil 1.8 mg/kg; 5.5 < pH ≤ 6.5), but it is noteworthy that some sites in China had Hg concentrations exceeding the risk control value (2.5 mg/kg). Pb, a naturally occurring metal, contaminates the environment through industrial activities and urban development. Pb accumulates in the human body via absorption, bioconcentration, and biomagnification, posing significant health risks (Collin et al., 2022). In most regions, the Pb concentrations were below the soil environmental quality standard (GB 15618-2018) (paddy soil, 100 mg/kg; other soil, 90 mg/kg; 5.5 <pH ≤ 6.5) but much higher than the background value (26 mg/kg). High Pb concentrations are notably present in South, Central, and East China (Zhang et al., 2019). As, a metalloid widely distributed in the environment, is highly toxic, non-biodegradable, and prone to bioaccumulation. As predominantly enters the food chain through contaminated water and food sources, posing substantial health risks (Khan et al., 2021). The highest As concentrations in farmed soils were reported in South China (mean 18.8 mg/kg), followed by Northeast China (mean 15.8 mg/kg) (Zhou et al., 2018). The prevalence of heavy metal contamination in Chinese soils has far-reaching implications for environmental health and food safety. Regional variations in metal concentrations necessitate targeted remediation strategies and stricter regulatory measures. In China, the Ministry of Ecology and Environment has implemented soil environmental quality risk control standards for soil contamination of agricultural land to manage heavy metal pollution.

2.2.2 Organic emerging pollutants

POPs constitute a major environmental concern in agricultural soils across China due to their remarkable durability, bioaccumulation potential, and toxicity (Yu et al., 2020). POPs encompass a range of substances including pesticides, industrial chemicals, and their by-products, which are known for their resistance to environmental degradation and their propensity to accumulate along the food chain, thereby posing substantial risks to both ecosystems and human health (Li et al., 2024b). OCPs, a prominent subgroup of POPs, have historically been employed to manage agricultural pests. These pesticides have been widely produced and used globally to enhance agricultural productivity, leading to their widespread presence in Chinese farm soils (Yu et al., 2020). Despite their effectiveness, OCPs exhibit significant environmental persistence and bioaccumulation, resulting in potential health risks for humans throughout the food chain (Chen et al., 2023).

PAHs are another critical class of POPs. These compounds, characterized by their multiple benzene ring structures, are ubiquitous in soil, sediment, plants, food, water, and air. PAHs are particularly concerning due to their carcinogenic, teratogenic, and mutagenic properties (Liu et al., 2023). In China, the distribution of PAH concentrations in soil follows a regional gradient, with higher levels observed in the Northeast, North, and East, and decreasing towards the South and West (Zhang & Chen, 2017). PAEs, commonly used as plasticizers, are also prevalent environmental pollutants. They have been frequently detected in farmlands across various regions in China, with higher concentrations noted in Central, Southern, and Eastern regions (Sun et al., 2018). The widespread use and persistence of PAEs underscore their potential impact on soil and water quality.

PFASs are another group of POPs extensively used in industrial processes and consumer products for their resistance to stains, chemicals, and heat. PFAS contamination has been detected in soil, surface water, and groundwater across China, with significant adverse effects on human health, particularly concerning liver and immune system functions. Notably, PFAS levels are considerably higher along the east coast of China compared to inland areas, with Tianjin identified as the province with the highest PFAS contamination (Wang et al., 2024b). Dioxins and furans, toxic by-products of industrial processes and waste incineration, pose severe health risks such as immunotoxicity and developmental disorders. In China, soil contamination with dioxins and furans is a concern, particularly near industrial facilities and waste disposal sites (Song et al., 2022; Xu et al., 2020).

The pervasive presence and persistence of these POPs in Chinese agricultural soils highlight critical environmental and public health issues. The ongoing accumulation of POPs exacerbates the risks of long-term ecological damage and human health impacts, particularly in regions with higher concentrations. Addressing these challenges requires comprehensive monitoring and remediation strategies to mitigate the adverse effects of POPs and safeguard both environmental and human health.

Pharmaceuticals and personal care products (PPCPs) are increasingly recognized as significant threats to soil quality and agricultural safety in China (Lyu et al., 2022). Unlike traditional pollutants, these contaminants are often not monitored or regulated, yet they pose substantial risks due to their persistence, potential for bioaccumulation, and negative impact on soil ecosystems (Wang et al., 2024a). Pharmaceuticals and PPCPs are primarily detected in farmed soils due to the application of wastewater, which may contain antibiotics, hormones, and various synthetic chemicals (Liu et al., 2020a). Research indicates that concentrations of these contaminants are particularly high in the East and South regions of China, including Jiangsu, Zhejiang, Guangdong, and Guangxi provinces, largely due to intensive livestock farming and industrial and urban wastewater discharge (Wang et al., 2022).

2.2.3 Microplastics

Microplastics (MPs), defined as plastic particles less than 5 millimeters in diameter, are increasingly detected in farmed soils across China (Zhang et al., 2022). These particles originate from various sources, including plastic waste, agricultural plastics, and synthetic fibers from textiles. The distribution of MPs in farmed soils in China is significantly influenced by regional industrial activities, agricultural practices, and waste management systems. Regions such as Jiangsu, Zhejiang, Guangdong, Guangxi, Henan, Shandong, and Shanghai exhibit elevated levels of MPs in soil, largely due to their high industrial output and extensive agricultural operations (Yu et al., 2021; Tang, 2023). The presence and distribution of MPs in farmed soils in China highlight the need for improved monitoring and regulatory frameworks. Addressing these issues requires a comprehensive approach involving better waste management practices, stricter regulations on industrial emissions, and advancements in wastewater treatment technologies. Effective management of various contaminants is crucial for safeguarding soil health, food safety, and overall environmental quality.

3 Environmental and agricultural challenges in China

3.1 Impact of contaminants on soil quality and crop production

Soil pollution in China is predominantly caused by heavy metals and metalloids, which account for 82.4% of the total soil contamination, with Cd being

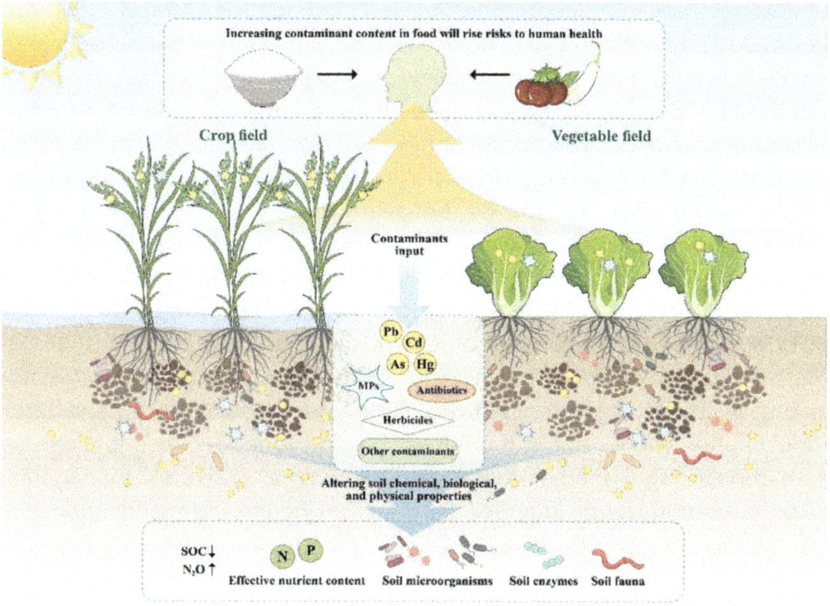

Figure 1 The impact of soil contamination on soil quality and crop production, along with the associated health risks to humans. MPs: microplastics, SOC: soil organic carbon, Pb: lead, Cd: cadmium, As: arsenic, Hg: mercury.

particularly problematic (Zhao et al., 2015). This contamination has detrimental effects not only on the soil itself but also on the microorganisms, animals, and plants that rely on it (Fig. 1). Heavy metals such as Cd, Pb, Zn, and Cu can bind with organic matter in the soil, forming metal-organic complexes (Abdu et al., 2017). These complexes reduce the availability of organic carbon, which is crucial for microbial activity. Consequently, this limitation impairs microbial decomposition processes and disrupts soil carbon sequestration, leading to decreased nutrient availability for plants (Xu et al., 2021).

In addition to heavy metals, other contaminants exacerbate soil pollution. Enrofloxacin, one kind of antibiotic, has been shown to increase nitrous oxide (N_2O) emissions by inhibiting the microbes that convert N_2O to nitrogen (Lin et al., 2022a). MPs disrupt soil nutrient dynamics by reducing ammonium (NH_4^+) content and interfering with the distribution of soil hydrolysis enzymes, making nutrients less effective (Tong et al., 2022). MPs also negatively impact soil fauna, such as microarthropods and nematodes, which in turn affects soil microbial functioning (Lin et al., 2020). Additionally, plastic pollution has been linked to significant reductions in crop growth, with estimates showing a 9%–21% reduction in cotton root weight, a 6%–10% decrease in plant height, and a 6%–10% reduction in cotton yield in Xinjiang Province due to prolonged plastic film coverage (Zhang et al., 2020).

Pesticide use also contributes to soil degradation. Atrazine, the most widely used herbicide in maize fields in China, has been shown to reduce soil enzyme activity and bacterial α-diversity, and negatively affect soybean growth (Gao et al., 2024). The fungicides tebuconazole and carbendazim, which are commonly used in agriculture, can inhibit soil respiration and enzyme activity, with potentially adverse effects on soil health (Wang et al., 2016). Earthworms (*Eisenia fetida*) show avoidance behavior toward low concentrations of neonicotinoid insecticides such as imidacloprid and cause reproductive damage to earthworms, which is not conducive to the sustainable development of soil ecosystems (Ge et al., 2018). According to predictions, when the cumulative risk quotient for soil antibiotic contamination exceeds 8.30–9.98, there is a significant reduction in crop yields for maize, rice, wheat, and vegetables (Zhao et al., 2023). Beyond direct crop yield losses, contaminated agricultural products often cannot enter the market, leading to indirect economic losses. These impacts highlight the urgent need for comprehensive soil management and pollution mitigation strategies to protect soil health and ensure agricultural productivity.

3.2 Human health risks associated with contaminated farmed soils

Crops and vegetables are critical components of human diets, yet their contamination by heavy metals and metalloids poses significant health risks (Fig. 1). Contaminants such as Cd, As, Hg, Pb, and Zn are prevalent in vegetables and paddy fields (Huang et al., 2019). Notably, Cd and Pb tend to accumulate in rice grains (Xie et al., 2017). A substantial portion of the Cd in rice is present as Cd–thiolate complexes (66%–92%), likely bound to thiol-rich proteins, which enhances its bioavailability in the acidic environment of the human gastrointestinal tract (Gu et al., 2020). For residents of Shifang, a periurban agricultural area in the Chengdu Plain in Sichuan Province, Cd intake from cereal grains represents a major health risk (Liu et al., 2022b). Long-term exposure to Cd can increase the risk of several diseases, including kidney disease, osteoporosis, and cancer (Qin et al., 2021).

The health risks associated with cereal intake are compounded by contaminants in vegetables, particularly raw ones. Raw edible vegetables have been found to contain various contaminants, including MPs in lettuce (Wang et al., 2023c), PAE-di-(2-ethylhexyl) phthalate (DEHP) (Xiang et al., 2022b) and 2,2′,4,4′-tetrabrominated diphenyl ether (BDE-47) in carrots (Xiang et al., 2019). Additionally, antibiotic-resistance genes from manure have been detected in vegetables (Mei et al., 2021). Soil particles also contribute to the contamination risk; a recent study on Hainan Island reported that the average daily exposure dose of MPs was 1.68×10^{-2} item MPs·kg $BW^{-1} \cdot d^{-1}$ (based on abundance) and

7.23×10^{-2} mg MPs·kg $BW^{-1}·d^{-1}$ (based on mass), with the main source of soil (hand-to-mouth) ingestion (>99%) (Lin et al., 2024). Once introduced into the human body, MPs may exert adverse effects on respiratory, gastrointestinal, and cardiovascular systems. These effects may result from the toxicity of MPs themselves and the toxicity of chemicals transported by MPs (Yu et al., 2022).

4 Case studies from China

4.1 Policy approaches to farmland soil pollution – challenges and progress

The National Soil Pollution Survey Report issued by the Ministry of Environmental Protection of China reveals that 19.4% of surveyed farmland soil sites exceed national pollution standards (Ministry of Environmental Protection of the People's Republic of China, 2014a). The survey was based on the Technical Regulations for the Evaluation of National Soil Pollution Conditions, and the evaluation standards were taken in principle from the Soil Environmental Quality Standards (GB15618-1995), the Farmland Environmental Quality Evaluation Standards for Edible Agricultural Products (HJ332-2006), and the Environmental Quality Evaluation Standard for Farmland of Greenhouse Vegetables Production (HJ333-2006) (Ministry of Environmental Protection of the People's Republic of China, 2014a). The primary contaminants identified include heavy metals (Cd, Cu, As, Hg, Pb, Ni,), dichlorodiphenyltrichloroethanes (DDTs), and PAHs (Ministry of Environmental Protection of the People's Republic of China, 2014b). Additionally, research into emerging contaminants–often neglected in traditional assessments–has gained momentum, revealing new challenges for environmental management (Wang et al., 2024a). A comprehensive list of contaminants detected in Chinese farmed soils is presented in Table 1.

In response to the growing problem of soil pollution, the Chinese government has taken active measures and achieved remarkable results. The measures included (1) conducting soil pollution surveys to assess the quality of the soil environment; (2) promoting soil pollution prevention and control legislation, and establishing and improving the regulatory standards system; (3) implementing classification management of agricultural land to ensure the environmental safety of agricultural production; (4) strengthening the protection of uncontaminated soil and strictly controlling new soil pollution; (5) enhancing pollution source supervision and ensuring effective soil pollution prevention efforts; (6) undertaking pollution control and remediation to improve regional soil quality; (7) increasing R&D efforts to promote the development of the environmental protection industry; (8) emphasizing the government's leading role in constructing a soil environmental management system; (9) strengthening goal assessment and enforcing strict accountability (The State Council of the

Table 1 The contaminants detected in farmed surface soils in China

Contaminants	Surveyed sites	Average concentration	Background	References
Heavy metals	Jinqu Basin, Zhejiang Province	Cd: 0.37 mg/kg Pb: 32.1 mg/kg Ni: 15.2 mg/kg Cr: 24.2 mg/kg Zn: 98.7 mg/kg Cu: 16.7 mg/kg	Cd: 0.17 mg/kg Pb: 25.6 mg/kg Ni: 22.3 mg/kg Cr: 54.6 mg/kg Zn: 69.0 mg/kg Cu: 17.8 mg/kg	(Guo et al., 2020)
	Lanzhou, Gansu Province	Pb: 20.96 mg/kg Cd: 0.24 mg/kg Cr: 0.05 mg/kg Hg: 6.40 mg/kg As: 8.36 mg/kg Cu: 48.57 mg/kg Zn:101.92 mg/kg Ni: 34.34 mg/kg	Pb: 22.0 mg/kg Cd: 0.14 mg/kg Cr: 59.85 mg/kg Hg: 0.03 mg/kg As: 10.6 mg/kg Cu: 22.9 mg/kg Zn: 67.1 mg/kg Ni: 26.5 mg/kg	(Chai et al., 2021)
Organochlorine pesticides (OCPs)	Pearl River Delta	α-HCH: 1.09 μg/kg γ-HCH: 1.40 μg/kg HCB: 0.40 μg/kg	α-HCH: 0.023 μg/kg γ-HCH: 0.206 μg/kg HCB: 0.076 μg/kg	(Yao et al., 2022)
Insecticides	Yangtze River Delta	Fenpropathrin: 4.92 μg/kg Cypermethrin: 1.10 μg/kg Deltamethrin: 0.89 μg/kg Cyhalothrin: 0.20 μg/kg	/	(Deng et al., 2020)
Polycyclic aromatic hydrocarbons (PAHs)	Changzhou, Jiangsu Province	∑15NPAHs: 35.6 μg/kg ∑16PAHs: 437 μg/kg		(Ma et al., 2023)
	Changchun, Jilin Province	∑7CarPAHs: 4.847 μg/kg ∑16PAHs: 13.791 μg/kg	∑7CarPAHs: 1.22 μg/kg ∑16PAHs: 3.09 μg/kg	(Ma et al., 2023)
Pharmaceuticals and personal care products (PPCPs)	Beijing	SMX: 0.86 μg/kg Ac-SMX: 10.92 μg/kg SX: 1.15 μg/kg SCP: 0.92 μg/kg CAP: 2.20 μg/kg IP: 1.16 μg/kg TCS: 20.22 μg/kg	/	(Liu et al., 2020a)
Microplastics	Shouguang, Shandong Province	Polytunnels: 2506.9 items/kg, Greenhouses: 2986.1 items/kg Woodland: 1243.6 items/kg Grassland: 1610.5 items/kg Maize: 1700.1 items/kg Wheat: 1343.8 items/kg	/	(Li et al., 2023)

People's Republic of China, 2016). By 2020, the safe utilization rate of farmed soils surpassed 90%, effectively curbing the trend of worsening soil pollution and fundamentally controlling associated risks. Legislative and regulatory frameworks have also been strengthened, evidenced by the enactment of the Law of the People's Republic of China on Prevention and Control of Soil

Contamination and the introduction of several regulations concerning the environmental management of farmed soil (China, 2021). These initiatives are supported by the Soil Pollution Prevention and Control Action Plan, which sets a target of achieving over 95% safe utilization rate for contaminated farmlands by 2030. By mid-century, the plan aims for a substantial enhancement in soil quality, fostering a beneficial ecological cycle (China, 2016).

4.2 Case studies in managing heavy metal contamination in farmed soils

Soil contamination with heavy metals has been identified as a critical factor affecting the safety of agricultural products. In 2013, the well-known incident of "Cd-tainted rice" highlighted concerns about the safety of grain crops and heavy metal contamination in farmed soils. A major source of rice with excessive Cd levels is Hunan Province, a major rice-growing region in China. Hunan Province is also the main production area of non-ferrous metals in China, and the wastewater, waste gas, and waste residue discharged by industrial and mining enterprises around the paddy field are the main sources of Cd contamination in the farmed soil (News, 2013). The study of Cd concentrations in farmed soils in Hunan Province found that the background concentration of Cd increased from 0.08 mg/kg in 1990 to between 0.5 mg/kg and 0.6 mg/kg in 2018 (Wang et al., 2018).

The government has responded with a suite of policies aimed at mitigating heavy metal pollution, achieving significant progress. Moreover, China has developed a strategic plan–Opinions on Further Strengthening the Prevention and Control of Heavy Metal Pollution to establish a robust heavy metal pollution control system by 2035, aimed at improving treatment capacity, environmental risk prevention, and regulatory enforcement (China., 2022). Various remediation technologies have been employed in China to address soil contamination, including bioremediation, chemical remediation, and physical remediation. The remediation of farmland in Guixi City, Jiangxi Province, is a typical case: Cultivated land contaminated with heavy metals in Jiangxi accounted for about 14.2% of the total cultivated land in the province, and heavy metal pollution was particularly severe in the areas around the mine processing areas and smelters. The case involved a comprehensive remediation strategy for heavy metal contamination: (1) Physicochemical stabilization of heavy metal using improved materials with a proportional combination of micron hydroxyapatite, ordinary apatite powder, lime, and biomass ash; (2) remediation via a combined physicochemical and biological approach; (3) restoration with the establishment of vegetation to restore soil ecological functions; and (4) efficiency enhancement to maximize the ecological, economic, and social benefits of the rehabilitated land. After two years of remediation, the remediation materials and the plant *Pennisetum* sp. resulted in a reduction of more than 50% of both Cu and Cd (0.1 mol/l $CaCl_2$ extraction) in the active state of the contaminated soil (Zhou & Cui, 2014). For heavily contaminated soils, the cost of soil remediation was low, US$10–20/t, and the *pennisetum* sp. could be used as a raw material for biomass power generation. For mildly contaminated farmland, rice yields after remediation increased by more than 32.8% compared to the control, and the Cu and Cd levels in rice were lower than the national threshold value in food. In conclusion, the remediation technology has been completed over 141 hm^2 of contiguous contaminated soil in the project area, has high environmental, economic, and social benefits (Zhou & Cui, 2014).

4.3 Case studies in managing pesticide contamination in farmed soil

In the northeastern region of China (Liaoning, Jilin and Heilongjiang provinces), the iconic black soil zone has become a hotspot for pesticide usage (Ren et al., 2022). According to the National Bureau of Statistics, in 2019, the northeastern region used 190,000 tons of pesticides, accounting for 13.7% of the national total, predominantly herbicides (China, 2020). A comprehensive survey of 1.09 million hectares of farmed surface soil in the northeastern black soil region detected residues of 56 different herbicides, with total concentrations ranging from 1.0 μg/kg to 1558.1 μg/kg (average 227.5 μg/kg). Notably, atrazine, its degradation products deethyl atrazine (DEA) and diisopropyl atrazine (DIA), trifluralin, and butachlor were the most frequently detected (Li et al., 2024a). The "Outline of Northeast Black Soil Protection Plan (2017–2030)" emphasizes the reduction of pesticide use and pollution in these critical farmed soils.

Regarding the remediation of pesticide pollution, bioremediation has emerged as a mature technology. This method involves isolating bacteria or fungi capable of degrading herbicides from the environment and introducing them into the contaminated soil. For example, Chinese researchers screened atrazine-degrading bacteria *Paenarthrobacter sp.* W11 from a corn field in Yanbian city, Jilin Province, where atrazine had been applied for a long time. After 49 days of incubation, *Paenarthrobacter sp.* W11 increased the degradation rate of atrazine in the black soil from 75.7% to 96.0% (Chen et al., 2021). The consortia of microbial strains have been used to increase degradation efficiency, such as the co-degradation of atrazine by *Arthrobacter aurescens* AT5, *Halobacillus sp.* NY15 (Xu et al., 2019b). To increase the stability of strains in the soil, carrier materials are often used to immobilize them and make them easy to apply. Bacterially loaded biochar significantly increased atrazine removal rates in black soils and soybean biomass (Harindintwali et al., 2024). Microbial agents prepared with the carrier materials (peat 1 g:

attapulgite 0.5 g: bentonite 1g) could achieve a degradation rate of atrazine up to 95.89% within 28 days (Cao, 2017).

5 Solutions and management strategies in China

Soil pollution has significant impacts on human health, crop growth, and ecosystem stability, necessitating the urgent implementation of effective management strategies and remediation measures to mitigate its adverse effects.

5.1 Sustainable soil management practices and remediation techniques

5.1.1 Sustainable soil management approaches

(1) Actively promote the development of local soil pollution prevention laws: Establishing legislation for soil pollution prevention is fundamental to addressing both current and long-term soil pollution issues in China. This ensures legal support for the prevention, control, remediation, and management of soil pollution.

(2) Permit local adaptation of soil environmental quality standards: Encourage the formulation of soil heavy metal quality standards tailored to diverse soil types across different regions. Develop standards for natural soil, agricultural soil (including farmland, forests, grasslands, vegetable fields, and orchards), and industrial land soil based on soil properties, conditions, and usage. For areas with geochemical anomalies and high background levels, create local soil environmental quality standards based on bioavailability and environmental risk.

(3) Strengthen technological innovation for zoned soil pollution remediation and safe utilization: Support research on soil pollution formation mechanisms, monitoring, and early warning systems, risk control, remediation technologies, materials, and equipment tailored to different zones. Develop integrated solutions and industrial models for regional soil pollution prevention, green sustainable remediation, and large-scale demonstration applications in typical areas to achieve comprehensive economic, social, ecological, and health benefits.

5.1.2 Remediation techniques

Based on the nature and characteristics of the pollutants, soil pollution remediation technologies applied in China include but are not limited to adsorption, bioremediation, phytoremediation, and immobilization techniques (Fig. 2). The following examples will illustrate the practical applications of these remediation technologies.

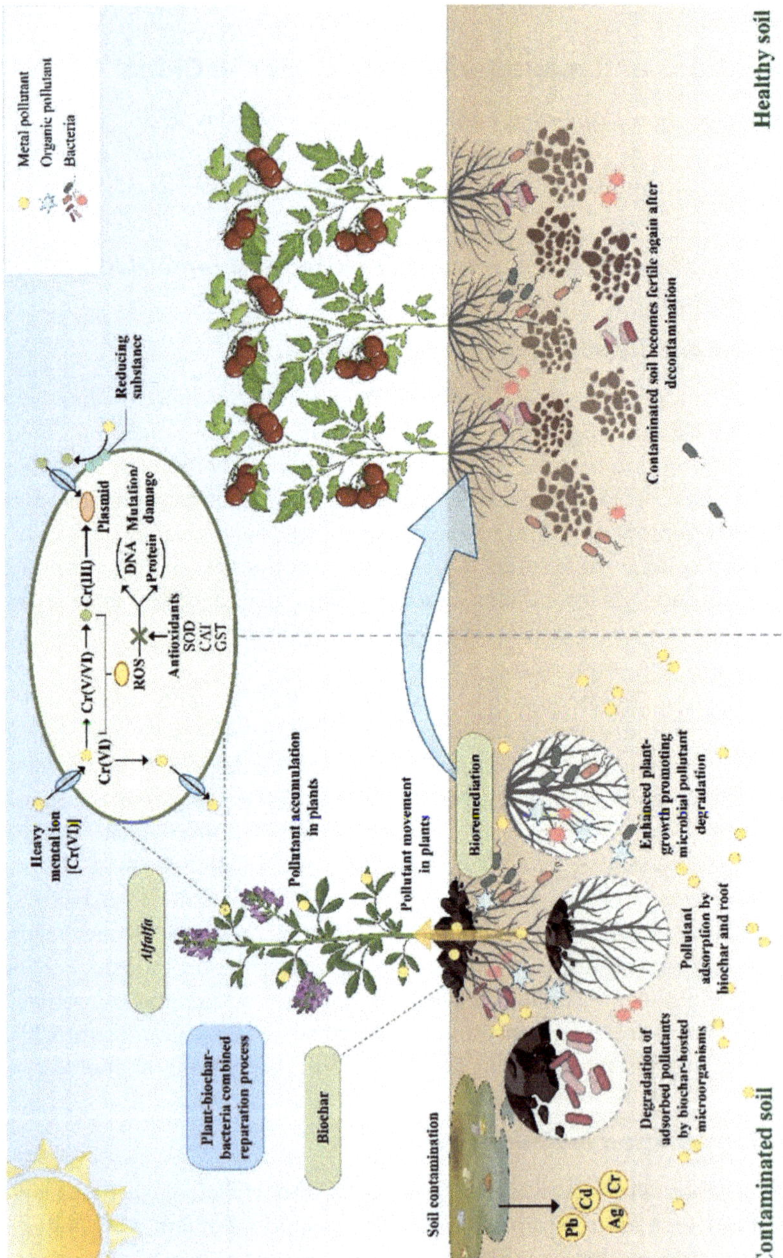

Figure 2 The remediation processes of contaminated soil via bioremediation, phytoremediation, and immobilization techniques.

Adsorption: Using carbon-based materials such as activated carbon, carbon nanotubes, graphene, and biochar exhibit high adsorption capacities due to their porous structures and surface functional groups. These materials can effectively remove or immobilize pollutants from the soil. For instance, using carbon-containing materials in agricultural soils can reduce the migration of antibiotics to groundwater and/or rivers, thereby mitigating secondary pollution in aquatic environments (Liu et al., 2022a). Furthermore, biochar serves as an excellent soil amendment for agricultural fields, not only enhancing the soil's water retention capacity but also improving nutrient cycling during crop growth. This, in turn, reduces the uptake of bioavailable heavy metals by plants (Fu et al., 2021). For example, the treatment of Cr-contaminated agricultural soil with magnetic porous biochar spheres resulted in a Cr(VI) removal rate of 90.77% (Liu et al., 2020d). Wheat straw biochar is 42 times more effective than soil at adsorbing hexachlorobenzene, and it can significantly reduce the bioavailability of various OCPs in contaminated soils (Ali et al., 2019; Song et al., 2012). The addition of biochar made from rice husks accelerates the dissipation rate of herbicides in farmland, reducing the half-life of atrazine and oxyfluorfen by half. This mitigates the phytotoxicity of herbicide pollution on soybeans and corn, enhancing the yield and quality of agricultural products. Utilizing the unique adsorption properties of attapulgite clay, Cd in the soil can be rendered inactive, effectively preventing its transfer to crops and ensuring the safety of agricultural products.

Bioremediation: Earthworm activity, including burrowing, feeding, and excretion can enhance the degradation of antibiotic-resistance genes in the soil by up to 43% through improved soil properties (Pu et al., 2020). Previous studies have also found that the addition of carbon sources (such as biochar and sodium acetate) can enhance the biodegradation efficiency of antibiotics (up to 78%) by enriching the biomass of microbial strains (Liang & Hu, 2019). Two urease-producing bacteria, *Enterobacter bugandensis,* and *Bacillus megaterium*, were isolated from the rhizosphere soil of lettuce in Cd- and Pb-contaminated farmland. These bacteria mitigated the toxicity of Cd and Pb through extracellular adsorption, promoted lettuce growth, improved quality, and reduced Cd and Pb uptake by lettuce (Wang et al., 2020b). *Pseudoalteromonas* sp. SCSE709-6 removes 96% of Cd(II) through the mechanisms involving extracellular glycoproteins and the carboxyl, amino, and hydroxyl groups of extracellular proteins. Chen et al. (2021) investigated the removal efficacy of atrazine by the strain *Paenarthrobacter* sp. W11 and the results demonstrated that *Paenarthrobacter* sp. W11 increased the degradation rate of atrazine in soil from 75.7% to 96.0%, significantly enhancing the degradation of atrazine in soil and mitigating its phytotoxic effects on wheat growth.

Phytoremediation: *P. vittata* is the first known As hyperaccumulator that is fast-growing, highly regenerative, and resistant to plant pests and diseases (Ma et al., 2001). Following a 2-year remediation project utilizing *P. vittata* in Huanjiang Maonan autonomous county in southwestern China, available arsenic levels were reduced by 55.3%, and local agricultural products have met national standards (Wan et al., 2016). *Sedum plumbizincicola* is a hyperaccumulator of cadmium and zinc. A field trial conducted in Hangzhou, Zhejiang Province, demonstrated that the removal rates of cadmium could reach 21.1% after 13 months when the planting density of *Sedum plumbizincicola* was 4.4×10^5 plants per hm^2 (Liu et al., 2009). Planting ramie in Hg-contaminated paddy fields resulted in a soil Hg purification rate of 41%, reducing the soil purification period by 8.5 times compared to planting rice (Liu et al., 2020b). Several plants, including Brassica juncea, *Lolium multiflorum*, *Mirabilis jalapa* L., *Tagetes patula* L., *Lythrum salicaria*, and *Chrysopogon zizanioides*, exhibit high capabilities for antibiotic removal due to their long root systems that create favorable conditions for microorganisms (Cui et al., 2021). Furthermore, research has shown that planting Pennisetum can increase the degradation rate of atrazine from 15.22% to 51.46% (Lin et al., 2018). Substantial elimination of total PAE compounds of over 87% was attained using alfalfa both in monoculture and in different intercropping systems under field conditions of the PAE-polluted soil (Ma et al., 2012).

Immobilization treatment: The 2-year field demonstration confirmed sepiolite's stability in immobilizing Cd contaminants in paddy soil, significantly reducing Cd levels in brown rice and soil without additional sepiolite in the second year (Liang et al., 2016). Furthermore, applying zeolite and biochar in cadmium-contaminated farmland soil can reduce both the content and transfer coefficient of Cd in maize plants and grains, suitable for immobilization remediation in Cd-contaminated farmland (Zhan et al., 2019). The sepiolite mineral material modified with nano-zero-valent iron was used as a stabilizer to immobilize cadmium, lead, and chromium in soil for cabbage cultivation, reducing their soil bioavailability and thereby decreasing plant uptake (Xu et al., 2019a).

Combined remediation: A field experiment found that inoculating alfalfa with *Glomus caledonium*, *Rhizobium meliloti*, or both reduced soil PCB concentrations by up to 43.5% over 180 days, while also increasing alfalfa yield and PCB accumulation in shoots, indicating enhanced phytoremediation potential through synergistic interactions (Teng et al., 2010). Phytoremediation using oilseed crops, which does not disrupt soil ecosystems or structure, improves soil quality and offers cost-effective, high-return solutions without secondary pollution risks. Furthermore, A field trial demonstrated that using a bio-preparation of immobilized and modified peanut shell powder with *Mycobacterium gilvum* significantly enhanced the degradation of PAHs in contaminated vegetable fields, especially high-molecular-weight PAHs, increasing the degradation rate by 16.5% to 43.5% compared to untreated soil (Ma et al., 2018a).

5.2 Contrasting approaches to soil contamination management in different regions of China

The situation of soil pollution prevention and control in China is severe. Currently, the increasing risk of soil pollution threatens the safety of agricultural product quality, the living environment, and ecological security in China (Luo & Teng, 2018). Soil pollution in China spans seven major regions: Southwest, Central China, South China, East China, North China, Northwest, and Northeast, covering provinces, cities, and districts within these regions (Luo & Teng, 2020). It is spatially distributed across urban, suburban, rural, and natural environments (e.g. areas with high geological background levels), and includes land use types such as agricultural land, construction land, mining areas, oilfields, and military sites (Zhu et al., 2011). Therefore, there are certain differences in soil pollution management methods across different regions in China, mainly reflecting in regulations and policies, monitoring systems, and remediation technologies. The following is a comparative study of soil pollution management methods in several representative major regions, including Beijing in North China, Shanghai in East China, Guangdong in South China, Liaoning in Northeast China, Henan in Central China, and Sichuan in Southwest China (Table 2).

Overall, the soil remediation techniques across Beijing, Shanghai, Guangdong, Henan, Liaoning, and Sichuan reflect the unique environmental challenges and remediation goals of each region. Beijing and Shanghai utilize advanced technologies like chemical oxidation and thermal desorption to address industrial contamination and urban redevelopment. In contrast, Guangdong and Henan focus on agricultural soil contamination, employing phytoremediation and organic amendments. Liaoning combines physical methods with phytoremediation to tackle industrial pollution, while Sichuan emphasizes cost-effective bioremediation and phytoextraction using local plant species. This diversity in approaches highlights the adaptation of remediation strategies to regional pollution sources and environmental conditions.

6 Future trends and conclusion

6.1 Challenges for contaminants control in farmed soils

Widespread chemical pollution poses a major threat to soil safety. Soil has become a potential long-term source of environmental pollution and an important link in the transport and transformation of pollutants. Currently, the main challenges for pollution prevention and control in the soil sector

Table 2 Soil contamination management in different regions of China

Surveyed site	Policy	Policy orientation	Monitoring system	Remediation method	Reference
Beijing	*Beijing Soil Pollution Prevention and Control OrdinanceBeijing Soil Pollution Prevention and Control Work Plan*	Emphasis on source control and prevention, particularly strict management of industrial enterprises and construction activities.	A comprehensive soil environment monitoring network covering the entire city, focusing on industrial areas, agricultural zones, and urban construction land	Utilizes a combination of physical, chemical, and biological remediation techniques.	(Congress, 2022)(Beijing Municipal People's Government, 2016)
Shanghai	*Implementation Plan for the Soil Pollution Prevention and Control Action PlanShanghai Soil Pollution Prevention and Control Ordinance*	Focus on soil pollution prevention during industrial restructuring and urban renewal, with particular attention to soil remediation in the transformation of old industrial areas.	Several monitoring stations have been set up throughout the city, especially in industrial zones, chemical parks, and urban redevelopment zones	Emphasizes safety and sustainability in remediation processes, promoting green remediation technologies such as phytoremediation and microbial remediation.	(Shanghai Municipal People's Government, 2016) (Congress, 2023a)
Guangdong	*Implementation Plan for the Soil Pollution Prevention and Control Action Plan*	Focus on regional joint prevention and control, especially soil pollution issues in the Pearl River Delta region.	A soil environment monitoring network has been established at the provincial, city, and county levels, focusing on agricultural land, industrial parks, and polluted sites	Uses a combination of various remediation techniques, such as thermal treatment, soil washing, and stabilization/solidification, based on the type and extent of pollution.	(Guangdong Provincial People's Government, 2016)

Liaoning	*Liaoning Province Soil Pollution Prevention and Control Work PlanLiaoning Province Construction Land Soil Pollution Risk Control and Remediation Management Measures (Trial)*	Focus on industrial pollution prevention and agricultural land protection.	Covers industrial zones, mining areas, and agricultural regions across the province, with a focus on heavy industrial zones and heavily polluted mining areas	Frequently employs physical, chemical, and biological remediation techniques, focusing on heavy metals and organic pollutants.	(Liaoning Provincial People's Government, 2016) (Liaoning Provincial Department of Ecology and Environment, 2019)
Henan	*Henan Province Clean Soil Action PlanHenan Province Soil Pollution Prevention and Control Ordinance*	Focus on agricultural land protection and rural environmental governance.	A three-tiered monitoring network at the provincial, municipal, and county levels, focusing on agricultural land and areas surrounding chemical enterprises	Emphasizes sustainability and promotes green remediation technologies such as phytoremediation and microbial remediation.	(Henan Provincial People's Government, 2017) (Environment, 2021)
Sichuan	*Sichuan Province Soil Pollution Prevention and Control Action Plan Work PlanSichuan Province Soil Pollution Prevention and Control Ordinance*	Emphasis on both ecological protection and pollution prevention, particularly soil protection in mountainous and hilly areas.	A three-tiered monitoring network at the provincial, municipal, and county levels, focusing on agricultural land, industrial zones, and mountainous ecological protection areas	Uses a combination of various techniques such as thermal treatment, soil washing, and stabilization/solidification.	(Sichuan Provincial People's Government, 2016) (Congress, 2023b)

in China include: (1) the diversity of pollution sources: the sources of soil pollutants are complex, and in addition to direct inputs from agriculture, the trans-media transport processes of pollutants, such as atmospheric deposition, and wastewater irrigation, can also cause soil pollution (Zhao et al., 2015). The challenge of accurately identifying the sources of soil contaminants is great in order to address these different sources. (2) Persistence of contaminants: many soil contaminants, such as heavy metals and POPs, are persistent in the environment over the long term (Tufail et al., 2022). Effective remediation strategies must comprehensively address these persistent contaminants. (3) Technical and economic constraints: although technological advances have provided new solutions for soil contamination management, many innovative technologies are either still at the experimental stage or too costly to be widely adopted. (4) Monitoring and enforcement: establishing robust monitoring systems and ensuring compliance with regulations requires significant resources and strong political commitment.

6.2 Opportunities for further research and monitoring in soil contamination management

Monitoring and managing soil contamination is important for agricultural productivity and environmental integrity. Existing remediation approaches have encountered challenges in preserving soil structure and function throughout the remediation process and addressing the complexity of soil contamination at regional scales. Viable solutions include monitoring and modeling soil processes, using big data to connect micro-scale and macro-scale processes (Chu & Zhu, 2024), and predicting soil contamination by using machine learning models to simulate complex non-linear relationships between various environmental conditions and soil contamination (Zhang et al., 2023). Meanwhile, remote sensing, geographic information systems, intelligent sensing, and other technologies are being used to support the detection and monitoring of soil contaminants (Wang et al., 2023b; Shi et al., 2021). In addition, an in-depth understanding of the interaction mechanisms between soil and pollutants, and the mutual effects of compound pollution is also crucial for soil pollution management. Establishing collaborative networks among academic institutions, research organizations, and government agencies can effectively improve the understanding and management of soil pollution.

6.3 Policy recommendations for enhancing soil quality and agricultural sustainability in China

To address soil pollution and promote sustainable agricultural practices in China, there is a need to strengthen the regulatory framework by imposing stringent regulations on the use of pesticides, fertilizers, and industrial chemicals, which can optimize resource utilization and reduce environmental pollution (Yin & Cui, 2024; Tang et al., 2021). Promoting sustainable agricultural practices and encouraging the adoption of sustainable agricultural technologies, such as the adoption of diversified crop rotations (Yang et al., 2024) and organic and conservation agriculture practices (Xu et al., 2022), provide synergistic benefits for crop productivity, soil health, and environmental protection. At the same time, investment in research and development of new technologies for interdisciplinary research, technological innovation, and sustainable agricultural practices should be increased, and a collaborative chain of research institutes, companies, and governments should be established to maintain sustainable soil health. In addition, the introduction of economic incentives and support to provide financial incentives and subsidies to farmers who adopt sustainable farming practices can accelerate the transition to environmentally friendly agriculture.

6.4 Conclusion

Proactive management of soil contaminants is critical to the sustainable development of Chinese agriculture, and healthy soils are the cornerstone of a productive and resilient agricultural system. Therefore, comprehensive research on the behavior and fate of contaminants in soil ecosystems is particularly important. In addition, the purification and recycling efficiency of key pollutants, such as nitrogen and phosphorus, must be further improved to ensure that environmental pollution is reduced while agricultural production is increased. Actively addressing soil contamination will ensure long-term food security, protect public health, and mitigate environmental degradation. Achieving these goals will require a combination of advanced research, robust policies, and sustainable agricultural practices.

Proactive soil health management will not only benefit China but also promote sustainable agriculture and environmental protection worldwide. China's soil pollution management strategies and policies can serve as models for other countries, and the international exchange of knowledge and best practices can strengthen global soil management. Collaborative research programs, policy development, and technology transfer between countries can render global soil health strategies more coherent and effective. The establishment of international standards for soil quality and contamination levels will contribute to better monitoring and management practices on a global scale.

7 Acknowledgements

This work was supported by the Strategic Priority Research Program of the Chinese Academy of Sciences (XDA28030501), National Key R&D Program of

China (2024YFC3713900), the National Natural Science Foundation of China (41991333, 42307048), the Institute of Soil Science, Chinese Academy of Sciences (ISSAS2419), and International Atomic Energy Agency Coordinated Research Project (D15022). Fang Wang was partly supported by the fellowship of Alexander von Humboldt for experienced researchers.

8 Where to look for further information

8.1 Good review articles or textbook

- Food & Agriculture Organization (FAO) has published a book *Soil Pollution: A Hidden Reality*, it highlights the need to assess the extent of soil pollution globally and develop soil management practices that enhance agricultural resilience to extreme climate events by reducing soil degradation processes. (https://openknowledge.fao.org/server/api/core/bitstreams/3f7e6959-db0b-44d3-971e-109bcfe78195/content)
- The review article *China's Soil Pollution: Farms on the Frontline* revealing China's farmland soil pollution was published in the journal Science. (https://www.science.org/doi/10.1126/science.344.6185.691-a)

8.2 Professional Organizations

- The FAO Soils Portal provides global soil health data and policy reports. (https://www.fao.org/soils-portal/zh/)
- The International Union of Soil Sciences publishes soil pollution research trends and conference information. (https://www.iuss.org/)
- Department of Soil Ecology and Environment, Ministry of Ecology and Environment of the People's Republic of China is responsible for the supervision and management of soil, groundwater and other pollution prevention and ecological protection across the country, organizing and guiding rural ecological environment protection, and supervising and guiding agricultural pollution control.(https://www.mee.gov.cn/zjhb/bjg/trhj/)

8.3 Key journals

- *Environmental Science & Technology* (IF=10.9): A flagship journal in the field of environment, it frequently publishes research on the mechanisms of migration and transformation of soil pollutants and emerging pollutants (such as microplastics). https://pubs.acs.org/journal/esthag
- Soil Biology and Biochemistry (IF=9.8): A top-tier, high-impact journal in soil science, publishing rigorous research on soil organisms, biochemistry,

and ecosystem processes. https://www.sciencedirect.com/journal/soil-biology-and-biochemistry

- Pedosphere (IF=5.2): A journal focuses on soil science, publishing high-quality research in environmental, agricultural, and ecological fields with rigorous peer review. https://www.sciencedirect.com/journal/pedosphere

8.4 Conferences

- World Congress of Soil Science
- National Conference on Soil Remediation in China
- International Conference on Soil Pollution and Remediation
- International Cleanup Conference

8.5 Research Centres

- Institute of Soil Science, Chinese Academy of Sciences. https://www.issas.ac.cn/

9 References

Abdu, N., et al. 2017. Heavy metals and soil microbes. *Environmental Chemistry Letters*, 15, 65–84.

Alavanja, M.C.R. 2009. Introduction: pesticides use and exposure extensive worldwide. *Reviews on Environmental Health*, 24, 303–310.

Ali, N., et al. 2019. Influence of biochars on the accessibility of organochlorine pesticides and microbial community in contaminated soils. *Science of the Total Environment*, 647, 551–560.

Atafar, Z., et al. 2010. Effect of fertilizer application on soil heavy metal concentration. *Environmental Monitoring and Assessment*, 160, 83–89.

Beijing Municipal People's Government. 2016. Beijing soil pollution prevention and control work program[EB/OL]. https://sthjj.beijing.gov.cn/bjhrb/index/xxgk69/zfxxgk43/fdzdgknr2/zcfb/szfgfxwj/10939747/index.html.

Cai, L., et al. 2012. Source identification of eight hazardous heavy metals in agricultural soils of Huizhou, Guangdong Province, China. *Ecotoxicology and Environmental Safety*, 78, 2–8.

Cai, Z., et al. 2015. Intensified soil acidification from chemical N fertilization and prevention by manure in an 18-year field experiment in the red soil of southern China. *Journal of Soils and Sediments*, 15, 260–270.

Cao, B. 2017. Construction of functional bacterial agents based on atrazine-degrading bacteria and their in situ remediation of contaminated soil. Northeast Agricultural University.

Chai, L., et al. 2021. Pollution characteristics, spatial distributions, and source apportionment of heavy metals in cultivated soil in Lanzhou, China. *Ecological Indicators*, 125, 107507.

Chen, L., et al. 2023. A national-scale distribution of organochlorine pesticides (OCPs) in cropland soils and major types of food crops in China: Co-occurrence and associated risks. *Science of The Total Environment*, 861, 160637.

Chen, S., et al. 2021. Soil bacterial community dynamics following bioaugmentation with Paenarthrobacter sp. W11 in atrazine-contaminated soil. *Chemosphere*, 282, 130976.

Chen, Z., et al. 2019. Characteristics, sources and environmental implications of atmospheric wet nitrogen and sulfur deposition in Yangtze River Delta. *Atmospheric Environment*, 219, 116904.

Chu, C., and Zhu, L. 2024. Paving the way toward soil safety and health: current status, challenges, and potential solutions. *Frontiers of Environmental Science & Engineering*, 18, 74.

Clunes, J., et al. 2021. MARTgineeringl safety and health: current status, challenges, and potential solutions in charge of soils[EB/OL]. *Ecological Indicators*, 139, 108969.

Collin, M.S., et al. 2022. Bioaccumulation of lead (Pb) and its effects on human: a review. *Journal of Hazardous Materials Advances*, 7, 100094.

Cui, E., et al. 2021. Ryegrass *Lolium multiflorum* and Indian mustard *Brassica juncea* intercropping can improve the phytoremediation of antibiotics and antibiotic resistance genes but not heavy metals. *Science of the Total Environment*, 784, 147093.

Deng, F., et al. 2020. Contamination of pyrethroids in agricultural soils from the Yangtze River Delta, China. *Science of The Total Environment*, 731, 139181.

Fu, H., et al. 2021. Hierarchically porous magnetic biochar as an efficient amendment for cadmium in water and soil: Performance and mechanism. *Chemosphere*, 281, 130990.

Gao, T., et al., Wang, F. 2024. Characteristics of bacterial community and extracellular enzymes in response to atrazine application in black soil. *Environmental Pollution*, 343, 123286.

Ge, J., et al. 2018. Sub-lethal effects of six neonicotinoids on avoidance behavior and reproduction of earthworms (Eisenia fetida). *Ecotoxicology and Environmental Safety*, 162, 423–429.

Gu, Y., et al. 2020. Chemical speciation and distribution of cadmium in rice grain and implications for bioavailability to humans. *Environmental Science & Technology*, 54, 12072–12080.

Guangdong Provincial People's Government. 2016. Implementation plan for soil pollution prevention and control action plan of Guangdong Province[EB/OL]. https://www.gd.gov.cn/gkmlpt/content/0/145/post_145728.html#7.

Guo, B., et al. 2020. Health risk assessment of heavy metal pollution in a soil-rice system: a case study in the Jin-Qu Basin of China. *Scientific Reports*, 10, 11490.

Guo, J., et al. 2022. Farmland scale and chemical fertilizer use in rural China: new evidence from the perspective of nutrient elements. *Journal of Cleaner Production*, 376, 134278.

Harindintwali, J.D., et al. 2024. Synergistic mitigation of atrazine-induced oxidative stress on soybeans in black soil using biochar and Paenarthrobacter sp. AT5. *Journal of Environmental Management*, 359, 120951.

He, G 2014. The soil pollution crisis in China: a cleanup presents daunting challenge[EB/OL].https://e360.yale.edu/features/the_soil_pollution_crisis_in_china_a_cleanup_presents_daunting_challenge.

Henan Provincial People's Government. 2017. Henan clean soil action plan[EB/OL]. https://www.henan.gov.cn/2017/05-15/239729.html.

Henan Provincial Department of Ecology and Environment. 2021. Henan Province soil pollution prevention and control regulations[EB/OL].https://sthjt.henan.gov.cn/2021/07-27/2190044.html.

Huang, S., et al. 2009. Multivariate analysis of trace element concentrations in atmospheric deposition in the Yangtze River Delta, East China. *Atmospheric Environment*, 43, 5781–5790.

Huang, Y., et al. 2019. Current status of agricultural soil pollution by heavy metals in China: a meta-analysis. *Science of The Total Environment*, 651, 3034–3042.

Khan, I., et al. 2021. Arsenic behavior in soil-plant system and its detoxification mechanisms in plants: a review. *Environmental Pollution*, 286, 117389.

Khan, S., et al. 2008. Health risks of heavy metals in contaminated soils and food crops irrigated with wastewater in Beijing, China. *Environmental Pollution*, 152, 686–692.

Li, J., et al. 2023. Vertical distribution and characteristics of soil microplastics under different land use patterns: a case study of Shouguang City, China. *Science of The Total Environment*, 903, 166154.

Li, R., et al. 2024a. Occurrence, distribution and ecological risk assessment of herbicide residues in cropland soils from the Mollisols region of Northeast China. *Journal of Hazardous Materials*, 465, 133054.

Li, X., et al. 2020. Status of mercury accumulation in agricultural soils across China (1976–2016). *Ecotoxicology and Environmental Safety*, 197, 110564.

Li, Y.-F., et al. 2024b. Persistent organic pollutants in global surface soils: distributions and fractionations. *Environmental Science and Ecotechnology*, 18, 100311.

Li, Z., et al. 2014. A review of soil heavy metal pollution from mines in China: pollution and health risk assessment. *Science of The Total Environment*, 468–469, 843–853.

Li, Z., et al. 2018. Organic contaminants in the effluent of Chinese wastewater treatment plants. *Environmental Science and Pollution Research*, 25, 26852–26860.

Liang, D.H., and Hu, Y. 2019. Simultaneous sulfamethoxazole biodegradation and nitrogen conversion by *Achromobacter* sp. JL9 using with different carbon and nitrogen sources. *Bioresource Technology*, 293, 122061.

Liang, X., et al. 2016. Two-year stability of immobilization effect of sepiolite on Cd contaminants in paddy soil. *Environmental Science and Pollution Research*, 23, 12922–12931.

Liaoning Provincial People's Government. 2016. Liaoning provincial soil pollution prevention and control work program[EB/OL]. https://www.ln.gov.cn/web/zwgkx/lnsrmzfgb/2016n/zk/zk13/szfwj/D4AF1F3541234435BFFF514F34B8E5A5/index.shtml.

Liaoning Provincial Department of Ecology and Environment. 2019. Administrative measures for risk control and remediation of soil pollution on construction land in Liaoning Province (for trial implementation)[EB/OL].https://sthj.ln.gov.cn/sthj/tc/zwgkzdgz/xyzcyjd/xgzcjd/7E3B4867BDD846B3AABD8546E3A99B38/index.shtml.

Lin, B., et al. 2024. Health assessment based on exposure to microplastics in tropical agricultural soil. *Journal of Hazardous Materials*, 465, 133372.

Lin, D., et al. 2020. Microplastics negatively affect soil fauna but stimulate microbial activity: insights from a field-based microplastic addition experiment. *Proceedings of the Royal Society B: Biological Sciences*, 287, 20201268.

Lin, H., et al. 2022a. Plants mitigate nitrous oxide emissions from antibiotic-contaminated agricultural soils. *Environmental Science & Technology*, 56, 4950–4960.

Lin, H., et al. 2022b. Technologies for removing heavy metal from contaminated soils on farmland: a review. *Chemosphere*, 305, 135457.

Lin, Z., et al. 2018. Isolation of urease-producing bacteria and their efflaterite soils of (*Pennisetum alopecuroides* L).Spreng. *Environmental Science and Pollution Research*, 25, 12407–12418.

Liu, L., et al. 2009. [Effect of planting densities on yields and zinc and cadmium uptake by Sedum plumbizincicola]. *Environmental Science & Technology*, 30(11), 3422–3426.

Liu, J., et al. 2022a. Peroxymonosulfate activation by cobalt particles embedded into biochar for levofloxacin degradation: efficiency, stability, and mechanism. *Separation and Purification Technology*, 294, 121082.

Liu, J., et al. 2023. Abiotic transformation of polycyclic aromatic hydrocarbons via interaction with soil components: a systematic review. *Critical Reviews in Environmental Science and Technology*, 53, 676–699.

Liu, L., et al. 2021. Excessive application of chemical fertilizer and organophosphorus pesticides induced total phosphorus loss from planting causing surface water eutrophication. *Scientific Reports*, 11, 23015.

Liu, Q., et al. 2022b. Health risk assessment of heavy metals in soils and food crops from a coexist area of heavily industrialized and intensively cropping in the Chengdu Plain, Sichuan, China. *Front Chem*, 10, 988587.

Liu, X., et al. 2020a. Occurrence and human health risk assessment of pharmaceuticals and personal care products in real agricultural systems with long-term reclaimed wastewater irrigation in Beijing, China. *Ecotoxicology and Environmental Safety*, 190, 110022.

Liu, X., et al. 2020b. Review on phytoextraction of heavy metals from contaminated agricultural land. *Environmental Pollution & Control*, 42, 507–513.

Lu, Y., et al. 2015. Addressing China's grand challenge of achieving food security while ensuring environmental sustainability. *Science Advances*, 1, e1400039.

Liu, Y., et al. 2020c. Atmospheric deposition of hazardous elements and its accumulation in both soil and grain of winter wheat in a lead-zinc smelter contaminated area, Central China. *Science of The Total Environment*, 707, 135789.

Liu, Y., et al. 2020d. A magnetic macro-porous biochar sphere as vehicle for the activation and removal of heavy metals from contaminated agricultural soil. *Chemical Engineering Journal*, 390, 124638.

Luo, Y., and Teng, Y. 2018. Regional differences in soil pollution in China and zoning treatment and remediation strategies. *Proceedings of the Chinese Academy of Sciences*, 33, 145–152.

Luo, Y., and Teng, Y. 2020. Research progresses and prospects on soil pollution and remediation in China. *Acta Pedologica Sinica*, 57(5), 1137–1142.

Lyu, S., et al. 2022. Effects of reclaimed wastewater irrigation on soil-crop systems in China: a review. *Science of The Total Environment*, 813, 152531.

Ma, L., et al. 2015. Heavy metal contamination of agricultural soils affected by mining activities around the Ganxi River in Chenzhou, Southern China. *Environmental Monitoring and Assessment*, 187, 1–9.

Ma, L., et al. 2018a. Bioremediation of PAH-contaminated farmland: field experiment. *Environmental Science and Pollution Research*, 25, 64–72.

Ma, L.Q., et al. 2001. A fern that hyperaccumulates arsenic. *Nature*, 411, 438–433.

Ma, T., et al. 2012. Removal of phthalic esters from contaminated soil using different cropping systems: a field study. *European Journal of Soil Biology*, 50, 76–82.

Ma, T., et al. 2023. Inventory, source and health risk assessment of nitrated and parent PAHs in agricultural soils over a rural river in Southeast China. *Chemosphere*, 329, 138688.

Ma, W.-L., et al. 2018b. PAHs in Chinese atmosphere, part I: concentration, source and temperature dependence. *Atmospheric Environment*, 173, 330–337.

Mei, Z., et al. 2021. Bioaccumulation of Manure-borne antibiotic resistance genes in carrot and its exposure assessment. *Environment International*, 157, 106830.

Meng, W., et al. 2016. Heavy metals in soil and plants after long-term sewage irrigation at Tianjin China: a case study assessment. *Agricultural Water Management*, 171, 153–161.

Ministry of Environmental Protection of the People's Republic of China. 2014a. National soil contamination survey report[EB/OL]. https://www.mee.gov.cn/gkml/sthjbgw/qt/201404/t20140417_270670.htm

Ministry of Environmental Protection of the People's Republic of China. 2014b. The Ministry of Environmental Protection and the Ministry of Land and Resources and other relevant persons in charge of the National Soil Pollution Status Survey answered reporters' questions[EB/OL]. https://www.mee.gov.cn/gkml/sthjbgw/qt/201404/t20140417_270671.htm.

Ministry of Natural Resources, P. 2021. The third national land survey main data bulletin[EB/OL].https://www.gov.cn/xinwen/2021-08/26/content_5633490.htm.

Mu, H., et al. 2023. Ecological risk assessment of pesticides on soil biota: an integrated field-modelling approach. *Chemosphere*, 326, 138428.

National Bureau of Statistics of China. 2020. China rural statistical yearbook-2020[EB/OL]. https://www.stats.gov.cn/sj/ndsj/2020/indexch.htm.

National Bureau of Statistics of China. 2023. China statistical yearbook 2023[EB/OL]. https://www.stats.gov.cn/sj/ndsj/2023/indexeh.htm.

News, BBC. 2013. Guangdong announces 120 batches of rice with excessive cadmium levels.https://www.bbc.com/zhongwen/simp/china/2013/05/130523_china_rice_scandal

Pan, M., and Chu, L.M. 2018. Occurrence of antibiotics and antibiotic resistance genes in soils from wastewater irrigation areas in the Pearl River Delta region, Southern China. *Science of The Total Environment*, 624, 145–152.

Pan, M., et al. 2014. Distribution of antibiotics in wastewater-irrigated soils and their accumulation in vegetable crops in the Pearl River Delta, Southern China. *Journal of Agricultural and Food Chemistry*, 62, 11062–11069.

Pu, Q., et al. 2020. Enhanced removal of ciprofloxacin and reduction of antibiotic resistance genes by earthworm *Metaphire vulgaris* in soil. *Science of the Total Environment*, 742, 140409.

Qin, G., et al. 2021. Soil heavy metal pollution and food safety in China: effects, sources and removing technology. *Chemosphere*, 267, 129205.

Qing, X., et al. 2015. Assessment of heavy metal pollution and human health risk in urban soils of steel industrial city (Anshan), Liaoning, Northeast China. *Ecotoxicology and Environmental Safety*, 120, 377–385.

Ren, W., et al. 2022. Progress and prospects of research on herbicide pollution processes and abatement technologies in agricultural fields of Black soil in Northeast China. *Soil Journal*, 59, 888–898.

Sarkar, D., et al. 2021. Organic interventions conferring stress tolerance and crop quality in agroecosystems during the United Nations Decade on Ecosystem Restoration. *Land Degradation & Development*, 32, 4797–4816.

Shanghai Municipal People's Government. 2016. Shanghai soil pollution prevention and control action plan implementation program[EB/OL]. https://www.shanghai.gov.cn/nw39327/20200821/0001-39327_50912.html.

Shao, L., et al. 2021. Organic interA review of current perspectives and environmental implications. *Journal of Cleaner Production*, 347, 131048.

Shi, T., et al. 2021. Mapping lead concentrations in urban topsoil using proximal and remote sensing data and hybrid statistical approaches. *Environmental Pollution*, 272, 116041.

Sichuan Provincial People's Government. 2016. Soil pollution prevention and control action plan work program of Sichuan Province[EB/OL]. https://www.sc.gov.cn/10462/c103044/2017/1/3/7419ab233eef49ac851b2620b72b6eae.shtml

Song, A., et al. 2022. Polybrominated dibenzo-p-dioxins/furans (PBDD/Fs) in soil around municipal solid waste incinerator: a comparison with polychlorinated dibenzo-p-dioxins/furans (PCDD/Fs). *Environmental Pollution*, 293, 118563.

Song, Y., et al. 2012. Bioavailability assessment of hexachlorobenzene in soil as affected by wheat straw biochar. *Journal of Hazardous Materials*, 217, 391–397.

Standing Committee of the Beijing Municipal People's Congress. 2022. Beijing soil pollution prevention and control regulations[EB/OL].http://www.bjrd.gov.cn/rdzl/dfxfgdxb/202210/P020221013332665003499.pdf.

Standing Committee of the Sichuan Provincial People's Congress. 2023a. Regulations on prevention and control of soil pollution in Sichuan Province[EB/OL].https://www.scspc.gov.cn/flfgk/scfg/202304/t20230403_43772.html.

Standing Committee of the Shanghai Municipal People's Congress. 2023b. Shanghai soil pollution prevention and control regulations[EB/OL].https://sthj.sh.gov.cn/hbzhywpt1013/hbzhywpt1042/20230810/cbf3402b1d3d46eb9999d060cb6ff180.html.

The State Council of the People's Republic of China. 2016. Action plan on prevention and control of soil pollution[EB/OL]. https://www.gov.cn/zhengce/content/2016-05/31/content_5078377.htm.

The State Council of the People's Republic of China. 2021. Soil, groundwater and rural ecological environment protection plan of the 14th five-year plan[EB/OL]. https://www.gov.cn/zhengce/zhengceku/2022-01/04/content_5666421.htm.

The State Council of the People's Republic of China. 2022. Opinions on further strengthening the prevention and control of heavy metal pollution. https://www.mee.gov.cn/xxgk2018/xxgk/xxgk03/202203/t20220315_971552.html

Sun, J., et al. 2018. Organic contamination and remediation in the agricultural soils of China: a critical review. *Science of the Total Environment*, 615, 724–740.

Tang, F.H.M., et al. 2021. Risk of pesticide pollution at the global scale. *Nature Geoscience*, 14, 206–210.

Tang, K.H.D. 2023. Microplastics in agricultural soils in China: sources, impacts and solutions. *Environmental Pollution*, 322, 121235.

Teng, Y., et al. 2010. Influence of arbuscular mycorrhiza and rhizobium on phytoremediation by alfalfa of an agricultural soil contaminated with weathered PCBs: A field study. *International Journal of Phytoremediation*, 12, 516–533.

Textor, C. 2023. Use of chemical fertilizers in China between 2012 and 2022[EB/OL]. https://www.statista.com/statistics/280131/use-of-chemical-fertilizers-in-china/#:~:text=The%20statistic%20shows%20the%20use,fertilizer%20were%20used%20in%20China.

Tong, Y., et al. 2022. Microplastics affect activity and spatial distribution of C, N, and P hydrolases in rice rhizosphere. *Soil Ecology Letters*, 5, 220138.

Tufail, M.A., et al. 2022. Recent advances in bioremediation of heavy metals and persistent organic pollutants: a review. *Science of The Total Environment*, 850, 157961.

Wan, X., et al. 2016. Cost-benefit calculation of phytoremediation technology for heavy-metal-contaminated soil. *Science of The Total Environment*, 563, 796–802

Wang, C., et al. 2016. Individual and combined effects of tebuconazole and carbendazim on soil microbial activity. *European Journal of Soil Biology*, 72, 6–13.

Wang, C.-C., et al. 2023a. Heavy metal(loid)s in agriculture soils, rice, and wheat across China: status assessment and spatiotemporal analysis. *Science of The Total Environment*, 882, 163361.

Wang, F., et al. 2024a. Emerging contaminants: a One Health perspective. *The Innovation*, 5, 100612.

Wang, L., et al. 2024b. Spatial distribution, compositional characteristics, and source apportionment of legacy and novel per- and polyfluoroalkyl substances in farmland soil: a nationwide study in mainland China. *Journal of Hazardous Materials*, 470, 134238.

Wang, M., et al. 2023b. Pollution level mapping of heavy metal in soil for ground-airborne hyperspectral data with support vector machine and deep neural network: a case study of Southwestern Xiong'an, China. *Environmental Pollution*, 321, 121132.

Wang, P., et al. 2018. Cadmium risk and mitigation in paddy systems in China. *Journal of Agro-Environment Science*, 37, 1409–1417.

Wang, P., et al. 2019. Cadmium contamination in agricultural soils of China and the impact on food safety. *Environmental Pollution*, 249, 1038–1048.

Wang, Q., et al. 2020a. Review on perfluoroalkyl and polyfluoroalkyl substances (PFASs) in the Chinese atmospheric environment. *Science of The Total Environment*, 737, 139804.

Wang, R., et al. 2022. Industrial source discharge estimation for pharmaceutical and personal care products in China. *Journal of Cleaner Production*, 381, 135129.

Wang, T., et al. 2020b. Isolation of urease-producing bacteria and their effects on reducing Cd and Pb accumulation in lettuce (*Lactuca sativa* L.). *Environmental Science and Pollution Research*, 27, 8707–8718.

Wang, Y., et al. 2023c. Transcriptomic and metabolomic changes in lettuce triggered by microplastics-stress. *Environmental Pollution*, 320, 121081.

Wu, Y., et al. 2018. Policy distortions, farm size, and the overuse of agricultural chemicals in China. *Proceedings of the National Academy of Sciences of the United States of America*, 115, 7010–7015.

Xiang, L., et al. 2022a. Integrating biochar, bacteria, and plants for sustainable remediation of soils contaminated with organic pollutants. *Environmental Science & Technology*, 56, 16546–16566.

Xiang, L., et al. 2019. Reducing plant uptake of a brominated contaminant (2,2INKLEB-tetrabrominated diphenyl ether) by incorporation of maize straw into horticultural soil. *Science of The Total Environment*, 663, 29–37.

Xiang, L., et al. 2022b. Visualizing the distribution of phthalate esters and plant metabolites in carrot by matrix-assisted laser desorption/ionization imaging mass spectrometry. *Journal of Agricultural and Food Chemistry*, 70, 15311–15320.

Xie, L.H., et al. 2017. The cadmium and lead content of the grain produced by leading Chinese rice cultivars. *Food Chemistry*, 217, 217–224.

Xu, C., et al. 2020. Polychlorinated naphthalenes, polychlorinated dibenzo-p-dioxins and dibenzofurans, and polychlorinated biphenyls in soils in an industrial park in Northwestern China: levels, source apportionment, and potential human health risks. *Ecotoxicology and Environmental Safety*, 188, 109895.

Xu, C., et al. 2019a. Immobilization of heavy metals in vegetable-growing soils using nano zero-valent iron modified attapulgite clay. *Science of The Total Environment*, 686, 476–483.

Xu, M., et al. 2021. Heavy metal pollution increases soil microbial carbon limitation: evidence from ecological enzyme stoichiometry. *Soil Ecology Letters*, 3, 230–241.

Xu, P., et al. 2022. Role of organic and conservation agriculture in ammonia emissions and crop productivity in China. *Environmental Science & Technology*, 56, 2977–2989.

Xu, X., et al. 2019b. Modeling microbial communities from atrazine contaminated soils promotes the development of biostimulation solutions. *The ISME Journal*, 13, 494–508.

Yang, Q., et al. 2018. A review of soil heavy metal pollution from industrial and agricultural regions in China: pollution and risk assessment. *Science of The Total Environment*, 642, 690–700.

Yang, X., et al. 2024. Diversifying crop rotation increases food production, reduces net greenhouse gas emissions and improves soil health. *Nature Communications*, 15, 198.

Yao, S., et al. 2022. Levels, distribution and health risk assessment of organochlorine pesticides in agricultural soils from the Pearl River Delta of China. *International Journal of Environmental Research and Public Health*, 19, 13171.

Yin, Y., and Cui, Z. 2024. Fertilizer application improvements in China. *Nature Food*, 5, 351–352.

Yu, H., et al. 2020. Assessment of the spatial distribution of organochlorine pesticides (OCPs) and polychlorinated biphenyls (PCBs) in urban soil of China. *Chemosphere*, 243, 125392.

Yu, H., et al. 2022. Microplastics as an emerging environmental pollutant in agricultural soils: effects on ecosystems and human health, 10, 855292.

Yu, L., et al. 2021. Distribution characteristics of microplastics in agricultural soils from the largest vegetable production base in China. *Science of The Total Environment*, 756, 143860.

Z, Y 2020. Contaminated land in China – tThe legal regime and its weakest link. University of Pennsylvaniaasian Law *Review*, 16, 150.

Zhan, F., et al. 2019. Field experiment on the effects of sepiolite and biochar on the remediation of Cd- and Pb-polluted farmlands around a Pb-Zn mine in Yunnan Province, China. *Environmental Science and Pollution Research*, 26, 7743–7751.

Zhang, D., et al. 2020. Plastic pollution in croplands threatens long-term food security. *Global Change Biology*, 26, 3356–3367.

Zhang, P., and Chen, Y. 2017. Polycyclic aromatic hydrocarbons contamination in surface soil of China: A review. *Science of The Total Environment*, 605–606, 1011–1020.

Zhang, Y., et al. 2015. Contamination of phthalate esters (PAEs) in typical wastewater-irrigated agricultural soils in Hebei, North China. *PLOS One*, 10, e0137998.

Zhang, Y., et al. 2019. Lead contamination in Chinese surface soils: source identification, spatial-temporal distribution and associated health risks. *Critical Reviews in Environmental Science and Technology*, 49, 1386–1423.

Zhang, Y., et al. 2023. Spatial prediction of soil contamination based on machine learning: a review. *Frontiers of Environmental Science & Engineering*, 17, 93.

Zhang, Y., et al. 2018. Legacies of Organochlorine Pesticides (OCPs) in soil of China: a review, and cases in Southwest and Southeast China. In: *Environmental Geochemistry (Second Edition)*, (Eds.) B. De Vivo, H.E. Belkin, A. Lima. Amsterdam: Elsevier, pp. 543–565.

Zhang, Y., et al. 2022. Effects of land use and landscape on the occurrence and distribution of microplastics in soil, China. *Science of The Total Environment*, 847, 157598.

Zhao, F., et al. 2023. Reducing risks of antibiotics to crop production requires land system intensification within thresholds. *Nature Communications*, 14, 6094.

Zhao, F.-J., et al. 2015. Soil contamination in China: current status and mitigation strategies. *Environmental Science & Technology*, 49, 750–759.

Zhao, L., et al. 2014. Occurrence, sources, and potential human health risks of polycyclic aromatic hydrocarbons in agricultural soils of the coal production area surrounding Xinzhou, China. *Ecotoxicology and Environmental Safety*, 108, 120–128.

Zhou, J., and Cui, H. 2014. Engineering and prospect for remediating large-scale arable land contaminated by heavy metals with "demonstration project of soil remediation on the periphery of guixi semlter" as example. *Bulletin of Chinese Academy of Sciences*, 29, 336–343.

Zhou, K., et al. 2023. Soil pollution prevention and control law. In: *Environmental and Resource Protection Law*, (Eds.) K. Zhou, L. Yi, X. Su, Y. Sun. Singapore: Springer Nature, pp. 139–171.

Zhou, Y., et al. 2018. Arsenic in agricultural soils across China: distribution pattern, accumulation trend, influencing factors, and risk assessment. *Science of The Total Environment*, 616–617, 156–163.

Zhu, Y., et al. 2011. Heavy metal contamination of shooting range soils and their biological effectiveness. *Journal of Environmental Science*, 31, 594–602.

Index

www.ingramcontent.com/pod-product-compliance
Ingram Content Group UK Ltd.
Pitfield, Milton Keynes, MK11 3LW, UK
UKHW021824100926
8556IPUK00003B/145

9 781801 468787